W0090726

Advances in Dermatological Sciences

Issues in Toxicology

Series Editors:
Professor Diana Anderson, *University of Bradford, UK*
Dr Michael D Waters, *Integrated Laboratory Systems, Inc, N Carolina, USA*
Dr Martin F Wilks, *University of Basel, Switzerland*
Dr Timothy C Marrs, *Edentox Associates, Kent, UK*

How to obtain future titles on publication:
A standing order plan is available for this series. A standing order will bring delivery of each new volume immediately on publication.

For further information please contact:
Book Sales Department, Royal Society of Chemistry, Thomas Graham House, Science Park, Milton Road, Cambridge, CB4 0WF, UK
Telephone: +44 (0)1223 420066, Fax: +44 (0)1223 420247
Email: booksales@rsc.org
Visit our website at www.rsc.org/books

Advances in Dermatological Sciences

Edited by

Robert Chilcott
*Health Protection Agency, Centre for Emergency Preparedness and Response,
Porton, UK.
Email: rob.chilcott@hpa.org.uk*

Keith Brain
*University of Oxford, Department of Pharmacology, Oxford
Email: keith.brain@pharm.ox.ac.uk*

RSC Publishing

Issues in Toxicology No. 20

ISBN: 978-1-84973-398-4
ISSN: 1757-7179

A catalogue record for this book is available from the British Library

© The Royal Society of Chemistry 2014

All rights reserved

Apart from fair dealing for the purposes of research for non-commercial purposes or for private study, criticism or review, as permitted under the Copyright, Designs and Patents Act 1988 and the Copyright and Related Rights Regulations 2003, this publication may not be reproduced, stored or transmitted, in any form or by any means, without the prior permission in writing of The Royal Society of Chemistry or the copyright owner, or in the case of reproduction in accordance with the terms of licences issued by the Copyright Licensing Agency in the UK, or in accordance with the terms of the licences issued by the appropriate Reproduction Rights Organization outside the UK. Enquiries concerning reproduction outside the terms stated here should be sent to The Royal Society of Chemistry at the address printed on this page.

The RSC is not responsible for individual opinions expressed in this work.

Published by The Royal Society of Chemistry,
Thomas Graham House, Science Park, Milton Road,
Cambridge CB4 0WF, UK

Registered Charity Number 207890

For further information see our web site at www.rsc.org

FOREWORD

The skin is a complex organ which necessitates a multidisciplinary investigative approach encompassing studies ranging from isolated, individual molecules through to holistic, whole-body interactions. Hopefully, this broad spectrum of methodologies is reflected across the six sections of this book which aim to provide an overview of contemporaneous research in the dermatological sciences.

The chapters of this book originate from presentations at the most recent Perspectives in Percutaneous Penetration (PPP) conference, a long-established, biannual meeting which brings together leading researchers from the international community. Consequently, the chapters are representative of state-of-the-art developments across a number of very active research areas.

Collating and editing the contents of this book has been a challenging but very rewarding experience. It would have been unthinkable a decade ago to be measuring skin function using synchrotron radiation and mass spectroscopy imaging, yet the humble skin diffusion cell remains to be central in advancing our understanding of transport processes within the skin. The availability and juxtaposition of new technologies with traditional techniques is likely to be the key to improving our fundamental understanding of the integument in the future.

Overall, we hope that this collation of scientific work will provide a useful summary of the current direction, scope and attainment of research in the dermatological sciences.

R P Chilcott and K R Brain

ACKNOWLEDGEMENTS

First and foremost, we would like to thank all the authors and section editors who contributed to this work. The research summarised herein represents many thousands of hours spent in laboratories and clinical research environments across the world. Hopefully, the book reflects that effort which summarises current progress in the dermatological sciences. We would also like to acknowledge Neil Redding for his assistance in collating the final draft and the patience and assistance of commissioning staff at the Royal Society of Chemistry. Finally, we would like to thank Caitlin Blackly for the cover design which is based on a collage of several skin imaging techniques and is entitled "frontier of man".

R P Chilcott and K R Brain

CONTRIBUTORS

Abdalghafor, H — School of Pharmacy, University College London, London WC1N 1AX, United Kingdom.

Adams, R G — School of Engineering & Information Science, University of Hertfordshire, Hatfield, United Kingdom.

Anjum, K — School of Pharmacy, University of Hertfordshire, Hatfield, United Kingdom.

Ansaldo, A B — Dermatology Research Centre, The University of Queensland, School of Medicine, Translational Research Institute, Brisbane, Australia.

Azeke, J — Medical Toxicology Branch, US Army Medical Research Institute of Chemical Defense, Aberdeen Proving Ground, Maryland, United States of America.

Baraldi, A — Pharmaceutical Science Division, King's College London, United Kingdom.

Barbosa-Barros, L — Department of Chemical and Surfactant Technology, Institute of Advanced Chemistry of Catalonia, Barcelona, Spain and SmartNano, SL, Barcelona, Spain.

Barrier, G — Service Départemental d'Incendie et de Secours (SDIS) 06, 140 av Maréchal de Lattre de Tassigny, BP99, 06271 Villeneuve Loubet – Cedex, France.

Benson, H A E — Curtin Health Innovation Research Institute, School of Pharmacy, Curtin University, Perth WA, Australia.

Bifarella, R — Institut de Recherche Biomédicale des Armées, Antenne de La Tronche, Département de Toxicologie, Protection et décontamination de la peau, 24 avenue des Maquis du Grésivaudan, 38700 La Tronche, France.

Bolzinger, M A — Université de Lyon F-69008, Lyon, France; Université Lyon 1, Laboratoire de Dermopharmacie et Cosmétologie, F-69008, Lyon, France; UMR CNRS 5007, Laboratoire d'Automatique et de Génie des Procédés, F-69622, Villeurbanne, France.

Bonniol, V — Faculty of Medicine, Aix-Marseille University, France and Prenyl Bio, Marseille, France.

Boulton, S J — Institute of Cellular Medicine and Medical Toxicology Centre, Medical School, Newcastle University, United Kingdom.

Braue, E   Medical Toxicology Branch, US Army Medical Research
                            Institute of Chemical Defense, Aberdeen Proving Ground,
                            Maryland, United States of America.

Briançon, S   Université de Lyon F-69008, Lyon, France; Université Lyon
                            1, Laboratoire de Dermopharmacie et Cosmétologie, F-69008,
                            Lyon, France; UMR CNRS 5007, Laboratoire d'Automatique
                            et de Génie des Procédés, F-69622, Villeurbanne, France.

Brown, M B   MedPharm Ltd, 50 Occam Road, Surrey Research Park,
                            Guildford, Surrey, United Kingdom and Department of
                            Pharmacy and Centre for Topical Drug Delivery and
                            Toxicology Research, University of Hertfordshire, Hatfield,
                            United Kingdom.

Cantecor, B   Biopharmacy Laboratory, Faculty of Pharmacy, Aix-Marseille
                            University and Prenyl Bio, Marseille, France.

Caserta, F   School of Pharmacy, University of Hertfordshire, Hatfield,
                            Hertfordshire, UK and MedPharm Ltd, 50 Occam Road,
                            Surrey Research Park, Guildford, Surrey, United Kingdom .

Chevalier, Y   Université de Lyon F-69008, Lyon, France; Université Lyon
                            1, Laboratoire de Dermopharmacie et Cosmétologie, F-69008,
                            Lyon, France; UMR CNRS 5007, Laboratoire d'Automatique
                            et de Génie des Procédés, F-69622, Villeurbanne, France.

Chilcott, R P   Department of Pharmacy, Centre for Topical Drug Delivery
                            and Toxicology Research, University of Hertfordshire,
                            Hatfield, United Kingdom and Biomedical Sciences
                            Department, Defence Science and Technology Laboratory,
                            Porton Down, Salisbury, United Kingdom.

Chipman, J K   School of Biosciences, University of Birmingham,
                            Birmingham, United Kingdom.

Cócera, M   Department of Chemical and Surfactant Technology, Institute
                            of Advanced Chemistry of Catalonia, Barcelona, Spain.

Cruz, C   Institut de Recherche Biomédicale des Armées, Antenne de
                            La Tronche, Département de Toxicologie, Protection et
                            décontamination de la peau, 24 avenue des Maquis du
                            Grésivaudan, 38700 La Tronche, France.

Dalton, C H   Biomedical Sciences, Dstl Porton Down, Salisbury, United
                            Kingdom and School of Biosciences, University of
                            Birmingham, Birmingham, United Kingdom.

Davey, N   School of Engineering & Information Science, University of
                            Hertfordshire, Hatfield, United Kingdom.

de la Maza, A   Department of Chemical and Surfactant Technology, Institute
                            of Advanced Chemistry of Catalonia, Barcelona, Spain.

Edwards, A   School of Pharmacy, University of Hertfordshire, Hatfield,
                            Hertfordshire, United Kingdom.

Erdal, M S — Faculty of Pharmacy, Department of Pharmaceutical Technology, Istanbul University, Istanbul, Turkey.

Ernst, B — Institute of Molecular Pharmacy, University of Basel, Klingelbergstrasse 50, CH-4056 Basel, Switzerland.

Fernandes, D — The Renaissance Body Science Institute, 183 Bree Street, Cape Town 8001.

Fernández, E — Department of Chemical and Surfactant Technology, Institute of Advanced Chemistry of Catalonia, Barcelona, Spain.

Ford, G — Department of Metabolism, Quotient Bioresearch Ltd, Rushden, Northamptonshire, United Kingdom.

Franz, TJ — Dermal and Transdermal Research, PRACS Institute, Fargo, North Dakota, United States of America.

Frelichowska, J — Laboratoire d'Automatique et de Génie des Procédés, University of Lyon 1, Villeurbanne, France.

Ghosh, T — Office of New Drugs and Quality Assessment, Center for Drug Evaluation and Research, Food and Drug Administration, Silver Spring, MD, United States of America.

Graham, SJ — Biomedical Sciences, Dstl Porton Down, Salisbury, United Kingdom.

Gratieri, T — School of Pharmaceutical Sciences, University of Geneva & University of Lausanne, 30 Quai Ernest Ansermet, 1211 Geneva, Switzerland and Faculdade de Ciências da Saúde, Universidade de Brasília. Campus Universitário Darcy Ribeiro, s/n, 70910-900, Brasília, DF, Brazil.

Griffiths, D — School of Pharmacy, University of Central Lancashire, Preston, United Kingdom.

Gullick, D R — The University of Georgia College of Pharmacy, Athens, GA, United States of America.

Güngör, S — Faculty of Pharmacy, Department of Pharmaceutical Technology, Istanbul University, Istanbul, Turkey.

Hafeez, F — Dermatology Department, University of California, San Francisco, San Francisco, United States of America.

Heylings, J — Dermal Technology Laboratory Ltd, Keele, Staffordshire, United Kingdom.

Hutton, O J — Biomedical Sciences Department, Defence Science and Technology Laboratory, Porton Down, Salisbury, United Kingdom.

Ibrahim, S F — The Institute of Optics, University of Rochester, 275 Hutchinson Road, Rochester NY 14627, United States of America and

Jenner, J — Biomedical Sciences, Dstl Porton Down, Salisbury, United Kingdom.

Jewell, C	Institute of Cellular Medicine and Medical Toxicology Centre, Medical School, Newcastle University, United Kingdom.
Jones, S A	Pharmaceutical Science Division, King's College London, United Kingdom.
Josse, D	Institut de Recherche Biomédicale des Armées, Département de Toxicologie et Risques Chimiques, La Tronche, France and Service de Santé et de Secours Médical, SDIS06, Villeneuve-Loubet, France.
Judd, A	School of Pharmacy, Keele University, Staffordshire, United Kingdom.
Kalaria, D	School of Pharmaceutical Sciences, University of Geneva & University of Lausanne, 30 Quai Ernest Ansermet, 1211 Geneva, Switzerland.
Kalia, Y N	School of Pharmaceutical Sciences, University of Geneva & University of Lausanne, 30 Quai Ernest Ansermet, 1211 Geneva, Switzerland.
Khengar, R H	Pharmaceutical Science Division, King's College London, United Kingdom.
Kyriacou, T	School of Computing & Mathematics, Keele University, Keele, United Kingdom.
Labrador, A	MAX-lab, Lund University, Lund, Sweden.
Lane, M E	School of Pharmacy, University College London, London WC1N 1AX, United Kingdom.
Laredj-Bourezg, F	Laboratoire d'Automatique et de Génie des Procédés, University of Lyon 1, Villeurbanne, France.
Lee, K S	Center for Analytical Instrumentation Development, Korea Basic Science Institute, Daejeon, South Korea 305-806 and The Institute of Optics, University of Rochester, 275 Hutchinson Road, Rochester NY 14627, United States of America.
Lehman, P A	Dermal and Transdermal Research, PRACS Institute, Fargo, North Dakota, United States of America.
Liu, F	School of Pharmacy, University of Hertfordshire, Hatfield, Hertfordshire, United Kingdom.
López, O	Department of Chemical and Surfactant Technology, Institute of Advanced Chemistry of Catalonia, Barcelona, Spain.
Maibach, H I	Dermatology Department, University of California, San Francisco, San Francisco, United States of America.
Mansson, R	Detection Department, DSTL, Porton Down, Salisbury, United Kingdom.
Marti-Mestres, G	Faculty of Pharmacy, University of Montpellier I, France.

Matar, H	Health Protection Agency, Porton Down, Salisbury, United Kingdom and Division of Biochemical Sciences, University of Surrey, Guildford, United Kingdom.
McAuley, W J	School of Pharmacy, University of Hertfordshire, Hatfield, Hertfordshire, UK.
McEwen, A B	Department of Metabolism, Quotient Bioresearch Ltd, Rushden, Northamptonshire, United Kingdom.
McGarry, D	Institute of Cellular Medicine and Medical Toxicology Centre, Medical School, Newcastle University, United Kingdom.
Meemon, P	Institute of Science, Suranaree University of Technology, Nakhon Ratchasima, Thailand 30000 and Department of Dermatology, University of Rochester Medical Center, 400 Red Creek Drive, Suite 200, Rochester, NY 14623, United States of America.
Mistry, T	School of Pharmacy, University of Hertfordshire, Hatfield, United Kingdom.
Mohammed, Y H	Curtin Health Innovation Research Institute, School of Pharmacy, Curtin University, Perth WA, Australia.
Moss, G P	The School of Pharmacy, Keele University, Keele, United Kingdom.
Mostefa Side Larbi, MA	Biopharmacy Laboratory, Faculty of Pharmacy, Aix-Marseille University, France, Faculty of Medicine, Aix-Marseille University, France and Prenyl Bio, Marseille, France.
Murdan, S	Department of Pharmaceutics, School of Pharmacy, University of London, United Kingdom.
Mutch, E	Institute of Cellular Medicine and Medical Toxicology Centre, Medical School, Newcastle University, United Kingdom.
Owen, J D	School of Pharmacy, University of Hertfordshire, Hatfield, United Kingdom.
Özdin, D	Faculty of Pharmacy, Department of Pharmaceutical Technology, Istanbul University, Istanbul, Turkey.
Payne,O J	Biomedical Sciences, Dstl Porton Down, Salisbury, United Kingdom.
Pelletier, J	Laboratoire d'Automatique et de Génie des Procédés, University of Lyon 1, Villeurbanne, France.
Piccerelle, P	Biopharmacy Laboratory, Faculty of Pharmacy, Aix-Marseille University and IMBE, Aix-Marseille University, France.

Price, S C	Division of Biochemical Sciences, University of Surrey, Guildford, United Kingdom.
Prow, T W	Dermatology Research Centre, The University of Queensland, School of Medicine, Translational Research Institute, Brisbane, Australia.
Raney, S G	Dermal and Transdermal Research, PRACS Institute, Fargo, North Dakota, United States of America.
Ravi, V	Institute of Cellular Medicine and Medical Toxicology Centre, Medical School, Newcastle University, United Kingdom.
Rodrigues, L M	Universidade Lusófona - CBIOS (Research Center for Health Sciences & Technologies) and Universidade de Lisboa - Pharmacol Sciences Dept. (Lab Experimental Physiology - Faculty of Pharmacy) Lisboa, Portugal.
Rodríguez, G	Department of Chemical and Surfactant Technology, Institute of Advanced Chemistry of Catalonia, Barcelona, Spain.
Rolland, J P	The Institute of Optics, University of Rochester, 275 Hutchinson Road, Rochester NY 14627, United States of America.
Rolland, P	Université de Lyon, F-69622, Lyon, France, Université Lyon 1, Villeurbanne, CNRS UMR5007, Laboratoire d'Automatique et de Génie des Procédés and Université de Lyon, F-69373, Lyon, France, Université Lyon 1, Lyon, Laboratoire de Dermopharmacie et Cosmétologie, Institut des Sciences Pharmaceutiques et Biologiques.
Rovere, M R	Laboratoire des Substituts Cutanés, University of Lyon 1, Villeurbanne, France.
Rubio, L	Department of Chemical and Surfactant Technology, Institute of Advanced Chemistry of Catalonia, Barcelona, Spain.
Sabés, M	Centre d'Estudis en Biofísica, Universitat Autònoma de Barcelona, Bellaterra, Spain.
Savelli, M P	Biopharmacy Laboratory, Faculty of Pharmacy, Aix-Marseille University, France.
Scurr, D	The School of Pharmacy, University of Nottingham, Nottingham, United Kingdom.
Shetage, S S	Department of Pharmacy, Centre for Topical Drug Delivery and Toxicology Research, University of Hertfordshire, Hatfield, United Kingdom.
Smatti, B	Centre commun de Quantimétrie, Université de Lyon, Villeurbanne, France.
Spencer, P M	Detection Department, DSTL, Porton Down, Salisbury, United Kingdom.

Staff, K	School of Pharmacy and Medical Sciences, University of South Australia, Adelaide, Australia.
Stair, J	School of Pharmacy, University of Hertfordshire, Hatfield, United Kingdom.
Stewart, C A	School of Pharmacy, University of Hertfordshire, College Lane Campus, Hatfield, Herts, AL10 9AB, United Kingdom and MedPharm Ltd, 50 Occam Road, Surrey Research Park, Guildford, Surrey, GU2 7AB, United Kingdom.
Sun, Y	School of Engineering & Information Science, University of Hertfordshire, Hatfield, United Kingdom.
Tamarkin, D	Foamix Ltd., Weizmann Science Park, 2 Holzman Street, Rehovot 76704, Israel.
Traynor, M J	Department of Pharmacy, Centre for Topical Drug Delivery and Toxicology Research, University of Hertfordshire, Hatfield, United Kingdom.
Turner, R B	MedPharm Ltd, 50 Occam Road, Surrey Research Park, Guildford, Surrey, GU2 7AB, United Kingdom.
Valour, J P	Laboratoire d'Automatique et de Génie des Procédés, University of Lyon 1, Villeurbanne, France.
Wagner, B	Institute of Molecular Pharmacy, University of Basel, Klingelbergstrasse 50, CH-4056 Basel, Switzerland.
Wan, K W	School of Pharmacy, University of Central Lancashire, Preston, United Kingdom.
Wilkinson, S C	Institute of Cellular Medicine and Medical Toxicology Centre, Medical School, Newcastle University, United Kingdom.
Williams, F M	Institute of Cellular Medicine and Medical Toxicology Centre, Medical School, Newcastle University, United Kingdom.
Wilson, K	Department of Metabolism, Quotient Bioresearch Ltd, Rushden, Northamptonshire, United Kingdom.
Wood, D G	MedPharm Ltd, 50 Occam Road, Surrey Research Park, Guildford, Surrey, United Kingdom and School of Pharmacy, University of Hertfordshire, College Lane, Hatfield, Herts, AL10 9AB, United Kingdom.
Wood, S G	Department of Metabolism, Quotient Bioresearch Ltd, Rushden, Northamptonshire, United Kingdom.
Xiao, P	Faculty of ESBE, London South Bank University, London SE1 0AA, United Kingdom.

CONTENTS

SECTION I: CUTANEOUS BIOLOGY

SECTION II: DERMAL THERAPEUTICS

SECTION III: COSMECEUTICALS

SECTION IV: REGULATORY AND TOXICOLOGY

SECTION V: Experimental and Mathematical Models

SECTION VI: SKIN PROTECTION

SECTION I: CUTANEOUS BIOLOGY

F M Williams

Institute of Cellular Medicine and Medical Toxicology Centre, Medical School, Newcastle University, United Kingdom.

Understanding the biology of healthy skin and how it changes in disease is important for the development of new drugs to treat skin disease and for understanding and interpreting the effects of chemical exposure. Current developments in molecular medicine, use of *in vitro* models, *in vivo* imaging techniques integrated with clinical observation in dermatology are directed towards improving knowledge in this area. The chapters in this book section cover a wide range of such developing areas of biology and make a useful contribution to our current understanding of the biology of the skin.

The first two chapters in this section consider the nature of hydrolytic enzymes (carboxylesterases) and their role in the dermal metabolism of paraben esters. Esterases in the skin have traditionally received limited attention, as expression levels within the epidermis are relatively low compared to the liver and so they have been thought to have limited importance when compared to hepatic metabolism and subsequent initiation of systemic effects. However, low-level xenobiotic-metabolising enzymes such as cytochromes 450 (CYPs) and peroxidases have been shown to be important in development of DNA adducts which may lead to tumorigenesis. Also, the formation of activated protein adducts may result in hypersensitivity and inflammatory responses in the skin. A rate–limiting role for esterases has been dismissed by researchers as they are ubiquitous and highly expressed but, as indicated in the two presented papers, their variability in expression may influence the toxicity of esters during percutaneous penetration and carboxylesterase isoforms exhibit substrate specificity which influences metabolism of different chain-length parabens.

Both Williams *et al.* and Ford *et al.* addressed the local hydrolysis of paraben esters in skin using *in vitro* systems because of their potential oestrogenic effects and toxicity. Thus, these studies represent a positive contribution to the reduction, refinement and replacement of animals in research. The focus of the studies on parabens is highly relevant, as this class of compound is becoming ubiquitous in the environment and in the general population due to their application as anti-bacterial agents in a range of cosmetic, heath and household products. Esters are lipophilic and cross the stratum corneum but undergo hydrolysis in keratinocytes in the skin. Ford *et al.* also identified cross-esterification in whole skin *ex vivo*. This, in addition to the effects of inhibitors on the hydrolysis profile aid identification of the isoforms involved for specific molecules and so aid prediction of the hydrolysis profile for newly developed esters.

It is important to define the physiology and surface biology of healthy skin; how it varies between individuals and is influenced by age, skin treatments, sun exposure and chemical insults. Clearly, there may be a role for genetic polymorphisms in rate-limiting enzymes influencing skin biology as has recently been shown for fillagrin. The studies of paraben metabolism and carboxylesterases presented here are mainly in cultured skin cells and so metabolism profiles do not necessarily reflect those *in vivo* where the 3D architecture is intact and is more representative of a heterogenous genetic population. It is important to relate studies with cells in culture to 3D models such as skin equivalents and then to whole skin *ex vivo* and finally to the situation in humans, *in vivo*.

An area of cutaneous biology which tends to be overlooked is, perhaps surprisingly, the outer skin surface. Specifically, the material traditionally referred to as the acid mantle which arises from corneocyte debris and accumulated secretions of the sweat and sebaceous glands. In Chapter 4 by Shetage *et al.*, preliminary work is presented which aims to characterise the nature of this superficial layer, more correctly termed residual skin surface material (RSSC). Using a traditional method (cigarette paper), samples of RSSC were acquired from the forehead of volunteers representative of a number of ethnic groups. Qualitative gas chromatography mass spectrophotometry (GCMS) indicated that RSSC contains at least 49 components from 5 lipid classes. Based on these data, Shetage *et al.* have proposed that RSSC may have utility as a bio-monitoring matrix to assess environmental exposure to chemicals or the secretion of specific endogenous biomarkers. The use of RSSC appears to provide several practical advantages as can be collected non-invasively and reproducibly from individuals. Moreover, the general composition was not significantly influenced by age, sex or ethnicity. Clearly, further investigations of the profile of RSSC in healthy individuals and patients with diseases will be an important aspect in developing the potential clinical applications of RSSC as a biomarker. The proposal for using RSSC as a medium for detecting chemical exposure deserves further investigation, but this approach will need considerable evaluation.

Non-invasive skin imaging techniques with good cellular resolution offer the potential to monitor changes in the skin in relation to chemical and drug use and in disease states. Indeed, the application of non-invasive visualising techniques to whole skin *in vivo* in healthy volunteers or patients and to whole *ex vivo* skin is evolving rapidly and provides important information about the 3D architecture of normal and diseased skin. Chapter 5 by Rolland *et al.* reports the application of Gabor Domain Optical Coherence Microscopy (GD OCM) as a technique for producing 3D images of skin *in vivo* at several anatomical locations. The resulting images clearly reveal the cutaneous microstructure and the array of epidermal keratinocytes and skin appendages. The technique was also used to compare normal and diseased (melanoma) skin *in vitro*. The work by Rolland *et al.* clearly demonstrates the potential utility of the technique: it will be interesting to see future studies where measurements in patients are compared directly with histological images of *ex vivo* samples.

Visualisation of the skin traditionally relies on the differential absorbance and reflectance of relatively low-energy photons. In contrast, synchrotron radiation (SR) can be used as a source of high energy photons with small-angle X ray scattering (SAXS) to characterise the structure of soft tissues such as skin. Its use is restricted by the need for a synchrotron (cyclic particle accelerator). In Chapter 6, Cocera *et al.* describe the advantages of the technique compared to Fourier transform infrared spectroscopy (FITR) for characterising pig skin in relation to collagen and lipid structure and also the influence of age and disease using *ex vivo* dermatomed and heat treated human skin. The use of SAXS, although not widely available, could provide useful 3D structural information on skin *ex vivo* but has no current *in vivo* application due to the potential adverse health effects of the high energy electromagnetic radiation. The continued development and improvement of such *in vitro* and *ex vivo* methodology will undoubtedly improve our understanding of skin biology in health and disease.

2

FACTORS INFLUENCING EXPRESSION AND SPECIFICITY OF CARBOXYLESTERASES IN SKIN: IMPLICATIONS FOR LOCAL AND SYSTEMIC PARABEN ESTER TOXICITY

F M Williams, V Ravi, D McGarry, S J Boulton, E Mutch, C Jewell and S C Wilkinson

Institute of Cellular Medicine and Medical Toxicology Centre, Medical School, Newcastle University, United Kingdom.

1   INTRODUCTION

Carboxylesterases (CES) are present in a number of skin cells and have known physiological and exogenous substrates including drugs, chemicals, pro-drugs, carbamate and pyrethroid insecticides, cosmetic chemicals and environmental toxicants. A typical reaction scheme for CES is presented in Figure 1. During drug discovery and design, an ester linkage is frequently included to selectively target a pro-drug to a tissue or to improve dermal absorption and deliver a novel compound. Skin is an important tissue regulating uptake of environmental chemicals as well as a delivery site for drugs targeted locally to the skin and systemically.

Satoh and Hosokawa [1] classified five groups of CES enzymes (CES 1–5) on the basis of amino acid homology and substrate specificity, with the majority of identified CES enzymes belonging to either the CES-1 or CES-2 sub-families. Amino acid sequence homology between human carboxylesterase 1 (hCE-1; a member of CES-1 family) and human carboxylesterase 2 (hCE-2, CES-2 family) is 48%. However, the substrate selectivity of these two enzymes is different. The hCE-1 enzyme mainly hydrolyses substrates with low molecular weight alcohol groups and higher molecular weight acyl groups such as cocaine (methyl ester), meperidine, and delapril. In contrast to hCE-1, the hCE-2 enzyme efficiently hydrolyses compounds with high molecular weight alcohol groups and relatively smaller carboxylate groups such as 4-methylumbelliferyl acetate, heroin, and 6-acetylmorphine CPT-11 or SN38 [2]. The hCE-1 and hCE-2 genes encode the two major forms of human liver microsomal carboxylesterase enzymes.

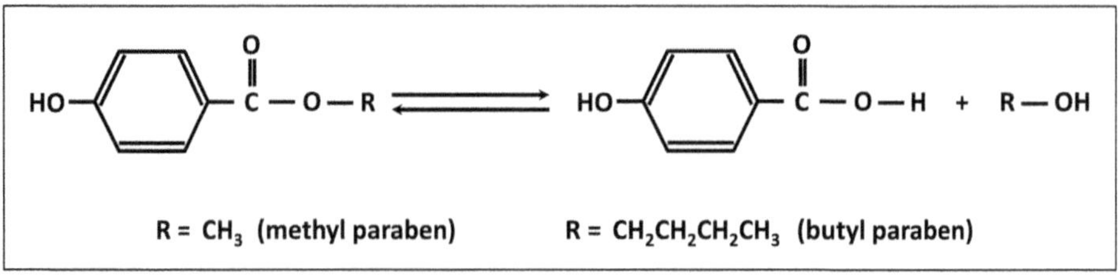

Figure 1 *General reaction scheme for paraben hydrolysis.*

Carboxylesterase activity has been found both in the microsomal and cytosolic fractions of the skin [3].

There are differences in the distribution of CES isoforms between extra-hepatic tissues (such as the small intestine, lung, skin) and the liver which results in differing hydrolysis profiles [4]. Staudinger *et al.* [5] summarised the differences in distribution between CES-1 and CES-2, but did not refer to skin. Hydrolysis profiles have been defined using isozyme specific substrates such as CPT-11 ironotecan or procaine for CES-2 and methyl phenidate for CES-1. Several nuclear receptor (NR) family members regulate drug-inducible expression and activity of CES-1 and 2 in mammalian liver and intestine but these have not been investigated in the skin [5].

Parabens (4-hydroxy benzoic acid esters) are substrates for CES enzymes. Parabens are widely used in topically applied cosmetic products and are known to penetrate the skin and have some oestrogenic effects *in vitro*. Paraben esters have been used as model substrates to identify carboxylesterase activity in human liver and skin [5]. Previous studies have shown that parabens are rapidly hydrolysed after oral intake through first pass metabolism by the liver [6]. The capacity for hydrolysis of the parabens in skin varies with the side-chain of the substrate. Lobemeier *et al.* showed that in human skin and subcutaneous fat tissue, paraben esters could be hydrolysed by 4 different carboxylesterase enzymes classified by their isoelectric points [7]. Paraben esters are hydrolysed to 4-OH benzoic acid by carboxylesterases in both skin microsomes and in cytosol by CES-1 and CES-2 differentially relating to the leaving group. Inhibition by loperamide (specific for CES-2) indicated that butyl paraben hydrolysis was more inhibited than methyl paraben suggesting involvement of CES-2 with higher molecular weight leaving groups [4].

In the liver, greater hydrolysis of methyl paraben is consistent with higher expression of CES-1. The profile is less easy to explain for skin where the hydrolysis rate for methyl paraben is greater than for butyl paraben, although previous studies had suggested lower expression of CES-1 in the skin. The degree of hydrolysis during absorption through the skin could potentially influence the efficacy and toxicity of esters and thus be influenced by variability in esterase expression in the skin. Absolute variability in activity of carboxylesterases in skin versus other enzymes (e.g. balance between hydrolysis and conjugation, oxidation or transport) may be important. CES are localised in keratinocytes so the absorbed ester needs access to enzymes and so higher lipid solubility might promote uptake into keratinocytes. We had postulated that transporters might be required for efflux of the metabolite from cells but found no evidence of this (unpublished data). It was previously shown for fluazifop butyl ester pesticides that inhibition of all serine esterases in skin by BNPP reduced absorption through skin in flow-through diffusion cell system [8].

The aims of the present study were to further define the activity and expression profile of carboxylesterases isoforms in human skin and to determine whether expression of carboxylesterases or hydrolysis of paraben esters could be induced by the steroid dexamethasone or 8-methoxypsoralen (8-MOP) which are applied to skin. Dexamethasone is used orally and applied topically to the skin in preparations for inflammation to produce a local effect, whilst 8-MOP is used orally for psoriasis and distributes to the skin from the systemic compartment. Studies were conducted *in vitro* using keratinocytes and human skin in culture compared with hepatocytes. Also the interaction between UV light and paraben ester local hydrolysis and toxicity in skin cells was examined.

2   METHODS

HaCat cells, primary human keratinocytes and HepG2 cells were maintained in culture. The cells were treated with dexamethasone (0-500µM) or 8-MOP (0-100µM) for 24-72 hours followed by methyl or butyl paraben (0-100 µM) treatment for 24 hours. Parabens were also applied to the stratum corneum surface of skin in short term culture for similar times. Formation of 4–OH benzoic acid was measured by HPLC as previously described [4]. Carboxylesterase protein expression (CES-1 and CES-2) was measured by western blotting and mRNA by RT PCR with CES-2 specific primers.

Epilife™ medium and human keratinocyte growth factor (HKGF) were obtained from Cascade Biologics. RNeasy mini kit was purchased from Qiagen, and cDNA synthesis kit from Bioline. The primers for PCR were from MWG biotech and Taq polymerase from Molzyme. GAPDH antibody was obtained from Abcam. Unless stated, all other chemicals were purchased from Sigma Aldrich Chemicals.

Cells were grown in 75 cm^2 vented Corning flasks, and maintained in the exponential phase of their growth by incubation in a humid atmosphere containing 5% CO_2 at a temperature of 37°C. Cells were routinely passaged on reaching a confluence of 80 – 90%. HaCaT cells were cultured in Dulbecco's Modified Eagle Medium (DMEM) supplemented with 1% penicillin streptomycin and 10% foetal calf serum (FCS). HepG2 cells were cultured in Williams Essential Media supplemented with 10% FCS and 1% glutamine. Primary Human Epidermal Keratinocytes (PEK) (kindly provided by Professor N Reynolds, Department of Dermatology, Newcastle University) were grown to the fifth passage in Epilife™ medium supplemented with 1% HKGF and 1% penicillin streptomycin.

For incubation studies, cells were sub-cultured by washing with Dulbecco's Phosphate Buffered Saline (DPBS), brought into suspension by incubating with trypsin, and neutralising with an equal volume of growth media. The cells were seeded out at a density of 80,000 HaCat or HepG2 or 60,000 cells PEK/ml media and dosed at 80-90% confluency.

Human skin was obtained from breast reduction surgery of healthy females. Patients gave informed consent and ethical approval was obtained from the relevant authority. Cleaned skin was stored at -80°C. Paraben hydrolysis was measured in skin in short term culture using Costar Netwell Supports with small glass rings (1 cm diameter) glued to the outer surface to contain the applied parabens.

Cells were treated for 24 - 72 hours depending on the experiment with a change of media containing the inducer every 24 hours. On the final day of treatment with the inducer, the cells were dosed with methyl paraben or butyl paraben in DMSO (5µL per ml medium) for 24 h. Hydrolysis was terminated by mixing medium with an equal volume of methanol containing 20µg/ml di-hydroxy benzoic acid (DHBA internal standard) and 1% (v/v) formic acid, and the supernatant retained for analysis. The cells on the surface of the culture plates were incubated with 0.5M sodium hydroxide for 10 minutes and scraped off to determine the protein concentration.

The concentration of the metabolite, 4-hydroxy benzoic acid (OHBA) formed from methyl paraben or butyl paraben was analysed on a Varian Prostar HPLC, using a C18 Gemini microbore column (5 μm, 250 mm × 2 mm, Phenomenex) by a modification of the method of Jewell et al. [4]. Enzyme activities were expressed as nmol 4-hydroxy benzoic acid formed per mg cellular protein per 24 h (n = 3, mean ± SEM). Activities in treated cells were compared to controls by Analysis of Variance (ANOVA) followed by the Dunnett's test using Prism (GraphPad Inc., San Diego).

Proteins (approximately 20 μg) isolated from the cells were denatured by treatment with lithium dodecyl sulphate (Invitrogen) and β-mercaptoethanol at 95°C, resolved on a bis-tris polyacrylamide gel (4 – 12 % gel, Invitrogen) and transferred onto a nitrocellulose membrane (Hybond-C extra, GE Biosciences) by electrophoresis. Non-specific binding sites on the membrane were blocked overnight at 4°C with 5% milk in TBS-T20 buffer following which the membranes were incubated with human anti CES-1 (1:2000 dilution) or anti CES-2 (1:1000 dilution) antibodies (gifts from Professor Satoh, Chiba University, Japan) for 90 minutes. The membranes were incubated with anti-rabbit antibody conjugated with horse radish peroxidise for 1 hour and were visualised using enhanced chemiluminiscent (ECL) reagent (GE Biosciences), whose signal was captured using a Syngene G:Box Chemi XL Gel Documentation System. For CES-2 the bands were quantified using Syngene densitometric analysis (Genetools, Syngene) and the data was expressed as a ratio of the absorbances of CES to GAPDH.

Reverse transcriptase-coupled polymerase chain reaction (RT-PCR) was performed on mRNA eluted from the cells using the RNA easy mini kit from Qiagen according to the manufacturer's instructions. The RNA samples were quantified by NanoDrop ND-1000 spectrophotometer (NanoDrop Technologies). The genomic DNA in the RNA samples was digested by DNase I recombinant, RNase-free (Roche). Approximately 2 μg RNA was reverse transcribed using the cDNA synthesis kit from Bioline according to the manufacturer's instructions. The DNA was amplified by 35 cycles of PCR in a 25 μl reaction mixture consisting of 0.25 μM specific primer, 0.1 mM dNTPs, 0.1 μg of cDNA as template, and 0.625U Taq polymerase in buffer (50 mM potassium chloride, 10 mM Tris-HCL pH 9.0, 0.1% (v/v) Triton X-100, 1.5 mM $MgCl_2$). The PCR program consisted of an initial denaturation for 15 seconds at 95°C, annealing at 50°C for 30 seconds and extension at 72°C for 45 seconds and was performed in a GeneAmp PCR System 9700 (PE applied biosystems, Foster city, CA). The gene specific primers used for CES-2 (product size: 415bp) were forward primer: 5' – AGCCTGTCCCTAGCATTGTT and reverse primer: 5' –GCCCCCAAAGAAACTTCTGA. The amplified DNA was separated by a 2% agarose gel containing ethidium bromide visualised by Bio-Rad Fluor S multi imager.

Protein concentrations were determined using the Bicinchoninic acid (BCA) assay or Bradford assay.

3   RESULTS

Protein expression: CES-2 and sometimes CES-1 protein were detected in cells and skin in short term culture by Western blot analyses using antibodies against CES-1 and CES-2. CES-2 was present in primary human keratinocytes and HaCats and in human skin cytosol and microsomes. Induction of CES-2 expression was detected with both dexamethasone and 8-MOP. CES-1 was detected in HaCats and human cytosol but not primary human keratinocytes induced by either dexamethazone or 8-MOP (Figure 2).

Both CES-1 and CES-2 were detected in HepG2 cells where CES-1 was more abundant, although surprisingly no induction of CES-1 or CES-2 was observed. Preliminary studies of mRNA expression confirmed CES-2 message present in HaCats. It had previously been shown using native PAGE and naphthyl acetate activity stain that two bands with esterase activity were present in human skin [4].

Rates of hydrolysis of methyl paraben and butyl paraben by skin cells and skin in short term culture differed when using similar concentrations on a μg basis with methyl paraben hydrolysis generally higher (methyl paraben [50uM, 6.0 h]; 15.0 ± 10.0 nmol in receptor fluid compared butyl paraben; 5.0 ± 2.0 nmol). Similarly, there was a differential effect of loperamide on the proportion of metabolite detected in the receptor fluid following application of methyl paraben or butyl paraben to the surface of skin in short term culture.

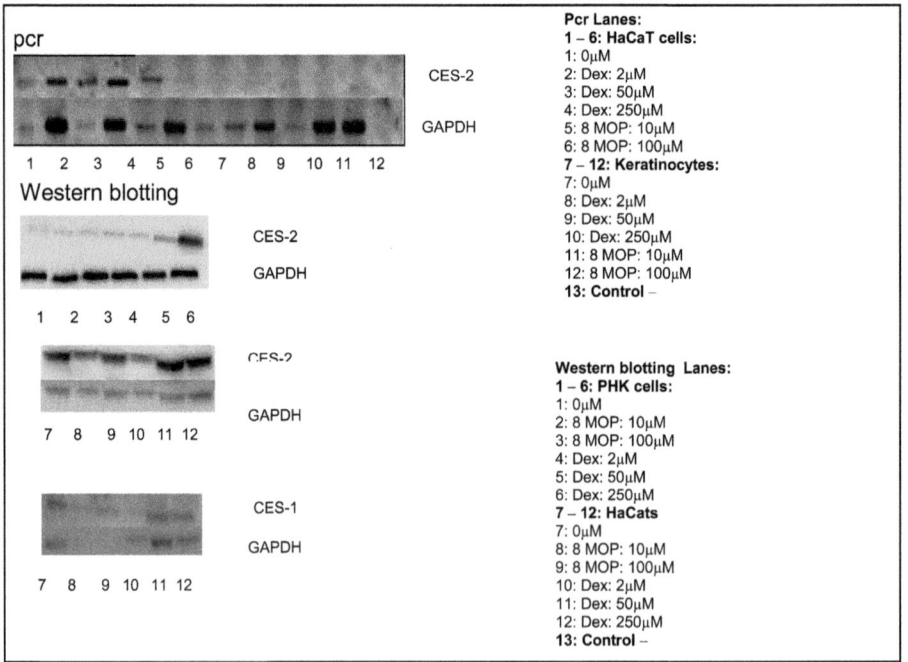

Figure 2 *Expression of carboxylesterases CES-2 and CES-1 in HaCat cells and normal human keratinocytes (NHK).*

Butyl paraben hydrolysis was increased by dexamethasone pre-treatment in skin cells but had little effect on methyl paraben hydrolysis (Figures 3a and 3b).

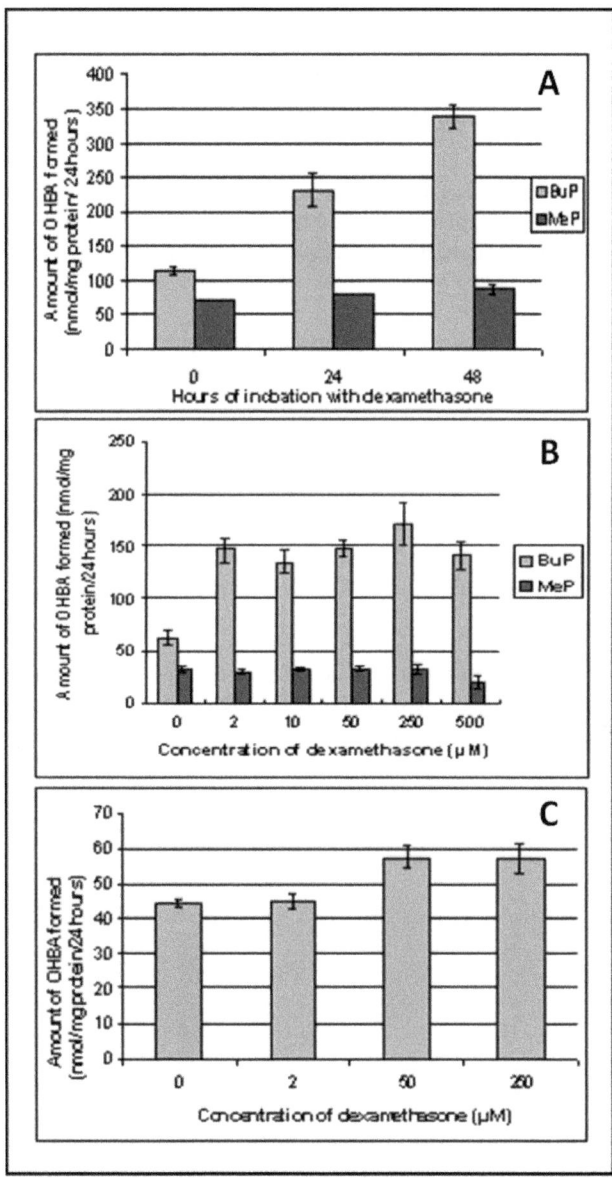

Figure 3 *Amount of 4-hydroxybenzoic acid (OHBA) formed (nmol/mg protein/24 h) following treatment of HaCaT cells (A and B) or primary human epidermal keratinocytes (PHK) (C) with dexamethasone and incubated with BuP or MeP or only BuP for PHK for 24 hs. Mean ± SEM (n = 3). MeP = Methyl paraben, BuP = Butyl paraben.*

Advances in Dermatological Sciences

The amount of 4-hydroxybenzoic acid (OHBA) formed by HaCat cells was greater from butyl paraben than from methyl paraben and an increase in the formation of OHBA from BuP was seen in those cells dosed with 50 μM dexamethasone (approximate two fold and three fold increases for 24 and 48 hour incubation periods, respectively). There was negligible increase in the formation of OHBA from methyl paraben with 24 hour treatment. When HaCaT cells were treated with a range of concentrations of dexamethasone for 48 hours followed by 24 hours incubation with butyl paraben there was a corresponding increase in OHBA up to 20 μM dexamethasone, after which further increases were minimal suggesting saturation of the induction pathway. There was no effect of dexamethasone on metabolism of methyl paraben. Primary human epidermal keratinocytes treated with dexamethasone and incubated with butyl paraben (50μM) for 24 hours showed significant induction of metabolism at 50μM and 250μM dexamethasone. (Figure 3c).

Pre-treatment with 8-MOP did not significantly increase butyl paraben hydrolysis whereas methyl paraben hydrolysis decreased in skin but increased in Hep G2 cells, suggesting that 8-MOP might induce CES-1. In HepG2 cells, hydrolysis rates were higher and there was a dose-dependent increase in butyl paraben hydrolysis with 8-MOP (results not shown). A proportion of the 4-OH benzoic acid formed was conjugated as the glucuronide, particularly in hepatocytes (released by β-glucuronidase treatment; results not shown).

Collectively, these data suggest that up-regulation of CES-2 activity occurs in skin cells treated with micromolar concentrations of dexamethasone but not 8-MOP. However, CES-1 was not inducible.

The rate of hydrolysis of parabens in HaCat cells related to the local effect on DNA as measured by the COMET assay. It was found that DNA damage was increased in cells exposed to both the highest methyl paraben concentration and UVB light compared to the control without any paraben or UVB light and to the cells with only exposure to UVB (Figure 4).

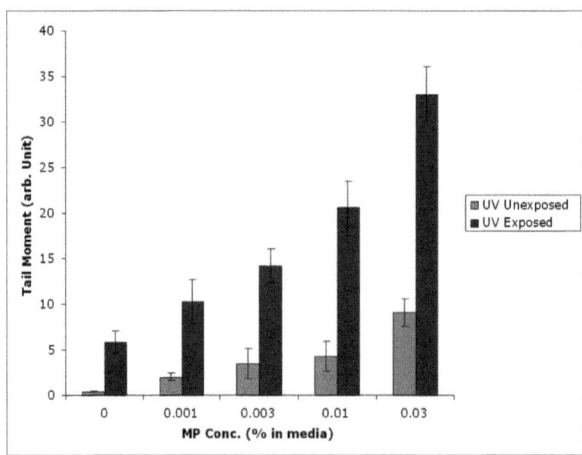

Figure 4 *DNA damage (tail moment) in HaCat cells treated with methyl paraben with and without UV light Mean ± SEM (n = 3).*

4   DISCUSSION

These results support our previous conclusion that CES-2 is the predominant isoform in human skin and that it hydrolyses butyl paraben and other long chain hydroxybenzoate esters. Methyl paraben is also metabolised in human skin, but hydrolysis by microsomes was previously shown not to be inhibited by loperamide and the effect was less in cultures of whole skin.

Current protein and RNA expression studies confirmed that expression of CES-1 in skin was lower than CES-2. CES-1 may be sufficient to carry out the metabolism of methyl paraben, or other esterases known to be expressed in skin might be involved or another (as yet undetected) CES isoform such as CES-3 (expressed in the brain). CES-2 has been shown to predominate in other extra-hepatic tissues which is in contrast to the liver (where CES-1 expression is dominant).   Thus, our observations in skin are consistent with previous findings. It has not been possible so far to define the kinetics of hydrolysis of parabens by purified CES isoforms or to conduct kinetic studies in induced skin cells with and without inhibitors.

We have successfully shown that CES-2 hydrolysis in skin is inducible by dexamethasone, which suggests that co-administered drugs might lead to induction of CES isoforms and so influence the proportion of paraben ester hydrolysed. This is the first time steroids have been shown to induce carboxylesterases in human skin, although dexamethasone-induced CES-1 and CES-2 in human hepatocytes [9] and dexamethasone-induced CES-2 in the rat have been previously demonstrated [10]. Studies by Zhu *et al.* [9] showed the involvement of PXR (pregnane X receptor) and the glucocorticoid receptor (GR, nuclear receptor NR3C1) in transcriptional regulation of expression and activity of carboxylesterases in both rat and human liver. In this present study, exposure of primary cultures of human hepatocytes to micromolar concentrations of dexamethasone induced hCE-1 and hCE-2 protein expression in a concentration-dependent manner. In contrast, exposure of rat hepatocytes to nanomolar concentrations of dexamethasone repressed the expression of carboxylesterase genes [11].

Pregnane X receptor has previously been shown to regulate drug metabolising enzymes such as some cytochrome P450 (CYP) genes, glutathione-S-transferase, sulphotransferase and uridinyl diphosphate glucuronyl transferases in liver. CAR (constitutive androgen receptor) was demonstrated to regulate phenobarbitone inducible expression of CYP 2B family but PXR and CAR can overlap. Esterase activity was shown to be induced by phenobarbitone in rat liver but not in skin [12]. Also, 8-MOP induced CES-2 in hepatocytes [13]. Induction of 8-MOP also involves the PXR receptor, skin expression of which is low compared to glucocorticoid receptors [14] and this would explain the differential effects of dexamethasone and 8-MOP in skin where GR predominates compared to liver. Carboxylesterase enzymes are also regulated by PPARα in liver but this may not be the case in skin.

The extent of local dermal hydrolysis of topically applied paraben esters to hydroxy benzoic acid will influence the relative amount of parent intact ester and metabolite entering the blood supply and being distributed systemically and hence potential to induce toxic effects. There is a discrepancy in the literature regarding the degree of toxicity exhibited following paraben exposure and the relative importance of parent and metabolite in the development of toxicity.

Recent studies have related the toxicity of paraben esters to effects at the oestrogen receptor and so provide a biologically plausible link between paraben exposure and breast cancer: Darbre [15] reported that parabens exhibited estrogenic activity *in vitro* in human breast cancer cell lines (MCF7) and *in vivo*, with such activity directly correlating with the length of the alkyl chain. Other work [15, 16] has showed upregulation of expression of an estrogen-responsive reporter gene and increased proliferation in breast cell lines and in murine uterine tissue *in vivo* in response to exposure to benzylparaben. Interestingly, isobutyl paraben had no effect on growth of estrogen-unresponsive MDA-MB-231 human breast cancer cells [15]. Importantly, when butyl paraben was compared to 17β-estradiol for its effects on MCF7 cell lines, the growth responses of the cells were similar. However, the gene expression induced by each agent was different, suggesting that downstream effects of parabens did not parallel those of oestrogen [17]. Methyl paraben administered to rats at a dose of up to 5700 mg/kg had no effects at the oestrogen receptor [18].

Administration of 8-MOP via oral or topical routes followed by UVA (PUVA) is used in the treatment of dermatological conditions such as psoriasis and sclerotic lesions. Reus *et al.* measured the toxicity of 8-MOP and UVA using the COMET assay [19]. More recently, parabens have been reported to potentiate oxidative stress, nitric oxide production and lipid peroxidation following ultraviolet (UV) light exposure in cultured skin keratinocytes [20]. Our data support these observations in confirming the augmented UV-induced DNA damage by parabens even after a single dose exposure. Methyl paraben has been shown to modulate Ca^{2+} concentration and histamine release in other types of cell and has some agonist activity at the ryanodine receptor, modulating Ca^{2+} signalling. Persistent elevation of Ca^{2+} in the cell can lead to toxicity and apoptosis [21]. It would be interesting to investigate exposure to MeP and UVB with regard to Ca^{2+} signalling and to assess the effect of dexamethasone and 8-MOP on this response.

Further work is required to relate metabolism and the effect of steroid inducers and 8-MOP to local end points such as DNA damage and oxidative stress in skin cells and to compare these with effects on oestrogen receptors in the skin. Also, it is important to determine whether parabens induce their own metabolism with repeated exposure as might occur with repeated use in a topical application product or interaction with UV as in a sun block. Unpublished data suggests that much of hydroxybenzoic acid formed by hydrolysis is conjugated before excretion *in vivo*. It is important to determine the role of phase 2 metabolism in inter-individual variability in uptake and responses to parabens.

In conclusion, expression and function of carboxylesterases in human skin is variable and differentially influenced by chemicals such as steroids. This may contribute to inter individual variability in the hydrolysis of ester substrates such as paraben esters and hence potential toxic effects.

5 REFERENCES

1. T. Satoh and M. Hosokawa, Structure, function and regulation of carboxylesterases, *Chem. Biol. Interact.*, 2006, **162**, 195–211.

2. C. Jewell, J. J. Prusakiewicz, C. Ackermann, N. A. Payne, G. Fate, R. Voorman and F. M. Williams, 'Hydrolysis of a series of parabens by skin microsomes and cytosol from human and minipigs and in whole skin in short-term culture', *Toxicol. Appl. Pharmacol.*, 2007, **225**, 221–8.

3. C. Jewell, C. Ackermann, N. A. Payne, G. Fate, R. Voorman and F. M. Williams, Specificity of procaine and ester hydrolysis by human, minipig, and rat skin and liver, *Drug Metab. Dispos.*, 2007, **35**, 2015–22.

4. C. Jewell, P. Bennett, E. Mutch, C. Ackermann and F. M. Williams, Inter-individual variability in esterases in human liver, *Biochem. Pharmacol.*, 2007, **74**, 932–9.

5. J. L. Staudinger, C. Xu, Y. J. Cui and C. D. Klaassen, Nuclear Receptor-Mediated Regulation of Carboxylesterase Expression and Activity Expert, *Opin. Drug Metab. Toxicol.*, 2010, **6**, 261–271.

6. C. L. Barker and R. H. Clothier, Human and keratinocyte cultures as models of cutaneous esterase activity, *Toxicology In vitro*, 1997, **11**, 637–640.

7. C. Lobermeier, C. Tschoetschel, S. Westie and E. Heymann, Hydrolysis of parabens by extracts from differing layer of human skin, *Biological Chemistry*, 1996, **377**, 647–651.

8. N. Clark, 'Cutaneous metabolism and its role in percutaneous absorption' PhD thesis, 2010, Newcastle University.

9. W. Zhu, L. Song, H. Zhang, L. Matoney, E. LeCluyse and B. Yan, Dexamethasone differentially regulated expression of carboxylase genes in humans and rats, *Drug Metab Dispos.*, 2000, **28**, 186–91.

10. T. Furihata, M. Hosokawa, A. Fujii, M. Derbel, T. Satoh and K. Chiba, Dexamethasone-induced methylprednisolone hemisuccinate hydrolase: its identification as a member of the rat carboxylesterase 2 family and its unique existence in plasma, *Biochemical Pharmacology*, 2005, **69**, 1287–1297.

11. D. Shi, J. Yang, D. Yang and B. Yan, Dexamethasone suppresses the expression of multiple rat carboxylesterases through transcriptional repression: evidence for an involvement of the glucocorticoid receptor, *Toxicology,* 2008, **254**, 97–105.

12. N. W.McCracken, P. G. Blain and F. M. Williams, Nature and role of xenobiotic metabolising enzymes in rat liver, lung, skin and blood, *Biochemical Pharmacology*, 1993, **45**, 31–36.

13. J. Yang and B. Yan, Photochemotherapeutic Agent 8-Methoxypsoralen Induces Cytochrome P450 3A4 and Carboxylesterase HCE2: Evidence on an Involvement of the Pregnane X Receptor, *Toxicological Sciences*, 2007, **95**, 13–22.

14. K. Chang., Q. Shen, G. Oh, S. Jelinsky, S. Jenkins, W. Wang, Y. Wang, M. LaCava, M. Yudt, C. Thompson, L. Freedman , J. Chung and S. Nagpal, Liver X Receptor Is a Therapeutic Target for Photoaging and Chronological Skin Aging, *Molecular Endocrinology*, 2008, **22**, 2407–2419.

15. P. D. Darbre, J. R. Byford, L. E. Shaw, R. A. Horton, G. S. Pope and M. J. Sauer, 'Oestrogenic activity of isobutylparaben *in vitro* and *in vivo*', *J. Appl. Toxicol.*, 2002, **22**, 219–226.

16. P. D. Darbre, J. R. Byford., L. E. Shaw, S. Hall, N. G. Coldham, G. S. Pope and M. J. Sauer, 'Oestrogenic activity of benzylparaben', *J. Appl. Toxicol.*, 2003, **23**, 43–51.

17. D. Pugazhendhi, A. J. Sadler and P. D. Darbre, Comparison of the global gene expression profiles produced by methylparaben, n-butylparaben and 17beta-oestradiol in MCF7 human breast cancer cells, *J. Appl. Toxicol.*, 2007, **27**, 67–77.

18. M. G. Soni, S. L. Taylor, N. A. Greenberg and G. A. Burdock, Evaluation of the health aspects of methyl paraben:a review of the published literature, *Food Chem. Toxicol.*, 2002, **40**, 1335–1373.

19. A. A. Reus, R. N. van Meeuwen, N. de Vogel, W. J. Maas and C. A. Krul, Development and characterisation of an *in vitro* photomicronucleus test using *ex vivo* human skin tissue, *Mutagenesis,* 2011, **26**, 261–8.

20. O. Handa, S. Kokura, S. Adachi, T. Takagi, Y. Naito, T. Tanigawa, N. Yoshida and T. Yoshikawa, Methylparaben potentiates UV-induced damage of skin keratinocytes, *Toxicology,* 2006, **227**, 62–72.

21. T. T Vo, E. M. Jung, K. C. Choi, F. H. Yu and E. B. Jeung, Estrogen receptor α is involved in the induction of Calbindin-D(9k) and progesterone receptor by parabens in GH3 cells: a biomarker gene for screening xenoestrogens, *Steroids,* 2011, **76**, 675–81.

3

METABOLIC TRANSFORMATIONS OF ALKYL PARABENS IN THE SKIN: SPECIFIC ESTERASE INHIBITION STUDY

G Ford, A B McEwen, K Wilson and S G Wood

Department of Metabolism, Quotient Bioresearch Ltd, Rushden, Northamptonshire, United Kingdom

1   INTRODUCTION

The skin plays a vital role in protecting the body from invading pathogens and provides a barrier against the absorption of hazardous chemicals. One of the key physical properties that enable xenobiotics to be absorbed in the skin, lipophilicity, also presents an obstacle to their elimination. The skin contains a number of enzymes capable of either activating or detoxifying absorbed xenobiotics [1 – 3]. Although the metabolic capacity of the skin is considerably lower than that observed in the liver it is still an important factor in determining the nature and extent of compounds entering the systemic circulation. The aim of this study was to examine the metabolism of octyl paraben during passage through human skin *in vitro*.

2   MATERIALS AND METHODS

Radiolabelled test compound, octyl paraben (specific activity 287 μCi mg^{-1}) was obtained from BioDynamics Research Limited (Figure 1). Esterase inhibitors (Figure 2) were purchased from Sigma Aldrich.

Two replicates of human skin for each test group were used from three different donors. The radiolabelled test compounds were prepared at a specific activity suitable for administration. Triplicate volumes for each dose solution were transferred to volumetric flasks prior to and following dose application in order to calculate the dose applied and to check for homogeneity of the dose formulations.

Testosterone was used as a reference compound to check the operation of the system and was applied in duplicate at 0.11 mg cm^{-2}. The radiolabelled testosterone was diluted with non-radiolabelled reference compound to meet the required specific activity.

TRA002492-OC01, 478 μm; TRA002488-OC02, 637 μm, Transkin, TCS Cellworks Ltd.) was used and the experiment conducted using Franz-type static diffusion cells with a nominal 12 ml receptor chamber volume and 1.77 cm^2 exposed skin area for dosing.

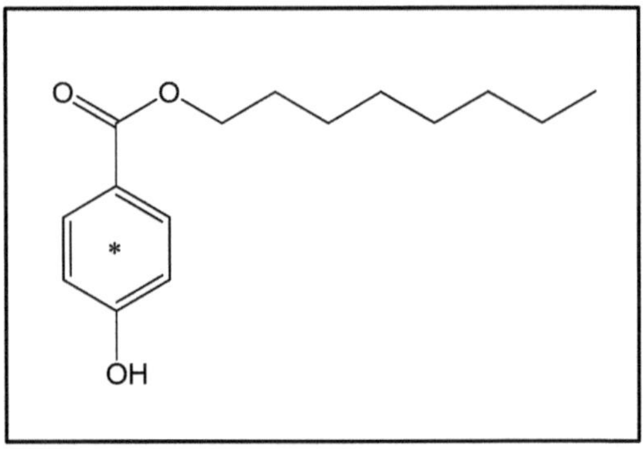

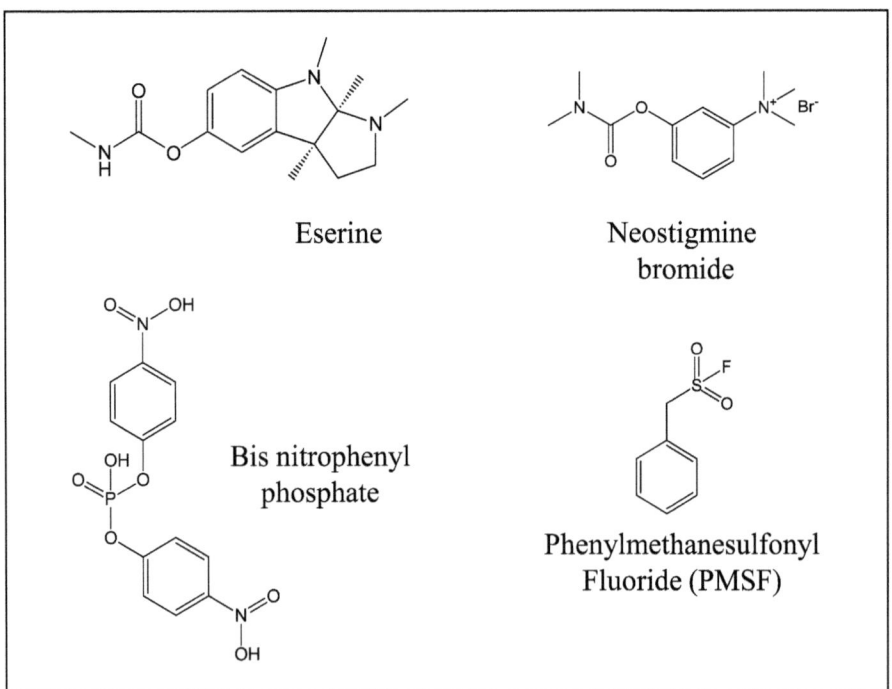

Figure 1 *Structure of octyl paraben (asterisk indicates position of radiolabel).*

Eserine

Neostigmine bromide

Bis nitrophenyl phosphate

Phenylmethanesulfonyl Fluoride (PMSF)

Figure 2 *Structure of enzyme inhibitors.*

Dermatomed (split thickness) human abdominal skin (TRA002490-OC04, 550 µm; The skin was visually checked for overt physical damage prior to being mounted in the test system. The total weight of the cell was recorded, the receptor fluid chamber filled with receptor fluid (0.9% saline + 5% Bovine serum albumin (BSA)) and the weight recorded again. An aliquot (50 µl) of tritiated water (7 µCi ml^{-1}) was dispensed onto the surface of the skin. Aliquots (2 x 50 µl) were removed from the receptor fluid for analysis at 0, 0.5, 1 and 2 hours after application. Each 50 µL aliquot was mixed with 2 mL Ultima Gold XR$^{®}$ (Perkin Elmer) scintillant and analysed using a liquid scintillation counter (Tri Carb 2300TR, Perkin Elmer) using an appropriate counting protocol. From the liquid scintillation count, the rate of penetration and permeability coefficient (Kp) was determined. The original volume of the receptor fluid was maintained by the addition of 0.9% saline + BSA at each sampling time-point. After the last sample time-point, the tritiated water was removed by rinsing with deionised water. The skin was maintained for use by filling the receptor chamber and standing in a covered water bath where the temperature was kept at 32 ± 2^{0}C. Skin samples were deemed viable if the rate of penetration was not greater than 2 µL cm^{-2} h^{-1} (equivalent to a permeability coefficient (Kp) >10 x 10^{-4} cm h^{-1}).

Skin samples were pre-treated with 1 mL of ethanol/water (1:1 v/v) containing an esterase inhibitor (1 mg mL^{-1} eserine, bis(p-nitrophenyl) phosphate or phenylmethanesulphonyl fluoride; PMSF). Several skin samples were treated with the vehicle only. The radiochemical purity of the dose solution was assessed using HPLC (Figure 3).

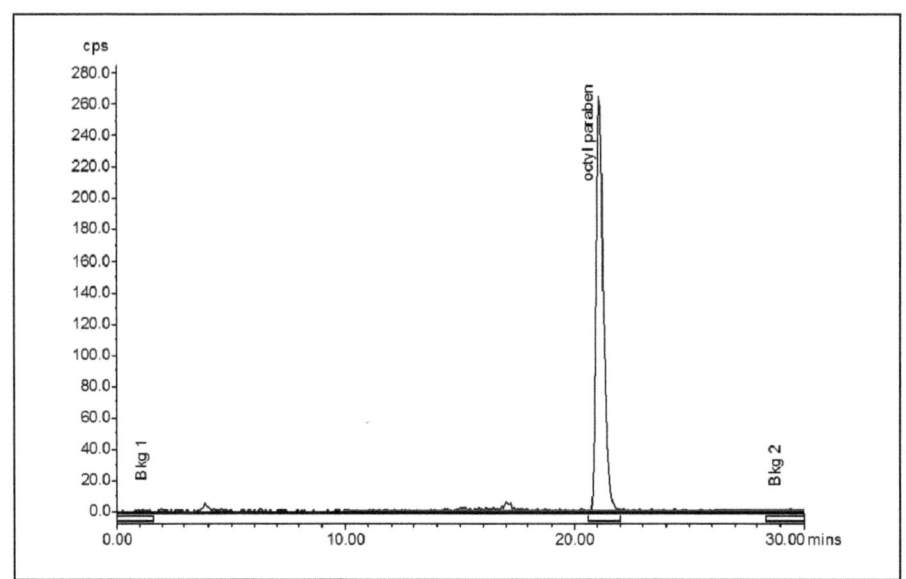

Figure 3 *HPLC radiochromatogram of octyl paraben application (dosing) solution.*

The test compound (octyl paraben) was applied to the skin at a dose of 10 µL cm^{-2} at a target activity of 2 µCi cm^{-2}. The reference compound (testosterone) was applied as a solution in ethanol/water (1/1 v/v). Compounds were, as far as reasonably possible, evenly distributed across the surface of the skin using a microman pipette. As the vapour pressure of the test item is negligible the donor chamber was not covered during the exposure period. The receptor fluid temperature was recorded at each sampling time point and was generally 32 ± 1°C.

Skin was exposed to the test compound for 8 hours after which the dose formulation was removed from the test skin and this area washed using mild detergent solution using a plastic disposable Pasteur pipette. The washings were retained for recovery analysis. Samples of receptor fluid were acquired at 11 sampling points over a 24 hour period (*viz.*, 0.5, 1, 2, 4, 6, 8, 10, 12, 18 and 24 hours).

Duplicate aliquots (2 x 50µl) were removed for sampling and replaced with an equal volume of receptor fluid. The testosterone reference cells were sampled at the same time points as the test compound cells. The removal of any residual testosterone from the surface of the skin using mild detergent solution was carried out after the 24 hour sample had been removed and the washings retained for recovery analysis.

Upon termination of the study all components of the test system were analysed to determine a full system dose recovery. The residual receptor fluid was transferred into an appropriate container and stored at *ca.* -20°C for possible metabolite profiling. The skin was removed and skin fractionation was performed on the skin samples at the termination of the study as described in OECD 428 (section 19) and the OECD guidance document (No 28) section 72. Tape strips (Standard D-Squame, CuDerm Corporation), were used to attempt to remove *stratum corneum* layer of the skin samples from each test cell. Twenty individual strips were used and the first ten strips analysed separately. The remaining skin was solubilised using Solvable® (Perkin Elmer) and analysed by liquid scintillation counting.

The remaining components of each Franz cell (comprising donor chamber, receptor chamber and cell clamp) were placed in separate, pre-weighed containers to which an appropriate volume of 1:1 v/v ethanol:water was added. Each container was sonicated for around 10 minutes and the cell components removed. The solution and container were then re-weighed. Duplicate aliquots (1 ml) of the ethanol water solution were removed for quantitative radiochemical analysis (QRA).

The amount of radioactivity associated with the integrity samples, dosing formulations and receptor fluid were determined by liquid scintillation counting of known volumes/weights of sample. Dosing formulation was diluted with dosing vehicle prior to mixing triplicate aliquots of each solution with scintillant (Ultima Gold XR, Perkin Elmer) prior to liquid scintillation counting.

Skin samples were solubilised using Solvable® (Perkin Elmer). Samples were solubilised at 50°C and mixed with scintillant prior to liquid scintillation counting.

Radioactivity in samples was quantified directly by liquid scintillation counting using a Perkin Elmer Tri-Carb 2300TR with automatic external quench correction. Detected counts per minute (cpm) were converted to disintegrations per minute (dpm) using quench correction. The quench curves were prepared using standards purchased from Perkin Elmer Life and Analytical Sciences using stock solutions calibrated against National Institute of Standards and Technology (NIST) reference materials. The validity of the curves was checked throughout the experiments. Radioactivity with less than twice background counts (<2 x background = 37.9 dpm) were considered to be below the limit of quantification.

The parent compound (octyl paraben) and metabolites were identified by reverse-phase HPLC using both radiochemical detection and UV detection at 254 nm. A Prodigy C18 reverse phase HPLC column was used (250 x 4.6 mm i.d., Phenomenex, Cheshire, UK), which was fitted with a C18 pre-column (Phenomenex). The compounds were eluted using a gradient of (0.45 µm) filtered mobile phase: mobile phase A was methanol, whilst mobile phase B was (40 mM aqueous) ammonium acetate:methanol (45:55, v/v). The gradient was held at 100% B for 5 minutes, lowered to 90% B at 10 minutes and then 20% B at 20 minutes. Analysis was carried out at room temperature with a mobile phase flow rate of 1.0 ml min^{-1}. Injections (200 µl) of each sample were eluted and metabolites and/or parent compounds were identified by comparing retention times with those of authenticated standards.

3   RESULTS

The radiochemical purity of the octyl paraben used for application was > 97% (Figure 3).

The percutaneous absorption data for octyl paraben through skin at 24 hours is shown in Table 1. Test compound was applied to untreated skin and skin pre-treated with esterase inhibitors. Total recoveries were 90.8 – 93.2% of applied radioactivity. Radioactivity recovered in the receptor fluid following application to untreated skin accounted for 9.1% of applied radioactivity and similar proportions of radioactivity (9.8 and 10.3% of applied radioactivity) were observed in skin samples treated with bis (p-nitrophenyl) phosphate and neostigmine bromide, respectively. In samples of skin pre-treated with eserine and PMSF, the proportions of radioactivity recovered in the receptor fluid were notably lower (4.9 and 1.4% of applied radioactivity, respectively).

Concentrations of radioactivity in samples of receptor fluid were also measured to provide flux data for octyl paraben in each cell (Table 2). The lag time for permeation was in the range of 0.91 – 1.94 hours. Concentrations of test compound in receptor fluid from untreated skin and skin pre-treated with bis (p-nitrophenyl) phosphate and neostigmine bromide were similar at 0.44, 0.48 and 0.51 µg mL^{-1} whilst concentrations of radioactivity observed in receptor fluid from skin pre-treated with serine and PMSF were notably lower, being 0.24 and 0.06 µg mL^{-1}, respectively.

Permeability coefficients calculated for octyl paraben in receptor fluid from untreated skin and skin pre-treated with bis (p-nitrophenyl) phosphate and neostigmine bromide were similar, being 1.11 x10^{-4}, 1.12 x 10^{-4} and 1.26 x 10^{-4} cm h^{-1}, respectively. In contrast, permeability coefficients for skin pre-treated with serine and PMSF were 2.37 x 10^{-4} and 0.14 x 10^{-4} cm h^{-1}, respectively.

Table 1 *Recovery of radioactivity from skin 24 hours (expressed as percentage of applied dose) after application of octyl paraben from untreated (control) skin and following pre-treatment with the inhibitors Bis (p-nitropheny)l phosphate (BPNPP), eserine, neostigmine bromide (NB) and phenylmethylsulfonyl fluoride (PMSF).*

Compartment	Inhibitor				
	Control	BPNPP	Eserine	NB	PMSF
Receptor fluid	9.11	9.82	4.87	10.32	1.41
Skin Wash	19.33	35.08	27.83	26.31	49.33
Skin Residue	52.76	40.63	47.08	48.71	35.47
Cell wash	9.61	7.33	12.01	7.86	5.76
Total recovery	90.81	92.86	91.79	93.21	91.97

Table 2 *Permeation parameters for octyl paraben through untreated (control) skin or following pre-treatment with inhibitor (Bis (p-nitrophenyl) phosphate (BPNPP), eserine, neostigmine bromide (NB) and phenylmethylsulfonyl fluoride (PMSF).*

Inhibitor	Parameter			
	Absorbed Dose (24 h) $\mu g\ mL^{-1}$	Flux $(\mu g\ cm^{-2}\ h^{-1})$	Kp $(cm\ h^{-1}\ x\ 10^{-4})$	Lag time (h)
Control	1.61	0.44	1.11	1.48
BPNPP	1.56	0.48	1.12	0.91
Eserine	0.79	0.24	2.37	1.19
NB	1.61	0.51	1.26	1.38
PMSF	0.20	0.06	0.14	1.94

Receptor fluid samples obtained at 24 hours were taken for analysis using HPLC and proportions of radioactive components determined (Table 3). Receptor fluid obtained following application of octyl paraben to untreated skin contained three main radioactive components identified as octyl paraben (3.75 % AR), ethyl paraben (0.76% AR) arising from trans-esterification with the ethanol solvent used for dose application and para-hydroxy benzoic acid, 4.06 % AR (Figure 4A). Proportions of radioactive components in receptor fluid from skin pre-treated with esterase inhibitors are also provided in Table 3 and Figures 4B - D.

Table 3 *Proportion of radioactive compounds in receptor fluid 24 hours after application of octyl paraben, expressed as concentration (μg mL⁻¹), with corresponding percentage of total recovered radioactivity in parenthesis. Skin was either untreated (control) or pre-treated with inhibitor (Bis-paranitrophenyl phosphate (BPNPP), eserine, neostigmine bromide (NB) and phenylmethylsulfonyl fluoride (PMSF).*

Compound	Inhibitor				
	Control	BPNPP	Eserine	NB	PMSF
p-OH benzoic acid	4.06 (45)	1.65 (17)	3.04 (63)	2.83 (27)	0.02 (1)
ethyl paraben	0.76 (8)	0.38 (4)	0.61 (13)	0.48 (5)	0.02 (1)
octyl paraben	3.75 (41)	7.51 (76)	1.03 (21)	6.71 (65)	1.25 (89)

Proportions measured in receptor fluid from skin pre-treated with bis (p-nitrophenyl) phosphate and neostigmine bromide were similar to those observed in untreated skin; octyl paraben (6.71 – 7.51 % AR), ethyl paraben (0.38 – 0.48% AR) and para-hydroxy benzoic acid (1.65 – 2.83 % AR), whilst the proportion of radioactive components observed in receptor fluid from skin pre-treated with serine and PMSF indicated that the extent of both percutaneous permeation and biotransformation was notably lower; octyl paraben (0.02 – 0.61 % AR), ethyl paraben (0.76% AR) and para-hydroxy benzoic acid, (0.02 – 3.04 % AR).

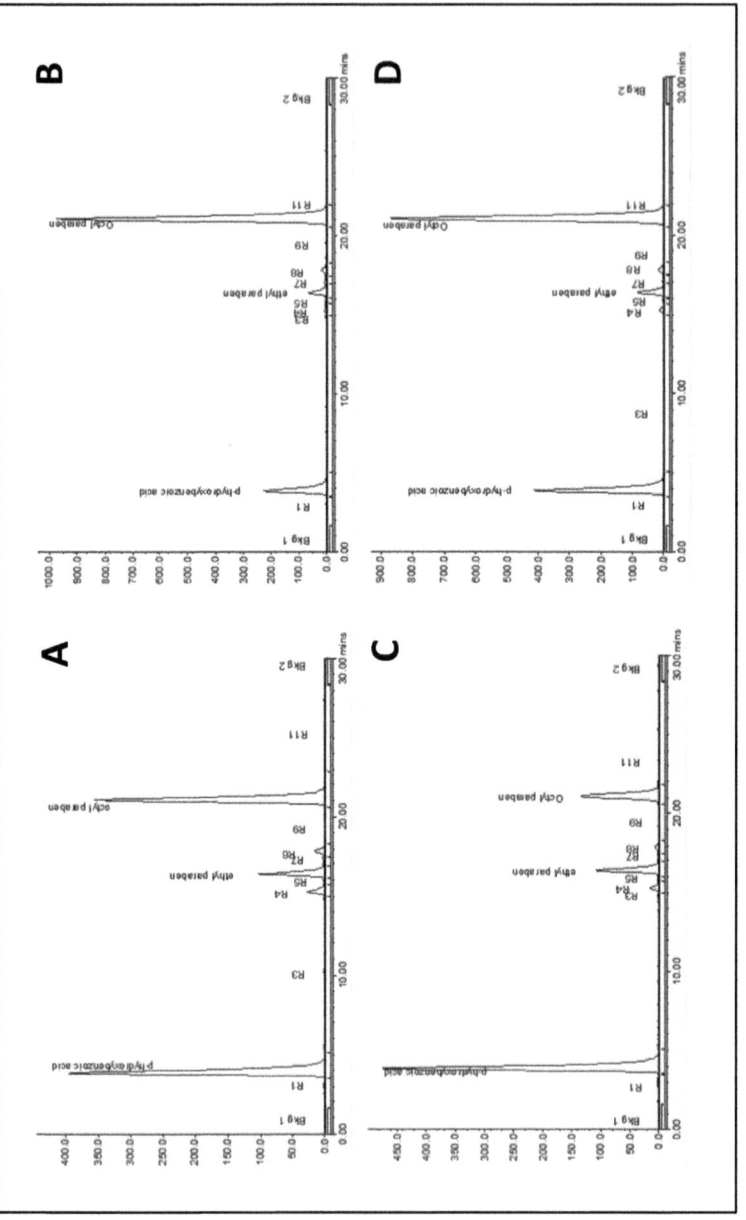

Figure 4 HPLC radio-chromatogram of receptor chamber fluid from diffusion cells containing untreated skin (A) or skin pre-treated with (B) bis para nitrophenyl, (C) Eserine and (D) neostigmine bromide prior to application of octyl paraben.

4   CONCLUSIONS

The test compound (octyl paraben) was shown to be metabolised during passage through the skin. Both hydroxybenzoic acid and ethyl paraben were present in receptor fluid indicating that trans-esterification had occurred. The results from pre-treatment with bis (p-nitrophenyl) phosphate, eserine and neostigmine bromide showed a decrease in both trans-esterification and conversion to hydroxybenzoic acid indicating partial/reversible inhibition of esterases present in the skin. Following pre-treatment with phenylmethanesulphonyl fluoride (a smaller, less specific molecule that binds irreversibly to serine proteases) the major component observed in the receptor fluid was the parent compound.

Whilst the skin has long been considered an organ with low capacity for drug metabolism, it should be noted that most drug metabolising enzymes are present in the skin, albeit at low activity. These can determine whether a topically applied material reaches the systemic circulation as parent material or metabolite. As metabolism can lead to both activation and inactivation of xenobiotic materials an understanding of the distribution, activity and metabolic capability of these skin enzymes is important when performing a dermal risk assessment.

5   REFERENCES

1. N. Ahmad, R. Agarwal and H. Murkhtar, Cytochrome P-450-dependent drug metabolism in skin, *Clinics in Dermatology*, 1996, **14**, 407–413.

2. F. Oesch, E. Fabian, B. A. Oesch-Bartlomowicz, C. Werner and R. Landsiedel, Drug-Metabolizing Enzymes in the Skin of Man, Rat and Pig, *Drug-Metabolism Reviews*, 2007, **39**(4), 659–69.

3. C. Svensson, 'Biotransformation of Drugs in Human Skin', *Drug Metabolism and Disposition*, 2009, **37**(2), 247–253.

4

COLLECTION AND CHARACTERISATION OF RESIDUAL SKIN SURFACE
COMPONENTS OBTAINED FROM SEBUM, SWEAT AND EPIDERMAL LIPIDS

S S Shetage[1], M J Traynor[1], M B Brown[1,2] and R P Chilcott[1]

[1]Department of Pharmacy, Centre for Topical Drug Delivery and Toxicology Research, University of Hertfordshire, Hatfield, United Kingdom. [2]MedPharm Ltd, 50 Occam Road, Surrey Research Park, Guildford, Surrey, United Kingdom

1   INTRODUCTION

The outermost surface of human skin, the *stratum corneum*, is covered with a superficial layer known as the acid mantle which consists of sebum, sweat and corneocyte debris [1]. Sebum is secreted by holocrine, muiltilobular sebaceous glands which are mainly associated with hair follicles and are present all over the body except the palmer-plantar regions [2]. Freshly secreted sebum is a clear, oily substance and mainly consists of squalene, wax esters and triglycerides, free fatty acids with small amount of cholesterol and cholesterol esters [3-5], the composition of which varies according to anatomical location [6]. Free fatty acids present in sebum on the skin surface impart low pH to the skin surface and are associated with hydrolysis of triglycerides by lipases activity of *P. acnes* [7]. The role of sebaceous glands in humans has been a controversial subject [8]. However, the control of sebaceous glands by complex hormonal mechanism and their development in particular areas such as face, upper back and chest imply that these glands are not vestigial [9]. In fact, sebum may contribute in maintaining the skin surface [10], thermoregulation [11], control of gram positive colonisation [12], delivery of Vitamin E (superficial antioxidant) to the skin surface [13] and excretion of lipophilic compounds [14-16].

Collection and analysis of skin surface lipids or sebum has been reported in several previous studies where Sebutape™ [17], polyurethane foam [18], cigarette paper [19, 20] and organic solvents [6, 21, 22] were used to acquire samples, analysis of which was performed by thin layer chromatography [23, 24], gas chromatography [25, 26] and infra-red spectroscopy [27]. However, there is great variation in the reported quantity of sebum present on the skin surface as well as its composition [28]. As sweat and sebum are present on the skin surface at the same time in the form of a mixture [1], material collected from the skin surface may not necessarily be representative of sebum but rather a mixture of sweat, sebum and the lipids of the *stratum corneum*. Thus, the phrase 'residual skin surface components' (RSSC) is a more appropriate term for the mixture of components recovered from the skin surface. The aims of this human volunteer study were to identify an optimal collection medium for RSSC, to investigate the quantity of RSSC collected with regards to age, gender and ethnicity and to determine the composition of RSSC using gas chromatography – mass spectrometry (GC-MS).

2   MATERIALS AND METHODS

Cigarette paper (Rizla+™ red, density: 17.5 gm^{-2}, thickness: 27 µm, composition: 14% calcium carbonate, 86% eucalyptus cellulose fibers), absorbent cotton (Boots UK Ltd), polyurethane foams (Boots UK Ltd) and Scotch™ tape (3M UK Ltd) were purchased from a local supplier. Sebutape™ was purchased from CuDerm Corporation, Dallas, USA. Diethyl Ether and Hexane (GC grade) were obtained from Sigma-Aldrich UK. A Mettler Toledo AX205 series balance was used for gravimetric analysis and gas chromatography analysis was performed on Varian 450 GC with Varian 240 MS.

All volunteers provided fully informed consent prior to participating in the study. Personal information and details of health conditions of volunteers were collected with the help of a questionnaire. Volunteers were categorised into different ethnic groups based on the self-declared information according to the 'Household Questionnaire, Census 2011' [29]. For the purpose of this study, the 'Asian' group comprised individuals from India, Pakistan and Bangladesh. A total of 230 volunteers participated in the study. Ethical permission to perform this study was approved by the School of Pharmacy and Postgraduate Medicine Ethics Committee with Delegated Authority (ethics approval number: PHAEC/10-25).

Identification of Suitable Collection Media.

Suitable collection material for RSSC acquisition was determined on three volunteers (1 male and 2 females, Age: 29 ± 4 years, mean age ± SD) by comparing six products: cigarette paper, polyurethane foams type A (pore size: 300 µm) and type B (pore size: 75 µm), absorbent cotton, Scotch™ tape and Sebutape™. Each sampling medium was cut into 2 cm x 2 cm squares, soaked in diethyl ether and fully dried before use, with the exception of Sebutape™ and Scotch™ which were not solvent treated. Collection of RSSC was performed by applying each medium to five sites on the forehead which was held in place with an elasticised headband for three hours. Each sampling media was dried and weighed before and after RSSC collection and the difference in weight was ascribed to the amount of RSSC collected which was expressed as mg cm^{-2}.

Characterisation of Cigarette Paper Collection Method.

The cigarette paper sample acquisition methods for RSSC collection [19] was characterised using 10 volunteers (5 males and 5 females, 31 ± 11 years old, mean age ± SD) by collecting samples on two different occasions. Cigarette papers were pre-treated before application as described above and were subjected to dehydration by passive evaporative loss in plastic sample cups covered with pierced Parafilm™ in a fume cupboard for a minimum period of two hours [30] after removal from the forehead. At the end of this period, each cigarette paper was reweighed to determine the quantity of RSSC collected.

Characterisation of RSSC

Using the optimised (cigarette paper) sample acquisition method described above, RSSC was collected from 230 volunteers. Dried, pre-weighed cigarette papers were applied to the left and right side of the forehead for a period of one hour. For volunteers ≤ 17 years old, RSSC was collected for one hour while for volunteers ≥ 18 years old, the papers were replaced with a fresh papers each hour up to 3 hours. All the samples were collected at ambient room temperature (18 – 25° C) and relative humidity (50 – 60 %). Papers were reweighed after collection period as described above. Gravimetric analysis was performed to determine the quantity of RSSC collected from each volunteer.

For GC-MS analysis, all the papers from an individual volunteer were placed in a glass vial containing 4 ml of hexane [31]. The vials were shaken vigorously for 5 minutes and allowed to stand for 20 minutes. The papers were then removed and the extract filtered using a 0.2 µm PTFE membrane. The extract was concentrated by purging with nitrogen gas until approximately 1 ml of sample remained. Chromatographic separation was performed on Varian 450 GC, using 5% phenyl-95% dimethyl polysiloxane capillary column (length 30 m, i.d. 0.25 mm, film thickness 0.25 µm) and helium (1 ml min^{-1}) as a carrier gas. The column was heated to 50° C for 2 minutes following sample injection and subsequently increased (10° C min^{-1}) to 330° C which was maintained for a further 5 minutes. Detection was performed using a Varian 240 MS, on full ion scan mode (range 40 - 1000 Da) following electron impact ionisation.

3   RESULTS

The quantity of RSSC collected from cotton was significantly higher than Sebutape™, cigarette paper, PU foam type A and B and Scotch™ tape ($p < 0.05$, Kruskal-Wallis ANOVA) and the coefficient of variation (CV) for Sebutape™, cigarette paper, cotton and Scotch tape™ was generally less than 0.26. In contrast, both type A and B foams exhibited high variability (CV exceeding 0.5; Figure 1).

The cigarette paper method for RSSC sample acquisition was reproducible: there was no statistically significant difference in the quantity of RSSC collected on two occasions from the same volunteer (p>0.05, paired t-test) and was reproducible between individuals (r = 0.86)

No statistically significant differences were observed in the amount of RSSC collected in the first hour from males (0.12 ± 0.06 mg cm^{-2}, Mean ± SD) and females (0.13 ± 0.07 mg cm^{-2}, p>0.05, Mann Whitney). A comparison of ethnic groups showed that RSSC accumulation was not significantly different between African, White or Asian population (p>0.05, Kruskal-Wallis; Figure 2). Similarly, there was no significant correlation between RSSC recovery and age (r = 0.27; Figure 3). However, recovery of RSSC was dependent on the collection time, with accumulation being significantly higher in first hour as compared to the second and third hour (Figure 4).

Qualitative (GC-MS) analysis of human RSSC indicated the presence of forty nine compounds which were classified in five lipid classes: squalene, cholesterol and cholesterol esters, wax esters, free fatty acids and triglycerides. The major components in each of these classes were squalene; cholesterol; cholest-5-en-3-ol (3β)-, acetate; Z-hexadec-9-enoic acid icosyl ester; *cis*-9-Octadecenoic acid and trilinolein, respectively.

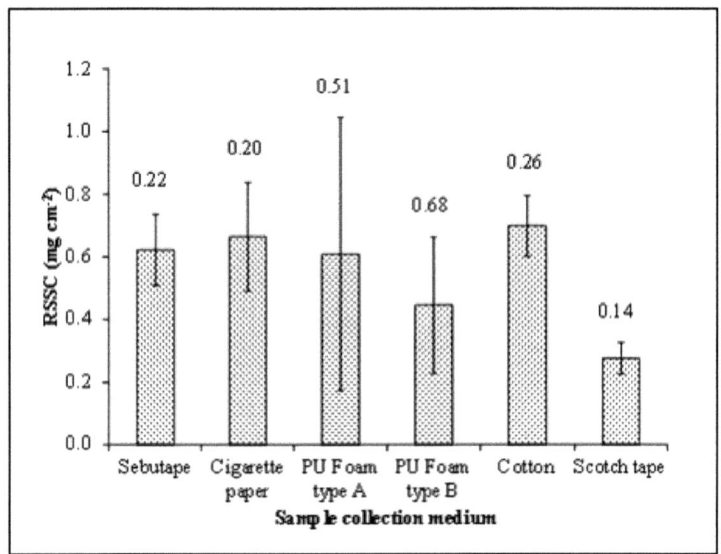

Figure 1 *Amount of RSSC collected (n = 5) from a volunteer (Volunteer A: African, Female, 26 years old) using six collection media. Number on the top of each bar represents the coefficient of variation (CV) of RSSC collection among three volunteers (Volunteer B: Asian, Male, 29 years old and Volunteer C: African, Female, 34 years old)*

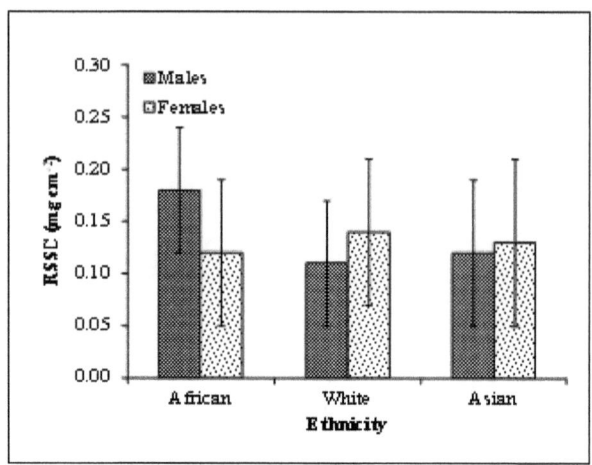

Figure 2 *Collection of RSSC from three different populations. There was no statistically significant difference in the amount of RSSC collected from any of these groups (p<0.05), Kruskal-Wallis).*

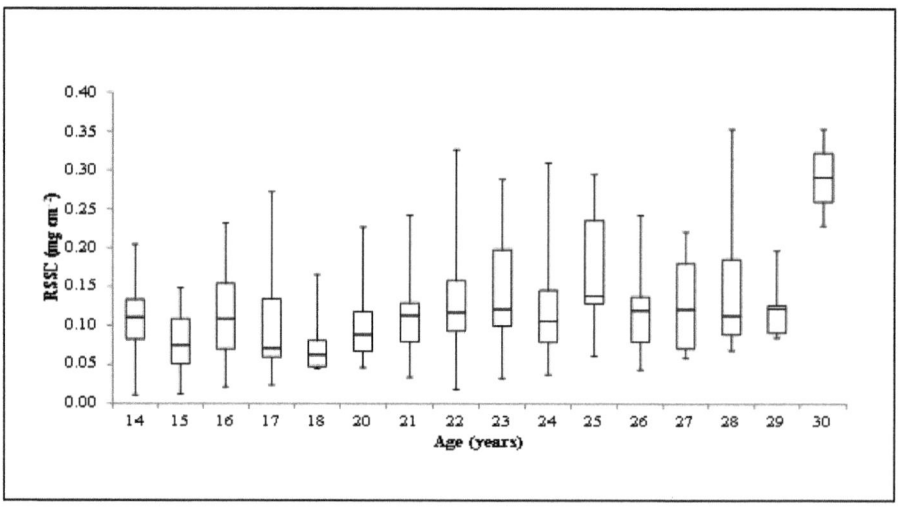

Figure 3 *Quantity of RSSC collected from human volunteers. Each box represents interquartile range with median, while minimum and maximum values of RSSC collection are shown by bars. There was no correlation between the age of volunteer and the quantity of RSSC collected.*

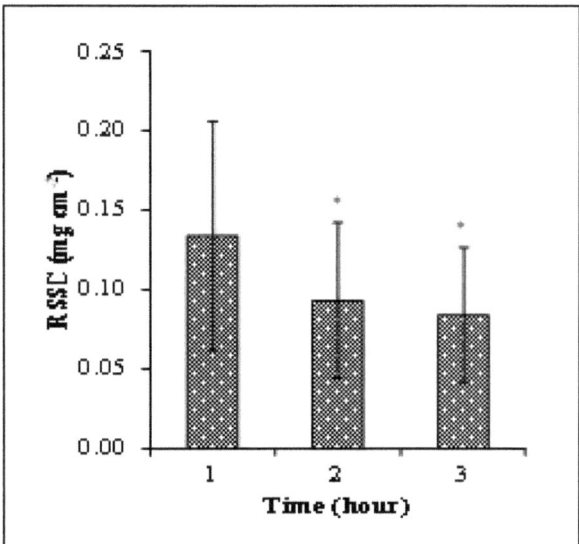

Figure 4 *Amount of RSSC collected in each hour using cigarette paper method. A statistically significant difference was observed in RSSC accumulation in the first hour compared to the second and third hour (p<0.05, as indicated by asterisks).*

4   DISCUSSION

Six collection media were compared to identify a suitable acquisition medium for RSSC (Figure 1). Cotton (wool) collected the highest quantities of RSSC and variability in the replicates was less than for the two polyurethane foams. However, retaining integrity of the cotton after application on the forehead was found to be practically difficult, as a proportion adhered to the application strap and thus prevented full recovery for gravimetric analysis. The polyurethane foams exhibited the highest variability in RSSC collection, making it unsuitable as an acquisition medium. The adhesive on the collecting surface of Sebutape™ and Scotch tape™ was likely to collect a large proportion of corneocytes (thus overestimating sebum recovery) and could potentially interfere with chemical analysis due to extraction of adhesive along with RSSC [31]. Additionally, RSSC collection from Sebutape™ was more variable than cigarette paper. In contrast, cigarette paper retained its integrity, contained no adhesive component and so presented the purest form of RSSC sample [32, 33] and so was selected as the sampling medium for RSSC.

The reproducibility of the cigarette paper method was determined by collecting RSSC on two separate occasions from the same volunteers: the method was highly reproducible as no significant difference was observed between samples. Therefore, these data confirmed that the cigarette paper method for acquisition of RSSC was robust and reproducible.

Overall, there was no significant difference in the total amount of RSSC collected from the forehead of male or female volunteers. This is in accordance with previous studies [34, 35]. Cotterill *et al.* reported that sebum secretion rate starts to increase around the age of 15, is higher in males than females and continues to increase gradually until 40 years after which sebum secretion decreases to a greater degree in females than males [36, 37]. These changes in RSSC were not observed in this current study (Figure 3), although the age range of volunteers in our study was limited to 30 years.

It has been reported that sebum excretion varies according to age and gender as a consequence of hormonal signals [38-40]: In prepubescent males, testosterone levels are relatively low, rise during puberty, peak in the third decade of life and then decline thereafter. Similarly, oestrogen levels are low in early childhood of females, rise during puberty and remain at this level until the menopause [41]. The outcomes of this study do not support the suggestion that there is a (hormone-mediated) correlation between age and RSSC secretion, at least within the 14 to 30 year old range of volunteers used in this study.

A previous study on differences in ethnic skin types showed an increasing order of sebum production from Hispanics to Caucasians to African-Americans. It was also reported that East Asians have lower sebum secretion than African Americans [37]. No statistically significant differences in RSSC production were identified between the African, White and Asian volunteers in this present study (Figure 2). Moreover, it has been suggested that African individuals may have higher RSSC accumulation as a result of larger sebaceous glands [42, 43]. Again, the lack of a significant difference between the ethnic groups in this present study does not support this hypothesis and so putative factors such as pore size and size of sebaceous glands do not appear to affect RSSC accumulation based on skin type [44].

The rate of RSSC accumulation on the skin was found to be highest over the first hour and then subsequently decreased (Figure 4). Eberhardt *et al.* [45] observed similar changes in sebum secretion, ascribing this effect to a 'feed-back' mechanism in which surface lipids physically inhibit further excretion [45]. Alternatively, the higher initial rate of sebum secretion (following skin cleansing) may be simply due to follicular depletion [46].

A relatively sensitive (GC-MS) method of analysis was employed in this study to identify the individual components of RSSC. Small sample quantity is a frequent technical issue in skin surface lipid analysis [33] and so the use of GC-MS provided a suitably sensitive technique. Compounds from all five reported chemical classes of sebum were detected, albeit in varying proportions. Previous studies have evaluated sebum composition using gas chromatography following the time-consuming process of sample derivatisation [26, 47]. The direct method developed in the present study involved no derivatisation steps and minimal sample preparation. Therefore, this method could be used as a rapid method for RSSC analysis.

Overall, whilst some variability in the composition of RSSC was noted, the reproducible recoveries between gender, age and ethnic groups suggest that RSSC has some potential for use as a non-invasive method for monitoring exposure to exogenous chemicals. Whilst further work is required to provide a more comprehensive analysis of individual components, it is conceivable that RSSC may also provide a means of detecting disease states in patients where lipid metabolism has been altered (such as diabetes) through changes in the relative amounts of individual residual skin surface components.

5 CONCLUSIONS

The results of the present study confirm that RSSC can be collected using a simple method based on the topical application of cigarette paper and subsequently analysed by GC-MS with minimal sample preparation. Gender, age and ethnicity had no significant impact on the quantity of RSSC recovered. Overall, these data imply that RSSC can be subject to a simple, non-invasive method of analysis and so may have application as a bio-monitoring matrix to assess exposure to environmental chemicals.

6 REFERENCES

1. A. B. Stefaniak, C. J. Harvey and P.W. Wertz, Formulation and stability of a novel artificial sebum under conditions of storage and use, *Int. J. Cosmetic Sci.*, 2010, **32**(5), 347–355.

2. P. W. Werts and B. B. Michniak, *Cosmeceuticals and Active Cosmetics, Drugs vs. Cosmetics.*, 2nd edn, ed. P. Elsner and H. I. Maibach, 2004, New York: Marcel and Dekker Inc, Chapter 5.

3. D. T. Downing, *et al.*, Skin lipids: an update, *J. Investigative Dermatology*, 1987, **88**, 2–6.

4. M. E. Stewart and D.T. Downing, Measurement of sebum secretion rates in young children, *J. Investigative Dermatology*, 1985, **84**(1), 59–61.

5. M. L. Rosenthal, *Cosmetics and the skin*, ed. I. Wells and F.V. Libowe, 1964, Reinhold, XX.

6. R. S. Greene, *et al.*, Anatomical variations in the amount and composition of human skin surface lipid, *J. Investigive Dermatology*, 1970, **54**(3), 240–247.

7. P. Clarys and A. Barel, Quantitative evaluation of skin surface lipids, *Clinics in Dermatology*, 1995, **13**(4), 307–321.

8. A. M. Kligman, The Uses Of Sebum, *British Journal of Dermatology*, 1963, **75**(8–9), 307–319.

9. A. J. Thody and S. Shuster, Control and function of sebaceous glands, *Physiological Reviews*, 1989, **69**(2), 383–416.

10. B. Rode, U. Ivens, and J. Serup, Degreasing method for the seborrheic areas with respect to regaining sebum excretion rate to casual level, *Skin Research and Technology*, 2000, **6**(2), 92–97.

11. A. M. W. Porter, Why do we have apocrine and sebaceous glands?, *Journal of Royal Society of Medicine*, 2001, **94**(5), 236.

12. A. Kydonieusb, J. J. Wille, Palmitoleic Acid Isomer (C16: 1¢ 6) in Human Skin Sebum Is Effective against Gram-Positive Bacteria, *Skin Pharmacol. Appl. Skin Physiol.*, 2003, **16**, 176–187.

13. J. J. Thiele, S.U. Weber, and L. Packer, Sebaceous gland secretion is a major physiologic route of vitamin E delivery to skin, *Journal of Investigative Dermatology*, 1999, **113**(6), 1006–1010.

14. J. Faergemann, *et al.*, Levels of terbinafine in plasma, stratum corneum, dermis-epidermis (without stratum corneum), sebum, hair and nails during and after 250 mg terbinafine orally once per day for four weeks, *Acta dermato-venereologica*, 1993, **73**(4), 305.

15. T. Iida, *et al.*, Recent trend of polychlorinated dibenzo-p-dioxins and their related compounds in the blood and sebum of Yusho and Yu-Cheng patients, *Chemosphere*, 1999, **38**(5), 981–993.

16. G. Cauwenbergh, *et al.*, Pharmacokinetic profile of orally administered itraconazole in human skin, *Journal of the American Academy of Dermatology*, 1988, **18**(2), 263–268.

17. A. M. Kligman and D. L. Miller, Sebutape: A Device for Visualising and Measuring Human Sebaceous Secretion, *J. Soc. Cosmet. Chem*, 1986, **37**, 369–374.

18. P. Ramasastry, *et al.*, Chemical composition of human skin surface lipids from birth to puberty, *Journal of Investigative Dermatology*, 1970, **54**(2), 139–144.

19. J. S. Strauss and P.E. Pochi, The Quantitative Gravimetric Determination of Sebum Production, *Journal of Investigative Dermatology*, 1961, **36**(4), 293–298.

20. W. Cunliffe and S. Shuster, The rate of sebum excretion in man, *British Journal of Dermatology*, 1969, **81**(9), 697–704.

21. I. Hodgson-Jones and V. Wheatley, Studies of Sebum. 3. Methods for the collection and estimation of small amounts of sebum, *Biochemical Journal*, 1952, **52**(3), 460.

22. D. T. Downing, J. S. Strauss and P. E. Pochi, Variability in the Chemical Composition of Human Skin Surface Lipids 1, *Journal of Investigative Dermatology*, 1969, **53**(5), 322-327.

23. D. T. Downing, M. E. Stewart and J. S. Strauss, Estimation of sebum production rates in man by measurement of the squalene content of skin biopsies, *Journal of Investigative Dermatology*, 1981, **77**(4), 358–360.

24. S. C. Green, M. E. Stewart and D. T. Downing, Variation in sebum fatty acid composition among adult humans, *Journal of Investigative Dermatology*, 1984, **83**(2), 114–117.

25. A. James and V. Wheatley, Studies of sebum. 6. The determination of the component fatty acids of human forearm sebum by gas–liquid chromatography, *Biochemical Journal*, 1956, **63**(2), 269.

26. R. Michael-Jubeli, J. Bleton and A. Baillet-Guffroy, High-temperature gas chromatography-mass spectrometry for skin surface lipids profiling, *Journal of Lipid Research*, 2011, **52**(1), 143–151.

27. A. Anderson and J. Fulton, Sebum: analysis by infrared spectroscopy, *Journal of Investigative Dermatology*, 1973, **60**(3), 115–120.

28. G. W. Lu, *et al.*, Comparison of artificial sebum with human and hamster sebum samples. *International Journal of Pharmaceutics,* 2009, **367**(1–2), 37–43.

29. *Household Questionnaire- England Census 2011*. 2011 [cited 2011 18 Dec 2011]; Available from: http://www.ons.gov.uk/ons/guide-method/census/2011/the-2011-census/2011-census-questionnaire-content/index.html.

30. D. Lookingbill, and W. Cunliffe, A direct gravimetric technique for measuring sebum excretion rate, *British Journal of Dermatology*, 1986, **114**(1), 75–81.

31. H. Vaule, S. Leonard and M. Traber, Vitamin E delivery to human skin: studies using deuterated [alpha]-tocopherol measured by APCI LC-MS, *Free Radical Biology and Medicine*, 2004, **36**(4), 456–463.

32. W. Cunliffe, J. Cotterill, and B. Williamson, Variations in skin surface lipid composition with different sampling techniques—I*, *British Journal of Dermatology*, 1971, **85**(1), 40–45.

33. N. Nicolaides and R. Kellum, *S*kin lipids. I. Sampling problems of the skin and its appendages, *Journal of the American Oil Chemists' Society*, 1965, **42**(8), 685–690.

34. K. P. Wilhelm, A. B. Cua, and H. I. Maibach, Skin aging: effect on transepidermal water loss, stratum corneum hydration, skin surface pH, and casual sebum content, *Archives of Dermatology*, 1991, **127**(12), 1806.

35. U. Jacobi, *et al.*, Gender-related differences in the physiology of the stratum corneum, *Dermatology*, 2007, **211**(4), 312–317.

36. J. Cotterill, *et al.*, Age and sex variation in skin surface lipid composition and sebum excretion rate, *British Journal of Dermatology*, 1972, **87**(4), 333–340.

37. G. G.Hillebrand, M. J. Levine, and K. Miyamoto, The age dependent changes in skin condition in African Americans, Asian Indians, Caucasians, East Asians & Latinos. *IFSCC*, 2001, **4**, 259–266.

38. S. Nouveau, *et al.*, Effects of topical DHEA on aging skin: A pilot study. Maturitas, 2008, **59**(2), 174–181.

39. E. Giltay and L. Gooren, Effects of sex steroid deprivation/administration on hair growth and skin sebum production in transsexual males and females, *Journal of Clinical Endocrinology & Metabolism*, 2000, **85**(8), 2913.

40. C. Piérard-Franchimont and G. Piérard, Postmenopausal aging of the sebaceous follicle: a comparison between women receiving hormone replacement therapy or not, *Dermatology*, 2000, **204**(1), 17–22.

41. W. J. Germann and C. L. Stanfield, *Principles of Human Physiology,* 2nd edn, 2005, San Francisco, Daryl Fox.

42. M. Roh, *et al.*, Sebum output as a factor contributing to the size of facial pores, British Journal of Dermatology, 2006, **155**(5), 890–894.

43. A. V. Rawlings, Ethnic skin types: are there differences in skin structure and function? 1, *International Journal of Cosmetic Science*, 2006, **28**(2), 79–93.

44. S. Marrakchi and H. I. Maibach, Biophysical parameters of skin: map of human face, regional, and age related differences, *Contact Dermatitis*, 2007, **57**(1), 28–34.

45. H. Eberhardt, The regulation of sebum excretion in man, *Archives of Dermatological Research*, 1974, **251**(2), 155–164.

46. J. L. Millns and H. I. Maibach, Mechanisms of sebum production and delivery in man. *Archives of dermatological research*, 1982, **272**(3), 351–362.

47. E. Haahti, E. Horning, and O. Castrén, Microanalysis of Sebum and Sebum-Like Materials by Temperature Programmed Gas Chromatography, *Scandinavian Journal of Clinical and Laboratory Investigation*, 1962, **14**(4), 368–372.

5

GABOR DOMAIN OPTICAL COHERENCE MICROSCOPY OF HUMAN SKIN

J P Rolland[1], K S Lee[1-2], P Meemon[1-3], and S F Ibrahim[4]

[1]The Institute of Optics, University of Rochester, 275 Hutchinson Road, Rochester NY 14627, United States of America. [2]Center for Analytical Instrumentation Development, Korea Basic Science Institute, Daejeon, South Korea 305-806 [3]School of Laser Technology and Photonics, Institute of Science, Suranaree University of Technology, Nakhon Ratchasima, Thailand 30000. [4]Department of Dermatology, University of Rochester Medical Center, 400 Red Creek Drive, Suite 200, Rochester, NY 14623, United States of America.

1 INTRODUCTION

The skin is the largest organ of the human body. Its key functions are to cover the internal organs, serve as a barrier to external pathogens and elements, and control haemostasis through thermoregulation and prevention of water loss. Because it is in constant contact with physical, chemical, and biological factors, the skin is susceptible to the development of various benign and malignant tumours. Non-melanoma skin cancers (NMSC), which include basal cell carcinoma (BCC) and squamous cell carcinoma (SCC), are the most prevalent human cancers, with numbers exceeding that of all other malignancies combined [1]. Recent reports estimate that the number of skin cancers in the United States has reached epidemic proportions, with an annual incidence as high as 3,000,000 and this figure is increasing world-wide [2]. Also referred to as keratinocyte carcinomas, they derive from keratinocytes in the epidermis of the skin, thus facilitating their detection, treatment and evaluation of response to treatment [3].

Even though many cutaneous diagnoses can be assessed through physical examination by a dermatologist, skin biopsy remains the gold standard for confirmation of clinical findings. However, biopsy of the skin is time-consuming, costly and has inherent risks such as bleeding, infection, and scarring. Hence, there is an unmet demand for non-invasive skin imaging with cellular resolution for diagnosis of diseases and monitoring of changes over time in response to various therapies.

To date, several non-invasive techniques using both optical and non-optical sources have been reported but have yet to replace the diagnostic role of standard histologic methods. Among non-optical approaches, high-frequency ultrasound and microscopic magnetic resonance imaging offer limited resolution but satisfactory imaging depth [4,5].

The optical properties of skin such as scattering and absorption are nonhomogeneous as a result of different refractive indices and absorption coefficients of skin layers and microstructures. These heterogeneous optical properties can be used to create contrast in imaging of the skin. Among non-invasive and optical imaging methods, dermoscopy is the most widely used in the clinical setting given its ease of use and cost-effectiveness at differentiating skin tumours with spectral reflectivity. However, dermoscopy has low resolution and does not support depth resolution. Reflectance confocal scanning microscopy (RCSM) and label-free two-photon fluorescence microscopy (TPFM) have provided cellular-level resolution and depth sectioning. They have been used widely in clinical investigations to identify NMSC to an imaging depth of 200-300 µm. Depth sectioning capability using a confocal detection mechanism degrades quickly as a result of scattering of light as it passes through skin, resulting in out-of-focus photons that are transmitted through the pinhole onto the detector. This rapidly decreasing contrast limits the usable depth, and hence utility, of imaging. In addition, axial resolution is intrinsically limited in confocal microscopy and thus requires an extremely high numerical aperture (NA) that is sensitive to optical aberrations as the light is focused deeper into tissue. On the other hand, the high NA significantly limits the depth of focus. In label-free TPFM, the imaging depth is limited by intrinsically weak auto-fluorescence signals whose intensity is further reduced by absorption as a function of depth.

Optical coherence tomography (OCT) is a technology applicable to non-invasive medical imaging that allows high-resolution visualisation of microstructural morphology of tissues [6]. Many investigations using OCT have been performed to image subsurface layers and structures of skin including the epidermis, dermal-epidermal junction, dermis, hair follicles, blood vessels and sweat ducts [7 – 12]. A number of clinical studies have suggested that OCT might be useful for non-invasive diagnosis of skin disorders such as inflammatory and bullous skin diseases and to assess wound healing [13,14]. Several reports to date have examined the use of OCT imaging of NMSC. Previous studies have demonstrated OCT images of BCC, SCC and actinic keratoses (AK) and have correlated these findings with standard histology [15 – 23]. The ability of OCT to differentiate NMSC lesions from normal skin has also been assessed in the clinical setting using observer-blinded evaluation by dermatologists and pathologists [24]. To date however, most of the clinical investigations in skin disease using OCT have used two dimensional images or relatively low resolution.

Initial reports of using OCT for skin imaging revealed significantly improved resolution over high frequency ultrasound imaging, as well as better imaging depth and field of view than with RCSM or TPFM. Nonetheless, conventional OCT does not provide histology-level resolution compared to RCSM and TPFM. To overcome this weakness, OCT can use a high NA objective similar to confocal microscopy to achieve cellular lateral resolution. In this imaging configuration, when further combined with phase modulation, en face *in vivo* cellular imaging is possible at specific depths, which is called optical coherence microscopy (OCM) [25,26]. However, this accomplishment does not enable *in-vivo* volumetric imaging ranging from the surface of the skin through the dermis given the limited depth of focus and acquisition speed [27]. Fourier domain OCM which is much faster than time domain OCM does not enable cellular volumetric imaging in sufficient depth of skin without fast refocusing and fusing. Full-field OCT is an approach to achieve cellular-level lateral resolution *in-vivo* volumetric imaging. However, full-field OCT has ~20 dB lower sensitivity compared to conventional FD-OCT. As such, full-field OCT is limited in its application to weakly scattering media such as the Xenopus tadpole or the cornea, or *ex-vivo* imaging [28].

To achieve additional contrast of specific cutaneous structures, several functional extensions to conventional OCT have been applied: phase-resolved Doppler OCT to measure blood flow [29, 30], polarisation-sensitive OCT to visualise birefringence properties of skin [31] (particularly connective tissue) and elastographic OCT to image the local variations of stiffness [32]. The combination of multiple imaging modalities can also be considered to achieve complimentary contrast, which may be helpful to improve the sensitivity and specificity of skin tumour diagnosis. The combination of OCT and TPFM is capable of providing biochemical specificity in addition to skin morphology [33]. Photoacoustic tomography (PAT) detects acoustic signals generated by optical absorption of nanosecond laser pulses within tissue with a spatial resolution in the order of tens of micrometres and a penetration depth of several millimetres. PAT is efficient at visualising vasculature containing haemoglobin, which is an optical absorber in skin. Recently, OCT combined with PAT has demonstrated high resolution structure together with cutaneous vasculature of *in-vivo* human skin [34]. Regardless of the type of extension or multimodality, the improvement in capability of OCT itself will enhance the overall performance of the combined system.

In OCT, axial and lateral resolution required for volumetric reconstruction is determined by the source coherence length and the numerical aperture of the objective lens, respectively. Axial resolution can be improved by use of a broadband light source such as that of an ultra-short pulse laser or a super-continuum source. On the other hand, the improvement in lateral resolution is not straightforward given that there is a trade-off between lateral resolution and depth of focus. Various methods have been investigated to achieve more depth of focus than the confocal parameter of the focusing Gaussian beam by the objective optics. Among current methods, Gabor domain optical coherence microscopy (GD-OCM) has been reported as a solution to achieve histology grade volumetric images for *in-vivo* clinical applications [35]. GD-OCM uses 0.2 NA imaging optics with high-speed variable focus capability within a FD-OCM experimental setup [36, 37].

In this study, we show cellular resolution 3D images of different anatomic locations of *in-vivo* normal skin such as the nail-fold, forearm, and fingertip. We also applied volumetric GD-OCM at cellular resolution to the morphologic investigation of *ex-vivo* NMSC, comparing cellular OCM images with 3D features of normal and abnormal skin (i.e. BCC and SCC).

2 METHODS

For normal skin imaging, healthy skin located in the nail-fold, fingertip, and forearm of two volunteers were imaged *in-vivo*. For skin cancer imaging, one biopsy-proven BCC from the nose and one biopsy-proven SCC from the ear were obtained as discarded specimens from Mohs micrographic surgical excisions. Two normal skin samples from equivalent locations (i.e. nose and ear) were used as comparisons for the two NMSCs. Patient information was unknown with the exception of gender and age. The excised samples were immediately placed on ice and transported to the imaging laboratory at the University of Rochester within 3 hours of excision. All imaging was conducted by applying the objective lens to the skin after application of ultrasound gel for index matching. Use of discarded skin samples was approved by the University of Rochester Research Subjects Review Board.

The imaging system with a micron-class resolution of 2 μm axially and laterally in skin tissue consisted of a Titanium:Sapphire femtosecond laser centred at 800 nm with 120 nm FWHM (Integral, Femtolasers Inc.), a custom liquid-lens-based three-dimensional scanning microscope [37], a broadband custom-made fiber coupler (NSF-DARPA/PTAP), a custom dispersion compensator [38] and a custom spectrometer with a high speed CMOS line camera (spl8192-70km, Basler Inc) [39]. The microscope was designed for clinical application as a handheld probe with a liquid lens embedded for refocusing with no moving parts. The microscope enabled 2 μm resolution at an imaging volume of 2 mm × 2 mm × 2 mm in the system [37]. A photograph of the microscope is shown in Figure 1. Spectra were acquired with an exposure time of 48 μs and a readout speed of 10,000 spectra s^{-1}. For *ex-vivo* skin cancer imaging, we acquired six volumes from the same 1000 μm × 1000 μm × 600 μm portion of a skin tissue with a sampling interval of 1 μm, which corresponds to 6 × 1000 × 1000 A-scans for total spectra. The six volumes were taken relative to a shifted focal plane with 100 μm separation. Using the 100 μm separation between focal planes, over a depth of 60 μm there is 2 μm lateral resolution and the remaining imaging depth (i.e. 40 μm) supports 3 μm lateral resolution measured as a minimum of 20% contrast criterion [40]. The six volume images were reconstructed in post processing into one volume using the Gabor-based fusion technique [36]. All acquisition and post processing including interpolation, fast Fourier transform, and fusion were done using Labview software based on 64 bit Windows 7.

Figure 1 *Custom, liquid-lens embedded, dynamic focusing microscope.*

3   RESULTS

In-Vivo 3D High Resolution Images of Normal Skin

In-vivo human fingertip skin was imaged (Figure 2). Epidermis, papillary dermis, stratum corneum, and a sweat gland are shown in a 300 μm × 200 μm × 400 μm volume in Figure 2a. The epidermis consists of the stratum corneum and viable epidermis. The stratum corneum is a thick and almost transparent layer except for the highly scattering sweat ducts as shown in Figure 2a. The viable epidermis consists of the stratum granulosum (SG), stratum spinosum (SS), and stratum basale (SB) compared to the stratum corneum that has dead cells without nuclei. The highly scattering sub-volume enclosed within the dashed lines in Figure 2a is the viable epidermis that is shown in *en face* views every 12 μm in depth in Figure 2b-i. Cellular morphology is seen in the ridge of the papillae in the bottom of Figure 2b-c, where the cellular size is larger than the observed cellular size in Figure 2g-i. The size of the cells designated by an arrow in Figure 2b, g, and i were estimated to be about 30 μm, 20 μm, and 10 μm, respectively.

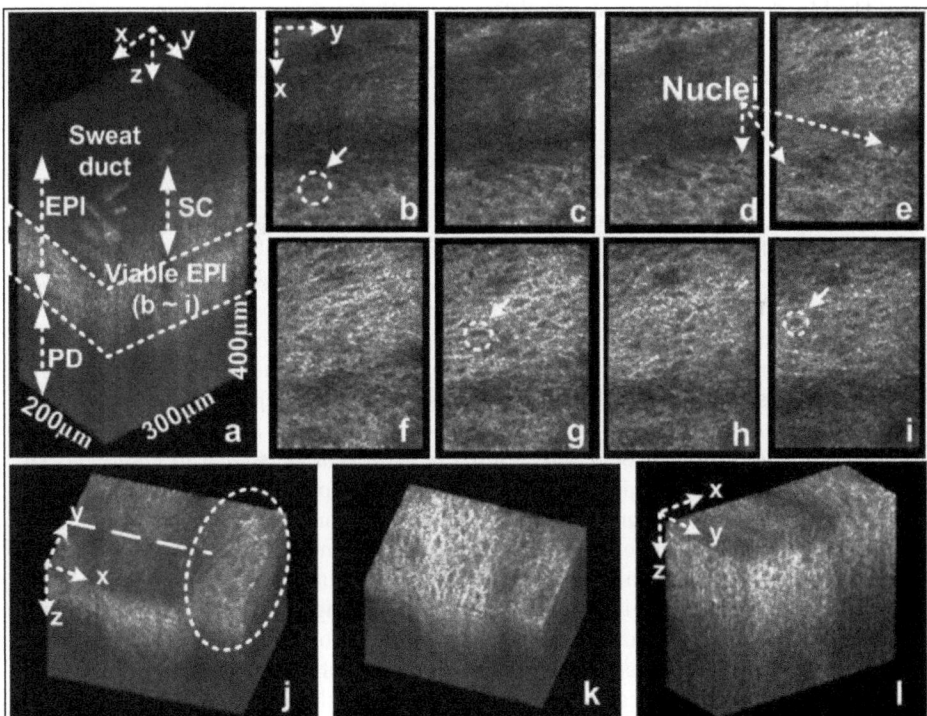

Figure 2 *(a) A volume (300 μm × 200 μm × 400 μm) of in-vivo human finger skin, (b)-(i) en face images every 12 μm in depth in viable epidermis, (j)-(k) volumetric images cut at the depth of (b), and (g) respectively, and (l) a half cut of volume (j) along a long dash line. EPI: epidermis, SC: stratum corneum, and PD: papillary dermis.*

According to the known location and cellular size from established histology and confocal microscopy of skin [41], we deduced that keratinocytes at different maturation states in SG, SS, and SB were successfully imaged. Figure 2j, and 2k show the cut images in the volume of Figure 2a at the depth of Figure 2b, and 2g, respectively. The volumetric cell structures are seen in the dashed region in Figure 2j. Figure 2l is the half cut of the volume 2j, showing cells in vertical cross section.

Figure 3a shows three orthogonal views in a 3D volume of *in-vivo* normal fingertip skin in a large field of view (i.e. 2 mm x 1 mm x 0.6 mm). In the cross sectional GD-OCM images of x-z plane and y-z plane (i.e. left-up and right-up images in Figure 3a), the stratum corneum of the fingertip is clearly visible with the border between the viable epidermis and stratum corneum. The *en face* image (i.e. left-bottom image in Figure 3a) shows dead keratinocytes together with some bright spots of sweat ducts in the stratum corneum on the ridge of the fingerprint at the corresponding depth indicated in the x-z cross sectional image in Figure 3a. The *en face* image in Figure 3b at the depth of viable epidermis shows viable keratinocyte aggregates on the ridge of the fringe depicted as brighter areas. The fringe comes from the undulations formed by dermal papillae. In Figure 3c, two white streaks correspond to the viable epidermis, where keratinocytes are seen. The blood vessels appear as signal-free round features. Figure 3d shows a snap shot of a 3D volume rendered with maximum intensity. The spiral sweat ducts within the epidermis (acrosyringium) are clearly visualised by their strong scattering.

Three orthogonal views in a 3D volume of *in-vivo* normal nail-fold skin acquired using GD-OCM are shown in Figure 4a. The imaged area of the *in-vivo* nail-fold corresponds to a line box in the schematic located in the right-bottom corner of Figure 4a. The cross sectional GD-OCM images of the x-z plane and the y-z plane (i.e. left-up and right-up images in Figure 4a) clearly delineate the features of papillary capillary loops as well as the cornified layer, germinal layer of the epidermis, and sub papillary venules. The *en face* image of x-y plane (i.e. left-bottom image in Figure 4a) shows dead keratinocytes in the cornified layer at the corresponding depth indicated in the x-z cross sectional image in Figure 4a. The *en face* image in Figure 4b at the depth of viable epidermis shows viable keratinocyte aggregates. Capillary loops extending vertically in the dermis are observed as small black spots in Figure 4c. The black streaks in Figure 4d represent the sub papillary venules in the dermis.

Three orthogonal views of *in-vivo* normal forearm skin are shown in Figure 5a. The cross sectional images of x-z plane and y-z plane (i.e. left-up and right-up images in Figure 5a) clearly delineate some signal-free cavities and long structures corresponding to blood vessels in the dermis. A lattice pattern of fingerprint lines and keratinocytes are seen in the *en face* image (i.e. left-bottom image in Figure 5a) at the corresponding depth indicated in the x-z cross sectional image in Figure 5a. The *en face* image in Figure 5b at the depth of dermal-epidermal junction shows small black spots corresponding to blood vessels. Figure 5c and 5d at the corresponding depths indicated in Figure 5a in the dermis show the extracellular matrix, predominantly composed of collagen, elastic fibres, and blood vessels.

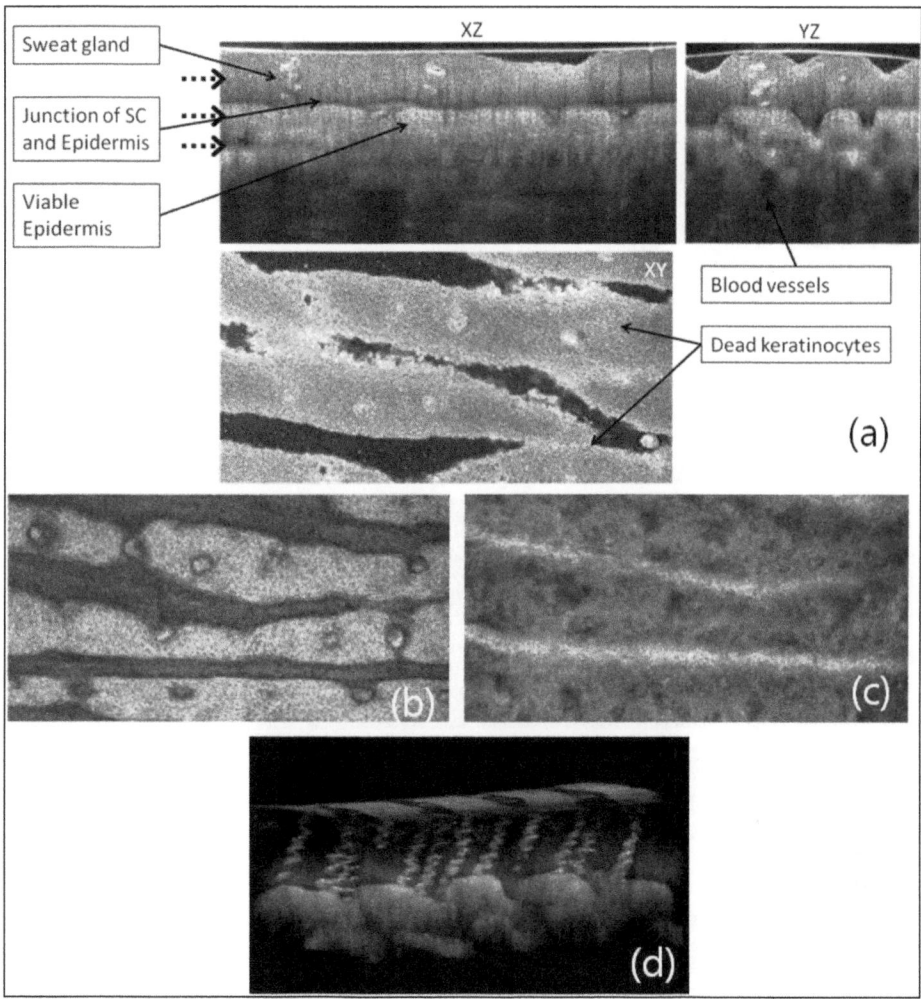

Figure 3 *(a) Three orthogonal views in a 3D volume of in-vivo normal fingertip skin (left-up: cross section in a x-z plane of 2 mm by 0.6 mm, right-up: cross section in a y-z plane of 1 mm by 0.6 mm, left-bottom: en face image in a x-y plane of 2 mm by 1 mm at the depth of the top dash arrow in (a), and right-bottom: (b) and (c) are the en face images at the corresponding depth indicated with the second and third dash arrows in the left-up image of Figure 3(a), respectively. (d) a snap shot of 3D volume rendered with maximum intensity rendering.*

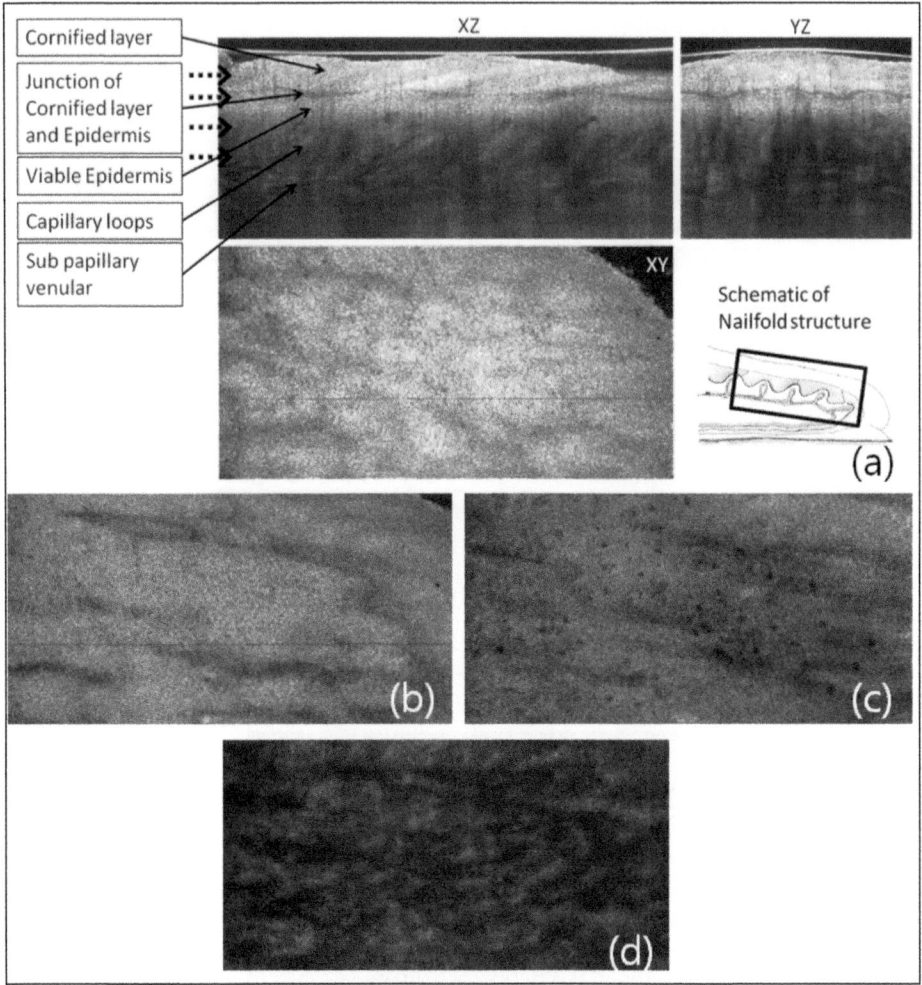

Figure 4 *(a) Three orthogonal views in a 3D volume of in-vivo normal nail-fold skin (left-up: cross section in a x-z plane of 2 mm by 0.6 mm, right-up: cross section in a y-z plane of 1 mm by 0.6 mm, left-bottom: en face image in a x-y plane of 2 mm by 1 mm, and right-bottom: schematic of nail-fold with a line box that indicates the imaged area. (b), (c), and (d) are the en face images at the corresponding depth indicated with the dash arrows in the left-up image of Figure 4(a), respectively.*

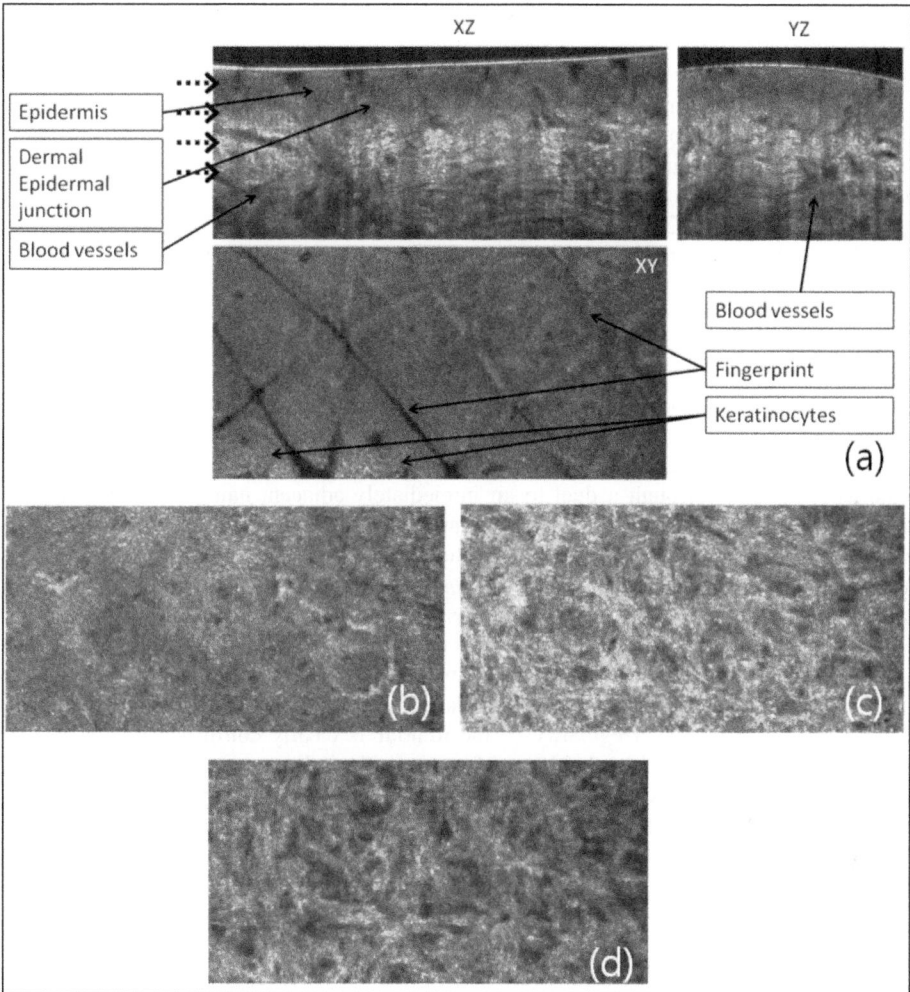

Figure 5 *Three orthogonal views in a 3D volume of in-vivo normal forearm skin (left-up: cross section in a x-z plane of 2 mm by 0.6 mm, right-up: cross section in a y-z plane of 1 mm by 0.6 mm, left-bottom: en face image in a x-y plane of 2 mm by 1 mm, and right-bottom: (b), (c), and (d) are the en face images at the corresponding depth indicated with the dash arrows in the left-up image of Figure 5(a), respectively.*

Ex-Vivo 3D High Resolution Images of Normal Skin and NMSC.

The cross sectional GD-OCM image shown in Figure 6a of normal nose skin clearly distinguishes the epidermis from the dermis while the BCC image in Figure 6e does not have the characteristic, organised layering. In addition, dark lobules consistent with nests of tumour cells were found in the BCC image, which is indicated by the white arrow in Figure 6e. Features such as disruption of skin layers and dark lobules have been observed in OCT images of NMSC in previous studies and used to differentiate NMSC from normal skin [42,43]. Furthermore, higher signal attenuation in depth was observed in Figure 6e compared to the corresponding image of normal skin [13].

In the *en face* image shown in Figure 6b at the corresponding depth indicated by the arrows in Figure 6a, the numerous keratinocytes surrounding the hair follicle are clearly observed. The *en face* image in Figure 6f of a BCC at the corresponding depth indicated by the arrows in Figure 6e does not show the organised cellular array. The *en face* image in Figure 6c of normal nose skin in the papillary dermis shows capillaries and hair follicles. The round object indicated with the black arrow in Figure 6c is consistent with a sebaceous gland as it connects through a duct to an immediately adjacent hair follicle. Clusters of dark cells, sometimes surrounded by lighter areas, are clearly observed (indicated by the white arrows in Figure 6g) and are classic features of BCC and its surrounding stroma seen in histology. The *en face* images of normal nose skin and BCC are shown at deeper depth in the dermis in Figure 6d and 6h, respectively.

The cross-sectional image in Figure 7a of normal ear skin shows the typical skin layers of the stratum corneum, epidermis, and dermis. The well-defined layers are not found in Figure 7e of the SCC image but replaced by irregularly shaped masses extending into the dermis [44]. The irregularity of the tumour is clearly confirmed in the *en face* images in Figure 7f, g and h at the three different depths indicated in Figure 7e. However, the *en face* image in Figure 7b at a depth in the epidermis in normal ear skin shows keratinocytes in the overall image except a hair follicle and the two sweat ducts. Blood vessels were imaged as small black spots due to light absorption by haemoglobin in Figure 7c and d.

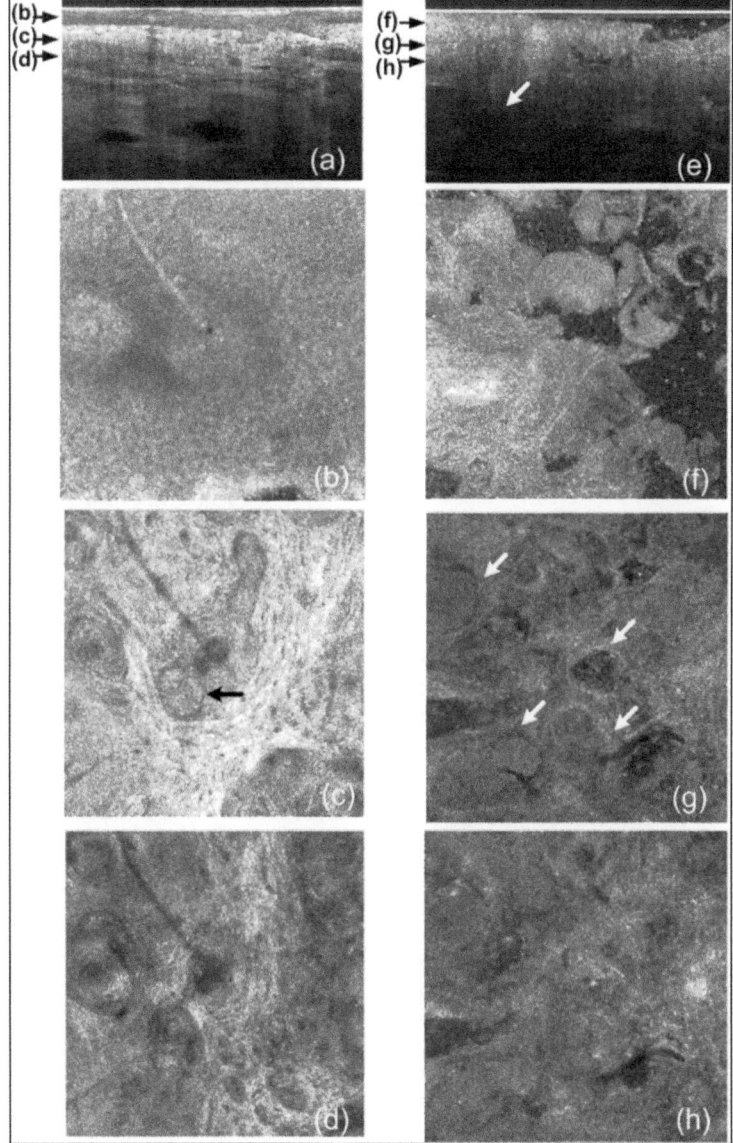

Figure 6 *(a) A cross sectional GD-OCM image of a normal nose skin. (b), (c), and (d) are the en face images at the corresponding depth indicated with the arrows in Figure 6(a), respectively. (e) A cross sectional GD-OCM image of BCC located in nose skin. (f), (g), and (h) are the en face images at the corresponding depth indicated with the arrows in Figure 6(e), respectively. The size of the cross sectional images (a) and (e) is 1 mm by 0.6 mm, and the size of the en face images is 1 mm by 1 mm.*

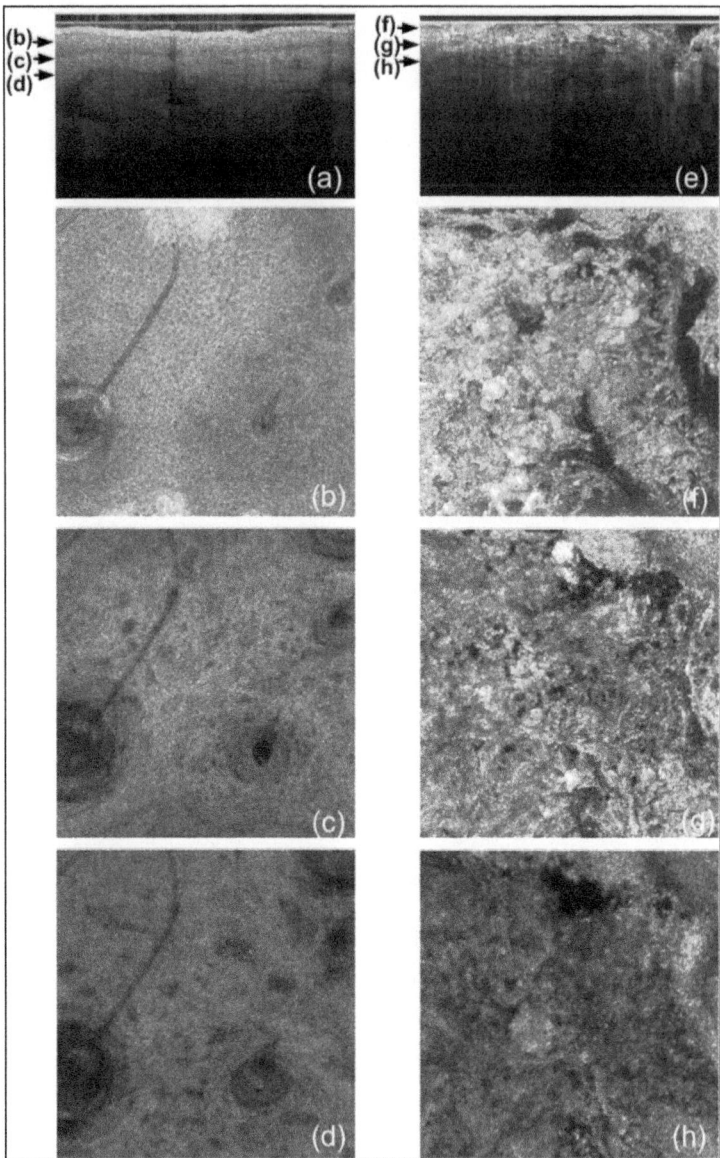

Figure 7 *(a) A cross sectional GD-OCM image of a normal ear skin. (b), (c), and (d) are the en face images at the corresponding depth indicated with the arrows in Figure 7(a), respectively. (e) A cross sectional GD-OCM image of SCC located on the ear. (f), (g), and (h) are the en face images at the corresponding depth indicated with the arrows in Figure 7(e), respectively. The size of the cross sectional images (a) and (e) is 1 mm by 0.6 mm, and the size of the en face images is 1 mm by 1 mm.*

4   CONCLUSION

In this study, the cellular resolution imaging technique of GD-OCM was applied to normal skin *in vivo* from different anatomic locations: nail-fold, fingertip, and inner forearm. The acquired images clearly reveal the array of epidermal keratinocytes at the three locations including skin appendages such as hair follicles, sebaceous glands, ducts of sweat glands, blood vessels, and the extracellular matrix of connective tissue. In addition, we investigated morphological differences between NMSC and normal skin in 3D images with cellular resolution using GD-OCM. Features of BCC such as the disruption of normal skin layers and nests of densely packed cells were clearly shown in cross section and *en face* images at different depths. Additionally, we also observed the disruption of the organised array of keratinocytes in BCC compared to normal skin. Disorganised arrays of keratinocytes were also observed in SCC. Follow-up clinical studies are planned for *in-vivo* imaging of NMSC with direct correlation to histology.

5   REFERENCES

1.   J. Allali, F. D'Hermies, G. Renard, Basal cell carcinoma of the eyelids, *Ophthalmologica*, 2005, **219**, 57–71.

2.   J. A. Neville, E. Welch, D. J. Leffell, Management of nonmelanoma skin cancer, *Nat. Clin. Pract. Oncol.*, 2007, **4**, 462–469.

3.   H. W. Rogers, M. A. Weinstock, A. R. Harris, *et al.*, Incidence estimate of nonmelanoma skin cancer in the United States, 2006, *Arch. Dermatol.*, 2010, **146**, 283–287.

4.   H. J. Buchwald, A. Muller, J. Kampmeier, *et al.*, Optical coherence tomography versus ultrasound microscopy of conjunctival and eyelid lesions, *Klin. Monatsbl Augenheilk*, 2003, **220**, 822–829.

5.   H. Gufler, F. E. Franke, W. S. Rau, High-resolution MRI of basal cell carcinomas of the face using a microscopy coil, *Am. J. Roentgenol*, 2007, **188**, 480–484.

6.   D. Huang, *et al.*, Optical Coherence Tomography, *Science*, 1991, **254**, 1178–1181.

7.   F. G. Bechara, T. Gambichler, M. Stucker, *et al.*, Histomorphologic correlation with routine histology and optical coherence tomography, *Skin Res. Technol.*, 2004, **10**, 169–173.

8.   J. Welzel, *et al.*, Optical coherence tomography in dermatology: a review, *Skin Res. Technol.*, 2001, **7**, 1–9.

9.   M. C. Pierce, J. Strasswimmer, B. H. Park, *et al.*, Advances in optical coherence tomography imaging for dermatology, *J. Invest. Dermatol.*, 2004, **123**, 458–463.

10. J. Welzel, *et al.*, Optical coherence tomography of the human skin, *J. Am. Acad. Dermatol.*, 1997, **37**, 958–963.

11. R. Steiner, *et al.*, Optical coherence tomography: clinical applications in dermatology, *Med. Laser Appl.*, 2003, **18**, 249–259.

12. A. Alex, *et al.*, Multispectral in vivo three-dimensional optical coherence tomography of human skin, *Journal of Biomedical Optics*, 2010, **15**, 026025.

13. V. R. Korde, G. T. Bonnema, W. Xu, *et al.*, Using optical coherence tomography to evaluate skin sun damage and precancer, *Lasers Surg. Med.*, 2007, **39**, 687–695.

14. J. Welzel, *et al.*, OCT in dermatology, in *Optical Coherence Tomography: Technology and Applications*, ed. W. Drexler and J. G. Fujimoto, Springer, Berlin, 2008, pp. 1103–1121.

15. J. Olmedo, K. Warschaw and J. Schmitt, Optical coherence tomography for the characterization of basal cell carcinoma *in vivo*: a pilot study, *J. Am. Acad. Dermatol.*, 2006, **55**, 408–412.

16. J. M. Olmedo, K. E. Warschaw, J. M. Schmitt, *et al.* Correlation of thickness of basal cell carcinoma by optical coherence tomography *in vivo* and routine histologic findings: a pilot study, *Dermatol. Surg.*, 2007, **33**, 421–425.

17. T. Gambichler, A. Orlikov, R. Vasa, *et al.*, *In vivo* optical coherence tomography of basal cell carcinoma, *J. Dermatol. Sci.*, 2007, **45**, 167–173.

18. J. K. Barton, K. W. Gossage, W. Xu, *et al.*, Investigating sun-damaged skin and actinic keratosis with optical coherence tomography: a pilot study, *Technol. Cancer Res. Treat*, 2003, **2**, 525–535.

19. P. Wilder-Smith, T. Krasieva, W. G. Jung, *et al.*, Noninvasive imaging of oral premalignancy and malignancy, *J. Biomed. Opt.*, 2005, **10**, 051601.

20. W. B. Armstrong, J. M. Ridgway, D. E. Vokes, *et al.*, Optical coherence tomography of laryngeal cancer, *Laryngoscope,* 2006, **116**, 1107–1113.

21. T. Hinz, *et al.*, Preoperative Characterization of Basal Cell Carcinoma Comparing Tumour Thickness Measurement by Optical Coherence Tomography 20-MHz Ultrasound and Histopathology, *Acta. Derm. Venereol.*, 2012, **92**, 132–137.

22. R. Ponerantz, *et al.*, Optical Coherence Tomography used as a Modality to Delineate Basal Cell Carcinoma prior to Mohs Micrographic Surgery, *Case Reports in Dermatology*, 2011, **3**, 212–218.

23. M. Mogensen, *et al.*, OCT imaging of skin cancer and other dermatological diseases, *J. Biophotonics*, 2009, **2**, 442–451.

24. M. Mogensen, *et al.*, Assessment of Optical Coherence Tomography Imaging in the Diagnosis of Non-Melanoma Skin Cancer and Benign Lesions Versus Normal Skin: Observer-Blinded Evaluation by Dermatologists and Pathologists, *Dermatol. Surg.*, 2009, **35**, 965–972.

25. J. A. Izatt, M. R. Hee, G. M. Owen, E. A. Swanson, and J. G. Fujimoto, Optical coherence microscopy in scattering media, *Optics Letters*, 1994, **19**, 590–592.

26. A. Aguirre, P. Hsiung, T. Ko, I. Hartl, and J. Fujimoto, High-resolution optical coherence microscopy for high-speed, in vivo cellular imaging, *Opt. Lett.*, 2003, **28**, 2064–2066.

27. M. Huzaira, *et al.*, Topographic variations in normal skin, as viewed by in vivo reflectance confocal microscopy, *J. Invest. Dermatol.*, 2001, **116**, 846–852.

28. M. Akiba and K. Chan, *In vivo* video-rate cellular-level full-field optical coherence tomography, *J. Biomed. Opt.*, 2007, **12**, 064024.

29. Y. Zhao, Z. Chen, C. Saxer, S. Xiang, J. F. deBoer, J. S. Nelson, Phase-resolved optical coherence tomography and optical Doppler tomography forimaging blood flow in human skin with fast scanning speed and high velocity sensitivity, *Opt. Lett.*, 2000, **25**, 114–116.

30. M. C. Aalders, M. Triesscheijn, M. Ruevekamp, M. deBruin, P. Baas, D. J. Faber , *et al.*, Doppler optical coherence tomography to monitor the effect of photodynamic therapy on tissue morphology and perfusion, *J. Biomed. Opt.*, 2006, **11**, 044011.

31. J. Strasswimmer, M. C. Pierce, B. H. Park, *et al.*, Polarization-sensitive optical coherence tomography of invasive basal cell carcinoma, *J. Biomed. Opt.*, 2004, **9**, 292–298.

32. B. F. Kennedy, *et al.*, *In vivo* three-dimensional optical coherence elastography, *Optics Express*, 2011, **19**, 6623–6634.

33. K. Konig, *et al.*, Clinical optical coherence tomography combined with multiphoton tomography of patients with skin diseases, *J. Biophotonics,* 2009, **2**, 389–397.

34. E. Z. Zhang, *et al.*, Multimodal photoacoustic and optical coherence tomography scanner using an all optical detection scheme for 3D morphological skin imaging, *Biomedical Optics Express*, 2011, **2**, 2202–2215.

35. K. Lee, K. P. Thompson, P. Meemon, and J. P. Rolland, Micron-class resolution optical coherence microscopy with high acquisition speed for *in vivo* human skin cell imaging, 2011, *Optics Letters*, **36**, 2221–2223.

36. J. P. Rolland, *et al.*, Gabor-based fusion technique for Optical Coherence Microscopy, *Opt. Express*, 2010, **18**, 3632–3642.

37. S. Murali, *et al.*, Three-dimensional adaptive microscopy using embedded liquid lens, *Optics Letters*, 2009, **34**, 145–147.

38. K. Lee, *et al.*, Dispersion control with Fourier-domain optical delay line in a fiber optic imaging interferometer, *Applied Optics*, 2005, **44**, 4009–4022.

39. K. Lee, *et al.*, Broadband astigmatism correction Czerny-Turner spectrometer, *Opt. Express*, 2010, **18**, 23378–23384.

40. S. Murali, *et al.*, Assessment of a liquid lens enabled in vivo optical coherence microscope, *Appl. Opt.*, 2010, **49**, D145–D156.

41. M. Huzaira, *et al.*, Topographic variations in normal skin, as viewed by in vivo reflectance confocal microscopy, *J. Invest. Dermatol.*, 2001, **116**, 846–852.

42. T. Gambichler, R. Matip, G. Moussa, *et al.*, *In vivo* data of epidermal thickness evaluated by optical coherence tomography: effects of age, gender, skin type, and anatomic site, *J. Dermatol. Sci.*, 2006, **44**, 145–152.

43. F. G. Bechara, T. Gambichler, M. Stucker, *et al.*, Histomorphologic correlation with routine histology and optical coherence tomography, *Skin Res. Technol.*, 2004, **10**, 169–173.

44. P. A. Khavari, Modelling cancer in human skin tissue, *Nature Reviews Cancer*, 2006, **6**, 270–280.

6
SYNCHROTRON RADIATION FOR DIAGNOSIS OF SKIN CONDITIONS

M Cócera[1], G Rodríguez[1], L Rubio[1], E Fernández[1], L Barbosa-Barros[1,2], A Labrador[3], M Sabés[4], A de la Maza[1], and O López[1]

[1]Department of Chemical and Surfactant Technology, Institute of Advanced Chemistry of Catalonia, Barcelona, Spain. [2]SmartNano, SL, Barcelona, Spain. [3]MAX-lab, Lund University, Lund, Sweden. [4]Centre d'Estudis en Biofísica, Universitat Autònoma de Barcelona, Bellaterra, Spain.

1   INTRODUCTION

The skin acts as a physical barrier at the interface with the external environment. This barrier is designed to protect the organism against insults, including desiccant, mechanical, chemical, and microbial damage [1]. It is primarily composed of three layers: the epidermis, dermis, and hypodermis. The epidermis, which contains the stratum corneum (SC), works as a barrier, preventing water loss from the body and protecting it from the environment. The dermis contains fibroblasts as the predominant cell type within a matrix of structural proteins (collagen or elastin), proteoglycans, nervous fibres and sebaceous glands. The hypodermis acts as a thermal and mechanical insulator [2]. Skin contains non-crystalline material, such as collagen, SC lipids and fat, with characteristic scattering patterns that are altered under specific conditions [3, 4]. These changes can be useful for understanding the structure and state of the tissue.

Collagen accounts for ~ 75% of the total dry weight of skin. This macromolecule provides tissue integrity (tissue repair, migration and adhesion) and facilitates tissue morphogenesis and platelet aggregation [5, 6]. Changes in the orientation and arrangement of collagen during ageing and disease progression are partly responsible for alteration in skin morphology and mechanical properties. These alterations are manifest as wrinkling, loss of elasticity, stretch marks and impairment in barrier function [7, 8]. Thus, a systematic analysis of dermal collagen characteristics such as discolouration, distribution, and orientation can provide insights into skin function and specific alterations [9]. From a structural point of view, collagen molecules are formed by three α-polypeptide chains folded together that form a triple helical structure with a characteristic length and diameter [10]. On the basis of the assembly between these triple helix molecules, collagen can be classified into fibrillar and non-fibrillar groups. In the fibrillar mode, collagen molecules are packed against one another in a quasi-hexagonal close-packed array [11]. Specific intermolecular cross-linking forces the molecules to form a regular staggered structure that induces periodic variations of electron density visible in electron microscopy as cross-striations and by X-ray diffraction as sharp Bragg peaks [12, 13]. Abnormalities of collagen molecular structure may adversely affect the packing of collagen molecules, thus producing structurally weaker connective tissues [5, 14]. Various studies have found that the presence of an aberrant form of collagen is associated with malignant breast tumours [15, 16]. Melanoma is a skin cancer arising from melanocytes, which are the cells responsible for melanin biosynthesis. Some investigations have associated melanoma with the degradation of collagen fibres [17, 18]. Collagen alterations may also be related to changes from ageing, exercise, processes of wound healing and degenerative conditions [19, 20].

Synchrotrons consist of an evacuated storage ring in which high-energy electrons circulate at relativistic velocities and incorporate "beamlines" that divert synchrotron light emitted by the electrons at a tangent to their orbital path at predefined positions using bending magnets and insertion devices. Synchrotron radiation (SR) is a very powerful source of high-intensity photons that allows a thorough characterisation of the structure and dynamics of soft matter. In addition to conventional X-ray diffraction, a wide range of other methods can be utilised such as small angle x-ray scattering, infrared-microscopy and other spectroscopic methods. Small angle X-ray scattering (SAXS) using SR is a useful technique for studying non-crystalline structures from biological tissues, such as skin [21, 22]. In particular, SAXS using conventional and SR sources is an excellent tool for determining the structural organisation of tissues that are endowed with collagen [4, 23]. It has been consistently demonstrated by Lewis *et. al.* [24], Fernández *et. al.* [16] and more recently by Conceição *et. al.* [25] that the supra-molecular structure of collagen can be studied using SAXS. These studies evaluated collagen to determine whether tissue samples had benign or malignant lesions. In skin research, X-ray analysis has yielded interesting information about the organisation of lipids from skin SC [26]. This technique has also proven useful in the study of the mechanism of a nephrogenic systemic fibrosis disease that causes disorders in the skin and systemic tissues [27]. Our group has evaluated the skin structures in several situations, and the influence of lipid nanostructured systems on the SC lipid organization by SAXS [28-30]. Moreover, other SR-based techniques such as infrared spectroscopy have demonstrable utility in investigating the physical state of SC lipids [31].

The present work presents the SAXS technique using SR to study the non-crystalline material of skin. Namely, collagen fibres and the lipid lamellar organisation of skin. Furthermore, we also report a chemical map of the skin before and after topical application of lipid vehicles (perdeuterated bicelles) using synchrotron infrared micro-spectroscopy. The use of a synchrotron source provides a photon beam between two and three orders of magnitude brighter than the conventional source, allowing Fourier Transform Infrared (FTIR) spectra to be recorded with a much higher lateral resolution. Moreover, the brightness of SR permit a confocal configuration to couple a microscope for imaging skin. The application of SR for the analysis of soft tissue represents a promising tool for the characterisation, diagnosis and treatment of skin diseases. In particular, evaluation of collagen provides information about underlying pathology of skin tissue, with the systematic study of different samples resulting in scattering signatures that are characteristic of each skin layer [32]. This study could help us understand the changes of supra-molecular structure of collagen and their possible correlation with the changes in the healthy and pathological state of skin tissues. Likewise, the evaluation of lipid carriers on the skin may have application for designing safe and more efficient delivery systems. The use of the SR opens up new possibilities in dermatological or cosmetic fields and augments our current range of biophysical instrumentation.

2   MATERIALS AND METHODS

Pepsin from porcine gastric mucosa powder, 800-2.500 units mg^{-1} and acetic acid were purchase by Sigma-Aldrich Chemie GmbH (Steinheim, Germany). Purified water was obtained using an ultrapure water system, Milli-Q plus 185 (Millipore, Bedford, MA). The lipids 1,2-dipalmitoyl(d62)-*sn*-glycero-3-phosphocholine (d-DPPC), and 1,2-dihexanoyl-*sn*-glycero-3-phosphocholine (DHPC), purity greater than 99%, were purchased from Avanti Polar Lipids (Alabaster, Al). For washing purposes sodium lauryl ether sulfate (SLES) solution at 0.5% w/v from Sigma-Aldrich Chemie GmbH (Steinheim, Germany) was used.

Sections of fresh human skin were excised during surgical procedures at the Department of Dermatology, University Hospital "Príncipes de España" (Barcelona, Spain). To minimise inter-individual variation, skin from younger patients was chosen as the probability of changes related to solar radiation, topical therapeutics and cosmetics was lower than from older patients. In some cases, skin samples exhibited stretch marks; a disorder which may occur when the skin suffers from rapid growth or stretching. Although the skin is usually fairly elastic, when it is overstretched, the normal production of collagen is disrupted, and stretch marks are formed [33]. Therefore, such tissue samples were used for determining possible collagen alterations.

The skin samples were washed with tap water and then dermatomed to 500 ± 50 μm thickness using an Aesculap dermatome GA 630 (Braun, Tuttlingen, Germany). The skin was placed in water at 65°C for 4-5 minutes and separated into sheets of epidermis and dermis via cleavage of the dermo-epidermal junction [34].

Diseased (melanoma) skin samples were surgically excised from patients undergoing surgery at the "Hospital Clínic" (Barcelona, Spain). The samples comprised skin biopsies from frozen samples provided by the Archive of the Dermatological Department. Thin sections (~1 mm) were cut with a scalpel. All samples were used with the consent of the patients.

To compare the degradation of collagen in skin invaded by melanoma, an additional skin sample was treated with pepsin (proteolytic enzyme). This enzyme solubilises the collagen into different fractions [35] producing a loss of order in the fibrillar structure. Skin collagen was digested with 0.1 mg ml^{-1} pepsin for 12 hours in 0.5 M acetic acid at 4° C [36].

Thus in total, four skin types were used: (1) young adult skin, (2) young, stretch-marked skin, (3) skin invaded with melanoma and (4) pepsin-digested skin. To obtain significant data, no less than ten samples for each of the four skin conditions were acquired from different individuals. Also, a minimum of five measurements were performed at several points of the samples to avoid radiation damage. The average of all measurements for each condition and the standard deviation were calculated using the software Origin Lab 8.

Porcine skin was obtained $2 - 3$ hours post mortem from the back of experimental animals in the Department of Dermatology, University Hospital Clinic of Barcelona

(Spain). The hair was carefully removed by close clipping and then the skin was washed with tap water. The excised skin was dermatomed to 500 ± 50 μm thickness.

Small pieces of skin were wrapped in aluminium foil to prevent contamination when embedded into OCT (Optimal Cutting Temperature, Tissue-Tek, Zoeterwoude, NL), frozen in liquid N_2 and finally placed in a cryostat (-30°C). Sections (~6 μm thickness) were obtained using a cryo-microtome. The resulting sections were then deposited onto CaF_2 windows.

Bicelles were prepared using perdeuterated lipids (d-DPPC/DHPC). Two disks of whole skin with dimensions of approximately 2.5 cm^2 were used to carry out the experiment (which was performed in triplicate). The integrity of each sample was checked by determining rates of transepidermal water loss (TEWL) using a Tewameter TM210 (Courage-Khazaka, Köln, Germany) following the protocol described for absorption experiments [29]. The tissue was placed in a Petri dish at 37 °C in order to maintain hydration. Treatment comprised four applications (every 1 h) of 13 μL of the bicellar system or water on the skin surface. Samples were washed after each application with Milli-Q water and left to dry before the following application. At the end of the treatment, each skin sample was washed with SLES solution (at 0.5% w/v) rinsed with Milli-Q water.

The SAXS profile of each skin layer was analysed to characterise the organisation of the non-crystalline material of the skin. Skin samples were analysed using two configurations (1.4 and 5 meters) using each of the four skin conditions. Additionally, the dermis and the epidermis from young skin were isolated and analysed individually. Prior to use, each skin sample was cut into sections (1 mm thick, 10 mm diameter) and placed in a holder (sandwiched between two aluminium plates each with a hole for transmission of SR) with the skin surface perpendicular to the beam.

Diffraction experiments were performed at the Spanish station BM16 of the European Synchrotron Radiation Facility (ESRF). The energy of the beam was 12 keV. The scattering patterns were recorded with a SAXS 2D detector MARCCD (Marresearch GmbH, Norderstedt, Germany) with a 165 mm diameter aperture with single exposure times of about 60 seconds. Two sample-to-detector distances at 5 (long) and 1.4 meters (short configuration) were used. Calibration was performed using silver behenate. Spacing was determined from axially integrated 2D images using the FIT2D program (http://www.esrf.eu/computing/scientific/FIT2D) from ESRF.

The typical axial periodicity of collagen is the result of the displacement of each molecule with respect to the adjacent molecules in a particular direction. The displacement generates gaps (less electron dense) and overlaps. The SAXS technique provides information about the larger structural units in the sample, namely, the repeat distance of one structure. The scattering intensity (I, in arbitrary units) was measured as a function of the scattering vector Q (in reciprocal nm), whose modulus is defined as:

$$Q=(4\pi sin\theta)/\lambda$$

where θ is the scattering angle and λ the wavelength of radiation (0.9795 Å).

The position of the diffraction peaks are directly related to the repeat distance of the molecular structure, as described by Bragg's law:

$$2dsin\theta=n\lambda$$

in which n and d are the order of the diffraction peak and the repeat distance, respectively. The different orders of peaks are calculated by:

$$Q_n=2\pi n/d$$

where Q_n is the position of the nth order reflection.

Analysis of the annular distribution from the 1st reflection of collagen was performed using FibreFix software (http://www.small-angle.ac.uk/small-angle/Software.html) from the Collaborative Computational Project for Fiber Diffraction and Solution Scattering (CCP13) for both young and aged skin. This analysis results in the annular distribution of scattered intensity from circular regions of the scattering pattern.

IR measurements on reflection (attenuated total reflection, ATR) and transmission mode were performed at SMIS beamline, at Soleil synchrotron. A Thermo Continuum XL IR microscope coupled to an FTIR Nicolet Nexus 5700 spectrometer was employed. The system was fitted with a mercury–cadmium–telluride (MCT) type A detector cooled by liquid nitrogen and a X–Y–Z motorised stage with incremental steps of 1 μm. In this study, a micro-slide on an ATR silicon crystal (refractive index n = 3.4) was directly connected to the objective with an aperture of 150 μm×150 μm, resulting in an area of investigation of ~ 40 × 40 μm. Optical micrographs were obtained on the microscope in the visible image mode in order to define the sampling positions. IR spectra were collected by gently lowering the ATR crystal onto the sample surface and applying pressure until optimum contact was achieved. Spectra were acquired at room temperature in the range of 4000–650 cm^{-1}, and a spectral resolution of 4 cm^{-1} and 128 scans were averaged. A Savitsky-Golay second derivative procedure was used to locate the peaks position using the Nicolet Omnic-Atlµs software. The penetration depth of IR in the skin is about 1 μm. For this reason, this is a useful technique for investigating the stratum corneum without necessitating physical isolation from the underlying skin layers. Combining this technique with tape stripping allowed the arrangement of lipids through the stratum corneum to be determined in normal and bicelle-treated skin.

The use of perdeuterated systems to treat the skin for FTIR experiments has the advantage that the CD mode appears in a region of the spectra where there is no vibrational feature associated with the skin (2000-2300 cm^{-1} range). This allows a differentiation of characteristic bands associated with the skin (such as intrinsic lipids (CH_2 vibration) or proteins (amide I and amide II vibrations)) and those linked to exogenous lipids (CD) [31].

For transmission mode, the IR microspectroscopic beamline was equipped with a Nic-Plan IR microscope coupled to a Magna 560 FT-IR spectrometer. The microscope operated in confocal mode, where the focusing Schwarzschild objective magnification was 32× (NA= 0.85) and the collection Schwarzschild objective magnification was 10× (NA= 0.71).

The area of illumination was defined with an adjustable aperture placed at an intermediate focal point and reimaged onto the sample. The upper and lower apertures were adjustable down to approximately 3×3 µm². The spectra were collected at either 8 or 4 cm^{-1} spectral resolution using AtlAs software (Thermo Nicolet Instruments). Area mapping was performed using a dual remote masking aperture to define the sample area for IR data collection and minimise diffraction. For each spectrum, at least 100 scans were accumulated. A Happ–Genzel apodisation function was applied for Fourier processing. Peak positions were determined using the Nicolet Omnic software (based on a polynomial least squares method) which allowed positions to be determined to within 1 cm^{-1}. Apertures of 6×6 µm² were used with mapping steps of 6 µm in both directions and a minimum of three mappings was collected for each treatment.

3   RESULTS

Comparison of dermatomed (whole) skin, dermis, and epidermis

Dermatomed skin, dermis, and epidermis were analysed at two sample-to-detector distances (Figure 1) with scattering profiles expressed as average (± standard deviation) of all measurements for the same condition. The standard deviation was consistently less than 5% and thus data reproducibility is very high. No differences were observed between frozen and fresh samples as the profiles of the samples were indistinguishable. The position of the diffraction peaks from collagen (at 5 and 1.4 m) is reported in direct space (d, nm; Table 1). The characteristic d-spacing of collagen was very similar for healthy skin samples (for both whole skin and dermis) but slightly lower for skin with stretch marks or subject to pepsin-treatment. The profiles of dermatomed skin and dermis were consistent with the typical scattering pattern of collagen.

Using the long configuration, the characteristic axial periodicity of collagen (63.4 ± 0.03 nm) from complete skin was clearly detected up to the 10th order reflection (Figure 1A). This curve also showed scattering features due to the cylindrical shape of the fibrils between the 1st and 3rd peak [37]. In the dermis profile, nine peaks were observed corresponding to the ten first orders of the periodical arrangement of the collagen molecules in fibrils (the 2nd reflection was not visualised). The SAXS profile of epidermis showed a band around 12 nm, corresponding to the long lamellar phase (LLP) of the SC lipid organisation as described Bouwstra and co-workers [26]. These authors propose the presence of two lamellar phases in the SC lipids, one with a repeat distance around 13 nm named LLP and the other with a repeat distance around 6 nm, which corresponds to the short lamellar phase (SLP).

For the short configuration (Figure 1B), the collagen diffraction pattern for complete skin went from the 3rd up to the 24th order reflection. The epidermis curve showed two bands (around 12 and 6.4 nm) corresponding to the LLP and the SLP of the lipid organisation in the SC [26, 38]. The influence of the scattering pattern from the SC lipids was undetectable in the pattern from complete skin at either configuration.

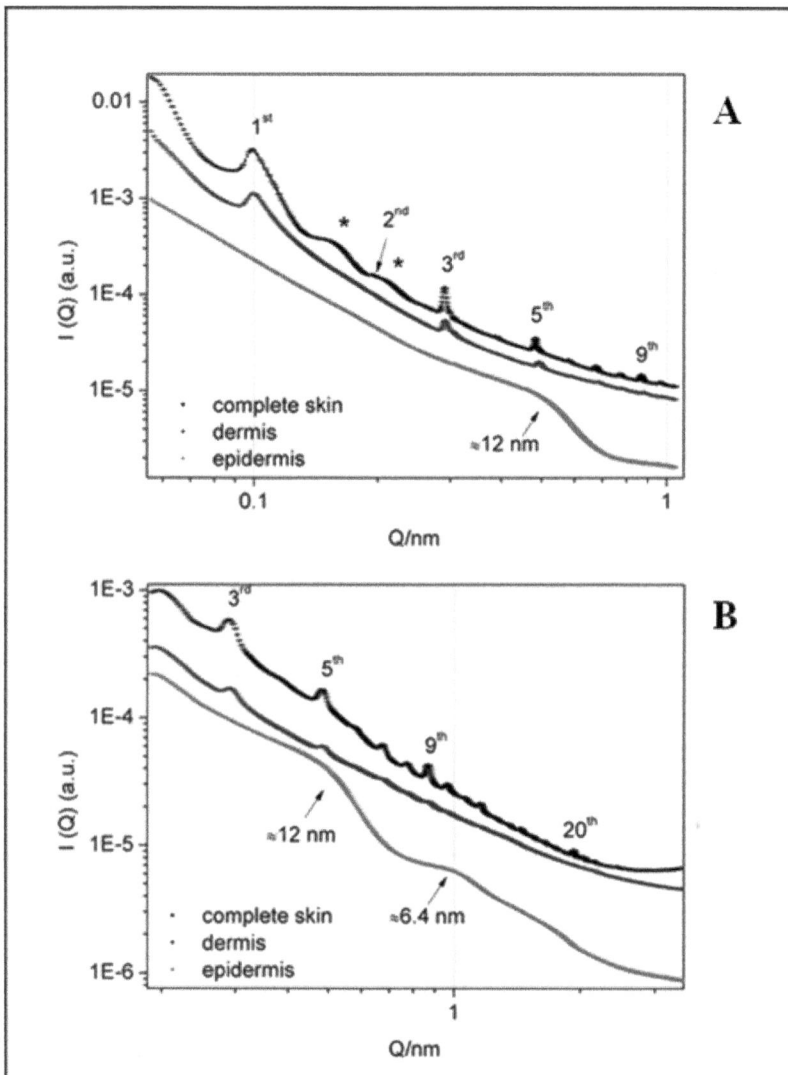

Figure 1 *SAXS profiles of skin layers (dermatomed skin, dermis, and epidermis) measured with two different sample-to-detector distances: A, long set-up (5 meters), and B, short set-up (1.4 meters). Graphs are plotted as scattering intensity (in arbitrary units) versus the modulus of scattering vector (inverse of nanometres) with standard deviation bars. The asterisks reflect the scattering features due to the cylindrical shape of the collagen fibrils.*

Peak	Reflection orders (d, nm)					
	Long			Short		
	Complete skin	Dermis	Stretch marks	Complete skin	Dermis	Pepsin treatment
1st	63.47±0.03	63.47±0.03	60.42±0.5			
2nd	33.07±0.05					
3rd	21.67±0.02	21.67±0.03	20.94±0.9	21.67±0.02	21.67±0.03	20.94±0.8
4th	16.53±0.03	16.11±0.4				
5th	13.37±0.02	13.09±0.3	12.82±0.4	13.09±0.2	13.09±0.03	12.82±0.3
6th	11.02±0.04	10.83±0.2		10.83±0.03		
7th	9.38±0.03	9.24±0.15		9.38±0.03	9.24±0.05	
8th	8.27±0.03	8.16±0.09		8.16±0.03	8.16±0.03	
9th	7.31±0.05	7.22±0.09		7.22±0.04	7.22±0.03	
10th	6.61±0.05	6.55±0.04		6.55±0.03	6.48±0.07	
11th				5.93±0.06		
12th				5.42±0.08	5.42±0.02	
20th				3.24±0.1	3.26±0.03	
21st				3.10±0.09		
22nd				2.95±0.2		

Table 1 *Reflection orders from the characteristic axial periodicity of collagen (d-spacing, nm) in complete skin, dermis, skin with stretch marks, and skin treated with pepsin samples. The sample-to-detector distances used were 5m (long set-up) and 1.4m (short set-up).*

Young and aged skin

The 2D scattering patterns from young (A) and aged (B) skin resulting from the average of several images performed with Fit2d software are shown in Figure 2. The skin pattern was dominated by a series of Bragg reflections, arising from constructive diffraction of X-rays from the regular repeating distance of 63.4 nm, characteristic of skin collagen [37, 39, 40]. The SAXS pattern from young skin presented a higher number of diffraction orders and a marked anisotropy in comparison with aged skin (Figure 2B). The annular analysis of the 1st reflection with the FibreFix software demonstrated that the SAXS pattern of young skin sample was formed by arcs, indicative of a preferential scattering in some areas of the detector. Scattering from aged skin samples showed rings as a result of a lower orientation of collagen structures than those in young skin.

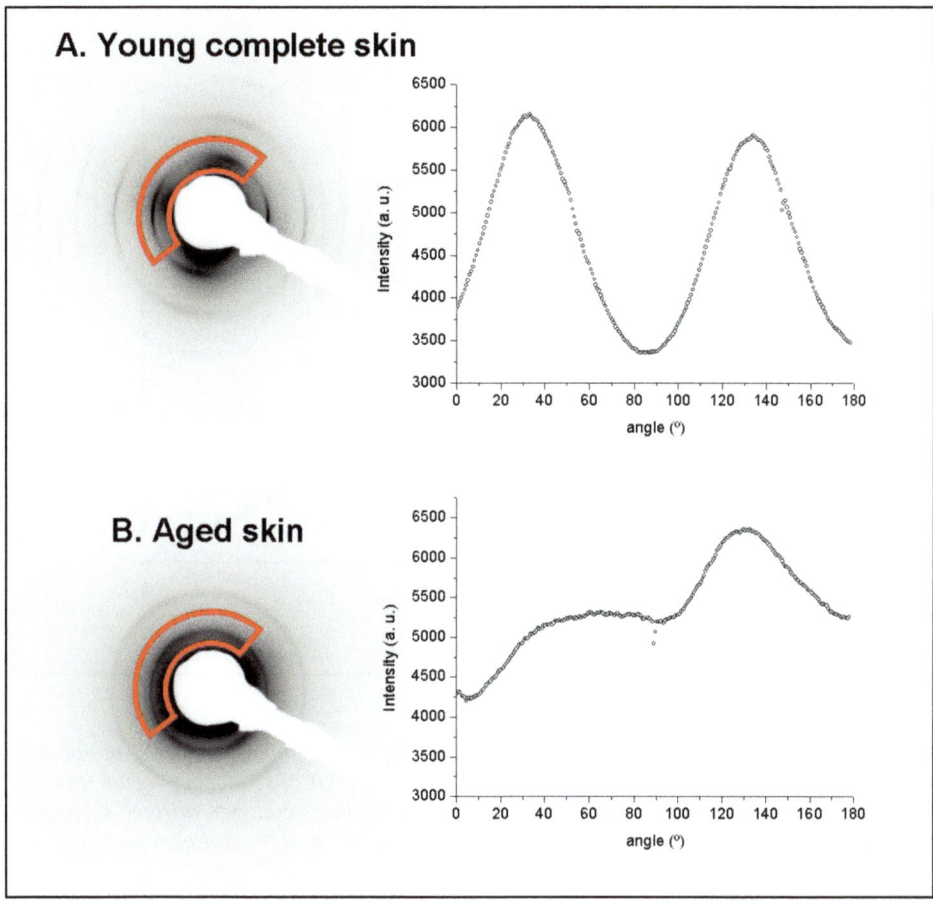

Figure 2 *2D SAXS patterns of A) complete young skin, and B) aged skin, using the long set-up (5 meters). The analysis of the annular distribution from the first reflection of collagen was conducted using the FibreFix software.*

Skin invaded by melanoma

Figure 3 illustrates the SAXS profile of skin affected by melanoma in comparison to healthy skin. The profile of the skin invaded by melanoma was clearly altered: the typical Bragg reflections from collagen were absent, leaving only a diffuse scattering band. The inset (Figure 3) demonstrated the effect of pepsin on collagen as an example of collagen degradation, whereby solubilised collagen produced different fractions.[35] In this case, the periodical arrangement of collagen was still detectable at very low intensity.

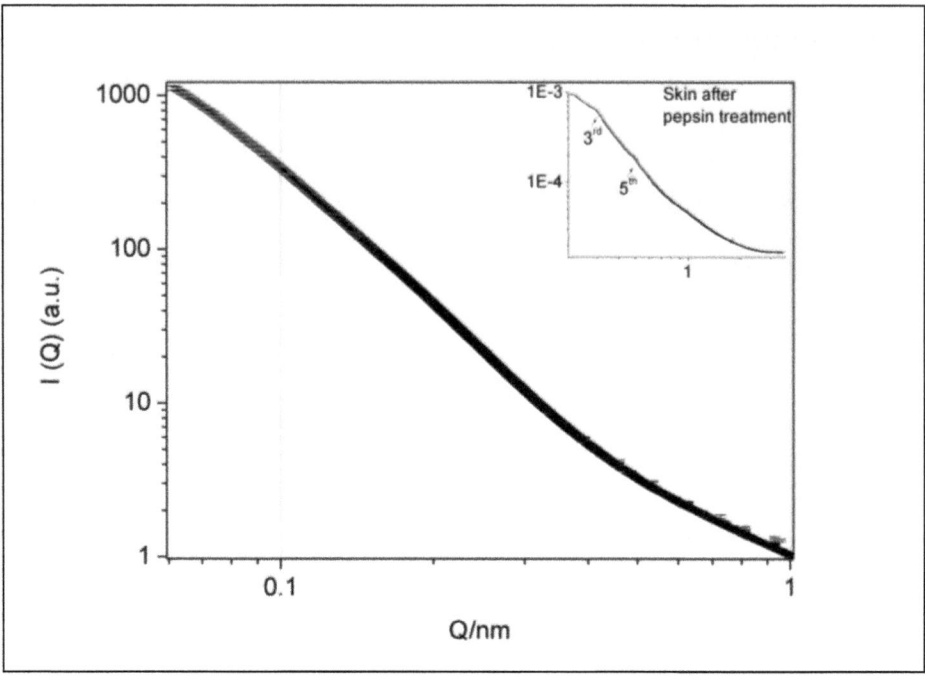

Figure 3 *SAXS profile of skin invaded by melanoma (measured at the long set-up) plotted as scattering intensity (in arbitrary units) versus the modulus of scattering vector (inverse of nanometres) with standard deviation bars. The inset shows collagen affected by pepsin treatment (proteolytic enzyme) for comparing the degradation of the collagen molecule.*

Skin with stretch marks

This technique proved useful for evaluating skin lesions such as stretch marks (striae), with the 1D SAXS profiles presented (with that of healthy skin) in Figure 4. Overstretching affects the hierarchical organisation of collagen molecules, leaving only some characteristic axial periodic reflections visible (1st, 3th, and 5th) with decreased intensity values.

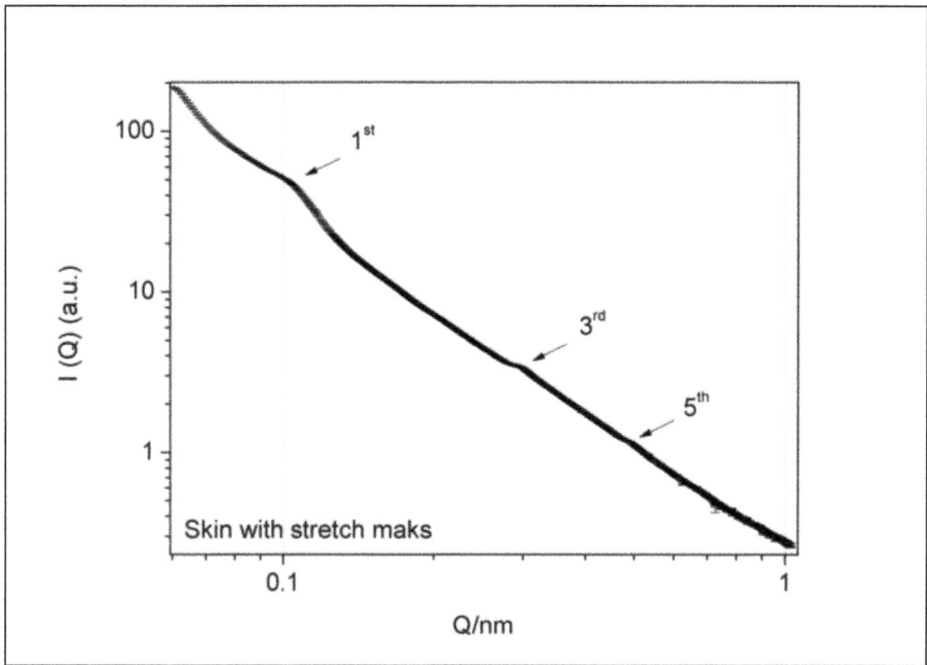

Figure 4 *SAXS profile of skin with stretch marks (measured at the long set-up) plotted as scattering intensity (in arbitrary units) versus the modulus of scattering vector (inverse of nanometres) with standard deviation bars. The numbers denote the order of the reflection from the characteristic axial periodicity.*

ATR-FTIR skin analysis

The absorption band from 3000 to 2700 cm^{-1} correspond to the C-H stretching motions of alkyl groups present in lipids (also present weakly in proteins) whereas the C-D stretching modes in the 2000-2300 cm^{-1} frequency range correspond to the deuterated lipids [31]. Figure 5A shows the IR spectra of the skin treated with d-DPPC/DHPC lipid system following tape stripping: a deuterated signal from the (exogenous) lipids were detected up to the 6th strip. The CH_2 stretching vibrations that report the physical state of the endogenous lipids of skin are shown in Figure 5B. The wavenumber associated to this vibration on the skin surface decreased up to the 6th strip. After this depth, the lipids are present in the hexagonal phase [28].

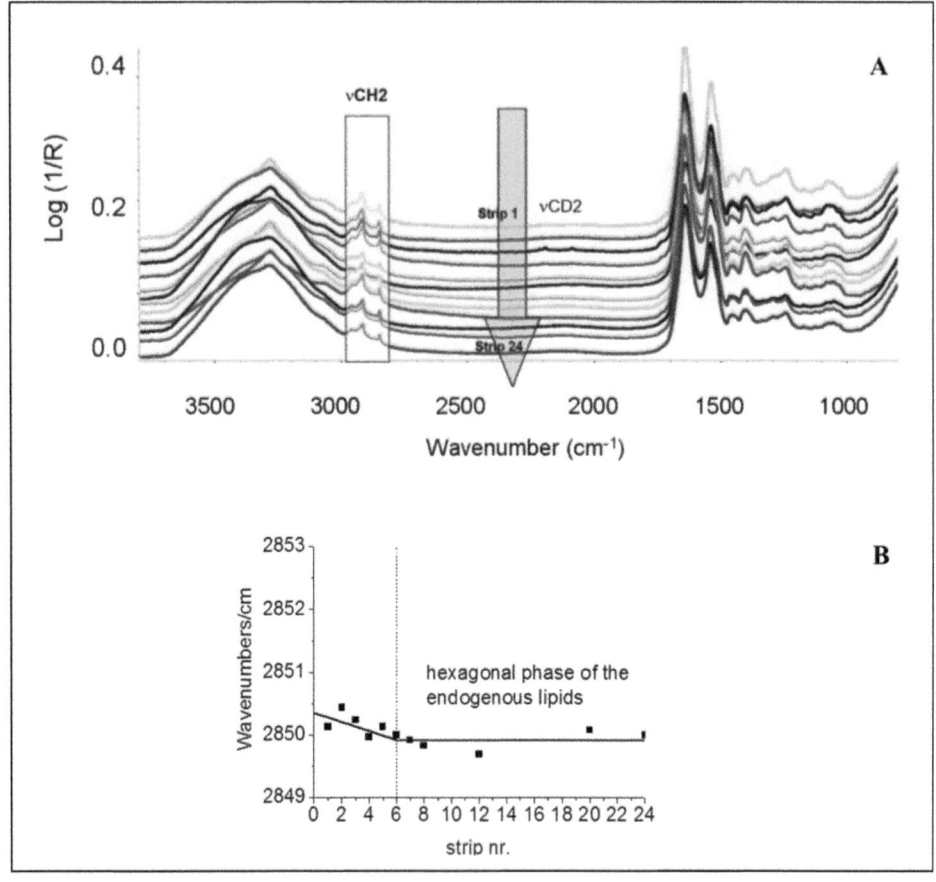

Figure 5 *ATR-FTIR scans of skin treated with bicelles d-DPPC/d-DHPC, from 1st (top) to 24th (bottom) strips (A), plotted as absorbance intensity versus the wavenumber. The shadowed square includes the CH$_2$ vibrations. The wavenumber of the CH$_2$ symmetric stretching vibrations versus the strip number are represented in B.*

Skin analysis in transmission FTIR mode

IR microspectroscopy allows selection of an area to perform several IR scans at different points (see dotted areas on micrographs; Figure 6). The colour within the dotted areas represents the relative distribution of vibrations selected in the IR spectra (Figure 6; vertical black line on IR spectra) that corresponded to the amide I of proteins and the CH$_2$ vibration of lipids [31]. The colour scale goes from blue (low intensity) to red (high intensity). Figure 7 shows the relative distribution of deuterated lipids (black line of the IR spectrum) on the skin micrograph derived from the IR spectrum performed at the asterisk position.

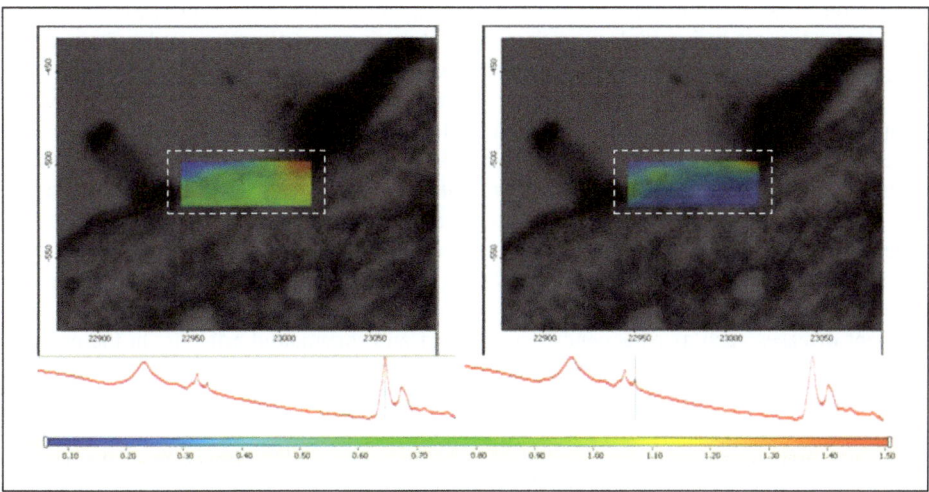

Figure 6 *Skin micrographs overlapped with the relative distributions of proteins (vibration of amide I, left) and lipids (vibration of CH$_2$, right) in skin. The relative distribution of amide I (right) and CH$_2$ (left) is marked with black lines in the IR spectra of skin (below the micrographs). The coloured scale (at bottom) represents the intensity (red, high; blue, low) of absorbance of IR radiation by each chemical bond.*

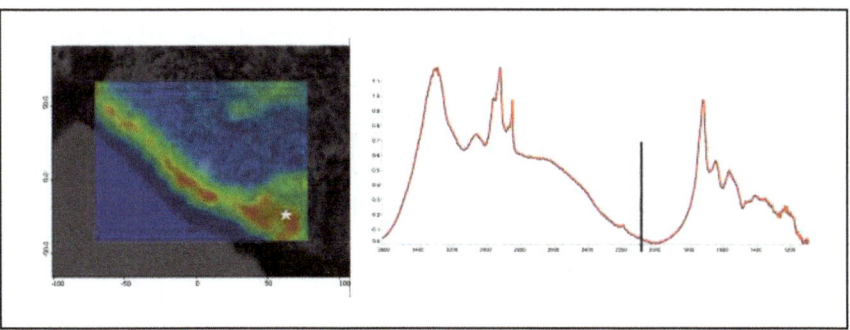

Figure 7 *IR microspectroscopy of skin showing the relative distribution of deuterated lipids (left). The IR spectrum corresponds with the asterisk position. The black line in the spectrum marks the vibrations of CD$_2$ bond (right).*

4   DISCUSSION

Synchrotron radiation (SR) represents an enhancement over techniques such as SAXS and IR due to the higher flux and coherence of photons. Consequently, measurement acquisition time is substantially reduced and spatial resolution is increased. This is important when analysing biological tissues, as there is generally a less ordered structure and greater sensitivity to radiation damage than with crystalline or inorganic materials.

The SAXS technique allows skin layers to be discriminated on the basis of composition and structure. The structural study of connective tissue matrices at the molecular level is a vital step in understanding the pathology and basis for possible treatment of many human diseases. Abnormalities of the molecular structure of collagen may adversely affect the packing of collagen molecules in fibrillar-forming types (e.g., bone, skin, and tendon) and so may reduce stability of these fibres, producing weaker and structurally inferior connective tissue. Clearly, an understanding the supra-molecular structure of healthy connective tissue is a prerequisite for elucidating the pathology of different diseases such as osteogenesis imperfecta and rheumatoid arthritis. In the present work, the application of the SAXS technique using SR proved adequate for studying ordered skin components such as collagen fibrils and SC lipids.

The SAXS profiles are modified only when collagen is altered. The axial periodicity of the collagen in our experiment was 63.4 ± 0.03 nm, slightly lower than the 65 nm associated with skin collagen [37, 39]. This small variation in the d-spacing could be explained by some variation in the water content of the sample [13]. Indeed, a comparison of skin collagen from other tissues such as tendons, bones or cartilage also results in lower d-spacing [39, 40]. However, the reasons for this are not immediately clear and may be due to a variety of factors such as the composition of the extracellular matrix, the proportion of different types of collagen molecules or water content of the sample. The biophysical properties of skin are different from those of tendons or bones in which strength and resistivity are important characteristics. In this sense, few studies have been focussed on the analysis of skin collagen by the SAXS technique. Cameron and co-workers explained this shortening of the d-spacing in heterotypic fibres of collagen (type III plus type I) as a consequence of the tilt angle of the fibre and the projection of this periodic spacing [41]. Stinson and Sweeney [39] suggested that the different glycosaminoglycan content of skin in relation to other collagen-rich tissues might be an important factor. Assuming that the collagen molecules have a net positive charge, these authors proposed that the glycosaminoglycan binding in skin could alter the charge distributions and water binding in such a way as to permit a somewhat shorter length for the molecule leading to a shorter d-period.

As would be expected, profiles from whole skin and dermis did not indicate a change the position of the peaks because both correspond to native collagen molecule (Figure 1). Correspondingly, these data infer that that the thermal treatment used to separate the skin layers did not affect the collagen. However, fibre coagulation has been described at a temperature of 73°C in murine skin [42]. The bands associated with scattering features due to the cylindrical shape of collagen fibrils (asterisked in Figure 1A) are only visible in complete skin. This could be the result of a loss of lateral organisation in collagen when the skin is manipulated for the experiment. The odd diffraction orders (1st, 3rd, and 5th) are clearly visible compared to the even orders. Fratzl and co-workers [43] associated a different behaviour for odd and even orders during the stretching process, although this condition has not been applied to skin samples. However, other authors have related the differential intensity in collagen reflections to the water content of the fibril: the higher the intensity of the odd peaks, the higher the degree of wetting of collagen [39, 44]. In this sense, although the water content in the upper layer of the skin (SC) is relatively low (*ca.* 30%), the dermis (where the collagen is located) contains more water (70%) [45] and this agrees fairly well with the differential intensities of the peaks. In contrast, the epidermis (Figure 1A and B) shows a pattern typical of a lipid lamellar organisation within the SC which is in agreement with other work [3, 38]. The two broad bands present in Figure 1B correspond to the two lamellar phases LLP and SLP (12 and 6.4 nm, respectively) as described by Bouwstra and co-workers [38]. These authors propose that the lipids in the lamellar phase form a predominantly crystalline lateral phase with a subpopulation of lipids forming a liquid phase. Our results do not show other components from the epidermis (e.g., crystalline cholesterol). This is possibly because the SC is included in the epidermal layer, where the ordered fraction (from the SC lipids) is surrounded by non-ordered material (corneocytes). As expected, the SC lipid pattern did not disturb the collagen pattern in the spectra of whole skin because the proportion of SC lipids contained in the skin (10%) is very low in comparison to the collagen content (75%) [8, 45].

Different ranges of the scattering vector value (Q) are achieved by varying the detector-to-sample distance: The higher the sample-to-detector distance, the lower the Q value. Thus, the 1st (up to the 10th) reflection and some features of lateral collagen organisation are visualised with the longer configuration. In contrast, more detailed organisation of SC lipids (larger d-spacings) and more than twenty reflections of collagen are better visualised with the shorter configuration.

Our results demonstrate that the SAXS technique is very useful for characterising different states of skin depending on the collagen and lipid structure. In this work, aged skin presented a loss of orientation in comparison with younger skin (Figure 2). Quaglino and co-workers reported that the diameter of fibres and the number of collagen bundles are decreased in people over 60 years old [46]. Also, the ratio of type III to type I collagen increases with age [8, 41, 46, 47]. Lavker compared skin from old (70–85 years) and young (19–25 years) individuals. This work showed that in young adults, collagen in the dermis forms a meshwork of randomly oriented thin fibres and small bundles [48]. Our results appear to indicate the opposite: a preferential fibre orientation in young individuals that is lost with age. The skin contains collagen type I and III, unlike the straight fibrils in tendons which comprise exclusively collagen type I. Local nonspecific "doping" of a type I fibrillar structure with type III molecules may disrupt crystalline-type interactions [41]. The fact that this occurs with ageing could explain the loss of orientation in the fibres from aged skin. Thus, changes in orientation and arrangement of this macromolecule during ageing could be partly responsible for age-related alterations in skin morphology and mechanical properties.

A comparison of the profiles from healthy skin (Figure 1) and skin invaded by melanoma (Figure 3) demonstrates clear structural differences. The characteristic reflections associated with the periodical axial distance of collagen were clearly present in healthy skin (Figure 1). However, they were not observed in the melanoma samples (Figure 3). This difference can be interpreted as a degradation (or disorder) of the collagen fibrils present in the tissue invaded by cancer cells. In diverse studies, Fernández and co-workers reported the characterisation of human breast tissue using the SAXS patterns of collagen [16, 30]. They found that regions far from the tumours were essentially different from those at the tumour sites. The axial period of collagen fibrils was slightly larger in cancer-invaded regions and the average intensity of scattering from cancerous regions was an order of magnitude higher than the intensity from healthy regions. In our experiment, the skin sample invaded by melanoma did not present any reflection associated with the characteristic d-spacing of collagen. The reason for this could be the advanced state of degeneration in the collagen molecule that is visualised as a disorganisation of the extracellular matrix in the melanoma. In this sense, this non-ordered material also produces diffuse scattering with high intensity.

The invasion and metastasis of melanoma have been shown to require proteolytic degradation of the extracellular environment, achieved primarily by enzymes of the matrix metalloproteinase (MMP) family. Increased enzyme activity has been localised at the border of tumour cells and adjacent peri-lesional connective tissue, emphasising the crucial role of tumour-stroma interactions in the regulation of MMP activity [15, 17]. Given these facts and with the aim of investigating the aggressiveness of melanomas on skin collagen, we undertook SAXS studies to compare skin melanoma and healthy skin submitted to proteolytic digestion with pepsin [35, 36]. This treatment was selected because the proteolytic process partially mimics the action of MMPs in melanoma (see the insert, Figure 3). Despite the strength of this enzymatic process, SAXS profiles of the insoluble residue of skin still exhibited a periodic arrangement of collagen structure, although with decreased intensity. As stated above, this periodic arrangement was not detected in the melanoma samples. In a previous study, the analysis of melanoma from the surrounding areas exhibited a gradual loss of order in the collagen, being substituted by very strong intensity bands [30].

Alterations in collagen were also detected in skin with stretch marks. The presence of stretch marks (Figure 4) affects the hierarchical organisation of the collagen molecule, with only slightly visible odd reflections indicative of a loss of order in the collagen molecular packing. The pathogenesis of striae is unknown, but is most likely related to changes in the structures that provide tensile strength and elasticity to the skin. Such structures are components of the extracellular matrix, including fibrillin, elastin and collagen. Thus, some workers have associated the presence of striae with an increase in glycosaminoglycan content, as well as with the reduction of vertical fibrillin fibres subjacent to the dermal-epidermal junction and elastin fibres in the dermis.[33] The increase in glycosaminoglycan content could explain the loss of order as a result of the increase of non-ordered material. As previously mentioned, some authors have related the high intensity of odd peaks with a wet state of collagen [39, 44] meaning that the physiological changes linked to the presence of stretch marks do not modify the water content associated to the collagen molecules. In this present study, we only detected indicative changes in collagen structure visible as a decrease (low intensity of the Bragg peaks) of the ordered material of the skin.

Ideally, transdermal delivery requires components or vehicles to traverse the skin barrier with minimum disruption. Bicellar systems are lipid-based vehicles of sufficiently small size to penetrate through the SC of skin [49]. The use of deuterated lipids allows one to distinguish signals in the IR spectra from both exogenous and endogenous source of lipid. This, combined with consecutive tape strips, facilitate the analysis of deeper skin tissue using ATR-FTIR (Figure 5). The deuterated lipids from d-DPPC/DHPC vesicles could be distinguished using the CD_2 vibration up to 6th tape strip (Figure 5A). Correspondingly, these data infer that the bicelles partitioned into the SC barrier but remained just near the surface. Indeed, the effect of the vehicle on the SC lipids (represented in Figure 5B as the wavenumber of CH_2 stretching vibration versus tape strip number) indicated a slight decrease in the wavenumber of this vibration close to the surface (up to the 6th tape strip) with no further detectable changes in subsequent tape strips. Cotte *et al.* previously demonstrated that the skin penetration of chemical enhancers was partially dependent on alkyl length chain [31]. It was observed in the current study that deuterated lipids from the vehicle and changes in endogenous lipids of SC occurred at the same level (6th tape strip). In this case, the small changes in the fluidity of the SC lipids could be influenced by the bicelles, thus confirming penetration of the vehicle into the superficial SC. Moreover, the wavenumber of the CH_2 stretching vibration corresponded to the gel state of the endogenous lipids (hexagonal phase) [28]. The typical lateral packing of SC lipids is gel phase: the more solid the lipid state is, the less permeable the barrier [38]. Hence, this bicellar system crosses the skin surface without altering skin barrier function. These data confirm those obtained in previous studies in which only the effect of the bicelles on the endogenous lipids was determined [28].

The evaluation by FTIR in transmission mode is represented in Figure 6, where the image of the microtomed skin section was superimposed with a colour scale representing the relative distribution of chemical bond vibrations associated with lipids (CH_2 stretching) or proteins (amide I). Thus, proteins would be distributed throughout the scanned area (Figure 6; left) while lipids were mainly detected at the surface (Figure 6; right). This agrees well the known lipid and protein distribution of the SC. Again, the deuterated signal (CD_2) from bicelles was detectable at the skin surface (Figure 7; left). The red colour in the relative distribution map indicated the presence of this specific vibration, inferred from the IR spectrum (Figure 7; right- marked with a black line). Thus, from the FTIR results it may be concluded d-DPPC/DHPC was able to partition into the SC layer and that the fluidity of SC lipids was not affected by the bicellar system (as it remained in the hexagonal phase). This is a potential advantage of the vehicle when compared to other systems that use chemical enhancers to disturb skin barrier function to achieve transdermal drug delivery [50, 51].

Overall, FTIR (comprising SR source) allowed qualitative and quantitative visualisation of the relative distribution of chemical components within the skin and so permitted differentiation between exogenous from endogenous sources of lipids. This combination reinforces the IR spectroscopy and enhances its applications in skin diagnosis.

4 CONCLUSION

In summary, the exhaustive structural analysis of skin by non-crystalline diffraction provides interesting results on lipid and protein organisation. In addition, IR microspectroscopy enhances the opportunities of skin diagnosis. The use of synchrotron radiation (SR) permitted the characterisation of healthy skin in terms of the structural properties of the barrier, the mechanical properties of the dermis skeleton and the chemical map of skin composition. Alterations such as ageing, stretch marks or illness modified these properties, resulting in a general disorganisation of skin structure. Interaction of

Advances in Dermatological Sciences

vehicles for topical delivery of substances may also modify the permeability of skin lipids. Consequently, SR facilitates the systematic evaluation of skin samples subject to chemical exposure and for evaluating pathological processes and so constitutes a powerful diagnostic tool.

5   ACKNOWLEDGEMENTS

The authors would like to thank the staffs of BM16 (ESRF), and SMIS (Soleil) for providing technical support. This work was supported by funds from CICYT (CTQ2010-16964), MICINN (proposals 16-02-13, 16-02-33), and FP7-ELISA. M.C. was funded by the JAE-Doc Program from CSIC (co-funded by FSE).

6   REFERENCES

1. D. J. Tobin, Biochemistry of human skin--our brain on the outside, *Chem. Soc. Rev.*, 2006, **35**(1), 52–67.

2. J. Kanitakis, Anatomy, histology and immunohistochemistry of normal human skin, *Eur. J. Dermatol.*, 2002, **12**(4), 390–401.

3. G. S. Pilgram, *et al.*, Aberrant lipid organization in stratum corneum of patients with atopic dermatitis and lamellar ichthyosis, *J. Invest. Dermatol.*, 2001, **117**(3), 710–717.

4. C. J. Moger, *et al.*, Regional variations of collagen orientation in normal and diseased articular cartilage and subchondral bone determined using small angle X-ray scattering (SAXS), *Osteoarthritis Cartilage*, 2007, **15**(6), 682–687.

5. T. R. Kyriakides, *et al.*, Mice that lack thrombospondin 2 display connective tissue abnormalities that are associated with disordered collagen fibrillogenesis, an increased vascular density, and a bleeding diathesis, *J. Cell Biol.*, 1998, **140**(2), 419–30.

6. S. A. Santoro and L.W. Cunningham, Collagen-mediated platelet aggregation: the role of multiple interactions between the platelet surface and collagen, *Thromb. Haemost.*, 1980, **43**(2), 158–62.

7. J. L. Contet-Audonneau, C. Jeanmaire and G. Pauly, A histological study of human wrinkle structures: comparison between sun-exposed areas of the face, with or without wrinkles, and sun-protected areas, *Br. J. Dermatol.*, 1999, **140**(6), 1038–47.

8. J. M. Waller and H. I. Maibach, Age and skin structure and function, a quantitative approach (II): protein, glycosaminoglycan, water, and lipid content and structure, *Skin Res. Technol.*, 2006, **12**(3), 145–54.

9. A. J. Singer, *et al.*, Standardized burn model using a multiparametric histologic analysis of burn depth, *Acad. Emerg. Med.*, 2000, **7**(1), 1–6.

10. B. Brodsky and A. V. Persikov, Molecular structure of the collagen triple helix, *Adv. Protein Chem.*, 2005, **70**, 301–39.

11. D. J. Hulmes and A. Miller, Quasi-hexagonal molecular packing in collagen fibrils, *Nature*, 1979, **282**(5741), 878–80.

12. F. Gobeaux, *et al.*, Cooperative ordering of collagen triple helices in the dense state. *Langmuir*, 2007, **23**(11), 6411–7.

13. T. J Wess and J. P. Orgel, Changes in collagen structure: drying, dehydrothermal treatment and relation to long term deterioration, *Thermochim. Acta*, 2000, **365**, 119–128.

14. T. Brandt, *et al.*, Pathogenesis of cervical artery dissections: association with connective tissue abnormalities. *Neurology*, 2001, **57**(1), 24–30.

15. S. Kauppila, *et al.*, Aberrant type I and type III collagen gene expression in human breast cancer *in vivo*, *J. Pathol.*, 1998, **186**(3), 262–8.

16. M. Fernández, *et al.*, USAXS and SAXS from cancer-bearing breast tissue samples, *Eur. J. Radiol.*, 2008, **68**(3 Suppl), S89–94.

17. T. D. McKee, *et al.*, Degradation of fibrillar collagen in a human melanoma xenograft improves the efficacy of an oncolytic herpes simplex virus vector, *Cancer Res.*, 2006, **66**(5), 2509–13.

18. V. J. James and N. Kirby, The connection between the presence of melanoma and changes in fibre diffraction patterns, *Cancers*, 2010, **2**, 1155–1165.

19. D. Skovgaard, *et al.*, Use of cis-[18F]fluoro-proline for assessment of exercise-related collagen synthesis in musculoskeletal connective tissue, *PLoS One*, 2011, **6**(2), e16678.

20. P. Martin, Wound healing-aiming for perfect skin regeneration, *Science*, 1997, **276**(5309), 75–81.

21. P. Willmott, An introduction to synchrotron radiation: techniques and applications, 1st edn, 2011, Wiley, Chichester, UK.

22. M. C. Martin, *et al.*, Recent applications and current trends in analytical chemistry using synchrotron-based Fourier-transform infrared microspectroscopy, *Trends Anal. Chem.*, 2010, **29**(6), 453–463.

23. C. Burger, *et al.*, Small-angle X-ray scattering study of intramuscular fish bone: collagen fibril superstructure determined from equidistant meridional reflections, *J. Appl. Cryst.*, 2008, **41**, 252–261.

24. R. A. Lewis, *et al.*, Breast cancer diagnosis using scattered X-rays, *J. Synchrotron Radiat.*, 2000, **7**(Pt 5), 348–52.

25. A. L. C. Conceição, M. Antoniassi and M. E. Poletti, Analysis of breast cancer by small angle X-ray scattering (SAXS), *Analyst*, 2009, **134**, 1077–1082.

26. J. A. Bouwstra, *et al.*, Structural investigations of human stratum corneum by small-angle X-ray scattering, *J. Invest. Dermatol.*, 1991, **97**(6), 1005–12.

27. S. J. George, *et al.*, Synchrotron X-ray analyses demonstrate phosphate-bound gadolinium in skin in nephrogenic systemic fibrosis, *Br. J. Dermatol.*, 2010, **163**(5), 1077–81.

28. G. Rodríguez, *et al.*, Conformational changes in stratum corneum lipids by effect of bicellar systems, *Langmuir*, 2009, **25**(18), 10595–603.

29. G. Rodriguez, *et al.*, A unique bicellar nanosystem combining two effects on stratum corneum lipids., *Mol Pharm*, 2012, **9**(3), 482–91.

30. M. Costa, *et al.*, Diagnosis Applications of Non-Crystalline Diffraction of Collagen Fibres: Breast Cancer and Skin Diseases, in *Applications of Synchrotron Light to Scattering and Diffraction in Materials and Life Sciences*, 2009, Springer, Berlin/Heidelberg, pp. 265–280.

31. M. Cotte, *et al.*, Synchrotron FT-IR microscopic study of chemical enhancers in transdermal drug delivery: example of fatty acids, *J. Control Release*, 2004, **97**(2), 269–81.

32. M. Cócera, *et al.*, Characterisation of skin states by non-crystalline diffraction, *Soft Matter*, 2011, **7**, 8605–8611.

33. R. E. Watson, *et al.*, Fibrillin microfibrils are reduced in skin exhibiting striae distensae, *Br. J. Dermatol.*, 1998, **138**(6), 931–7.

34. O. López, *et al.*, New arrangement of proteins and lipids in the stratum corneum cornified envelope, *Biochim. Biophys. Acta*, 2007, **1768**(3), 521–9.

35. D. W. Bannister and A. B. Burns, Pepsin treatment of avian skin collagen. Effects on solubility, subunit composition and aggregation properties, *Biochem. J.*, 1972, **129**(3), 677–681.

36. B. M. Kim, *et al.*, Collagen structure and nonlinear susceptibility: effects of heat, glycation, and enzymatic cleavage on second harmonic signal intensity, *Lasers Surg. Med.*, 2000, **27**(4), 329–35.

37. C. Mérigoux, *et al.*, *Supramolecular organisation of collagen fibrils in human tissues,* Newsletter, 1997, **29** 18–19.

38. J. Bouwstra, G. Gooris, and M. Ponec, The Lipid Organisation of the Skin Barrier: Liquid and Crystalline Domains Coexist in Lamellar Phases, *J. Biol. Phys.*, 2002, **28**(2), 211–223.

39. R. H Stinson and P. R. Sweeny, Skin collagen has an unusual d-spacing, *Biochim. Biophys. Acta*, 1980, **621**(1), 158-61.

40. L. J. Gathercole, J. S. Shah, and C. Nave, Skin-tendon differences in collagen D-period are not geometric or stretch-related artefacts, *Int. J. Biol. Macromol.*, 1987, **9**, 181–183.

41. G. J. Cameron, *et al.*, Structure of type I and type III heterotypic collagen fibrils: an X-ray diffraction study, *J. Struct. Biol.*, 2002, **137**(1-2), 15–22.

42. K. M. Kirsch, *et al.*, Ultrastructure of collagen thermally denatured by microsecond domain pulsed carbon dioxide laser, *Arch. Dermatol.*, 1998, **134**(10), 1255–9.

43. P. Fratzl, *et al.*, Fibrillar structure and mechanical properties of collagen, *J. Struct. Biol.*, 1998, **122**(1-2), 119–22.

44. C. A. Maxwell, T. J. Wess, and C. J. Kennedy, X-ray diffraction study into the effects of liming on the structure of collagen, *Biomacromolecules*, 2006, **7**(8), 2321–6.

45. G. Imokawa, *et al.*, Decreased level of ceramides in stratum corneum of atopic dermatitis: an etiologic factor in atopic dry skin?, *J. Invest. Dermatol.*, 1991, **96**(4), 523–6.

46. D. Quaglino, Jr., *et al.*, Ultrastructural and morphometrical evaluations on normal human dermal connective tissue–the influence of age, sex and body region, *Br. J. Dermatol.*, 1996, **134**(6), 1013–1022.

47. L. Vitellaro-Zuccarello, R. Garbelli, and V. D. Rossi, Immunocytochemical localization of collagen types I, III, IV, and fibronectin in the human dermis. Modifications with ageing, *Cell Tissue Res.*, 1992, **268**(3), 505–11.

48. R. M. Lavker, Structural alterations in exposed and unexposed aged skin, *J. Invest. Dermatol.*, 1979, **73**(1), 59–66.

49. L. Barbosa-Barros, *et al.*, Bicelles: lipid nanostructured platforms with potential dermal applications, *Small*, 2012, **8**(6), 807–18.

50. K. S. Warner, *et al.*, Structure-activity relationship for chemical skin permeation enhancers: probing the chemical microenvironment of the site of action, *J. Pharm. Sci.*, 2003, **92**(6), 1305–22.

51. N. Abla, *et al.*, Topical iontophoresis of valaciclovir hydrochloride improves cutaneous aciclovir delivery, *Pharm. Res.*, 2006, **23**(8), 1842–9.

SECTION II: DERMAL THERAPEUTICS

7

SECTION EDITORIAL – DERMAL FORMULATION DEVELOPMENT

M B Brown

MedPharm Ltd, Guildford, Surrey, United Kingdom and School of Pharmacy, University of Hertfordshire, Hatfield, United Kingdom

1 THE PHARMACEUTICAL DEVELOPMENT INDUSTRY

Since the mid-1990s, pharmaceutical R&D productivity has experienced what at best can be described as a downturn. From 1998 to 2008 the number of new molecular entities has been in general decline whereas attrition rates, development times and expenditure have all increased [1,2]. For example, over the past 40 years the US pharmaceutical industry's inflation-adjusted research and development spend has increased from $2.5 to $27 billion, which equates to a current average cost of $843 million per new molecular entity (NME). In addition, only 2 out of 10 marketed products end up providing a return on investment. During the same period the time to market has almost doubled, currently being around twelve to fourteen years. Many reasons have been put forward for this lack of productivity, including a more stringent and tightly controlled regulatory environment, increased competition and intellectual property issues. It is also important to note that the effects of the recent multinational mergers and the resultant so called 'NME pipeline consolidation' are yet to be seen. However, a striking observation is that this decline in productivity has coincided with multinational pharmaceutical companies beginning to search for, and spending enormous amounts of money on, the identification of increasingly complex druggable disease targets. This often relies on genomics and the resultant development and validation of techniques such as high-throughput screening, robotics, combinatorial chemistry and bioinformatics. In addition, although such techniques are producing an increasing number of potential drug candidates, the attrition rates of such molecules as a result of problematic physicochemical properties, safety and efficacy are also escalating. For example, at present it is estimated that for every 10,000 molecules screened during the discovery process only one will gain regulatory approval as a medicine. The pharmaceutical development industry often forgets that it is not a drug that is given to a patient but a medicine and the art of formulation development in producing a medicinal product that meets the relevant regulatory authority's criteria of acceptable quality, safety and efficacy is being lost. As such it is pleasing to see the strength and breadth of formulation development research described in this section on Dermal Therapeutics and this book series of *Advances in Dermatological Sciences*.

2   THE PHILOSOPHY OF FORMULATION DEVELOPMENT

When a formulation scientist starts the long and often painful process of developing a topical or transdermal formulation there are a plethora of issues that need to be considered. The pharmaceutical history is littered with examples of drugs that could have been the next blockbuster if only a formulation had been developed that delivered the drug, safely and efficaciously to the pathological site in a cost effective manner. Such problems are exacerbated even further in dermal drug delivery because patients care what they apply to the skin and often have a choice. As such, the consumer should often be at the forefront of the thoughts of a formulation scientist. The dermal products developed many years ago that were greasy, malodorous and stained clothes are no longer acceptable. Cosmetics and aesthetics of the final product are, in some cases, almost as important as the product's efficacy. As such the intricacies of dermal formulation development are considerably different to those for a tablet, for example. Consequently, this introductory chapter provides an overview of a dermal formulation development process encompassing pre-formulation development through to identification of the final lead formulation candidate and although generic in nature highlights the issues and possible pitfalls that need to be addressed.

3   *IN SILICO* DRUG CANDIDATE SELECTION

Drug candidate selection for dermal formulation development requires the use of rapid and cost effective methods applicable to a large number of samples. Over the last 25 years there has been much interest in the use of mathematical models in an attempt to predict dermal absorption *in silico* [3]. Formulation scientists have been tempted, especially in the pharmaceutical arena, to find the most promising compounds by investigating the relationship between percutaneous permeation and molecular parameters such as lipophilicity, hydrogen bonding, molecular size and/or melting point. Quantitative Structure Property-Activity Relationships (QSPR) studies have meant that, if a correlation is found, it is possible to screen any number of compounds, including those that have not been yet synthesised, for the selection of those structures with the required properties for dermal delivery. A range of non-linear methods have also been employed to improve predictions of skin absorption including artificial neural networks (ANN) [4], which in some cases have demonstrated high predictive power (for example, see Section 5 "Experimental and Mathematical Models" of this book). However, ANN's are a limited method in that they have a tendency to over-fit where large numbers of physicochemical descriptors exist, compared to the data points used. Such models are often weighted and are susceptible to over-training [5]. Gaussian process (GP) methods do not alleviate all these issues, but minimise them [6], reportedly providing better predictions of percutaneous absorption than existing models [7] perhaps suggesting that the approach of predicting skin absorption by means of a simple equation may have limited mechanistic value.

Whatever the opinion on the best model to use or even if any are actually valid, there are a plethora of *in silico* models [3] that exist for the formulation scientist to help identify the molecules that will be best absorbed. However, other factors need to be considered including predicted or experimentally derived molecule stability, solubility, irritancy, toxicity and potency before a final decision on candidate selection is made. Despite developments in *in silico* modelling, they are yet to replace actual skin permeation testing using the type of *in vitro* models described in the subsequent chapters and will probably remain as such until their reliability and correlation with real experimentally derived date can be confirmed. As often is the case, it is experience rather than theory that guides the best way forward.

4 PRE-FORMULATION STUDIES

Pre-formulation is a research and development stage where the drug's physicochemical properties and desired dosage form along with the drug's mechanism of action and target disease are considered. For the development of topical semi-solid drug products, pre-formulation studies typically involve initial solubility and compatibility studies. Such studies are conducted to identify any critical parameters which may affect the development of the final product and may include poor drug solubility, inherent drug instability, potential excipient/drug or excipient/excipient incompatibility amongst others. In addition, another aim of pre-formulation studies is to develop and explore methodologies to improve these defined issues such that the target profile of the formulation dosage form can be achieved. Thus, pre-formulation studies are conducted to form the basis for the rationale of formulation design.

The physicochemical properties of a drug that can influence its performance and manufacturability should be identified and considered during pre-formulation work and include log P, pKa and molecular weight. Such parameters play a key role in the inherent permeability of a drug across the skin as described previously. The Log P (partition coefficient) reflects how well a drug partitions between lipid (oil) and water whilst the pKa (dissociation constant) is a measure of the strength of an acid or base and allows the determination of charge on a molecule at any given pH. Both measurements are useful parameters for use in understanding the solubility and diffusivity and/or partitioning across the stratum corneum (SC).

The selection and identification of potential drug salt forms or the use of a salt or free acid or base is critical during pre-formulation studies since these selections obviously influence drug solubility, stability and ultimately drug absorption. Although the unionised form of a drug is generally thought to be a more suitable candidate for topical application due to its lower polarity and consequently higher partitioning into the SC, other considerations such as drug solubility and stability in the formulation have to be considered. These issues should be decided during the early development phase since changing the salt form at a later stage may force repetition of toxicological, formulation and stability studies, thus increasing development time and cost. The introduction of a new salt form at a late stage must also be evaluated for potential impurity changes, and its bioequivalence, pharmacokinetic equivalence, and toxicity equivalence to the previous salt form will have to be proven.

Given that the exact underlying cause of a lot of skin diseases may not be well defined or understood, the effective drug concentration required to reach the target area within the skin is also often not well defined. Thus, before a decision is made to take a drug into topical formulation development, several issues should be considered (if they have not already been addressed in the candidate selection process):

- Has the pharmacological activity of the drug been demonstrated or predicted and what is the IC_{50} or minimum concentration required to exert a therapeutic effect?

- Are the pharmacological models used in assessing and/or predicting the pharmacological activity of the drug appropriate?

- Is the target site known?

- Is the drug metabolised in the skin?

For example, when targeting a drug to the epidermis, a highly potent drug with a low permeability coefficient (Kp) may not necessarily be the most efficacious when compared to a less potent drug with a higher Kp as it is a combination of a drug's potency and ability to permeate the skin which is important. Thus, such parameters should be well defined and understood during drug selection and during any pre-formulation work since this information will allow proper evaluation of dosage form type, dose/drug concentration, and the selection of excipients.

Topical pharmaceutical formulations rarely contain a single excipient, with the number varying from of a few (e.g. aqueous gels or lotions) to greater than ten (e.g. emulsion systems). The simplest way to choose and utilise excipients is to select those with appropriate properties that are used in existing formulations that have received regulatory approval in the territories relevant to the product.

The legislation and nature of regulatory control varies from one country to another, with acceptable pharmaceutical excipients generally being listed in international pharmacopoeias and supported by extensive safety data. Notably, the use of existing excipients requires regulatory consideration when using an established excipient for an alternative delivery route. For example, the oral consumption of an excipient that has previously been used in a topical dosage form may well mean that additional toxicity studies are required. Equally, a wide range of novel excipients have been developed and described in the literature. However, for the reasons highlighted above, the use of such excipients in developing a pharmaceutical product is limited as they require supportive data similar to those required for a new drug, although such extensive supporting data may be reduced or eliminated if each excipient has associated toxicity data, previous approval for food use or oral administration or is already used cosmetically. On the other hand, regulation of cosmetic topical formulations, although controlled by legislation, has a less strict stance on excipients, thus a much wider range is available and acceptable than in the development of pharmaceutical topical formulations. With this in mind, it is obvious that the general concept of 'the less the better' holds true when developing a drug product.

Excipients typically make up more than 90% of a topical pharmaceutical product and are included to perform a variety of functional roles in such formulations. For the purpose of a semi-solid topical dosage form, such functional roles may include:

- Improvement of solubility to allow incorporation of the drug at the target concentration.

- Controlling drug release and permeation.

- Improving general aesthetics of the product to increase patient compliance.

- Improve drug skin permeability and/or deposition.

- Improve drug and formulation stability.

- Prevention of microbial growth and contamination.

The effective delivery of drugs into and through the skin is not trivial and in order for therapeutic quantities of drug to penetrate the skin, the barrier properties of the stratum corneum must be overcome. Current strategies to overcome the barrier properties of the stratum corneum may be divided into chemical or passive, such as the use of occlusion, penetration enhancers [8 – 10], colloidal [11, 12] and supersaturated systems [13] and physical or active strategies, such as iontophoresis, skin electroporation, ultrasound, magnetophoresis, thermophoresis [14] or microneedles [15, 16]. Currently, chemical/passive enhancement is the most successful approach and the choice and selection of enhancers depends upon the formulation type and nature of the drug. However, consideration should also be given to the enhancer's potential pharmacological activity, toxicity, duration of action, enhancing mechanism (and reversibility), stability and cosmetic acceptability [17]. Many of these strategies and issues are the subject of contemporary work reported in this and subsequent book sections [8 – 17].

During product use, most formulations undergo considerable physical changes once they are applied to the surface of the skin. For example, the effect of rubbing may decrease the viscosity of a formulation containing a thixotropic gelling agent and this in turn may have an effect on drug release from the formulation and permeation across the skin. If a volatile solvent is present, evaporation of a drug solvent within a formulation may reduce the solubility of the drug and result in precipitation or physical instability of the formulation. However, this effect has also been used by many formulation scientists to increase thermodynamic activity and thus increase drug release [13]. Consideration of contamination from frequent use may warrant the inclusion of a preservative within the formulation such that the microbial load throughout the product life remains below the recommended level.

Solubility plays an essential role as part of formulation development since inadequate drug solubility in a formulation will impede the development of certain drugs if the target dose cannot be achieved. Likewise, a very soluble drug may also pose issues such as poor drug release from a system. Thus the solubility of a particular drug may necessitate the selection of a narrow range of solvents which are only suitable for specific dosage forms.

For example, a highly lipophilic drug is less likely to be formulated as an aqueous gel but as a cream, ointment or non-aqueous gel. During pre-formulation study design, the selection of solvents is based on the initial dosage form of choice using a library of topically acceptable excipients. Some of the approaches used to dissolve drugs for topical formulation development include the use of co-solvents, pH adjustment, complexation, surfactants and a combination thereof. Of these approaches, the use of co-solvents is probably the most practical and commonly used method to solubilise poorly soluble compounds in aqueous systems. When sufficient solubility cannot be achieved in a single solvent (most likely aqueous), co-solvents are used as an alternative option to increase drug solubility. A variety of solvents can be used and it is important that the selection of co-solvents is based on the miscibility of each solvent to avoid phase separation. In addition, solvents and non-solvents are also used in co-solvent systems to increase the thermodynamic activity of highly soluble drugs at the desired drug concentration. Adjusting the pH of a vehicle is an effective means of increasing drug solubility, since most drugs are weakly ionisable acids or bases. Consequently, by considering the pH of the solution in combination with the pKa of the drug, solubility can be optimised. In addition, it should be emphasised that the pH of the final formulation should be in the region of 5 to 7 although slightly more acidic pH values (> pH 4) may still be acceptable for topical formulations depending on the application site, frequency and area. Surfactant systems have also been successfully used in the form of emulsions, micro emulsions and nano-emulsions to improve the delivery of drugs with poor aqueous solubility [11,12].

Excipients are the components of a formulation that in combination produce a 'successful' pharmaceutical product. Although the drug or active is of primary importance since it is responsible for the treatment of the disease, in order to 'present and deliver' the drug, excipients play an equally important role. Whilst traditionally, excipients have been used as 'inert' material such as bulking agents in dosage forms such as tablets, more recent development of drug delivery systems has used the benefit of excipient–drug interaction to produce formulations with specific properties and performance specifications. Thus any interaction between excipients and the drug substance must be well understood and is fundamental in developing any drug product.

Excipient/drug or excipient/excipient compatibility can be classified as physical, chemical or physiological and such interactions may have implications for drug stability, product manufacture, drug release, product efficacy, therapeutic activity and side-effect profiles. There are several approaches to conduct excipient compatibility screening and these can be dependent upon the dosage form, although the main approach is based on accelerated temperature studies. In this case, an experiment is designed to investigate the effect on compatibility of the drug and excipient under real time (25° C) and accelerated conditions (40 or 50° C). Usually, such experiments are designed using excipients comprising solutions of the drug in pure solvent to investigate experimental 'extremes'. Higher accelerated temperature conditions may be investigated but such experiments must be based on the assumption that any reaction rate is proportional to temperature and this is often not the case.

5   FORMULATION DEVELOPMENT

Dermal dosage forms have been generally classified as liquids (e.g. suspensions, sprays, lotions), semi solids (colloids, creams, gels, ointments, pastes and foams) and solids (e.g. patches, powders). The most conventional and probably well-known topical dosage forms are creams, gels and ointments, although classification of such topical dosage forms may be poorly defined and ambiguous depending on the literature referenced. Nevertheless, the initial selection of suitable dosage forms is most commonly dependent upon the drug's physicochemical properties, target disease type, required aesthetic and cosmetic properties, scalability, costs and the target product profile. The brand and its continual evolution can also be a significant consideration for over-the-counter (OTC) topical products.

For chronic dermatological disorders such as eczema and psoriasis, occlusive and hydrating formulations such as ointments or anhydrous systems are often preferred since these preparations have protective properties. However, such anhydrous mixtures are also usually very tacky and greasy and have poor aesthetic properties. Although such formulation types are extremely useful as emollients due to their occlusive properties, their value as topical products is limited by the poor solubility of many drug substances in them. In such cases drug solubility can only be enhanced by formulating them with hydrocarbon-miscible solvents. Anhydrous systems may also comprise pure polyethylene glycol systems or triglyceride derivatives in addition to the traditional hydrocarbon systems containing white soft paraffin (petrolatum). Alternatively, silicone-based formulations may also be used. However, the regulatory status of silicones for topical use is currently limited even though an extensive and positive safety profile is emerging.

Alternative monophasic systems to ointments are aqueous gel formulations. Such systems usually contain water based or alcohol based co-solvent systems with a thickening agent. Although these systems are more aesthetically pleasing than anhydrous systems or ointments, they lack the occlusive, hydrating properties of traditional formulations. Gels are usually used for the application of anti-inflammatories, anti-infectives or anti-histamines, or where facial application is required (for example, in the treatment of acne and rosacea). For such applications, occlusive properties are not desirable and gels are less likely to leave a cosmetically unacceptable (greasy) residue. To improve the occlusive properties of gels, 'Emugels' have been developed. Emugels, or emulsified gels, are essentially biphasic systems containing an aqueous gel dispersed with a lipid phase and are more closely related to creams.

Creams are dispersed systems comprising of an outer water continuous phase and an inner, oil-immiscible dispersed phase. Such systems are also known as o/w (oil-in-water) emulsions. Where the phases are reversed, the system is known as a w/o (water-in-oil) emulsion. The indication for which the product will be used is an important consideration in the choice of emulsion type. From a cosmetic acceptability point of view, o/w emulsions are generally less 'greasy' and thus more acceptable. The development of multiple emulsions (for example, oil-in-water-in-oil; o/w/o) can allow compartmentalisation of incompatible excipients/drugs with similar physicochemical properties. As with all multiple phase systems, their effect on drug release and permeation is complex and should always be considered.

More recently, the use of emulsifier free approaches has intensified in the development of pharmaceutical and cosmetic products. Creams or emulsion-based systems usually comprise one or more traditional emulsifiers (such as surfactants) to stabilise the formulation. However, due to the irritancy potential of surfactants [18] polymeric emulsifiers such as carbomers, celluloses and polyacrylates have been successfully used to stabilise emulsions [19]. The high molecular weight of such polymeric emulsifiers means that that they are less likely to penetrate the stratum corneum, thus minimising any of the unwanted effects often observed with surfactants. In addition, surfactants, like detergents and soaps, have a tendency to emulsify and remove natural lipids within the skin, leaving a 'dry skin' feel; a disadvantage not observed with polymeric emulsifiers.

Although the specifics of a target product profile for a topical formulation will vary depending on its ultimate purpose, there are key aspects of most target profiles that are the same for most formulations. The most basic of these is the use of approved excipients, where the type and concentration of excipient used should be acceptable from a regulatory perspective (as discussed earlier). It is also always the case that the excipients utilised must be suitable for use in the disease state for which the formulation is designed.

The extent and rate of release of drug from a formulation should be well understood [20]: *in vitro* release rates are a useful assessment of this parameter and can also serve as a valuable quality control (QC) tool in monitoring formulation changes on storage. Clearly, it should be demonstrated that the formulation can deliver the drug into or across the skin at the appropriate concentration at the required site of action. The cosmetic elegance and patient acceptability of any product is also of importance whilst the physical and chemical stability of the drug/formulation must yield adequate shelf life. It is also important to ensure that the developed formulation can be manufactured on commercial scales. Finally, the cost of goods for the product must satisfy the demands of the relevant market. Ultimately, it is always important to remember a general rule: the simpler a formulation, the fewer things there are to go wrong.

6   FORMULATION PERFORMANCE

The performance of a topical formulation can be assessed using a range of methods depending on factors such as formulation type, target disease, aesthetic requirements and application site. However, three parameters that are central to formulation development are stability, release and skin permeation characteristics of the drug from the formulation [20]. Ultimately, a formulation optimised for drug release and permeation is often more efficacious and may require a lower concentration of drug which may reduce the cost of the final product and potential toxicity.

Other performance assessments include aesthetic and cosmetic acceptability, the use of preclinical disease models and early stage toxicity assessment. In addition, stability in various packaging materials may be determined. The ultimate objective is to test the performance of candidate formulations in order to mitigate the risk of failure during clinical investigation. Once all such assessments and formulation optimisation has been completed, one lead (and preferably at least one back-up) formulation should be identified for full characterisation and stability testing.

7 FORMULATION CHARACTERISATION

Detailed characterisation of a candidate formulation is performed to define a provisional product specification and methods of measurement such that that the capability or performance of a product can be monitored during ICH [1] stability tests to ensure that they remain within the set specifications throughout its shelf life. For topical semi solids, some examples of typical parameters include-

- Macroscopic (visual) and microscopic appearance and odour.

- Drug content and uniformity, related substances and degradation products.

- Preservative content.

- Formulation pH.

- Rheology and viscosity.

- Microbial quality or microbial limit test (MLT) and preservative efficacy test (PET).

- Sterility.

Once characterised, the formulation(s) are placed on long-term accelerated and real-time stability studies performed under ICH conditions (25° C & 60% relative humidity (RH), 30° C & 65% RH and 40° C & 75% RH; [21]). Such a study may span a period of at least 2 years, with measurement of all parameters being performed at regular intervals (for example; 0, 1, 3, 6, 9, 12, 18 and 24 months). At the same time (if not already investigated), the primary packaging materials may also be selected and investigated for compatibility with the final formulation.

8 CONCLUSIONS

The development of any topical product is unique to the particular drug and dosage form used and various issues have to be considered during the development process. This brief overview of the processes and challenges involved with topical formulation development provides a background to which subsequent chapters in this section make a substantial contribution.

---

[1] International Conference on Harmonisation (of Technical Requirements for Registration of Pharmaceuticals for Human Use).

9   REFERENCES

1. F. Pammoli, L. Magazzini, and M. Riccaboni, The productivity crisis in pharmaceutical R and D, *Nature Reviews: Drug Discovery*, 2011, **10**, 428–438.

2. M. B. Brown, It is not a molecule you give to a patient but a medicine, the lost of science of formulation, *Drug Discovery Today*, 2005, **10**, 1405–1407.

3. M. B. Brown and S. T. Lim, Topical product development in *Transdermal and Topical Drug Delivery: Principles and Practice*, ed. H. Benson and A.C. Watkinson, Wiley, New Jersey, 2012, pp. 255–286.

4. I. T. Degim, J. Hadgraft, S. Ilbasmis, Y. Ozkan, Prediction of skin penetration using artificial neural network (ANN) modelling, *Journal of Pharmaceutical Sciences*, 2003, **92**, 656–664.

5. D. Neumann, O. Kohlbacher, C. Merkwirth, T. Lengauer, A Fully Computational Model for Predicting Percutaneous Drug Absorption, *Journal of Chemical Information and Modelling*, 2006, **46**, 424–429.

6. C. E. Rasmussen, C. K. I. Williams, Gaussian Processes for Machine Learning. The MIT Press, 2006.

7. G. P. Moss, Y. Sun, M. Prapopouloua, N. Davey, R. Adams, W. J. Pugh and M. B. Brown, The application of Gaussian processes in the prediction of percutaneous absorption, *J. Pharm. Pharmacol.*, 2009, **61**, 1147–1153.

8. S. Briançon, Y. Chevalier and M. A. Bolzinger, *Biopharmaceutical evaluation of various dosage forms intended for caffeine topical drug delivery*, Chapter 8.

9. A. Baraldi, R. H. Khengar, S. Murdan, M. J. Traynor, S. A. Jones and M. B. Brown. *The effect of human nail disulphide bond disruption on barrier integrity*, Chapter 9.

10. M. S. Erdal, D. Özdin and S. Güngör, *Synergistic effects of transcutol and terpenes as penetration enhancers: in vitro and in vivo ATR-FTIR spectroscopic imaging studies*, Chapter 10.

11. J. Frelichowska, M. A. Bolzinger, J. Pelletier, J. P. Valour and Chevalier, *Skin penetration from Pickering emulsions*, Chapter 11.

12. F. Laredj-Bourezg, M. A. Bolzinger, J. Pelletier, M. R. Rovere, B. Smatti and Y. Chevalier, *Pickering emulsions stabilised by biodegradable particles offer a double level of controlled delivery of hydrophobic drugs*, Chapter 12.

13. A. Edwards, F. Liu, M. B. Brown and W. J. McCauley, *Evaluation of methylphenidate permeation from Daytrana™ patches across silicone and human epidermal membranes*, Chapter 31.

14. F. Caserta, D. G. Wood, M. B. Brown, W. J. McAuley, *The effect of heat on diclofenac permeation through human skin*, Chapter 13.

15. T. W. Prow, Y. H. Mohammed, A. B. Ansaldo and H. A. E. Benson. *Topical microneedle drug delivery enhanced with magnetophoresis,* Chapter 14.

16. T. Gratieri, B. Wagner, D. Kalaria, B. Ernst and Y. N. Kalia, *Iontophoretic delivery of glycomimetics – a new approach for the treatment of inflammatory skin diseases,* Chapter 15.

17. D. Tamarkin, *Foam – a unique topical drug delivery system*, Chapter 16.

18. K.-P. Wilhelm, A. B. Cua, H. H. Wolff, and H. I. Maibach, Surfactant-induced Stratum Corneum hydration *in vivo*: Prediction of the irritation potential of anionic surfactants. *Journal of Investigative dermatology*, 1993, **101**, 310–315.

19. M. F. Bobin, V. Michel, E. Journet, and M. C. Martini, *Study of formulation and stability of emulsions with polymeric emulsifiers. Colloid and Surfaces*, 1999, vol 152, pp. 53–58.

20. P. A. Lehman, S. G. Raney and T. J. Franz. *Topical bioequivalence: a comprehensive approach using multiple surrogate methods*, Chapter 17.

21. ICH. 2003. Q1A(R2): Stability testing of New Drug Substances and Products. Proceedings of the International Conference on Harmonisation. US FDA Federal Register.

8

BIOPHARMACEUTICAL EVALUATION OF VARIOUS DOSAGE FORMS INTENDED FOR CAFFEINE TOPICAL DELIVERY

S Briançon, Y Chevalier, M-A Bolzinger

Université de Lyon F-69008, Lyon, France; Université Lyon 1, Laboratoire de Dermopharmacie et Cosmétologie, F-69008, Lyon, France; UMR CNRS 5007, Laboratoire d'Automatique et de Génie des Procédés, F-69622, Villeurbanne, France.

1 INTRODUCTION

The efficacy of actives applied by the topical route is limited not just by factors relating to the active ingredient(s) but also formulation design and biological factors such as skin permeability. Topical formulations have to be designed to ensure that a sufficient dose of the active molecule reaches its target site, be it the skin surface, the skin itself or the underlying cutaneous vasculature. In order to achieve this, the active substance has to overcome the natural barrier function of the skin. Several strategies have been studied to enhance skin penetration and so improve cutaneous bioavailability. The first approach is to disturb the skin barrier using chemical penetration enhancers or by increasing the hydration level [1]. Encapsulating the active substance in particulate systems such as polymeric or lipid nanoparticles or liposomes is another way to improve their skin penetration, as shown by many research groups [1–4]. Microemulsions are also recognised as formulations which promote drug penetration through the skin, resulting in permeation rates significantly higher than classical emulsions or other liquid vehicles [5–8].

The objective of our work was to evaluate the efficacy of several formulations to improve caffeine skin delivery. Caffeine is widely used in cosmetics as an active substance because of its slimming effect. Caffeine has also been widely studied as a model compound in transdermal delivery [1, 2, 9, 10] because it is not metabolised during transport through the skin [11] and its low $\log P$ value (-0.07) allows generation of high fluxes through the skin.

The site of action of caffeine is in the adipocytes located in the hypodermis, meaning that formulations have to be optimised to ensure a high penetration level and an adequate dose of caffeine delivered to the hypodermis. Three new formulations of caffeine were tested: polymeric microspheres, microemulsions and Pickering emulsions. Their skin absorption kinetics was compared to those of classical oil-in-water (o/w) or water-in-oil (w/o) emulsions, a hydrophilic gel and an aqueous caffeine solution. In the first part of the study, caffeine permeation from the different formulation was evaluated through full-thickness pig skin after removal of the subcutaneous fat in accordance with regulatory requirements [12]. In the second part, experiments were performed on full-thickness skin pieces with a 3 mm hypodermis in order to estimate the accumulation of caffeine at the intended target site.

2 MATERIALS AND METHODS

The composition of all formulations and the preparation protocols are listed below. Each formulation contained 0.8% (w/w) caffeine, except microspheres which contained 0.23%.

The microemulsion (ME) was prepared at pH 5.5 by mixing all the components in the same flask under stirring with a magnetic bar. The mixture comprised 13.3% PEG-8 Caprylic/Capric triglycerides (LAS®, Gattefossé, France), 6.67% Polyglycerol-6 dioleate (Plurol® oléique, Gattefossé, France), 4% isostearyl isostearate (ISIS®, Gattefossé, France), 2.3% cyclomethicone (DC®344, Dow Corning, Belgique), 1.6% diisopropyl adipate (Ceraphyl®230, ISP, France), 2% PPG-5 ceteth 20 (Procetyl AWS®, Croda, France), 2% propylene glycol (Cooper, France), 0.8% caffeine (Sigma, France) and purified water.

The Pickering Emulsion (PE) comprised 1% modified silica Wacker HDK®H20 (55% free OH), 47.7% cyclomethicone (DC®245 and DC®246, Dow Corning, Belgium), 0.8% caffeine (Sigma, France), purified water. The silica was first dispersed in the silicone oil using an ultrasound disperser (Sonics VibraCell, BioBlock Scientific, France) at 500 W for 30 s. The oil and aqueous phases were mixed with an UltraTurrax® device (Ika, Germany) at 22,000 rpm over 2 min to obtain a w/o emulsion.

Polymer microspheres (MS) comprised poly ε-caprolactone (M_w = 64,000, Aldrich), Poly(vinylalcohol) (Mowiol® 4–88, Fluka), Caffeine 0.23% (Sigma, France) and purified water. Caffeine-loaded microspheres were prepared by a $w_1/o/w_2$ multiple emulsion solvent evaporation method as previously described [13].

The oil-in-water emulsion (o/w CE) was prepared using 5% PEG-6 stearate PEG-32 stearate (Tefose®1500, Gattefossé, France), 3% stearic acid (Cooper, France), 10% isostearyl isostearate (ISIS®, Gattefossé, France), 5.75% cyclomethicone (DC®344, Dow Corning, Belgique), 4% diisopropyl adipate (Ceraphyl®230, ISP, France), 0.15% carbomer (Ultrez®10, Gattefossé, France), 0.3% xanthan gum (Rhodicare®D, Saci, France), 0.8% caffeine (Sigma, France) and purified water. The oil-in-water emulsion (w/o CE) was prepared using 0.1%, Bis-PEG/PPG-14/14 dimethicone (Abil®EM97, Evonik, France), 10% C24-28 alkylmethicone (Abil®Wax 9810, Evonik, France), 39.1% cyclomethicone (DC®245 and DC®246, Dow Corning, Belgium), 0.8% caffeine (Sigma, France) and purified water.

Both the w/o and o/w emulsions were prepared by heating the aqueous phase (water + thickening agents + caffeine) and the oil phase (lipophilic compounds + emulsifiers) to 70°C followed by mixing with a TurboTest® (Rayneri/VMI, Montaigu, France) at 1,000 rpm. On cooling (30°C), the pH was adjusted to 5.5 with sodium hydroxide.

The gel (G) was prepared by mixing all ingredients except the sodium hydroxide at room temperature with a Turbo-Test® at 500 rpm. Gelling was achieved by addition of sodium hydroxide up to pH 5.5. The initial mixture comprised 0.4% carbomer (Ultrez®10, Gattefossé, France), 0.3% NaOH, 0.3% hydroxypropylguar (Jagua®HP105, Saci, France), 5% glycerine (Cooper, France), 2% propyleneglycol (Cooper, France), 0.05% citric acid (Cooper, France), 0.8% caffeine), purified water.

The caffeine solution (CAF sol) comprised 0.8% caffeine (Sigma, France), phosphate buffer saline (PBS 0.01 M, pH 7.4, Sigma, France) and purified water.

Particle size distribution of emulsions and microspheres was analysed by laser scattering (MasterSizer® 2000, Malvern, UK). The average particle size was expressed as the volume-average diameter in micrometers. The shape and surface of the microspheres were observed using a scanning electron microscope (SEM, Hitachi S800, Japan).

The caffeine content in microspheres was assayed by HPLC after dissolving the dried microspheres in ethyl acetate. A Waters instrument was used with a reverse phase column (XTerra® RP8, 4.6×250 mm - 5 μm) and a Waters 2996 photodiode array UV detector at a wavelength of 271 nm. Elution with water/acetonitrile/acetic acid (85:15:1) solvent at 1 mL min^{-1} flow rate and 35°C gave a retention time of 5.0 min for caffeine. The calibration curve for quantitative analysis was linear up to 40 μg mL^{-1}.

The partition coefficient of caffeine between the aqueous and lipophilic phases of emulsions was measured for each emulsion containing 0.8% of caffeine in the absence of surfactant. The water and oil phases were mixed together with a magnetic stirrer during 24 h and equilibrated at rest during 24 h. The concentration of caffeine in each phase was measured by the same HPLC method as described above and it was checked that the partition coefficient was identical for the emulsion and microemulsion formulations.

The structure of microemulsions was determined by small angle neutron scattering measurements performed on the D22 spectrometer at the Institut Laue – Langevin (ILL) European facility at Grenoble (France). The samples contained in quartz cuvettes of 1 mm thickness were measured according to standard procedures available at the ILL and their structure was assessed from comparison with classical structural models [14]. Microemulsions were prepared with the same chemical composition as used for percutaneous penetration experiments, but normal water was substituted by deuterated water.

Percutaneous absorption studies were carried out with full thickness pig skin in accordance with OECD guidelines [12]. Skin integrity was assessed by measuring transepidermal water loss (TEWL) of skin samples using a Skin Station apparatus (La Licorne, France). Skin samples having TEWL rates greater than 10 g m^{-2} h^{-1} were considered damaged and eliminated from further study. The thickness of each section was measured with a micrometer (Mitutoyo). Individual skin pieces were mounted between the two (donor and receptor) chambers of a vertical (Franz-type) glass diffusion cells with the *stratum corneum* facing the (upper) donor chamber. The area available for diffusion was 2.54 cm^2.

Skin samples retaining hypodermis were sliced with a dermatome to a thickness of 4 mm, resulting in approximately 3 mm of hypodermis. The receptor phase (10 mL) was phosphate buffer saline solution (PBS, pH 7.4) which was continuously stirred. The caffeine solubility in the receptor fluid was checked before beginning the experiments to ensure sink conditions.

The formulations were applied to the skin surface using the same dose of caffeine (3.5 mg cm^{-2}). Diffusion experiments (six replicates) were conducted over a 24 or 72 h period. At each measuring time the entire receptor phase was collected, washed four times with phosphate buffered saline, filtered, and analysed by HPLC. At the end of the study (24 or 72 h), the formulation remaining in the donor compartment was removed and the caffeine quantified by HPLC. The epidermis was separated from the dermis by heat treatment (by immersion for 45 s in hot (60°C) water) and the amount of caffeine in each skin layer analysed by HPLC.

When microspheres were applied, a complementary analysis was performed to determine the encapsulated quantity of caffeine remaining within the formulation (in addition to that released). This was measured by dissolving the whole sample in ethyl acetate prior to HPLC analysis. The amount of caffeine still encapsulated was calculated by subtracting the released quantity from the total quantity. To investigate microsphere penetration into the skin, the stratum corneum was tape-stripped (Monaderm, Monaco 20 strips). The strips were observed using scanning electron microscopy (SEM, Hitachi S800, Japan).

Cumulative amounts of caffeine ($\mu g\,cm^{-2}$) penetrating through the skin were corrected to account for previous sample removal and plotted against time. The pseudo-steady state flux (J_{SS}) of *in vitro* caffeine permeation was estimated from the slope of the linear plot ($R^2 \geq 0.97$) from which the mean and standard deviation (S.D.) of $n = 6$ determinations were calculated. Statistical comparisons were made using Student's *t*-test (two-sample, assuming equal variances) and analysis of variance (ANOVA, single factor) with significance level of $p \leq 0.05$.

3 RESULTS AND DISCUSSION

The structure of microemulsions was assessed by small angle neutron scattering. Only the oil-in-water model of polydisperse spherical droplets [14] fit the experimental data, giving the mean droplet radius as 7.8 nm [15]. The visual appearance of the o/w microemulsion is shown in Figure 1A. The mean droplet sizes of the standard and Pickering emulsions were of the same magnitude, being 5.24 ± 0.08 μm for o/w CE, 10.9 ± 0.5 μm for w/o CE and 9.7 ± 0.5 μm for PE. The Pickering emulsion was of a water-in-oil type so it could be compared directly with the standard (w/o CE) emulsion: in both cases, caffeine was entrapped within the inner aqueous droplets. The mean diameter of the microspheres (MS) was 2.8 ± 0.2 μm and their caffeine loading was 2.3 ± 0.14 mg g^{-1} of particles (0.23%). MS have a porous inner structure as shown in Figure 1B-D; the pores correspond to the aqueous droplets of the primary water-in-oil emulsion.

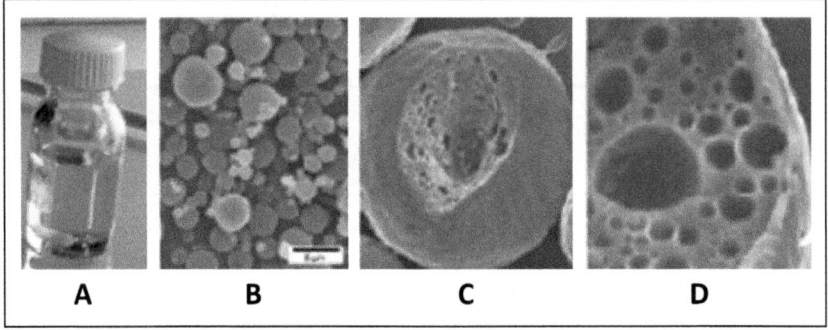

Figure 1 *Photos of the microemulsion (A) and scanning electron micrograph images of microspheres (MS) at increasing magnification (B – D).*

The initial studies used dermatomed skin in the absence of hypodermis in order to provide an OECD-compliant baseline measurement of the skin absorption kinetics of caffeine. The results are presented as a percentage of the applied dose recovered in the receptor compartment versus time (Figure 2).

The percutaneous penetration of caffeine was generally higher when delivered from the topical formulations in comparison with the caffeine solution (CAF sol), the exception being the w/o emulsion (w/o CE). The total amount of caffeine penetrating the skin at 24 h from the microspheres (MS) and microemulsion (ME) was more than double that of the caffeine solution. Thus, these two vehicles led to a significant enhancement of caffeine permeation with similar quantities of caffeine recovered in the receptor fluid after 24 h. However, the kinetic profiles were different: MS showed an initially high penetration commensurate with a more rapid partitioning of particles followed by rapid release of the caffeine. These data show that MS easily penetrate the skin and accumulate in the receptor fluid providing a continuous release of caffeine [16]. This particular behavior of polymeric particles has previously been demonstrated by other workers [17 – 22].

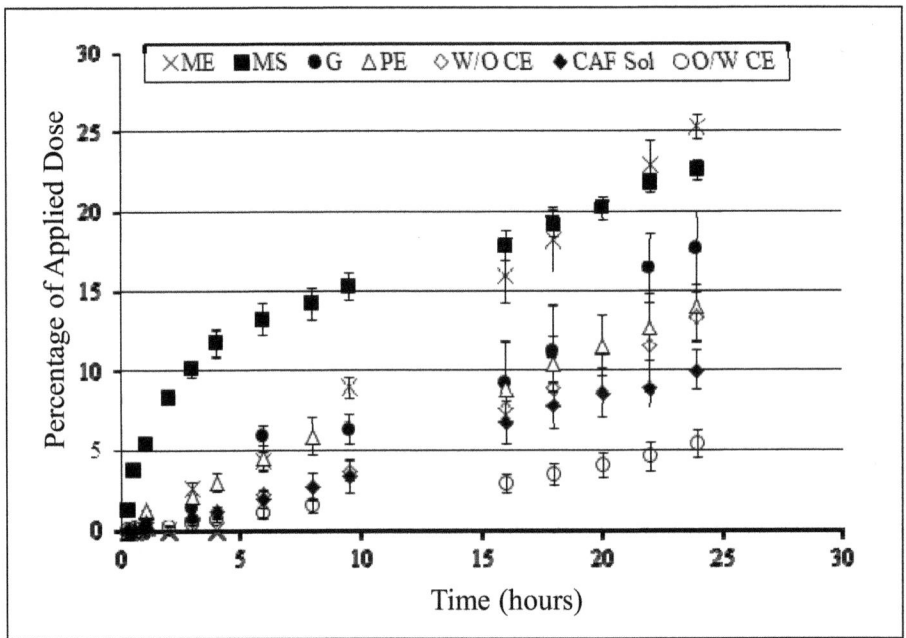

Figure 2 *Cumulative skin absorption of caffeine through dermatomed human skin (free of hypodermis) expressed as % of applied dose as a function of time for all formulations (ME = microemulsion; o/w CE = oil-in-water classical emulsion; PE = Pickering emulsion; w/o CE= water-in-oil classical emulsion; MS = microspheres; CAF sol = solution). All values are mean ± standard deviation of n = 6 replicates.*

Evidence of nano-carrier penetration and accumulation in the outermost layers of the stratum corneum (SC) and skin furrows has been obtained from similar penetration studies [21, 22] and indicate a size-dependent behaviour where smaller particles achieve higher penetration rates and accumulation in the *stratum corneum*. Particle deposition in hair follicles has also been previously described. For example, Alvarez-Román *et al.* observed the behaviour of nanoparticles after skin application by confocal microscopy and showed a size-dependent penetration over time in hair follicles and skin furrows [21, 22]. There is, however, no consensus about the optimal size to favour skin penetration: some workers have reported an optimal particle size of less than 300 nm [21 – 23] while others [17 – 20] have demonstrated the penetration of larger (μm range) particles as found in this present study. The follicular pathway should also be considered when dealing with particle penetration, as it appears that particles can at least reach the infundibulum where they accumulate and potentially constitute a drug reservoir [22 – 25]. A size-dependent penetration was established by Patzelt *et al.* who showed that both organic (poly(lactic-co-glycolic acid)) and inorganic (silica) nanoparticles penetrated deeper into the follicles when they had an optimum size of around 640 nm [26]. This could be important for caffeine which has been shown to use the follicular pathway even when applied as a solution; the study by Patzelt *et al.* indicated that approximately half of the permeated concentration of caffeine utilised the follicular pathway [27].

The amount of caffeine in the receptor fluid after 24 h from skin treated with the microemulsion (ME) was 1.3 greater than that obtained with the oil-in-water classical emulsion (o/w CE) and gel (G) and 2.5 greater than that obtained with the aqueous solution (CAF sol) (Figure 2). The enhanced permeation of caffeine by a microemulsion dosage form has previously been described [15, 28]; the lag time was significantly shorter for microemulsions when compared to an aqueous solution or classical emulsion. It has been suggested that the mechanism of enhancement may be linked to the emulsion microstructure, as well as caffeine partitioning within a vehicle and thermodynamic activity [28]. Other studies have indicated that permeation enhancement may be related to the water content of the vehicle [29]. An additional hypothesis for the mechanism of enhanced skin absorption relates to the use of surfactants which could dissolve or disrupt the arrangement of intercorneocyte lipids within the *stratum corneum*. Transepidermal water loss (TEWL) measurements before and after application of topical formulations can provide an indication of changes in skin permeability. Previous work in our laboratory has shown that TEWL rates do not significantly alter following application of different formulations [15]. Thus, incorporation of surfactant into microemulsions does not appear to be a dominant factor. Alternative hypotheses for the enhancement effect of microemulsions include improved drug solubility and reduced interfacial surface tension [30]. Thus overall, it appears that the microemulsion effect may be explained by several mechanisms linked to the microstructure but also to the water, oil and surfactant/co-surfactant content.

The Pickering emulsion (PE) behaved in an intermediate way (enhancement factor of 1.5 compared to caffeine solution). A comparison of the PE and standard w/o emulsion (w/o CE; which also contains caffeine in the inner droplets) indicates that the quantity of caffeine in the receptor fluid was 3 times higher with the PE. These data are in agreement with previous studies which have suggested that enhanced skin absorption of caffeine from

Pickering emulsions may be due to improved skin surface interactions and thus faster drug release into the *stratum corneum* [27, 28].

The second part of this study involved the use of skin with an intact layer of hypodermal tissue in order to further investigate the two most effective formulations identified above (microemulsion and microspheres) and so evaluate their potential for delivering caffeine to the proposed target site.

Microsphere permeation was not significantly affected by the presence of hypodermis: 20% of the applied dose of caffeine reached the receptor fluid after 24 h in the presence of hypodermis (Figure 3) against 22.6% of applied dose when the hypodermis was removed (Figure 2). This difference mainly corresponded to the amount stored in the hypodermis since very small amounts of caffeine were recovered in the epidermis and dermis. Conversely, the presence of the hypodermis significantly reduced caffeine permeation delivered from the microemulsion: the quantity of caffeine in the receptor fluid after 24 h was halved (12% of the applied dose versus 25% without hypodermis). Interestingly, the amount of caffeine released from the microspheres was very low, indicating that the microspheres were penetrating intact and accumulated in the hypodermis (from where they could potentially act as a reservoir for sustained caffeine delivery).

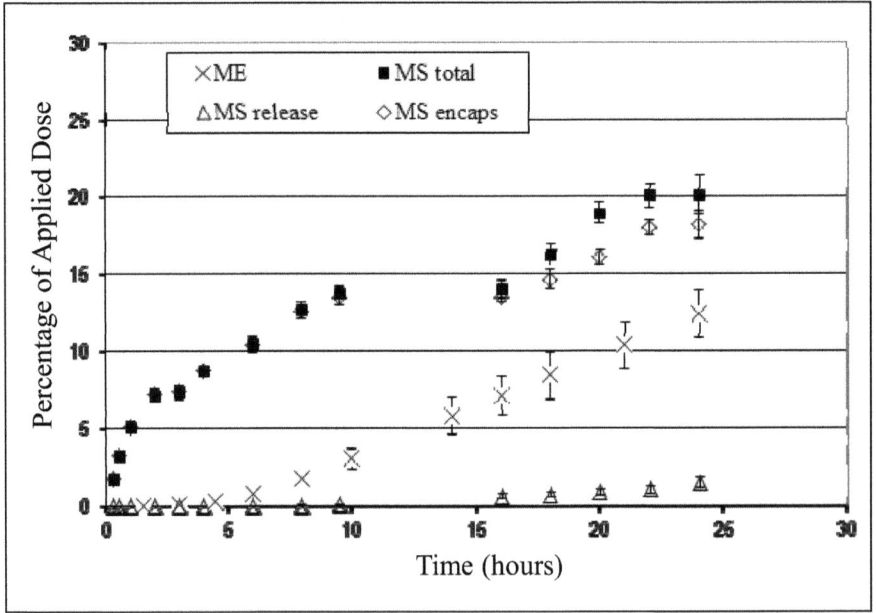

Figure 3 *Cumulative skin absorption of caffeine through human skin (with intact hypodermis) expressed as % of applied dose as a function of time for all formulations (ME = microemulsion; MStotal = microspheres; MSrelease = caffeine released from the microsphere; MSencaps = caffeine still entrapped in the microspheres). All values are mean ± standard deviation of n = 6 replicates.*

The distribution of caffeine recovered from the various skin layers after 24 h treatment with different vehicles is presented in Tables 1 (without hypodermis) and 2 (with hypodermis). In the case of the microspheres, caffeine distribution was measured after 72 h exposure to account for the extended sustained-release properties. It can be seen that the distribution of caffeine was dependent not just on the formulation but also the presence or absence of the hypodermal layer.

Skin Layer	ME	PE	O/W CE	MS*	G
Epidermis (%)	0.27 ± 0.01	0.30 ± 0.07	0.37 ± 0.05	0.08 ± 0.01	0.31 ± 0.03
Dermis (%)	1.48 ± 0.10	0.8 ± 0.1	0.92 ± 0.05	0.98 ± 0.2	1.06 ± 0.2
Receptor (%)	25.3 ± 1.6	12.7 ± 1.6	13.3 ± 1.5	52.08 ± 1.41	17.7 ± 1.6
Total absorbed (%)	27.0 ± 1.7	13.80 ± 1.64	14.6 ± 1.6	53.14 ± 1.59	19.1 ± 1.7

Table 1 *Caffeine distribution as a percentage of the applied dose in the skin layers after 24 h exposure (skin <u>without hypodermis</u>). Asterisk (*) indicates distribution after 72 h.*

Skin Layer	ME	O/W CE	MS*	G
Epidermis (%)	0.68 ± 0.06	0.62 ± 0.02	0,08± 0.01	0.51 ± 0.09
Dermis (%)			0,98± 0.2	
Dermis + hypodermis (%)	10.35 ± 1.3	7.8 ± 1.5		7.95 ± 2.3
Hypodermis (%)			14.52± 0.72	
Receptor compartment (%)	12.4 ±2	7.8 ± 0.8	42.46 ± 1.16	7.24 ± 2.1
Total absorbed (%)	23.4 ± 3	16 ± 2	58.03± 1.46	15.7 ± 2.5

Table 2 *Caffeine distribution as a percentage of the applied dose in the skin layers after 24 h exposure (skin* with hypodermis*). Asterisk (*) indicates distribution after 72 h.*

The distribution of caffeine in the skin layers showed that the major proportion of caffeine accumulated in the receptor fluid. This result was in accordance with several studies showing that hydrophilic and small molecules permeate well through the skin [13, 15]. The results revealed significant differences in the total penetration between the microemulsion (25% of applied dose) and the other formulations (classical emulsion, Pickering emulsion and gel; 15–20%). The quantity stored in the dermis was also higher with the microemulsion even though it was low for all other formulations (1.5% for the microemulsion against 0.8%, 0.92% and 1.06 for the Pickering emulsion, the classical emulsion and the gel, respectively). The amounts of caffeine in the epidermis were not statistically different. The use of microspheres resulted in a lower quantity of caffeine in the epidermis and dermis, comparable with that obtained with emulsions and gel.

The presence of the hypodermis did not significantly influence the total amount of penetrated material with ME, gel and MS, suggesting that the *stratum corneum* was the rate limiting barrier against penetration. However, the recovery of caffeine within the different skin layers differed according to the presence or the absence of the hypodermis. After 24 h exposure, caffeine was equally distributed between the dermal/hypodermal layer and the receptor chamber compartment for the three liquid formulations (ME, G and o/w CE), indicating that the hypodermis retained one half of the caffeine which had permeated.

The fraction of applied dose in the hypodermis and receptor compartment increased with the performance of the dosage form (microemulsion > emulsion and gel). It can be concluded that the presence of hypodermis facilitated the formation of a reservoir of caffeine in that layer (where it can exert its lipolytic activity).

When MS were applied, the quantity of caffeine in the epidermis and dermis was lower than with the other formulations even if it was measured after 72 h exposure. The distribution in the skin layers and the analysis of both fractions of caffeine (release and still encapsulated) showed that MS were present in the receptor fluid and that they were able to release their content for the whole experiment duration. The distribution in skin layers after 72 h in the presence or in the absence of hypodermis are shown in Figure 4. The results showed an accumulation of particles in the hypodermis when this layer was present, with 14% of the applied caffeine being recovered compared to less than 1% in the dermis when the hypodermis was absent. This quantity of caffeine in the hypodermis was release for one half, the other half being still encapsulated in the particles which play a role of active reservoir.

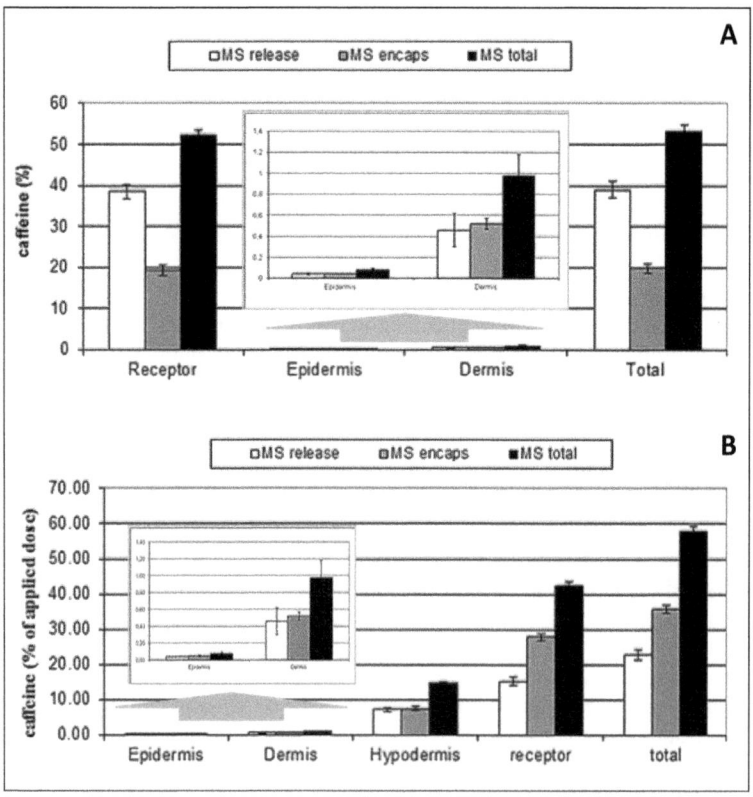

Figure 4 *Caffeine distribution in skin layers 72 h after MS application on skin without (a) or with (b) hypodermis (MStotal = microspheres; MSrelease = caffeine released from the microsphere; MSencaps = caffeine still entrapped in the microspheres).*

4 CONCLUSIONS

The choice of an appropriate vehicle is critical to improving the efficacy and cutaneous bioavailability of active substances. Ideally, a vehicle should be able to stabilise and protect the active drug inside the formulation, improve skin contact and adhesion on the skin surface and control the drug delivery in terms of the rate and extent of absorption. Many systems have been described to improve the performances of actives that are devoted to skin administration. The choice of formulation depends on the physicochemical properties of the active substance on the one hand and on the target site on the other hand.

As shown in this study, some formulations like microemulsions or particles are able to enhance skin penetration. The mechanism of action of these formulations are not well characterised, although key factors such as microstructure, water content and particle size are clearly factors. For an active such as caffeine (which has to penetrate deeper in the skin to reach the hypodermis in order to exert its required action), the microemulsions and particles used in this study represent promising candidates for delivery vehicles due to their ability to promote the skin penetration and accumulation in the hypodermis. Furthermore, microparticles can additionally provide sustained release of caffeine in the hypodermis.

Pickering emulsions, stabilised with solid particles, were able to improve skin penetration compared to classical emulsion stabilised with surfactant, but to a lower extent than microparticles. It was shown in this study that adjusting the type of formulation and its key properties such as particle size, water and surfactant content, is a way to control the skin permeation of a hydrophilic drug like caffeine. It is likely that these outcomes may be different when considering a hydrophobic drug, but the role of a vehicle to enhance diffusion through the skin barrier layer (*stratum corneum*) would remain.

5 REFERENCES

1. M. Förster, M.-A. Bolzinger, H. Fessi and S. Briançon, Topical delivery of cosmetics and drugs. Molecular aspects of percutaneous absorption and delivery, *Eur. J. Dermatol.*, 2009, **19**, 309–323.

2. G. Cevc and U. Vierl, Nanotechnology and the transdermal route. A state of the art review and critical appraisal, *J. Control. Release*, 2010, **141**, 277–299.

3. *Microspheres, microcapsules & liposomes*, Ed. R. Arshady, London: Citus books, 1999.

4. M.-A. Bolzinger, S. Briançon and Y. Chevalier, Nanoparticles through the skin: managing conflicting results of inorganic and organic particles in cosmetics and pharmaceutics, *WIRE Nanomed. Nanobiotechnol.*, 2011, **3**, 463–478.

5. A.C. Sintov and L. Shapiro, New microemulsion vehicle facilitates percutaneous penetration in vitro and cutaneous drug bioavailability *in vivo*, *J. Control. Release*, 2004, **95**, 173–183.

6. M.-A. Bolzinger-Thevenin, C. Carduner and M.-C. Poelman, Bicontinuous sucrose ester microemulsion: a new vehicle for topical delivery of niflumic acid, *Int. J. Pharm.*, 1998, **176**, 39–45.

7. M. Kreilgaard, Dermal pharmacokinetics of microemulsion formulations determined by in vivo microdialysis, *Pharm. Res.*, 2001, **18**, 367–375.

8. M. Kreilgaard, E.J. Pederson and W. Jarozewski, NMR characterization and transdermal drug delivery potential of microemulsion systems, *J. Control. Release*, 2000, **69**, 421–433.

9. F. Akomeah, T. Nazir, G.P. Martin, M.B. Brown, Effect of heat on the percutaneous absorption and skin retention of three model penetrants, *Eur. J. Pharm. Sci.*, 2004, 21, 337–345.

10. A. Mourgues, C. Charmette, J. Sanchez, G. Marti-Mestres and P. H. Gramain, EO/EP copolymer membranes as reservoir in a transdermal therapeutic system for caffeine delivery: Modeling and simulation, *J. Membr. Sci.,* 2004, **241**, 297–304.

11. R.L. Bronaugh, R.F. Stewart and J.E. Storm, Extent of cutaneous metabolism during percutaneous absorption of xenobiotics, *Toxicol. Appl. Pharmacol.*, 1989, **99**, 534–543.

12. OECD, Organisation for economic co-operation and development, Guideline for the testing of chemicals skin absorption: *in vitro* method, *OECD, Paris,* 2004.

13. L. Al Haushey, M.-A. Bolzinger, H. Fessi and S. Briançon, rhEGF microsphere formulation and in vitro skin evaluation, *J. Microencapsulation*, 2010, **27**, 14–24.

14. Y. Chevalier and T. Zemb, The structure of micelles and microemulsions, *Rep. Prog. Phys.*, 1990, **53**, 279–371.

15. M.-A. Bolzinger, S. Briancon, J. Pelletier, H. Fessi and Y. Chevalier, Percutaneous release of caffeine from microemulsion, emulsion and gel dosage forms, *Eur. J. Pharm. Biopharm.*, 2008, 68, 446–451.

16. S. Bourgeois, M.-A. Bolzinger, J. Pelletier, J.-P. Valour and S. Briançon, Caffeine microspheres – An attractive carrier for optimum skin penetration, *IFSCC Magazine*, 2009, **12**, 1–5.

17. A. Rolland, N. Wagner, A. Chatelus, B. Shroot and H. Shaefer, Site-specific drug delivery to pilosebaceous structures using polymeric microspheres, *Pharm. Res.*, 1993, **10**, 1738–1744.

18. S. Lombardi Borgia, M. Regehly, R. Sivaramakrishnan, W. Mehnert, H.C. Korting, K. Danker, B. Röder, K.D. Kramer and M. Schäfer-Korting, Lipid nanoparticles for skin penetration enhancement—correlation to drug localization within the particle matrix as

determined by fluorescence and parelectric spectroscopy, *J. Control. Release*, 2005, **110**, 151–163.

19. E.G. de Jalón, M.J. Blanco Prieto, P. Ygartua and S. Santoyo, PLGA microparticles: possible vehicles for topical drug delivery, *Int. J. Pharm.*, 2001, **226**, 181–184.

20. J. Shim, H. S. Kang, W.S. Park, S. H. Han, J. Kim and I. S. Chang, Transdermal delivery of mixnoxidil with block copolymer nanoparticles, *J. Control. Release*, 2004, **97**, 477–484.

21. R. Alvarez-Román, A. Naik, Y.N. Kalia, R.H. Guy and H. Fessi, Skin penetration and distribution of polymeric nanoparticles, *J. Control. Release*, 2004, **99**, 53–62.

22. R. Alvarez-Román, A. Naik, Y.N. Kalia, H. Fessi and R.H. Guy, Visualization of skin penetration using confocal laser scanning microscopy, *Eur. J. Pharm. Biopharm.*, 2004, **58**, 301–316.

23. J. Lademann, H. Richter, A. Teichmann, N. Otberg, U. Blume-Peytavi, J. Luengo, B. Weiß, U.F. Schaefer, C.M. Lehr, R. Wepf and W. Sterry, Nanoparticles—an efficient carrier for drug delivery into the hair follicles, *Eur. J. Pharm. Biopharm.*, 2007, **66**, 159–164.

24. R. Toll, U. Jacobi, H. Richter, J. Lademann, H. Schaefer and U. Blume-Peytavi, Penetration profile of microspheres in follicular targeting of terminal hair follicles, *J. Invest. Dermatol.*, 2004, **123**, 168–176.

25. F. Knorr, J. Lademann, A. Patzelt, W. Sterry, U. Blume-Peytavi, A. Vogt, Follicular transport route – Research progress and future perspectives, *Eur. J. Pharm. Biopharm.*, 2009, **71**, 173–180.

26. A. Patzelt, H. Richter, F. Knorr, U. Schäfer, C.M. Lehr, L. Dähne, W. Sterry, J. Lademann, Selective follicular targeting by modification of the particle sizes, *J. Control. Release*, 2011, **150**, 45–48.

27. S. Trauer, A. Patzelt, N. Otberg, F. Knorr, C. Rozycki, G. Balizs, R. Büttemeyer, M. Linscheid, M. Liebsch and J. Lademann, Permeation of topically applied caffeine through human skin – a comparison of *in vivo* and *in vitro* data, *Br. J. Clin. Pharmacol.*, 2009, **68**, 181–186.

28. W. Naoui, M.-A. Bolzinger, B. Fenet, J. Pelletier, J.-P. Valour, R. Kalfat and Y. Chevalier, Microemulsion microstructure influences the skin delivery of a hydrophilic drug, *Pharm Res*, 2011, **28**, 1683–1695.

29. J. Zhang and B. Michniak-Kohn, Investigation of microemulsion microstructures and their relationship to transdermal permeation of model drugs: Ketoprofen, lidocaine, and caffeine, *Int. J. Pharm.*, 2011, **421**, 34–44.

30. M. Kreilgaard, Influence of microemulsions on cutaneous drug delivery, *Adv. Drug Deliv. Rev.*, 2002, **54**, S77–98.

31. J. Frelichowska, M.-A. Bolzinger, J.-P. Valour, H. Mouaziz, J. Pelletier and Y. Chevalier, Pickering w/o emulsions: drug release and topical delivery, *Int. J. Pharm.*, 2009, **368**, 7–15.

32. J. Frelichowska, M.-A. Bolzinger, J.-P. Valour, J. Pelletier and Y. Chevalier, Topical delivery of lipophilic drugs from o/w Pickering emulsions, *Int. J. Pharm.*, 2009, **371**, 56–63.

9

THE EFFECT OF DISULPHIDE BOND DISRUPTION ON THE BARRIER INTEGRITY OF THE HUMAN NAIL

A Baraldi[1], R H Khengar[1], S Murdan[2], M J Traynor[3], S A Jones[1], M B Brown[3,4]

[1]Pharmaceutical Science Division, King's College London, United Kingdom. [2]Department of Pharmaceutics, School of Pharmacy, University of London, United Kingdom. [3]School of Pharmacy, University of Hertfordshire, United Kingdom. [4]MedPharm Ltd, Guildford, United Kingdom.

1 INTRODUCTION

Onychomycosis is the fungal infection of the nail and accounts for approximately 50% of all nail disorders [1]. The prevalence of onychomycosis is increasing and currently up to 10 % of people in the UK and US are reported to be affected by this condition [2,3]. It can be both physically and psychologically debilitating during the most progressive stages of the disease [4,5]. Fungi (most frequently *Trichophyton rubrum and Trichophyton mentagrophytes*), moulds (*Scytalidium* spp, *Scopulariopsis* spp, *Fusarium* spp, *Acremonium* spp, *Onychocola canadensis*) and yeasts (*Candida* spp) can be co-resident in a diseased nail and hence not every nail can be considered as being populated by the same community of species [6]. The infection, irrespective of the causative organisms, can be treated using oral medication, but systemic therapy has the disadvantages of causing adverse effects and drug interactions [7,8]. Topical therapy is more desirable as it avoids high levels of systemic drug loading but treatment times are longer (up to 12 months) and cure rates are low using currently marketed products [9-12]. Drug delivery to the nail via the topical route is a challenge because of the barrier properties of the nail plate [13]. The latter provides physical protection for the tips of fingers and toes and is highly impermeable to the ingress of foreign material. The low permeability of most drugs across the nail plate means that topical treatment is only recommended in early stages of disease if systemic therapy is contraindicated and therefore there remains a need to design new medicines that facilitate the passage of molecules across this barrier to allow the topical therapy of onychomycosis.

The chemical and mechanical resilience of the human nail plate is attributed to its high α-keratin content [14-15]. α-Keratin is composed of keratin filaments which are embedded in a non-filamentous matrix of proteins called keratin associated proteins [16-18]. Nail keratins have a large amount of the sulphur-rich amino acid, cysteine [19], which forms numerous chemical cross-links (disulphide bonds) between the filament and non-filamentous matrix [17] and it is this extensive molecular bonding that is believed to be, at least in part, responsible for the nail's barrier properties [15].

The use of physical and/or chemical means to promote drug permeation has been described [20]. Physical methods can be effective, but those which show the best results often involve removal of all or part of the affected nail and so is not an ideal strategy. Penetration enhancers (PEs), which break the chemical bonds responsible for the stability of nail keratin, allow the nail to remain largely intact. Reducing agents such as thiol-containing compounds have been proposed as promising nail penetration enhancers (PEs) [21-24]. Examples of such compounds include *N*-acetyl-cysteine (N-Ac), mercaptoethanol, thioglycolic acid (TA), *N*-(2-mercaptopropionyl) glycine (MPG) and the cysteine protease papain [23, 25-28]. The mechanism of action of these enhancers is thought to be via reduction of the nail disulphide bonds (-SS-) to sulfhydryl groups (-SH) [29] as shown in Equation 1.

$$\text{Nail-S-S-Nail} + \text{R-SH} \longleftrightarrow 2\,\text{Nail-SH} + \text{R-S-S-R} \tag{1}$$

Nogueiras-Nieto *et al.* observed that disruption of disulphide bonds upon treatment of nail clippings with *N*-acetyl-cysteine resulted in the formation of new pores in the nail, which may facilitate drug penetration [24]. However, sulphur containing compounds are irritant to the skin and they produce a strong odour which is not consumer friendly. Cleavage of disulphide bonds can also be accomplished by using strong oxidising agents which involves the formation of cystine oxide intermediates (monoxide, and dioxide) and terminates at the cysteic acid species (Equation 2) [30,31].

$$\text{Cy-S-S-Cy} \longrightarrow \text{Cy-S(O)-S-Cy} \longrightarrow \text{Cy-SO}_2\text{-S-Cy} \longrightarrow \text{Cy-SO}_3 \tag{2}$$

Oxidants like hydrogen peroxide (H_2O_2) or its stable adjunct urea hydrogen peroxide (urea H_2O_2) have previously shown potential in compromising α-keratin integrity via disulphide oxidation, but direct analysis of their solution state chemistry suggests that the mechanism by which this is achieved may be more complex than first thought [31,32].

Sequential application of thioglycolic acid (TA), a reducing agent, and urea H_2O_2, an oxidant, has been used to manipulate hair keratin during permanent waving or straightening [32]. During the permanent waving process, H_2O_2 is applied to re-oxidise disulphide bridges in hair keratin so that the new style is locked into place [33]. Despite the current applications of TA and urea H_2O_2 in the hair care industry, neither is currently used in a commercial product to enhance permeation into the human nail. Brown *et al.* studied the enhancing potential of these two agents on nail penetration and found that sequential application of TA and urea H_2O_2 produced the greatest nail penetration across a range of model permeants [22]. Traynor et al. subsequently demonstrated that the use of this novel permeation enhancing system was able to enhance the efficacy of existing topical formulations when applied topically [21]. In these studies, the concentrations of urea H_2O_2 (15-17.5 %) were considerably higher than those employed by the hair-care industry (2-3 %) therefore it is possible that further disulphide oxidation occurred upon -SS- bond reformation resulting in the production of cystine oxide intermediates and cysteic acid as principal end product.

The objective of this work was to further investigate the role of disulphide bonds in nail barrier integrity. First, an investigation into the change in the nail barrier properties changed when the nail disulphide content was altered was undertaken using rhodamine B and water as markers. This was followed by an investigation into the effect of nail disulphide bond disruption on the nail permeability to terbinafine hydrochloride.

2 MATERIALS AND METHODS

Human nail clippings were donated by healthy volunteers following approval by the King's College Research Ethics Committee (Study ref no. 04/05-126). Terbinafine hydrochloride (TBF) ($\geq$ 0.99 %) was purchased from QueMaCo Ltd (Nottingham, UK). Ethanol (HPLC grade) and sodium hydroxide was purchased from BDH (Dorset, UK). Thioglycolic acid (TA), urea hydrogen peroxide (urea H_2O_2), monopotassium phosphate, dipotassium phosphate, potassium hydroxide pellets, 95 % rhodamine B (RB), $\geq$ 99.5% 2-propanol, 95-98 % sulfuric acid were reagent grade and purchased from Sigma Aldrich (Dorset, UK). Phosphoric acid solution (85 % wt in water) and orthophosphoric acid (HPLC grade) was supplied by Fisher (Leicestershire, UK). Ringer's solution, Sabouraud dextrose agar (SDA), antimicrobial susceptibility blank discs and *C. albicans* culti-loops were from Oxoid (Hampshire, UK). Deionised water was used throughout the experiments. Validated TurChub permeation cells were kindly donated by MedPharm Ltd (Surrey, UK).

Experiments to measure nail swelling and rhodamine B uptake were conducted to investigate the link between disulphide bonds (-SS-) and nail barrier properties. Disulphide bond disruption was achieved by treatment of nails with thioglycolic acid and confirmed by Raman spectroscopy [34]. The assay was adapted from the method used by Potsch *et al* with hair keratin [35]. Briefly, nail clippings were weighed and incubated for 2 h at room temperature with a test solution containing 0.5 M thioglycolic acid in water (pH 2.0), or 0.5 M thioglycolic acid in phosphate buffer (pH 11.5) to break the nail -SS- bonds. After 2 h the nail clippings were briefly rinsed in water and re-weighed. The nails were incubated in 0.2 mM rhodamine B aqueous solution (pH 3.74) for 30 min at 32°C. After the incubation period (to allow rhodamine B uptake into the nail), the latter was briefly washed with ethanol and transferred to a solution of 4:1 v/v isopropanol:0.1 M sulphuric acid at 60°C for 1 h to allow extraction of rhodamine B from the nail. The extraction solution was removed and its absorbance was measured in a 1 ml quartz cuvette. Three consecutive aliquots of the extraction solution were analysed on a Perkin Elmer Lambda 25 UV/Visible spectrophotometer at an absorbance wavelength of 556 nm. For each tested condition, three nail clippings were used. Control nail clippings were treated with water at pH 6.7 and phosphate buffer at pH 11.5 and processed in an identical manner. Nail swelling was determined as the percentage increase in nail mass after treatment with the test solutions.

The TurChub assay was performed to evaluate the permeation of terbinafine hydrochloride across 5 μm nail sections following nail disulphide bond disruption by sequential pre-treatment of nails with thioglycolic acid and urea hydrogen peroxide (urea H_2O_2). The TurChub cell system uses a modified Franz-cell, where the receptor phase consists of agar gel in which fungus grows. The ability of the antifungal agent (placed in the donor compartment) to penetrate across the nail (positioned between the upper donor chamber and the lower receiver chamber) into the receptor chamber is evaluated by the production of areas of no growth of the organism (zones of inhibition, ZOI), as the permeated drug acts on the fungus. *Candida albicans* is reported to account for more than 50 % of candidal onychomycoses [36] and so was selected in the present study as a clinically relevant biological marker for its rapid growth compared to *T. rubrum* (a more common cause of onychomycosis) which requires a considerably longer incubation period of 14 - 21 days. Terbinafine hydrochloride, currently available worldwide in topical and oral dosage forms for the treatment of superficial mycoses [37] was chosen as a model drug for its broad-spectrum antifungal activity [38, 39].

Nail sections (5 μm thick) were generated by mounting a piece of full thickness nail onto the end of a 10 mm (length) by 6 mm (diameter) perspex rod with Araldite™ adhesive resin (Bostik Ltd. Stafford, UK) which was left overnight to harden. The rod with the attached nail was soaked in deionised water for ca. 10 min after which the nail was partially lifted away from the araldite and a small drop of cyanoacrylate glue (Bostik Ltd, Stafford, UK) introduced into the gap. The nail was pressed down and held until the superglue had set. For the sectioning procedure, the rod with the glued nail was soaked in distilled water for ca. 10 min, mounted in the standard chuck for the ultra-microtome (Riechart-Jung Ultracut E, Leica Microsystems Ltd, Milton Keynes, UK) and 5 μm sections cut using glass knives. The cut nail sections were air-dried for 1 h after which they were placed in sterile glass vials and stored at 4° C until use.

Penetration enhancer 1 (PE1) was 5% thioglycolic acid (w/w) prepared in 80:20 phosphate buffer (pH 2 or pH 8):EtOH. Penetration enhancer 2 (PE2) was 15% urea H_2O_2 (15 % w/v H_2O_2) prepared in pH 2 phosphate buffer. The saturated TBF solution was prepared in 50:50 phosphate buffer (pH 2):EtOH as previously described [22]. The pH 2 buffer consisted of phosphoric acid and monopotassium phosphate and was prepared at a concentration of 0.2 M, whilst the pH 8 buffer consisted of monopotassium phosphate and dispotassium phosphate and was prepared at 0.3 M. Where necessary, the pH of the solutions was adjusted using sodium hydroxide and orthophosphoric acid.

Isolated *C. albicans* (originating from *C. albicans* culti-loop cultures on Sabouraud dextrose agar) was transferred from a Sabouraud dextrose agar (SDA) plate incubated at 32 ± 3 °C for 36 h, to approximately 5 ml of sterile Ringer's solution. The density of the suspension was adjusted with a spectrophotometer (Shimazdu, Milton Keynes, UK) to 10^7 cfu ml^{-1} by diluting or concentrating the suspension as required with reference to a previously prepared calibration curve of cell density versus absorbance (measured at 610 nm). A viable plate count of the *C. albicans* cell suspension was performed by pipetting serial dilutions ($10^0 - 10^{-6}$) of fungal suspensions (50 μl) onto SDA plates. Viable count plates were then incubated at 32 ± 3 °C for 36 h after which time the colonies on the plate were counted to calculate the cfu in the original *C. albicans* suspension. A second *C. albicans* suspension (10^7 cfu ml^{-1}) was prepared, with 200 μl aliquots spread over the surface of SDA Petri dishes ($n = 6$). Plates were allowed to dry for 10 min inside a sterile cabinet after which time antimicrobial susceptibility discs (6 mm in diameter) impregnated with 20 μl of 5 % w/w TA (pH 2 or pH 8), 15 % urea H_2O_2 (15 % w/w H_2O_2, pH 2), saturated terbinafine hydrochloride (pH 2) and blank buffer/solvents were placed in the centre of the inoculated agar dishes using sterile forceps. Plates were incubated at 32 ± 3°C for 36 h after which time the inhibition of fungal growth was evaluated by measuring the diameter of the transparent inhibition zone around each disc.

The Turchub assay with 5 μm nail sections was performed as previously described [21] with minor modifications. Terbinafine hydrochloride was used as model antifungal drug and nail sections were incubated with terbinafine for 24 h after application of the penetration enhancers. A set of controls in which no penetration enhancer or no terbinafine hydrochloride was applied were prepared in the same way.

Results are presented as the mean ± standard deviation. The data was tested for normality (Shapiro–Wilk test) and statistical analysis was performed using the non-parametric Mann-Whitney test. In all cases a level of $p \leq 0.05$ was considered significant. All statistical analyses in the present study were performed using SPSS for Windows, version 11.0.

3 RESULTS AND DISCUSSION

The correlation between disulphide bonds and nail barrier integrity was investigated by modifying the nail plate disulphide bond content using thioglycolic acid and measuring nail swelling and rhodamine B uptake. Thioglycolic acid is a well-characterised reducing agent which has been shown to disrupt the disulphide bonds of hair keratin [35, 40-42] and of nail keratin [34]. Rhodamine B uptake and nail swelling in nails treated with thioglycolic acid (pH 2 and 11.5) and with the control solutions (water pH 6.7 and buffer pH 11.5) is shown in Figure 1. Nail incubation in water or phosphate buffer pH 11.5 resulted in a 15% and 19% weight gain, respectively. Such nail swelling upon incubation in aqueous solutions has previously been described [23].

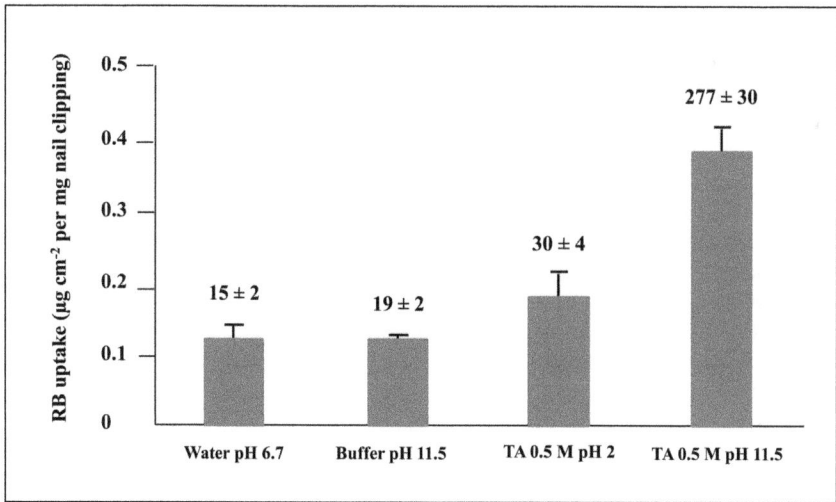

Figure 1 *Rhodamine B (RB) uptake (bars) and nail swelling (numbers above bars) in nails treated with the test (thioglycolic acid; TA) and control (water/buffer) solutions. All values are mean ± standard deviation of n=3 replicates.*

It is thought that water interacts with keratin filaments which swell as a result of disruption of keratin hydrogen bonds. Swelling results in an expansion of the pores within the keratin network which can increase the uptake of more water molecules and thereby allow the entry of rhodamine B. Greater nail swelling and rhodamine uptake upon incubation in alkaline buffer (pH 11.5) might have been expected, as the alkali could hydrolyse the peptide bonds among keratin chains and thereby increase nail porosity. For example, treatment of wool at high temperature in the presence of excess alkali has been reported to cause disulphide bond disruption [43]. The lack of significant differences in nail swelling and rhodamine B uptake between the two control solutions ($p > 0.05$) indicated that, in this study, alkali degradation of nail clippings was not prevalent during the 2 h incubation at room temperature.

When nail clippings were incubated with thioglycolic acid solutions a significantly greater ($p \leq 0.05$) uptake of water and rhodamine B was observed compared to the control solutions (Figure 1). This indicates that the chemical structure of the nail was damaged by thioglycolic acid. Reduction of disulphide bonds by thioglycolic acid may result in unfolding of the tightly packed keratin matrix so that the opportunity arises for water and rhodamine B to penetrate into regions of the protein which are normally inaccessible. The ability of thioglycolic acid to induce changes in nail keratin was found to be dependent on the pH. At acidic pH, thioglycolic acid did swell the nail, but the damage was comparatively lower than TA at the alkaline pH ($p \leq 0.05$). In particular, thioglycolic acid at alkaline pH increased nail swelling and rhodamine B uptake by ~ 9 and~ 2 fold, respectively, when compared to TA at acidic pH. The much greater difference in nail swelling compared to that in rhodamine B uptake indicate that the passage of large size molecules such as rhodamine B through the nail is constrained even after keratin damage induced by TA at pH 11.5. The pH-dependent efficacy of TA can be attributed to changes in its ionisation state. TA has previously been reported to break disulfide bonds most efficiently at pH above 10, as the active species is believed to be the thiolate ion, $-S-CH_2-COO-$, which is only present at alkaline pH [32]. Consequently, it can be inferred that at pH 2, where the majority of the molecules of TA are in the unionised form, the attack of the -SS- bonds by TA would be slower. Indeed, previous studies demonstrated the efficacy of TA at acidic pH when 20 h incubation was employed [23]. In this study, treatment duration was limited to 2 h to avoid keratin dissolution which occurred when the nail was treated with TA at pH 11.5 for longer periods.   Thus overall, these experiments demonstrate that the breakage of nail -SS- bonds reduces the barrier properties of the human nail plate.

The effect of nail disulphide bond (-SS-) disruption on nail permeability to terbinafine hydrochloride (TBF) was investigated *in vitro* in the TurChub efficacy model. Firstly, a spread plate inhibition experiment (without nail clippings) was conducted to ensure the antifungal activity of TBF preparations used in the TurChub assays against *C. albicans* and also to determine whether the penetration enhancers (PEs) or the solvents used in the TurChub assays inhibited fungal growth. The 5 % w/w TA preparations used in the TurChub assay did not display antifungal effects against *C. albicans* at acidic (pH 2) or alkaline (pH 8) conditions (Table 1). Conversely, 15 % urea H_2O_2 (15 % w/w H_2O_2) inhibited the growth of *C. albicans* on the agar plate to the same extent as TBF (inhibition zone of 3.5 cm in diameter). The inhibitory effects of urea H_2O_2 against *C. albicans* in the agar spread plate experiments were not surprising as H_2O_2 is a well-established bactericidal

[44] and antifungal agent used by industry as a preservative and as a disinfectant [45]. The toxicity of H_2O_2 is reported to result from oxidative damage to cell membranes and DNA modification [46]. The relatively high concentration of urea H_2O_2 intended for use in the TurChub assays (15 % w/w H_2O_2) dramatically impacted on *C. albicans* growth in the agar spread plate tests, as it was applied directly at a concentration above that which is reported to be sporicidal (> 7.5 - 10 % depending on the microorganism) [47, 48]. The antifungal effects of the solvents used to dissolve the PEs and the TBF in each TurChub assay were also investigated and did not cause *C. albicans* growth inhibition (Table 1).

Solution	Mean diameter of zone of inhibition (cm)
5 % w/w TA in 80:20 pH 2 buffer:EtOH	0.1 ± 0.05
5 % w/w TA in 80:20 pH 8 buffer:EtOH	0.0 ± 0.0
15 % urea H_2O_2 (15 % w/w H_2O_2) in pH 2 buffer	3.5 ± 0.1
Saturated TBF in 50:50 pH 2 buffer:EtOH	3.5 ± 0.2
pH 2 buffer	0.0 ± 0.0
50:50 pH 2 buffer:EtOH	0.0 ± 0.0
80:20 pH 2 buffer:EtOH	0.0 ± 0.0
80:20 pH 8 buffer:EtOH	0.0 ± 0.0

Table 1 *Zones of inhibition against Candida albicans on Sabouraud's dextrose agar (SDA) plates incubated with TA, urea H_2O_2, TBF solutions and the solvents of each solution (n=6). EtOH = ethanol. Data expressed as mean ± standard deviation of n=6 replicates.*

Secondly, the TurChub assay was performed using 5 μm nail sections to test for permeation. Sequential application of TA at pH 2 or pH 8 and urea H_2O_2 alone (without TBF) to 5 μm nail sections did not exert antifungal effects against *C. albicans* in the TurChub model over the time course of the present study, and no zone of inhibition was found (Table 2). The lack of antifungal activity of urea H_2O_2 in the TurChub studies was probably due to the amount of H_2O_2 penetrating nail sections being below a fungistatic dose for *C. albicans*. A small zone of inhibition was detected following single application of TBF for 24 h to the 5 μm nail sections (Table 2). *C. albicans* growth was also inhibited when nail sections were sequentially pre-treated with TA and urea H_2O_2 and the drug was applied for 24 h (Table 2). However, growth inhibition of *C. albicans* that occurred from the sequential PEs regime followed by TBF was not significantly different ($p > 0.05$) when compared to the control cells in which no PEs were used before application of TBF to 5 μm nail sections (Table 2). One reason for the lack of statistical significance observed may be that the TurChub assay required the use of two biological matrices – the human nail and a fungal marker, thus making this assay prone to a high degree of variability. In the 5 μm experiments for example, the % coefficient of variance (CV) of the TBF only controls was 73 %, making the discrimination between different PE treatment regimes speculative. In addition, the use of sectioned human nails may have compounded the issue of high variability in these experiments. Previous work has shown that the human nail is a tri-layered structure and that each layer has distinct permeability characteristics [49]. The nail sections used in this study were a random selection from the different layers and thus it is likely that this added to the variability observed.

Treatment	Stage 1 (20 h)	Stage 2 (20 h)	Stage 3 (24 h)	Mean zone of inhibition ± SD (cm)
Enhancer alone	TA pH 2	urea H_2O_2 pH 2	50:50 pH 2 buffer:EtOH	0.00 ± 0.00
	TA pH 8	urea H_2O_2 pH 2	50:50 pH 2 buffer:EtOH	0.07 ± 0.18
TBF alone	80:20 pH 2 buffer:EtOH	Buffer pH 2	TBF pH 2	0.70 ± 0.51
	80:20 pH 8 buffer:EtOH	Buffer pH 2	TBF pH 2	0.59 ± 0.47
Enhancer + TBF	TA pH 2	urea H_2O_2 pH 2	TBF pH 2	1.10 ± 0.41
	TA pH 8	urea H_2O_2 pH 2	TBF pH 2	0.30 ± 0.36

Table 2 *Zones of inhibition against Candida albicans in TurChub cells during control studies (enhancer alone or TBF alone) and upon application of the sequential PEs regime followed by TBF (enhancer + TBF) to 5 µm nail sections (n=3 -7). EtOH = ethanol.*

4 CONCLUSIONS

The data generated by this study supports the hypothesis that the human nail plate barrier arises, at least in part, from the integrity of disulphide (-SS-) bonds. The uptake study using rhodamine B and water as markers demonstrated that nail barrier permeation is increased upon cleavage of disulphide bonds. However, the terbinafine permeability study using the TurChub assay failed to show a significant difference between the quantities of drug transported across untreated nails or those treated with TA and urea H_2O_2 to damage disulphide links. This was mainly attributed to the combined use of two biological matrices in this model and thus a correspondingly high degree of variability.

5 REFERENCES

1. M. A. Ghannoum, R. A. Hajjeh, R. Scher, N. Konnikov, A. K. Gupta, R. Summerbell, S. Sullivan, R. Daniel, P. Krusinski, P. Fleckman, P. Rich, R. Odom, R. Aly, D. Pariser, M. Zaiac, G. Rebell, J. Lesher, B. Gerlach, G. F. Ponce-de-Leon, A. Ghannoum, J. Warner, N. Isham and B. Elewski, *J. Am. Acad. Dermatol.*, 2000, **43**, 641.

2. D. T. Roberts, Prevalence of dermatophyte onychomycosis in the United Kingdom - results of an omnibus survey, *Br. J. Dermatol.*, 1992, **126**, S39, 23–7.

3. D. T. Roberts, W. D. Taylor and J. Boyle, Guidelines for treatment of onychomycosis, *Br. J. Dermatol.*, 2003, **148**(3), 402–10.

4. A. K. Gupta, N. Konnikov, P. MacDonald, P. Rich, N. W. Rodger, M. W. Edmonds, R. McManus and R. C. Summerbell, Prevalence and epidemiology of toenail onychomycosis in diabetic subjects: a multicentre survey, *Br. J. Dermatol.*, 1998, **139** (4), 665–71.

5. L. A. Levy, Epidemiology of onychomycosis in special-risk populations, *J. Am. Podiatr. Med. Assoc.*, 1997, **87**(12), 546–50.

6. R. Baran, P. R. Dawber, E. Haneke, A. Tosti and I. Bristow I, in *A Text Atlas of Nail Disorders. Techniques in investigation and diagnosis*, 3rd edn, Martin Dunitz, Taylor & Francis Group, London, 2003, p. 197–220.

7. C. Ajit, A. Suvannasankha, N. Zaeri, and S. J. Munoz, Terbinafine-associated hepatotoxicity, *Am. J. Med. Sci.*, 2003, **325** (5), 292–5.

8. W. M. Chambers, A. Millar, S. Jain and A. K. Burroughs, Terbinafine-induced hepatic dysfunction, *Eur. J. Gastroenterol. Hepatol.*, 2001, **13** (9), 1115–8.

9. D. de Berker, Fungal Nail Disease, *N. Engl. J. Med.*, 2009, **360**(20), 2108–16.

10. A. K. Gupta, P. Fleckman and R. Baran, Ciclopirox nail lacquer topical solution 8% in the treatment of toenail onychomycosis, *J. Am. Acad. Dermatol.*, 2000, **43**(4), S70–80.

11. J. Lauharanta, Comparative efficacy and safety of amorolfine nail lacquer 2-percent versus 5-percent once weekly, *Clin. Exp. Dermatol.*, 1992, **17**, S1, 41–3.

12. E. Epstein, Fungus-free versus disease-free nails, *J. Am. Acad. Dermatol.*, 2004, **50**(1), 151–2.

13. S. Murdan, Drug delivery to the nail following topical application, *Int. J. Pharm.*, 2002, **236**(1–2), 1–26.

14. G. V. Gupchup, J. L. Zatz, Structural characteristics and permeability properties of the human nail: A review, *J. Cosmet. Sci.*, 1999, **50**(6), 363–85.

15. R. C. Marshall, D. F. G. Orwin and J. M. Gillespie, Structure and biochemistry of mammalian hard keratin, *Electron Microsc. Rev.*, 1991, **4**(1), 47–83.

16. B. Forslind, G. Nordstrom, D. Toijer and K. Eriksson, The rigidity of human fingernails - a biophysical investigation on influencing physical parameters, *Acta Derm.-Venereol.*, 1980, **60**(3), 217–22.

17. E. K. Levit and R. K. Scher, in *The biology of the skin.*, ed. R.K. Freinkel and D.T. Woodley, Parthenon Publishing Group, London, 2001, pp. 101–12.

18. H. H. Bragulla and D. G. Homberger, Structure and functions of keratin proteins in simple, stratified, keratinized and cornified epithelia, *J. Anat.*, 2009, **214**(4), 516–59.

19. G. Bulaj, Formation of disulfide bonds in proteins and peptides, *Biotechnol. Adv.*, 2005, **23**(1), 87–92.

20. S. Murdan, Enhancing the nail permeability of topically applied drugs, *Expert Opin. Drug Delivery*, 2008, **5**(11), 1267–82.

21. M. J. Traynor, R. B. Turner, C. R. G. Evans, R. H. Khengar, S. A. Jones and M. B. Brown, Effect of a novel penetration enhancer on the ungual permeation of two antifungal agents, *J. Pharm. Pharmacol.*, 2010, **62**(6), 730–7.

22. M. B. Brown, R. H. Khengar, R. B. Turner, B. Forbes, M. J. Traynor, C. R. G. Evans and S. A. Jones, Overcoming the nail barrier: A systematic investigation of ungual chemical penetration enhancement, *Int. J. Pharm.*, 2009, **370** (1–2), 61–7.

23. R. H. Khengar, S. A. Jones, R. B. Turner, B. Forbes and M. B. Brown, Nail swelling as a pre-formulation screen for the selection and optimisation of ungual penetration enhancers, *Pharm. Res.*, 2007, **24**(12), 2207–12.

24. L. Nogueiras-Nieto, J. L. Gomez-Amoza, M. B. Delgado-Charro and F. J. Otero-Espinar, Hydration and N-acetyl-l-cysteine alter the microstructure of human nail and bovine hoof: Implications for drug delivery, *J.Controlled Release*, 2011, **156**(3), 337–44.

25. Y. Kobayashi, M. Miyamoto, K. Sugibayashi and Y. Morimoto, Enhancing effect of N-acetyl-L-cysteine or 2-mercaptoethanol on the in vitro permeation of 5-fluorouracil or tolnaftate through the human nail plate, *Chem. Pharm. Bull.*, 1998, **46**(11), 1797–1802.

26. Y. Sun, J. C. Liu, E. S. Kimbleton and J. C. T. Wang, Anti fungal treatment of nails, *Off. Gaz. U. S. Pat. Trademark Off., Pat.*, 2000, **1232**(4).

27. G. G. Malhotra and J. L. Zatz, Investigation of nail permeation enhancement by chemical modification using water as a probe, *J. Pharm. Sci.*, 2002, **91**(2), 312–23.

28. D. Quintanar-Guerrero, A. Ganem-Quintanar, P. Tapia-Olguin, Y. N. Kalia and P. Buri, The effect of keratolytic agents on the permeability of three imidazole antimycotic drugs through the human nail, *Drug Dev. Ind. Pharm.*, 1998, **24**(7), 685–90.

29. Y. Sun, J. C. Liu, J. C. T. Wang and P. De Doncker, in *Percutaneous Absorption. Drugs–Cosmetics–Mechanisms–Methodology*, ed. R. L. Bronaugh and H. I. Maibach, Marcel Dekker Inc, New York, 3rd edn, 1999, p. 759–87.

30. F. J. Douthwaite, D. M. Lewis, U. Schumacherhamedat, Reaction of cystine residues in wool with peroxy compounds, *Text. Res. J.*, 1993, **63**(3), 177–83.

31. W. M. Marmer and R. L. Dudley, The oxidative degradation of keratin (wool and bovine hair), *J. Am. Leather Chem. Assoc.*, 2006, **101**(11), 408–15.

32. C. R. Robbins, in *Chemical and physical behavior of human hair*, ed. C. R. Robbins, Springer-Verlag GmbH and Co. KG, Berlin, 4th edn, 2002, pp. 105–343.

33. C. Bolduc, J. Shapiro, Hair care products: Waving, straightening, conditioning, and coloring, *Clin. Dermatol.*, 2001, **19**(4), 431–6.

34. A. Baraldi, S. Murdan, S. A. Jones, M. J. Traynor and M. B. Brown, A study on the structural integrity of healthy and diseased nail: the retention of disulphide bonds, Perspective in Percutaneous Penetration, Thirteenth International Conference, La Grande Motte, France, 11–14 April 2012.

35. L. Potsch and M. R. Moeller, On pathways for small molecules into and out of human hair fibers, *J. Forensic Sci.*, 1996, **41**(1), 121–5.

36. R. Segal, A. Kritzman, L. Cividalli, Z. Samra and M. David, Treatment of Candida nail infection with terbinafine, *J. Am. Acad. Dermatol.*, 1996, **35**(6), 958–61.

37. A. K. Gupta, D. N. Sauder, N. H. Shear, Antifungal agents - an overview 2, *J. Am. Acad. Dermatol.*, 1994, **30**(6), 911–33.

38. N. S. Ryder, Specific-inhibition of fungal sterol biosynthesis by sf 86-327, a new allylamine antimycotic agent, *Antimicrob. Agents Chemother.*, 1985, **27**(2), 252–6.

39. H. J. Schmitt, E. M. Bernard, J. Andrade, F. Edwards, B. Schmitt and D. Armstrong, Mic and fungicidal activity of terbinafine against clinical isolates of aspergillus spp, *Antimicrob. Agents Chemother.*, 1988, **32**(5), 780–1.

40. A. L. dos Santos Silva and I. Joekes, Rhodamine B diffusion in hair as a probe for structural integrity, *Colloids Surf., B.*, 2005, **40**(1), 19–24.

41. S. F. DeLauder, D. A. Kidwell, The incorporation of dyes into hair as a model for drug binding, *Forensic Sci. Int.*, 2000, **107**(1–3), 93–104.

42. E. A. Olsen, Methods of hair removal, *J. Am. Acad. Dermatol.*, 1999, **40**(2), 143–55.

43. J. M. Cardamone, Investigating the microstructure of keratin extracted from wool: Peptide sequence (MALDI-TOF/TOF) and protein conformation (FTIR), *J. Mol. Struct.*, 2010, **969**(1–3), 97–105.

44. S. Brul and P. Coote, Preservative agents in foods - Mode of action and microbial resistance mechanisms, *Int. J. Food Microbiol.*, 1999, **50** (1–2), 1–17.

45. T. M. Schreier, J. J. Rach and G. E. Howe, Efficacy of formalin, hydrogen peroxide, and sodium chloride on fungal-infected rainbow trout eggs, *Aquaculture*, 1996, **140**(4), 323–31.

46. A. J. Phillips, I. Sudbery and M. Ramsdale, Apoptosis induced by environmental stresses and amphotericin B in Candida albicans, *Proc. Natl. Acad. Sci. U.S.A.*, 2003, **100**(24), 14327–32.

47. J. L. Sagripanti and A. Bonifacino, Comparative sporicidal effects of liquid chemical agents, *Appl.Environ. Microbiol.*, 1996, **62**(2), 545–51.

48. A. E. Acosta-Gio, J. L. Rueda-Patino, L. Sanchez-Perez, Sporicidal activity in liquid chemical products to sterilize or high-level disinfect medical and dental instruments, *Am. J. Infect. Control*, 2005, **33**(5), 307–9.

49. Y. Kobayashi, M. Miyamoto, K. Sugibayashi and Y. Morimoto, Drug permeation through the three layers of the human nail plate, *J. Pharm. Pharmacol.*, 1999, **51**(3), 271–8.

10

SYNERGISTIC EFFECTS OF TRANSCUTOL AND TERPENES AS PENETRATION ENHANCERS: IN VITRO AND IN VIVO ATR-FTIR SPECTROSCOPIC IMAGING STUDIES

M S Erdal, D Özdin and S Güngör

Faculty of Pharmacy, Department of Pharmaceutical Technology, Istanbul University, Istanbul, Turkey

1 INTRODUCTION

The skin offers an ideal application site to deliver therapeutic agents for both topical and systemic effect. However, the integument is also a well-designed membrane to protect the organism from environmental factors, thus it is an effective obstacle for drug permeation. The main barrier for skin delivery is the stratum corneum (SC), the outermost layer of the epidermis. The SC is formed by dead and keratinized cells which present a unique barrier to passage of drugs across the skin [1, 2]. Several approaches are used to overcome the barrier property of SC in topical and transdermal administration of drugs, the most frequent of which is the inclusion of chemical penetration enhancers in formulations [3-6].

Over the last decade, biophysical techniques such as Fourier Transform Infrared (FTIR) spectroscopy have been employed to study lipid structure and organisation in the skin barrier [7, 8]. Attenuated Total Reflectance (ATR) FTIR is a well-established biophysical research tool used to characterise the molecular structure and properties of the SC and the influence of penetration enhancers [8 – 11]. It allows monitoring lipid conformation in the skin and has the advantage of well-defined spectra-structure correlations [12, 13]. Generally, the frequency of C-H stretching vibrations is a measure of the conformational order in the lipid alkyl chain domains in SC [14]. The absorption of the asymmetric and symmetric bands at 2920 and 2850 cm^{-1}, respectively, undergoes a blue-shift (shift to a higher wavenumber) as the degree of disorder of the lipid alkyl chains increases [8, 12, 14- 16].

ATR-FTIR spectroscopy facilitates an evaluation of the influence of solvents/penetration enhancers on the organisation of lipids and proteins within the skin. Some penetration enhancers may fluidise SC lipids and this can be deduced from the blue-shift of C-H stretching peaks as well as increment in bandwidth [17, 18]. The height and area of the peaks of asymmetric and symmetric C-H vibrations have been found to be proportional to the amount of the lipids present in SC. Solvent extraction of SC lipids results in a decrease in peak height and area of the C-H stretching bands [16, 17].

Terpenes obtained from natural sources are of low cutaneous irritancy and are considered to be safe and effective penetration enhancers. Indeed, the US Food and Drug Administration (FDA) have classified terpenes as "GRAS" (Generally Regarded as Safe). A variety of terpenes have been identified which can enhance the permeation of both hydrophilic and lipophilic drugs [19 – 22]. The mechanisms of action of terpenes are thought to be due to their chemical structure as well as the physicochemical properties of the drug. Terpenes increase the penetration of drugs through the skin by disrupting the intercellular packaging of SC lipids and/or increasing drug diffusivity within the SC [19, 23 – 25]. Topical application of terpenes results in a significant decrease in the height and area of asymmetric and symmetric C-H stretching peaks and the permeation of various drugs is enhanced due to overall extraction of SC lipids [17]. Terpene-induced alterations in SC barrier properties are generally reversible [26]. Examples of terpenes assessed for use in transdermal formulations include nerolidol, limonene and eucalyptol.

Nerolidol, a monoterpene (log p: 5.36; Figure 1A), is deemed safe and is widely used as penetration enhancer in transdermal delivery systems [22]. The effective permeation-promoting activity of nerolidol has been attributed to its amphiphilic structure which is suitable for alignment within the lipid lamellae and also for disruption of the highly organised packing of SC [21, 27, 28].

Limonene (*dl*-) is a hydrocarbon terpene (log P: 4.53; Figure 1B) [22]. A synergistic effect of ethanol and *dl*-limonene has been reported and it has been recognised that hydrocarbon terpenes provide a greater enhancement in permeation of lipophilic drugs [18].

The calculated log P value of eucalyptol is 2.82 (Figure 1C) [19]. It has been suggested that eucalyptol produces an enhancement effect through intercalation of lipids and proteins of the SC via hydrogen bonds [18].

The presence of various co-solvents can affect the interaction of terpenes with the SC due to additional interactions with SC [18]. Propylene glycol and ethanol have been found to produce a synergistic action when used in combination with terpenes in many studies [16, 17, 22, 29]. The combination of terpenes in ethanol and propylene glycol has been found to increase the accumulation of drugs in the skin as well as promoting a significant enhancement in percutaneous penetration [30-35]. Transcutol, a monoethyl ether of diethylene glycol (Figure 1D), is a hydroscopic liquid and is recognised as a co-solvent due to its miscibility with polar and nonpolar solvents [36 – 38]. Unlike terpenes, Transcutol has not been found to have a fluidising effect on SC intercellular lipids [7] but appears to increase the accumulation of topically applied drugs within the skin in the absence of a concomitant increase in transdermal permeation [36]. Indeed, ATR-FTIR studies have demonstrated that Transcutol affects the solubility and hence partitioning of the permeant into the SC [39, 40].

The novelty of our approach is to evaluate the possible synergistic action of terpenes and Transcutol combination through their effect on pig and human skin barrier properties by comparing skin barrier interaction of neat nerolidol (amphiphilic sesquiterpene), *dl*-limonene (cyclic terpene), eucalyptol (oxygen containing sesquiterpene) and Transcutol, with the activity of a terpene-Transcutol combination. ATR-FTIR spectroscopy was employed to elucidate the effect of terpenes, Transcutol and terpene-Transcutol system on molecular organisation of pig and human SC under *in vitro* and *in vivo* conditions, respectively.

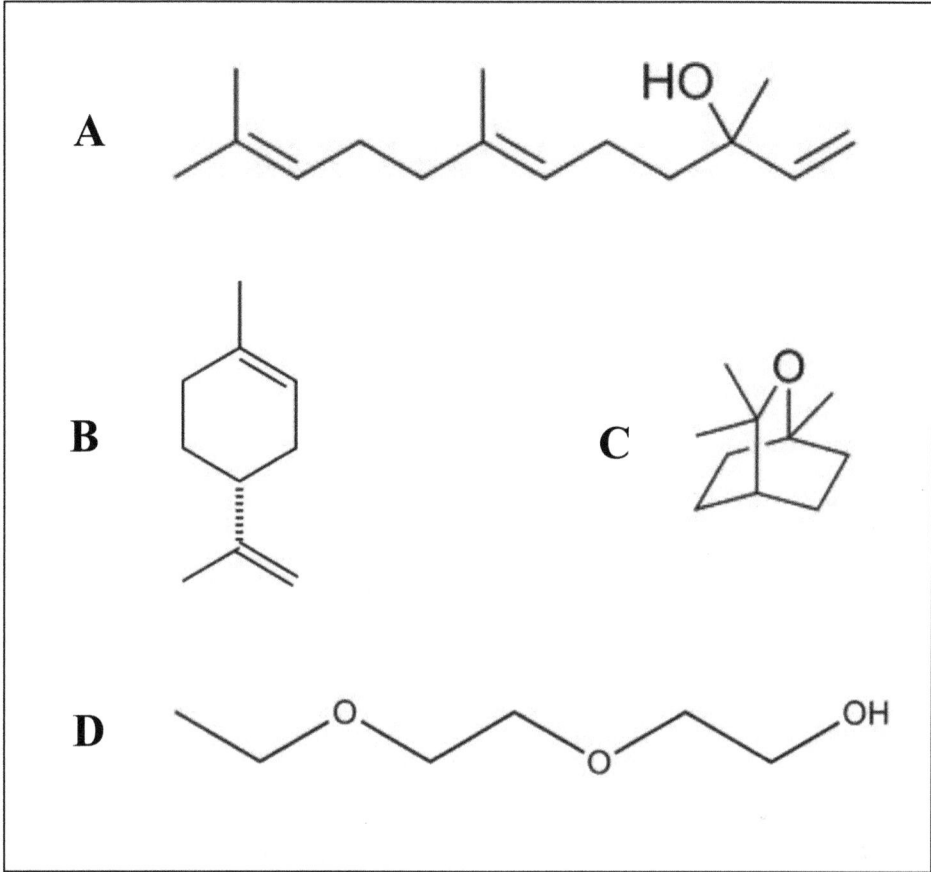

Figure 1 *Chemical structure of three terpenes (A – C) and Transcutol (D). A; nerolidol (C₁₅H₂₆O; mw 222.37). B; dl-limonene (C₁₀H₁₆, mw: 136.24). C; eucalyptol (C₁₀H₁₈O, mw: 154.25). D; Transcutol (diethylene glycol monoethyl ether - C₆H₁₄O₃, mw: 134.17).*

2 METHODS

Terpenes were used at a concentration of 3% (w/w) with the combination of Transcutol as co-solvent (Table 1). Mixtures were equilibrated for 30 minutes without stirring following 3 minutes mixing with vortex. At the concentrations used (3%), *dl*-limonene, nerolidol and eucalyptol were readily miscible with Transcutol.

Table 1 *Composition of terpene/co-solvent (Transcutol) systems.*

Treatments	Terpene/co-solvent mixtures
LM	*dl*-limonene
LMTC	*dl*-limonene + Transcutol
NR	Nerolidol
NRTC	Nerolidol + Transcutol
EU	Eucalyptol
EUTC	Eucalyptol + Transcutol
TC	Transcutol

ATR-FTIR spectroscopy studies were performed *in vitro* using pig skin. The subcutaneous fat was removed with a scalpel and hairs were close clipped. The skin was dermatomed to a nominal thickness of 750 μm with a Zimmer Dermatome (USA) and subsequently stored at -25°C (wrapped in Parafilm and packed in ZipLock bags) for not longer than 2 months prior to use.

Prior to collection of ATR-FTIR spectra, each skin sample was defrosted for 8 h at 4°C, after which the skin surface was wiped twice with a wet cotton swab, with any excess surface water removed by drying with paper tissue. The test vehicle (100 μl of terpene, Transcutol or terpene/Transcutol system) was applied to the surface for 3h at 25°C under non-occlusive conditions. Following the treatment period, all residual formulation was gently removed by blotting with paper tissue. Treated skin samples were subjected to ATR-FTIR analysis after vacuum drying.

In vivo studies were carried out in an examination room under constant environmental conditions (42% relative humidity, 25° C). Six healthy female volunteers, ranging in age from 25 to 40 years, with no history of dermatologic disease volunteered for the study which had been approved by the Istanbul University ethical committee. Written informed consent was obtained from all subjects. For each subject, two sites on the ventral aspect of both forearms were delineated using an adhesive tape. A non-treated skin site served as control (CNT) in the studies. Each (100 μl) treatment was applied to the test area under non-occlusive conditions.

All spectra were collected with a spectral resolution of 4 cm^{-1} in the 4000-650 cm^{-1} range using a Perkin Elmer Spectrum 100 FTIR spectrometer (USA) equipped with a diamond ATR crystal. Background spectra were acquired using the clean ATR crystal prior to use. Each spectrum comprised an average of 40 scans. *In vitro* skin samples, cut to dimensions 2x2 cm, were placed so that the SC side was facing the ATR crystal.

Attention was focused on characterising the occurrence of peaks near 2850 and 2920 cm^{-1} due to the symmetric (SSV) and asymmetric (ASSV) C-H stretching vibrations, respectively. For each skin sample, peak height and area was measured after enhancer treatment. A non-treated skin sample was served as control (CNT). The resulting spectra were analysed using Perkin Elmer Spectrum Version 6.0.2 software.

3 RESULTS AND DISCUSSION

Of particular interest were the peaks near 2920 and 2850 cm^{-1} caused by asymmetric and symmetric C-H bond stretching, respectively (Table 2). The shift of these peaks to a higher wave number is indicative of a trans- to cis-conformational change which implies fluidisation of SC lipids [16]. The variation of C-H stretching band frequencies provides a qualitative indication of alterations in chain packing as well as conformational order and 1-2 cm^{-1} frequency shifts or differences are indicating of significant changes [14, 41]. The C-H stretching peaks in the spectra of untreated and treated skin were analysed for frequency shifts and changes in peak height and areas in *in vitro* and *in vivo* studies, respectively. Our results are summarised in Table 2 and Figures 2 and 3.

The C-H symmetric stretching band frequency of 2851 cm^{-1} represents a state between gel and liquid crystalline. Frequencies of 2852 to 2854 cm^{-1} are found for a liquid crystalline phase [42]. Similar correlations were detected for the C-H asymmetric stretching mode which can range between 2915 and 2924 cm^{-1} [7]. Our ATR-FTIR data showed that treatment of pig skin with terpenes (*dl*-limonene, eucalyptol and nerolidol) or the mixtures (namely *dl*-limonene-Transcutol and eucalyptol-Transcutol) produced a blue shift of approximately 2-5 cm^{-1} in the asymmetric and symmetric stretching peak positions indicative of increased lipid chain disorder. The expected blue shift of stretching absorbance peaks due to reduced lipid order after treatment with nerolidol-Transcutol mixture and Transcutol alone was unremarkable in the *in vitro* study (Table 2).

Compared with Transcutol, treatment of human skin for 3 h with terpenes and terpene-Transcutol mixtures resulted in blue shifts to a higher frequency for the C-H asymmetric and symmetric absorbance peaks and had a great influence on the fluidity of the SC intercellular lipid matrix: the lipid chain disorder was increased. The synergistic effect of Transcutol can be explained by improved partitioning of terpenes into SC in the presence of Transcutol.

Several studies have shown that terpenes extract the SC lipids and thereby enhance transdermal drug permeation [23, 43, 44]. Extraction of lipids by solvents or enhancers results in decreased C-H stretching peak height and area [17].

In our *in vitro* studies, terpenes and terpene-Transcutol systems decreased the peak height and areas for asymmetric stretching absorbances in comparison to untreated control skin (Figure 2). Nerolidol and nerolidol-Transcutol mixture treatment produced the greatest decrease in symmetric stretching frequency peak height and areas due to an overall extraction of SC lipids. These data are in accordance with the finding that terpenes with larger log P values (like nerolidol) are more effective enhancers than those with smaller log P and this effect may be attributable to more extensive mixing with SC intercellular lipids resulting in extraction or lipid phase transition [18].

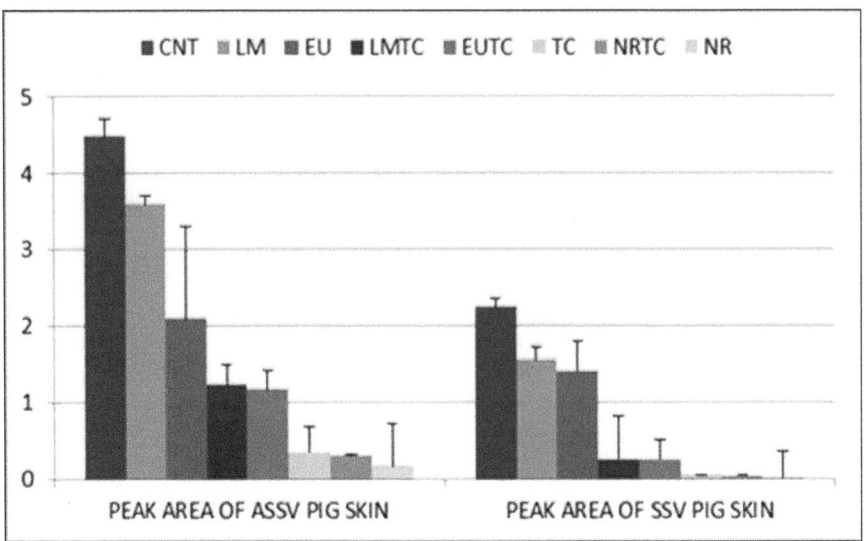

Figure 2 *Peak area of asymmetric and symmetric C-H stretching absorbances before and after treatment of ex vivo pig skin with terpene, Transcutol and terpene-Transcutol mixtures (ASSV: asymmetric stretching vibration, SSV: symmetric stretching vibration).*

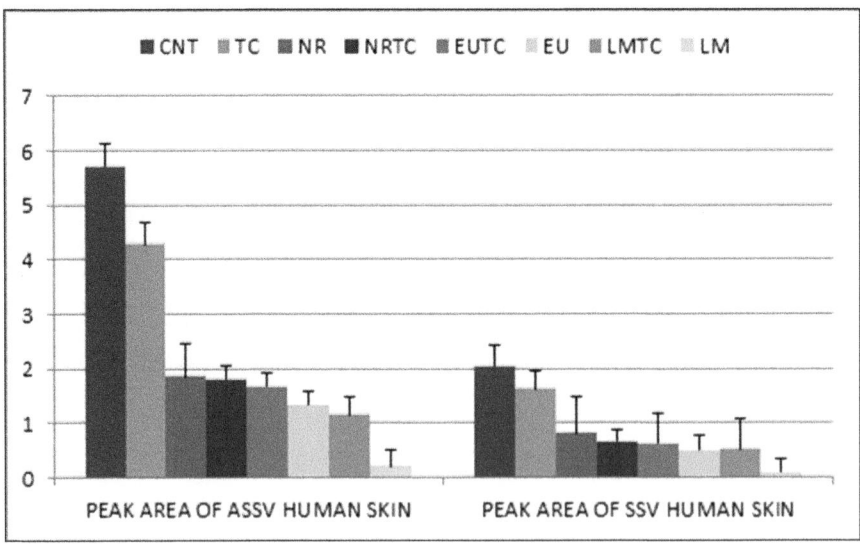

Figure 3 *Peak area of asymmetric and symmetric C-H stretching absorbances before and after treatment of in vivo human skin with terpene, Transcutol and terpene-Transcutol mixtures (ASSV: asymmetric stretching vibration, SSV: symmetric stretching vibration).*

Table 2 *Asymmetric and symmetric C-H stretching peak positions before and after treatment of pig and human skin with terpene, Transcutol and terpene-Transcutol mixtures (ASSV: asymmetric stretching vibration, SSV: symmetric stretching vibration).* Each value represents the mean ± SD of six samples/subjects.

Type of Treatment	Pig Skin		Human Skin	
	ASSV (cm^{-1})	SSV (cm^{-1})	ASSV (cm^{-1})	SSV (cm^{-1})
CNT	2918.97±1.68	2850.93±0.67	2918.25±1.15	2849.88±1.05
TC	2919.11±1.27	2851.65±1.08	2919.72±0.98	2851.21±0.70
LM	2922.16±0.41	2853.52±0.55	2921.40±0.41	2851.81±0.55
NR	2920.56±0.58	2852.06±1.12	2922.79±1.14	2852.74±1.18
EU	2921.66±0.28	2853.52±0.16	2922.63±1.45	2853.04±1.09
LM-TC	2922.75±0.89	2852.96±1.02	2920.54±0.04	2851.14±0.07
NR-TC	2919.34±1.03	2852.04±0.56	2921.56±1.80	2851.92±1.40
EU-TC	2923.09±0.71	2854.11±0.63	2919.60±1.21	2851.08±0.96

The human *(in vivo)* ATR-FTIR results revealed that the maximum extraction of SC lipids was observed after treatment with *dl*-limonene, as evidenced by the highest decrease in C-H stretching peak area (Figure 3). The lipid extraction effect and the percentage decrease in C-H stretching peak height and area in human subjects treated with other terpenes and with terpene-Transcutol systems were similar. Lipid extraction was higher with terpenes in Transcutol compared with that of skin treated with Transcutol alone. Vaddi and co-workers obtained similar results on the IR spectra of *ex vivo* human SC treated with propylene glycol-terpene systems [16].

Treatment with Transcutol caused a blue shift of approximately 2 cm^{-1} in the symmetric stretching frequency during *in vivo* studies, indicating a chain conformational disorder. In contrast, no remarkable decrease in area of C-H stretching absorbances was observed in untreated control skin regions.

4 CONCLUSIONS

Investigation of ATR-FTIR spectra of SC/skin is useful to investigate the mechanism of action of penetration enhancers and co-solvent-penetration enhancer combinations and to make a correlation with enhanced permeation of drug substances. Based on the *in vivo* data, Transcutol treatment resulted in a blue shift in the symmetric stretching frequency due to a chain conformational disorder, but C-H stretching absorbances in SC were not decreased significantly. The combination of terpenes with Transcutol increased SC lipid extraction compared to Transcutol treatment alone. The synergistic effect of Transcutol can be attributed to the increased partitioning of terpenes into SC in the presence of that co-solvent. Both *in vitro* and *in vivo* data also indicated that terpenes and terpene-Transcutol mixtures affect the lipid extraction from the SC.

It is well-known that extraction of SC lipids cause in enhanced drug permeation through the skin. For this reason, further studies will be run with the purpose of investigating transdermal permeation of hydrophilic and lipophilic model drugs in the presence of terpenes, Transcutol and terpene-Transcutol combination as well as investigation of their distribution among skin layers.

5 ACKNOWLEDGEMENT

This study was financially supported by Istanbul University Research Fund (Project number: 4583)

6 REFERENCES

1. K. Moser, K. Kriwet, A. Naik, Y.N. Kalia and R.H. Guy, Passive skin penetration enhancement and its quantification *in vitro*, *Eur. J. Pharm. Biopharm.*, 2001, **52**, 103.

2. A. Williams, *Transdermal and Topical Drug Delivery*, Pharmaceutical Press, 2003

3. H. Tromer and R.H.H. Neubert, Overcoming the stratum corneum: The modulation of skin penetration, *Skin Pharmacol. Physiol.*, 2006, **19**, 106.

4. R. H. Guy, Carriers in the Topical Treatment of Skin Disease, *Drug Delivery, Handbook of Experimental Pharmacology*, ed. M. Schäfer-Korting, Springer-Verlag, Berlin Heidelberg, 2010, p. 197.

5. H. A. Benson, Transdermal drug delivery: Penetration enhancement techniques, *Curr. Drug Deliv.*, 2005, **2**, 23.

6. T. Taner and R. Marks, Delivering drugs by the transdermal route: Review and comment, *Skin Res. Technol.*, 2008, **14**, 249.

7. R. Mendelsohn and D. J. Moore, Vibrational spectroscopic studies of lipid domains in biomembranes and model systems, *Chem. Physics Lipids*, 1998, **96**, 141.

8. D. J. Moore, M. Rerek and R. Mendelsohn, Role of ceramides 2 and 5 in the structure of the stratum corneum lipid barrier, *Int. J. Cosmetic Sci.*, 1999, **21**, 353.

9. M. V. L. B. Bentley, E. R. M. Kedor, R. F. Vianna and J. H. Collett, The influence of lecithin and urea on the in vitro permeation of hydrocortisone acetate through skin from hairless mouse, *Int. J. Pharm.*, 1997, **146**, 255.

10. S. Wartewig and R. H. H. Neubert, Pharmaceutical applications of Mid-IR and Raman spectroscopy, *Advanced Drug Del. Rev.*, 2005, **57**, 1144.

11. M. S. Erdal and A. Araman, Comprative evaluation of the structure vernix caseosa and human stratum corneum by transmission FT-IR spectroscopy, *Turkish J. Pharm. Sci.*, 2009, **2**, 73–83.

12. D. Bommannan, R. O. Potts and R. H. Guy, Examination of stratum corneum barrier function in vivo by infrared spectroscopy, *J. Inv. Dermatol.*, 1990, **95**, 403–408.

13. P. Garidel, B. Földing, I. Schaller and A. Kerth, The microstructure of the stratum corneum lipid barrier: mid-infrared spectroscopic studies of hydrated ceramide:palmitic acid:cholesterol model systems, *Biophysical Chem.*, 2010, **150**, 144.

14. H. E. Bodde, L. A. R. M. Pechtold, M. T. A. Subnel and F. H. N. de Haan, Monitoring *in vivo* Skin Hydration by Liposomes Using Infrared Spectroscopy in Conjunction with Tape Stripping, in *Liposome Dermatics*, ed. H.I. Maibach, Springer Verlag, Berlin Heidelberg, 1992, p. 137.

15. H. Schaefer and T.E. Redelmeier, Skin barrier. Principles of percutaneous absorption, *Skin Barrier*, Karger, Basel, 1996, pp. 76–77

16. H. K. Vaddi, P. C. Ho and S. Y. Chan, Terpenes in propylene glycol as skin-penetration enhancers: permeation and partition of haloperidol, Fourier transform infrared spectroscopy, and differential scanning calorimetry, *J. Pharm. Sci.*, 2002, **91**, 1639.

17. K. Babita, V. Kumar, V. Rana, S. Jain and A. K. Tiwary, Thermotropic and spectroscopic behavior of skin: Relationship with percutaneous penetration enhancement, *Current Drug Del.*, 2006, **3**, 95.

18. B. Sapra, S. Jain and A. K. Tiwary, Percutaneous permeation enhancement by terpenes: Mechanistic view, *AAPS Journal*, 2008, **10**, 120.

19. K. Cal, K. Kupiec and M. Sznitowska, Effect of physicochemical properties of cyclic terpenes on their *ex vivo* skin absorption and elimination kinetics, *J. Dermatol. Sci.*, 2006, **41**, 137.

20. S. Gao and J. Singh, *In vitro* percutaneous absorption enhancement of a lipophilic drug tamoxifen by terpenes, *J. Control. Rel.*, 1998, **51**, 193.

21. S. Güngör, A. Bektaş, F. I. Alp, B. S. Uydeş-Doğan, O. Özdemir, A. Araman and Y. Özsoy, Matrix-type transdermal patches of verapamil hydrochloride: in vitro permeation studies through excised rat skin and pharmacodynamic evaluation in rats, *Pharm. Dev. Techn.*, 2008, **13**, 283.

22. Y. S. R. Krishnaiah, S. M. Al-Saidan, D. V. Chandrasekhar and B. Rama, Effect of nerodilol and carvone on *in vitro* permeation of nicorandil across rat epidermal membrane, *Drug Dev. Ind. Pharm.*, 2006, **32**, 423.

23. A. C. Williams and B. W. Barry, Terpenes and the lipid-protein-partitioning theory of skin penetration enhancement, *Pharm. Res.*, 1991, **8**, 17.

24. A. C. Williams and B. W. Barry, Penetration enhancers, *Adv. Drug Del. Rev.*, 2004, **55**, 603.

25. K. Zhao and J. Singh, Mechanisms of percutaneous absorption of tamoxifen by terpenes: eugenol, *dl*-limonene and menthone, *J. Contr. Rel.*, 1998, **55**, 253.

26. J. R. Kunta, V. R. Goskonda, H. O. Brotherton, M. A. Khan and I. K. Reddy, Effect of menthol and related terpenes on the percutaneous absorption of propranolol across excised hairless mouse skin, *J. Pharm. Sci.*, 1997, **86**, 1369.

27. P. A. Cornwell, B. W. Barry, C. P. Stoddart and J. A. Bouwstra, Wide-angle X-ray diffraction of human stratum corneum: effects of hydration and terpene enhancer treatment, *J. Pharm. Pharmacol.*, 1994, **46**, 938.

28. A. F. El-Kattan, C. S. Asbill, N. Kim and B. B. Michniak, The effect of terpene enhancers on the percutaneous permeation of drugs with different lipophilicities, *Int. J. Pharm.*, 2001, **215**, 229.

29. K. K. Levison, K. Takayama, K. Isowa, K. Okabe and T. Nagai, Formulation optimization of indomethacin gels containing a combination of three kinds of cyclic monoterpene as percutaneous penetration enhancers, *J. Pharm. Sci.*, 1994, **83**, 1367.

30. J. S. Chang, Y. H. Tsai, P. C. Wu and Y. B. Huang, The effect of mixed-solvent and terpenes on percutaneous absorption of meloxicam gel, *Drug Dev. Ind. Pharm.*, 2007, **33**, 984.

31. D. A. Godwin and B. B. Michniak, Influence of drug lipophilicity on terpenes as transdermal penetration enhancers, *Drug Dev. Ind. Pharm.*, 1999, **25**, 905.

32. Y. S. R. Krishnaiah, V. Raju, M. S. Kumar, B. Rama, V. Raghumurthy and K. V. R. Murthy, Studies on optimizing *in vitro* transdermal permeation of ondansetron hydrochloride using nerodilol, carvone, and limonene as penetration enhancers, *Pharm. Dev. Technol.*, 2008, **13**, 177.

33. A. Nokhodchi, K. Sharabiani, M. R. Rashidi and T. Ghafourian, The effect of terpene concentrations on the skin penetration of diclofenac sodium, *Int. J. Pharm* 2007, **335**, 97.

34. T. Şenyiğit, C. Padula, Ö. Özer and P. Santi, Different approaches for improving skin accumulation of topical corticosteroids, *Int. J. Pharm.*, 2009, **380**, 155.

35. M. A. Yamane, A. C. Wiliams and B. W. Barry, Effects of terpenes and oleic-acid as skin penetration enhancers towards 5-fluorouracil as assessed with time-permeation, partitioning and differential scanning calorimetry, *Int. J. Pharm.*, 1995, **116**, 237.

36. D. A. Godwin, N. H. Kim and L. A. Felton, Influence of Transcutol CG on the skin accumulation and transdermal permeation of ultraviolet absorbers, *Eur. J. Pharm. Biopharm.*, 2002, **53**, 23.

37. M. Manconi, C. Caddeo, C. Sinico, D. Valenti, M.C. Mostallino, G. Biggio and A.M. Fadda, Ex vivo skin delivery of diclofenac by transcutol containing liposomes and suggested mechanism of vesicle skin interaction, *Eur. J. Pharm. Biopharm.*, 2011, **78**, 27.

38. P. Mura, M. T. Faucci, G. Bramanti and P. Corti, Evaluation of transcutol as a clonazepam transdermal permeation enhancer from hydrophilic gel formulations, *Eur. J. Pharm. Sci.*, 2000, **9**, 365.

39. J. E. Harrison, A. C. Watkinson, D. M. Green, J. Hadgraft and K. Brain, The relative effect of Azone and Transcutol on permeant diffusivity and solubility in human stratum corneum, *Pharm. Res.*, 1996, **13**, 542.

40. J. Hadgraft, Modulation of the barrier function of the skin, *Skin Pharm. Appl. Skin. Physiol.*, 2001, **14**, 72.

41. R. Mendelsohn and D. J. Moore, Infrared determination of conformational order and phase behavior in ceramides and stratum corneum models, *Methods in Enzymology*, 2000, **312**, 228.

42. C. L. Gay, R. H. Guy, G. M. Golden, V. H. W. Mak and M. L. Francoeur, Characterization of low temperature lipid transitions in human *stratum corneum, J. Inv. Dermatol.*, 1994, **103**, 233.

43. K. Takahashi, H. Sakano, M. Yoshida, N. Numata and N. Mizuno, Characterization of the influence of polyol fatty acid esters on the permeation of diclofenac through rat skin, *J. Contr. Rel.*, 2001, **73**, 351.

44. K. Zhao and J. Singh, *in vitro* percutaneous absorption enhancement of propranolol hydrochloride through porcine epidermis by terpenes/ethanol, *J. Control. Rel.*, 1999, **62**, 359.

11

SKIN PENETRATION FROM PICKERING EMULSIONS

J Frelichowska, M-A Bolzinger, J Pelletier, J-P Valour and Y Chevalier

Université de Lyon F-69008, Lyon, France; Université Lyon 1, Laboratoire de Dermopharmacie et Cosmétologie, F-69008, Lyon, France; UMR CNRS 5007, Laboratoire d'Automatique et de Génie des Procédés, F-69622, Villeurbanne, France.

1 INTRODUCTION

Pickering emulsions are especially attractive in pharmaceutical and cosmetic sciences because stabilisation of droplets by solid particles dramatically improves the stability of liquid droplets with respect to coalescence. Such remarkable behavior was originally identified by S.U. Pickering in 1907 [1]. The solid particles are adsorbed at the oil-water interface by means of partial wetting behaviour of the solid particles' surface by water and oil [2-4] (Figure 1).

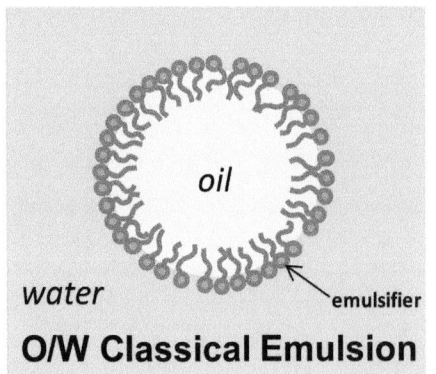
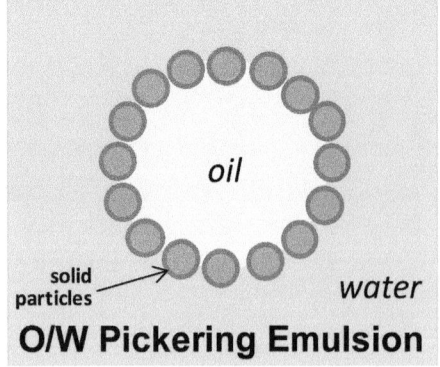

Figure 1 *Schematic of a classical (emulsifier-based) emulsion and Pickering emulsion. Solid particles stabilise the droplets in place of the emulsifier molecules. The solid particles are adsorbed at the oil-water interface when partial wetting conditions are fulfilled.*

Improved stability comes from the strong anchoring of the solid particles at the oil-water interface and the formation of a mechanically-rigid coating at the surface of the liquid droplets. Both oil-in-water (o/w) and water-in-oil (w/o) emulsions can be prepared depending on the hydrophobic character of the solid particles [5, 6]: hydrophobic particles stabilise w/o emulsions whereas hydrophilic particles stabilise the o/w type emulsions, in the same manner as hydrophobic and hydrophilic surfactants stabilise w/o and o/w emulsions, respectively (according to the Bancroft rule). In the case of silica particles, 'hydrophobised' particles are most often required in order to ensure the partial wetting conditions of the solid surface by oil and water. 'Hydrophobic' silica particles are prepared by grafting their surface with organic molecules (organosilanes); the hydrophobic character being controlled by the grafting density. However, bare silica particles (non-grafted) can be used in few instances for specific oils [7]. Resistance to coalescence allows the preparation of concentrated emulsions and stable coarse emulsions (millimeter-sized emulsion droplets) that surfactant-based stabilisation would not support. Pickering emulsion show strong similarities with classical emulsions (surfactant-based); emulsification processes are the same, the size of emulsion droplets is controlled by the emulsification process and by the amount of stabilising particles with respect to the amount of dispersed phase [8, 9].

Pickering emulsions are surfactant-free, which is an attractive characteristic regarding their application to cosmetic formulations given the potential toxicity of surfactants. Besides enhanced stability and reduced toxicity, the dense coating of solid particles around each droplet provides a barrier against diffusion of drugs [10, 11] and so controlled delivery can be expected in the same way as from encapsulation in solid particles. Such specific properties of Pickering emulsions make them potentially attractive for drug delivery applications to skin. Skin absorption from Pickering emulsions has recently been investigated for the first time [12, 13]. The research described in this chapter reports two comparative studies of skin transport of drugs by Pickering emulsions and classical emulsions stabilised with surfactants: the first deals with skin absorption of retinol as a model hydrophobic drug from o/w emulsions; the second addresses skin delivery of caffeine as hydrophilic drug from w/o emulsions. Beyond the comparative assessment of Pickering and classical emulsions, the underlying physicochemical mechanisms of such transport are touched on in the case of transport of caffeine by w/o emulsions.

2 MATERIALS AND METHODS

Ingredients were gifts from their suppliers: Polysorbate 85 emulsifier (Montanox®85) from Seppic, α-Tocopherol acetate from BASF, Caprylic/capric triglyceride (Labrafac WL®1349) from Gattefossé, Polyoxyethylene (20) oleyl ether (Brij®98) from Uniqema, the emulsifier Abil® EM97 and the wax Abil® Wax 9810 from Evonik, cyclomethicone oils DC®245 and DC®246 from Dow Corning, preservative Seppicide® HB from Seppic. *All-trans* retinol, caffeine and phosphate buffer pH = 7.4 were purchased from Sigma-Aldrich. Deionised water (18 MΩ cm resistivity) was used throughout the studies.

Hydrophobised silica's HDK® HKS D and HDK® H20 were kind gifts from Wacker Chemie (Germany). Such solid particles are fumed silica's that have been made partially hydrophobic by grafting dimethylsilyl groups at their surface. According to the producer's data [14, 15], HDK® H20 silica (used for the stabilisation of w/o emulsions) is grafted up to 1.1 wt% carbon content, leaving 55 % of free hydroxyl groups at the surface (Si-OH ≈ 3 μmol m^{-2}); it has a BET specific area of 170 m^2 g^{-1}. HDK® HKS D of BET specific area 200 m^2 g^{-1} contains 0.95 wt% carbon from elemental analysis, so that the silica surface comprises 71 % of free hydroxyl groups.

Full-thickness pig skin (mean thickness ± SE = 1.35 ± 0.05 mm) was used in the skin absorption experiments. The skins of 3 donor animals were excised from the flank of donor animals and washed. Subcutaneous fatty tissue was carefully removed and the skin pieces were stored flat at -20 °C until use. The compositions of the emulsions are given in Table 1.

Pickering emulsions were prepared in two steps: silica was first dispersed in the dispersing phase using an ultrasound disperser Sonics VibraCell at 500 W during 30 s; the oil and aqueous phases were mixed together with an UltraTurrax® at 22000 rpm for 2 min. For the o/w Pickering emulsions of silicone oil, HDK® HKS D silica was first dispersed water as the dispersing phase at a concentration of 8 % w/w; caprylic/capric triglyceride oil containing the *all-trans* retinol and α-tocopherol acetate as anti-oxidant was then emulsified into the aqueous phase in the dark. The final emulsion was a stable o/w emulsion with 10 % w/w dispersed phase. For the w/o Pickering emulsions in silicone oil, HDK® H20 silica was first dispersed in silicone oil at a concentration of 2 % w/w and an equivalent amount of water containing the caffeine and preservative was then emulsified into the oil phase. The final emulsion was a stable w/o emulsion with 50 % w/w dispersed phase.

The classical (emulsifier-stabilised) o/w emulsion was prepared by mixing together the oil and aqueous phases with an UltraTurrax® (Germany) at 13500 rpm for 2 min in the dark.

The classical w/o emulsion was prepared at 70 °C by slow addition of the aqueous phase into the silicone oil phase containing the emulsifiers while stirring with an UltraTurrax® at 11000 rpm. The emulsion was stirred with a TurboTest® (Rayneri/VMI) at 1000 rpm for 2 min and at 300 rpm during cooling to 25 °C (30 min).

Emulsion droplet sizes were measured by small angle light scattering (MasterSizer® 2000, Malvern) and by optical microscopy (Leica® DMLM optical microscope and image analysis using AnalySIS® software). Viscosity was measured at 20 °C using a Couette rheometer (Rheomat R180, Lamy, France) equipped with a mobile system N° 11 rotating at 200 rps. Contact angles and interfacial tensions were measured using a Drop Shape Analysis System DSA10 Mk2 (Krüss GmbH, Germany). Interfacial tensions between deionised water and cyclomethicone oil were measured by the pendant drop method (the densities used for the calculation were 0.9982 g cm^{-3} for water, 0.958 g cm^{-3} for cyclomethicone and 0.9593 g cm^{-3} for cyclomethicone containing silica particles). Contact angles between water, cyclomethicone and pig skin were determined by the sessile drop method. Samples of full-thickness pig skin stuck on glass slides were immersed in the oil phase, a drop of water (≈ 5 μL) was deposited on the skin surface with a syringe, and the picture of the drop shape was recorded and analysed for the contact angle.

Table 1 *Composition (% w/w) and physicochemical parameters of o/w and w/o emulsions. The ingredients of the aqueous phase are given in italics.*

Formulation Type	Ingredient / Parameter	Pickering emulsion (% w/w)	Classical emulsion (% w/w)
O/W emulsions	*all-trans Retinol*	0.1	0.1
	Caprylic/capric triglyceride	9.4	9.4
	α-Tocopherol acetate	0.5	0.5
	Polysorbate 85	-	*0.5*
	Hydrophobic silica HDK HKS D	7	-
	Water	*83*	*89.5*
	Mean droplet size (μm)	3 ± 1	3 ± 1
	Viscosity (mPa s)	6 ± 2	5 ± 2
W/O emulsions	Hydrophobic silica HDK® H20	1	-
	Abil® EM 97	-	0.1
	Abil® Wax 9810	-	10
	Cyclomethicones DC®245:DC®246 (1:1)	48.2	39.1
	Preservative Seppicide® HB	0.8	0.8
	Caffeine	0.8	0.8
	Water	49.2	49.2
	Mean droplet size (μm)	9.7 ± 0.5	10.9 ± 0.5
	Viscosity (mPa s)	550 ± 50	546 ± 50

Transepidermal Water Loss (TEWL) was measured using a Skin® Station (La Licorne, Meylan, France); skin samples with TEWL higher than 15 g m^{-2} h^{-1} were discarded. *In vitro* transdermal delivery of caffeine was determined using Franz-type diffusion cells containing full-thickness porcine skin (1.35 ± 0.05 mm thickness; mean ± S.E.) according to classical methods described in OECD guidelines (2004). Pig skin was mounted in the two-chamber glass diffusion cell of 2.54 cm^2 penetration area; the receptor compartment contained 10 mL of 0.9 % NaCl aqueous solution for sink conditions. Measurements were performed at 32 °C over a 24 h exposure period. Samples were collected at different exposure times and were analysed by HPLC for caffeine or retinol [12, 13]. After 24 h exposure, the cells were dismantled and the distribution of caffeine was measured in the donor compartment and the different skin layers. The mean and standard error (S.E.) of $n = 6$ samples were calculated and compared together using statistical analysis using the Student's t-test, analysis of variance (ANOVA); the differences were considered significant when $p \leq 0.05$.

Thin layers of the *stratum corneum* obtained by the skin stripping method were observed by scanning electron microscope (SEM) after 24 h exposure to Pickering emulsion. This was performed on three skin samples separate from those used for skin absorption studies. The skin samples were in Franz cells which were dismantled after 24 h exposure to Pickering emulsion, the emulsion was removed from the top of the skin by scraping, and the *stratum corneum* of the treated area was removed by 19 successive tape-strips [16 – 18] using D-Squame® (Monaderm) adhesive tapes (diameter 22 mm). The strips were stuck to a double-adhesive tape previously adhered to SEM aluminum stubs and sputter-coated with a thin gold/palladium layer using a cathodic pulverizer (Hummer II Technics). SEM images were obtained using a Hitachi S800 microscope working at 15 kV acceleration voltage (Centre Technologique des Microstructures, CTµ, University of Lyon, Villeurbanne, France).

3 RESULTS AND DISCUSSION

Skin Absorption of Retinol from O/W Pickering Emulsions.
There was no detectable amount of retinol in the receptor compartment: this was expected due to the hydrophobic character of retinol ($\log P = 5.68$). The storage of retinol within the skin layers was measured after 24 h exposure to the formulations (Figure 2). Retinol mainly remained as a reservoir inside the *stratum corneum* for the oil solution and the Pickering emulsion. Conversely, the emulsifier-based emulsion promoted deeper penetration to the viable epidermis and dermis. However, significant amounts reached the viable epidermis and the dermis from the two emulsions. More specifically, the balance of the absorbed retinol after 24 h exposure to the oil solution was 70 % in the *stratum corneum*; 19 % in the viable epidermis, and 11 % in the dermis (70/19/11). The same balance was 11/55/35 for the emulsifier-based emulsion and 56/33/11 for the Pickering emulsion. The difference between the two emulsions is remarkable: the Pickering emulsion caused a 5-fold higher accumulation of retinol in the *stratum corneum*.

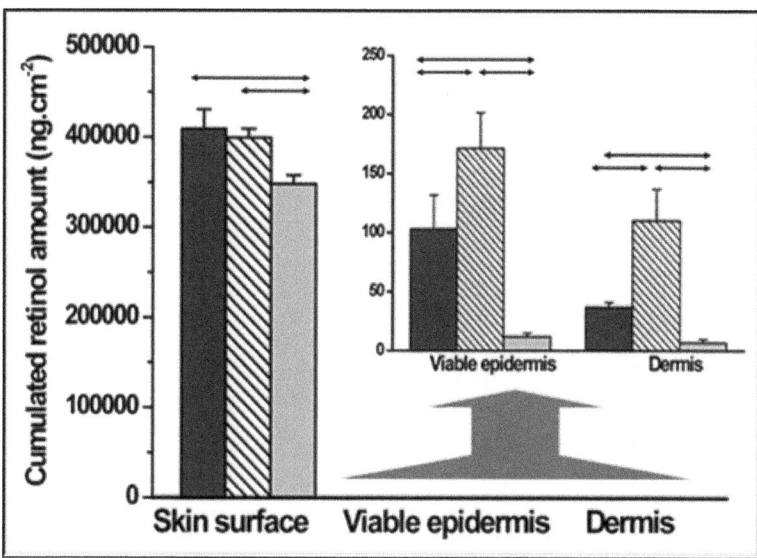

Figure 2 *Skin distribution of all-trans retinol in the skin after 24 h exposure to the o/w Pickering emulsion:* ■*, emulsifier-based emulsion:* ▨*, and oil solution:* ▨*. Mean ± SE, n = 6. Arrows indicate significant differences according to the ANOVA test with p < 0.05. Values for viable epidermis and dermis are expanded by a factor of 1000 in the upper right inset.*

The amounts of retinol stored inside the *stratum corneum* were measured by tape stripping [16-18]. The amount of recovered drug with respect to strip number is related to the depth in the *stratum corneum* and provides a picture of the penetration profile of the drug. Considering that the full *stratum corneum* was stripped off by 20 tape strips and its thickness was ~10 μm, each strip thus corresponds to a step of ~0.5 μm.

Figure 3 shows the penetration profiles of retinol from the two emulsions and the oil solution. The Pickering emulsion promoted the accumulation of retinol in the *stratum corneum*. Moreover retinol has been found in all *stratum corneum* sheets. Conversely, the emulsifier-based emulsion and the oil solution left the main parts of retinol in the upper layers of the *stratum corneum*.

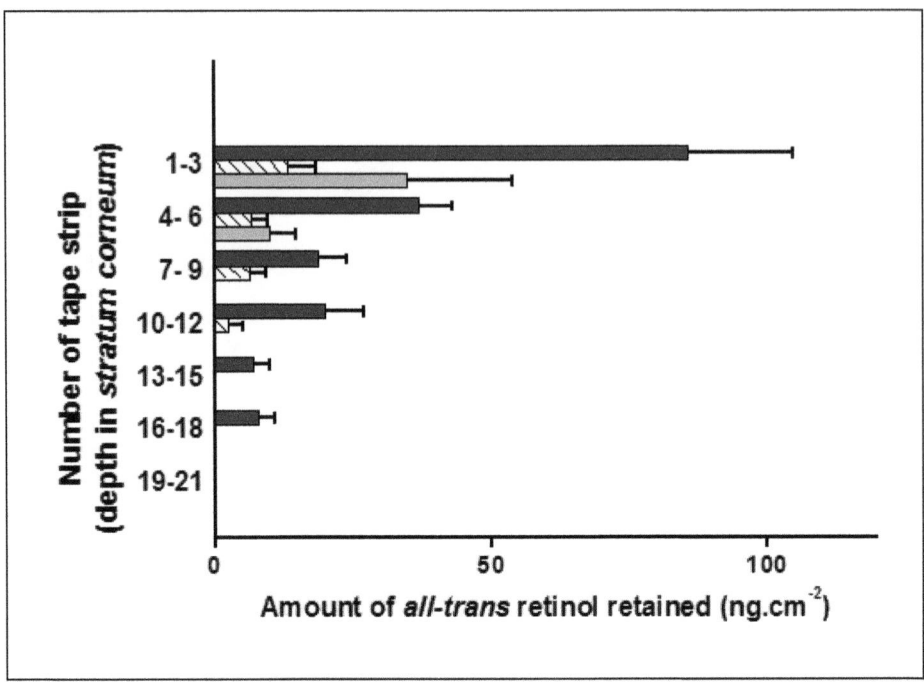

Figure 3 *Retinol distribution in stratum corneum as assessed from skin stripping experiments after 24 h exposure to the o/w Pickering emulsion:* ▇▇▇, *emulsifier-based emulsion:* ▨▨▨, *and oil solution:* ▨▨. *Mean ± SE, n = 6.*

The retinol penetration profile in the skin was significantly different for Pickering and classical emulsions. Retinol accumulated in the *stratum corneum* when it was driven by the rigid droplets of the Pickering emulsion. The emulsifier-based emulsion favored permeation towards the viable skin layers.

Skin Absorption of Caffeine from W/O Pickering Emulsions

Permeation of caffeine to the receptor compartment of the Franz cells showed the expected behavior of a permeation experiment carried out under infinite dose and occlusive conditions: A linear increase of the cumulated amount of caffeine was observed after a lag time when no caffeine was detected. Steady-state fluxes of caffeine were calculated from the slopes of the linear part of the release profiles (Figure 4).

The steady-state flux of caffeine was three-fold higher for the Pickering emulsion than for the classical emulsion and the lag time was twice as short for the Pickering emulsion (0.5 h) compared to classical emulsion (0.92 h). Both features indicate a faster penetration of caffeine from the Pickering emulsion. As expected, the amount of caffeine recovered in the skin layers (receptor fluid + dermis + epidermis) after 24 h exposure was twice as high from the Pickering emulsion compared to the classical emulsion (Figure 5).

Mechanisms of Skin Absorption from W/O Pickering Emulsions

The w/o Pickering emulsions increased the flux of caffeine through skin when compared to classical emulsions stabilised by surfactants. The skin permeation of caffeine as a model hydrophilic drug was investigated from a w/o Pickering emulsion and compared to a w/o classical emulsion having the same composition and physicochemical properties but stabilised with an emulsifier. The steady-state flux of caffeine through excised pig skin measured *in vitro* in Franz diffusion cells was 7.0×10^{-5} g m^{-2} s^{-1}, three-fold higher than from the classical emulsion (2.7×10^{-5} g m^{-2} s^{-1}). After 24 h exposure, the cumulated amounts of caffeine in the receptor fluid of the Franz cell and in the dermis were higher for the Pickering emulsion.

The mechanism of such accelerated skin absorption remains an open issue. The size of the particle is often suggested as the most relevant parameter controlling skin penetration [19]: It has been claimed that particles larger than 10 µm do not penetrate the skin, particles between 3 µm and 10 µm enter hair follicles and those smaller than 3 µm are able to diffuse through the *stratum corneum* [20, 21]. Such a view does not hold in the present case where droplets of Pickering emulsion and their reference surfactant-based emulsion had the same diameter. The mechanisms of skin absorption are manifold [22] and, in particular, surface chemistry of emulsion droplets are thought to make a major contribution to dermal absorption [23 – 25].

An investigation, based on the physical chemistry of the Pickering emulsions, is currently underway to ascertain the mechanism through which Pickering emulsions enhance the skin absorption of caffeine.

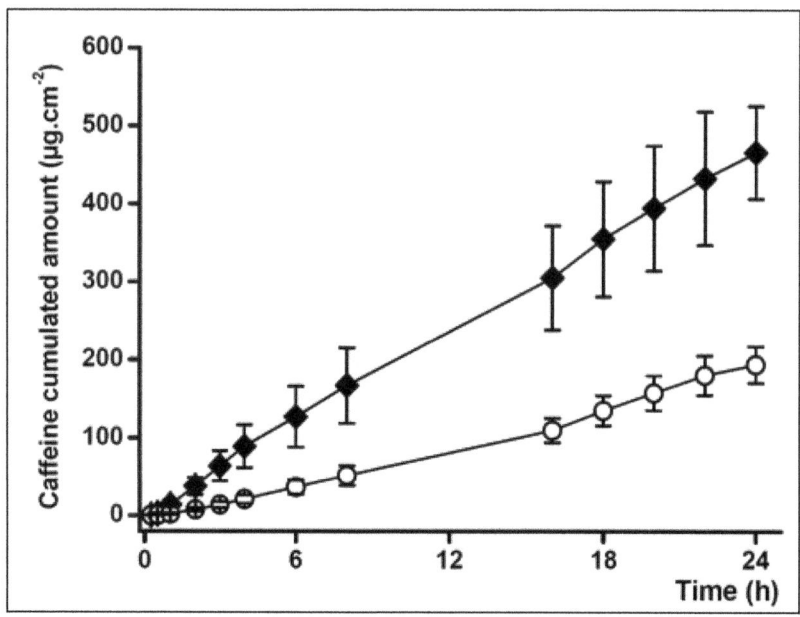

Figure 4 *Permeation profile of caffeine over 24 h exposure to Pickering emulsion (◆) and classical emulsion (O). Each point represents the mean ± S.E. of six determinations.*

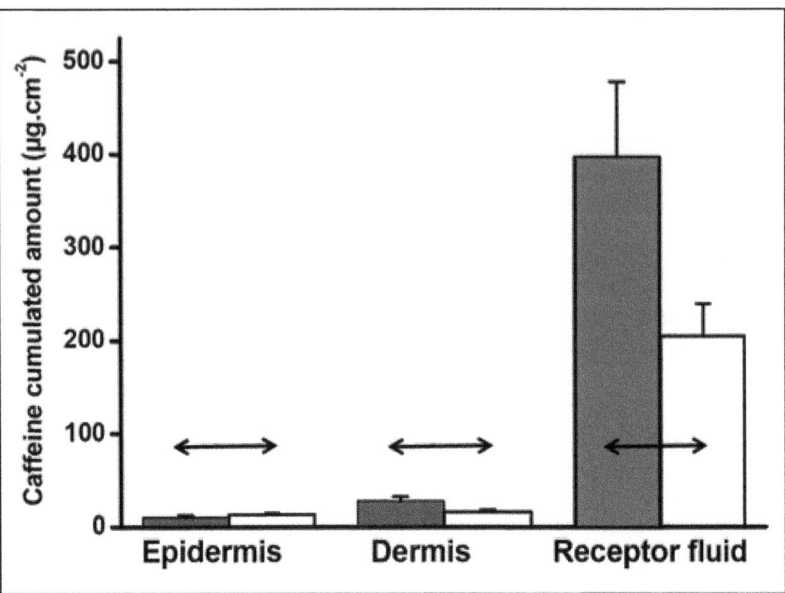

Figure 5 *Distribution of caffeine in the skin layers after 24 h to both emulsions. Pickering emulsion:* ■■■; *classical emulsion:* ☐. *Mean ± S.E., n = 6. Arrows indicate amounts showing significant difference according to the ANOVA test with p < 0.05.*

The path from the water droplets of the w/o emulsion to the deep layers of skin (dermis or deeper) involves several transfer processes that may be of relevance because they are potentially rate-limiting and controlled by the surface chemistry of the oil droplets. Caffeine is a very hydrophilic drug ($\log P$ = -0.07 for octanol-water [26]); 99.8 % of caffeine is solubilised in the water droplets of the w/o emulsion owing to its low oil-water partition coefficient ($\log P$ = -2.71 ± 0.02 for silicone oil [12]). The *stratum corneum* is a hydrophobic barrier against skin penetration. Once the *stratum corneum* has been passed, caffeine enters the relatively hydrophilic media of the viable epidermis and dermis where diffusion is relatively unimpeded. The rate-limiting step is the diffusion through the *stratum corneum*. Accordingly, two elementary stages are considered: *i*) release of caffeine from the water droplets to the outermost layer of skin; *ii*) transport of caffeine through the *stratum corneum*. Caffeine transfer from water droplets to skin (first step) may take place either by transfer through the oil phase or by direct transfer when water droplets are contacting the skin surface. Diffusion of caffeine to the skin surface through the oil continuous phase is slow because the solubility of caffeine in silicone oil is very low. The coating of water droplets acts as a barrier against release into the external phase. Such a sustained-release effect has previously been observed with surfactant-based emulsions [27 – 29] as well as Pickering emulsions [11]. In order to investigate this effect further, the kinetics of caffeine release from the water droplets to a bulk aqueous phase was measured for both emulsions at 32 °C under sink conditions as follows [30, 31]: 1 g of w/o emulsion was spread on the surface of 0.9 % NaCl aqueous solution (12 cm³) which was gently stirred with a magnetic bar so as not to mix the emulsion and saline. The caffeine concentration in the receptor was determined by HPLC at various time intervals. The release of caffeine from Pickering emulsion was 40 % slower than from the classical emulsion (Figure 6).

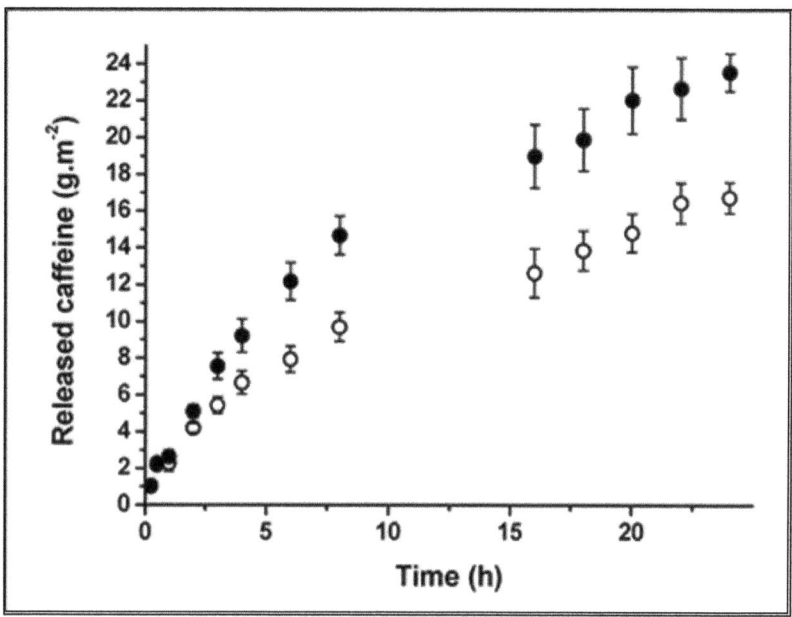

Figure 6 *Release profile of caffeine from Pickering (O) and classical (●) emulsion.*

The slow release of caffeine from the Pickering emulsion suggested that the dense shell of silica particles coating the emulsion droplets acted as a barrier against interfacial diffusion. A similar conclusion was reached from experiments performed on o/w emulsions of cross-linked polydimethylsiloxane particles coated with adsorbed silica particles [11]. This behavior is similar to encapsulation inside microcapsules made of a liquid core surrounded by a polymer shell. The typical size of fumed silica particles (20 nm) is 200-400 nm [9], making the silica shell relatively thick. In addition, there is complete coverage of the droplet surface with a dense layer of silica particles since the amount of silica was larger than required for full coverage (excess silica dispersed in silicone oil was detected as a sub-micronic population in the particle size distribution of the emulsions [9]).

The sustained release from Pickering emulsions does not provide a rationale to the skin absorption experiments because release out from the Pickering emulsion droplets was slower whereas the skin absorption was faster. An alternative mechanism is direct transfer to skin when water droplets are entering contact with the skin surface.

Direct transfer of caffeine to the skin from water droplets can take place when emulsion droplets are in contact with the skin surface. The efficiency of such process is controlled by the adhesion energy of the water droplets to the skin surface. Adhesion free energy was estimated from measurements of oil-water interfacial tension and contact angle of water droplets on skin immersed in silicone oil. The adhesion energy of a water droplet on the skin immersed in oil is defined from the three interfacial tensions [32]: skin-oil $\gamma_{(s-o)}$, skin-water $\gamma_{(s-w)}$, oil-water $\gamma_{(o-w)}$ as

$$E_{adh} = \gamma_{(s-o)} + \gamma_{(w-o)} - \gamma_{(w-s)}$$

(1)

E_{Adh} was calculated from combination of Eq (1) and Young's equation as:

$$E_{adh} = \gamma_{(w-o)} (1 + \cos \theta)$$

(2)

The interfacial tension between deionised water and silicone oil was measured by the pendant drop method (densities of phases were taken as 0.9982 g cm^{-3} for water, 0.958 g cm^{-3} for cyclomethicone and 0.9593 g cm^{-3} for cyclomethicone containing silica particles). Oil contained either 0.2 % w/w Abil® EM97 emulsifier or 0.1 % w/w HDK® H20 silica particles. Time-resolved experiments showed fast adsorption kinetics of both silica particles and emulsifier at the cyclomethicone/water interface. Adsorption of the Abil® EM97 emulsifier at the oil-water interface lowered the interfacial tension (Table 2). Conversely, adsorption of silica particles had no such effect. Lack of interfacial activity of adsorbed particles has previously been documented [33].

Table 2 *Interfacial tension γ, contact angle of water drop with skin θ, and adhesion free energy E_{Adh} for the skin - cyclomethicone - water system.*

Oil phase	γ (mN m^{-1})	θ (°)	E_{Adh} (mJ m^{-2})
Cyclomethicone	27 ± 1	151 ± 3	3.3
Cyclomethicone with emulsifier	1.4 ± 0.2	144 ± 1	0.27
Cyclomethicone with silica particles	27 ± 1	151 ± 1	3.3

The contact angle between water, cyclomethicone and pig skin was determined by the sessile drop method. Pieces of pig skin stuck on glass slides were immersed in silicone oil containing either dispersed silica particles or Abil® EM97 emulsifier. A picture of a small drop of water deposited on the skin surface was taken and analysed for the contact angle.

The surface of skin is hydrophobic because of the presence of intercellular lipids of the *stratum corneum*. The surface energy of human skin is of the order of 40 mJ m^{-2} and its dispersive component is 35 – 38 mJ m^{-2} [34, 35]. Partial wetting of water and silicone oil took place; contact angles larger than 90 ° indicated better wetting by silicone oil than water which was consistent with the hydrophobic nature of the skin surface (Figure 7).

Contact angle did not depend much on the type of surface coverage. However, the small differences were statistically significant (Table 2).

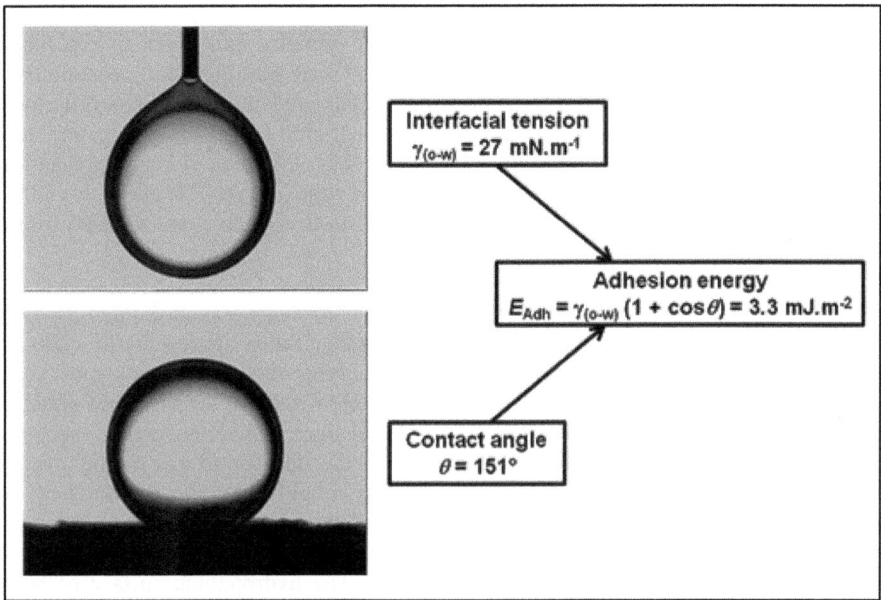

Figure 7 *Determination of the adhesion energy of water droplets (containing 2% HDK® H20 silica) to skin surface immersed in cyclomethicone oil. Oil-water interfacial tension, $\gamma_{(o-w)}$, was measured from the shape of the pendant drop of aqueous phase in cyclomethicone (top left). Contact angle was measured from a droplet of aqueous phase on skin immersed in cyclomethicone (bottom left).*

The adhesion energy of silica-coated water drops was of the same order of magnitude as that of bare water drops. The adhesion energy of a drop stabilised by the Abil® EM97 emulsifier was much lower. The larger adhesion free energy of Pickering emulsion droplets to the skin surface did not increase the contact area between water and skin (the contact angles were similar). The better adhesion of Pickering emulsion droplets makes the time spent at contact longer than for classical emulsions and so this mechanism may explain the enhanced caffeine absorption that was observed in the Franz cell experiments.

The *stratum corneum* is made of flat keratinised cells (corneocytes) imbedded inside an intercellular medium made of lamellar stacks of lipids. None of these compartments is hydrophilic; the overall water content of *stratum corneum* is 15 % (mostly in corneocytes). Two diffusion paths are possible: the intercellular path through the lipidic medium is the preferred path of hydrophobic penetrants whereas hydrophilic molecules preferably localise in corneocytes or at the interface between cell and lipids. Diffusion of hydrophilic caffeine in the *stratum corneum* is quite difficult because the lipidic intercellular medium is continuous; even an intracellular path requires that molecules pass through the intercellular lipidic medium when diffusing through the whole *stratum corneum*. Transport assisted by (nano)particles has often been advanced as a possible mechanism of skin penetration [21, 36, 37]. Thus, either full water droplets diffuse in the intercellular lipids or the emulsion breaks at the surface of skin freeing silica particles to travel through the *stratum corneum* with the adsorbed caffeine. Silica particles can enter lipidic media because they are made hydrophobic at their surface by silane grafting. Skin penetration of intact Pickering emulsion droplets can be discarded as the emulsion droplet size (10 μm) is much larger than the reported cut-off size for skin penetration [21, 36, 37]. Penetration of silica particles from broken emulsion droplets is thus the most likely explanation. Diffusion of caffeine requires that silica particles (i) penetrate the *stratum corneum* and (ii) caffeine adsorbs at the surface of hydrophobic silica. Both aspects were verified by the following experiments:

(i) The penetration of silica particles into the *stratum corneum* was monitored by SEM observations of thin layers acquired from tape stripping. Silica aggregates of ~200 nm diameter were visible on the outer surface of skin (top of Figure 8). The same typical silica aggregates were observed at the surface of corneocytes of the top strips. Silica particles were present until the 10th strip on which both bare and silica particle-covered corneocytes were observed. Silica particles were not detected in the deepest stripes of the *stratum corneum* (Figure 8). Thus, the penetration depth was roughly half the thickness of the *stratum corneum*, i.e. ~5 μm. Skin penetration of nanoparticles is currently well-documented and our observations are in accordance with several studies showing the limited penetration of inorganic particles [38 – 41]. The hydrophobic surface of silica particles helps penetration of the *stratum corneum* lipid medium. Therefore, the silica particles may act as vehicles for adsorbed caffeine; this appears as a likely mechanism that would accelerate caffeine delivery inside the deeper layers of the skin.

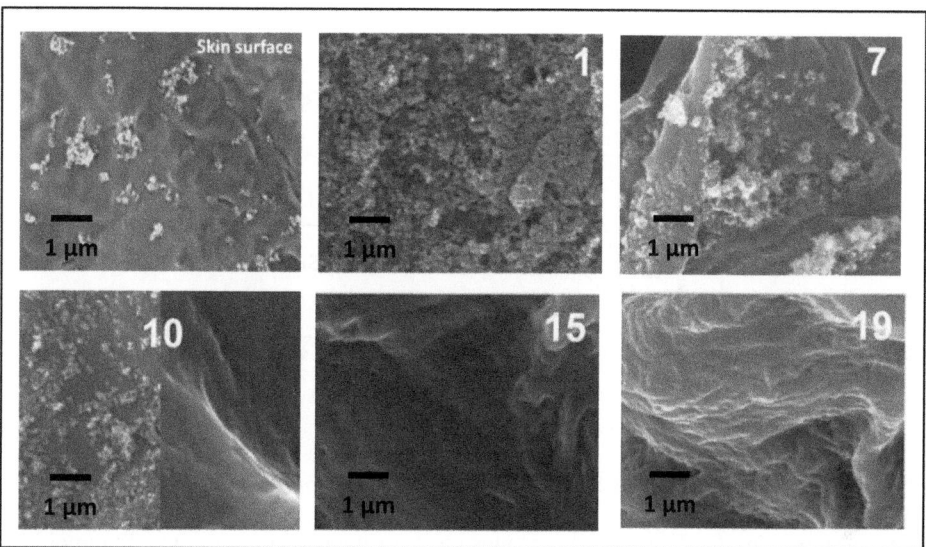

Figure 8 Representative *SEM pictures of stratum corneum layers after 24 h exposure to Pickering emulsion acquired from the skin surface and subsequent (1st, 7th, 10th, 15th and 19th) tape strips. Note that the 10th strip is a composite of two images and represents the boundary between recovery and non-recovery of the silica particles.*

(ii)   Caffeine transport driven by penetrating silica particles implies adsorption of caffeine on silica. Many polar molecules adsorb at the surface of silica particles [42]; water is one of these. Since water competes with caffeine for adsorption on silica, the adsorption of caffeine from aqueous solution was investigated. The role of hydrophobic dimethylsilyl grafts that may hinder adsorption was also investigated. Adsorption isotherms of caffeine were measured by the depletion method: Briefly; caffeine solution was equilibrated with silica particles and the supernatant collected by centrifugation and analysed for caffeine by UV absorption spectroscopy at 273 nm. The surface excess of caffeine was calculated from the mass balance of caffeine. Adsorption isotherms were measured for HDK N20 (Wacker) bare silica (100 % Si-OH) and dimethylsilyl-grafted silica HDK HKS D having 71 % residual Si-OH groups (Figure 9).

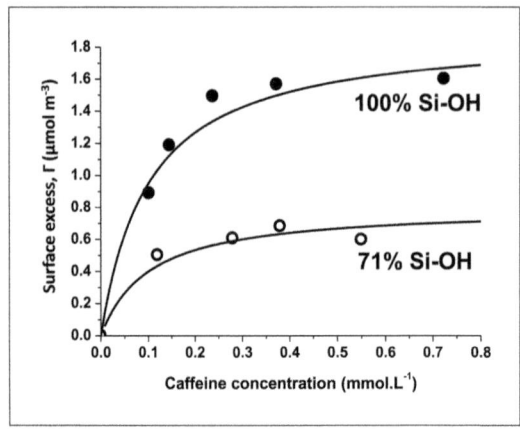

Figure 9 *Adsorption isotherms of caffeine from aqueous solution to silica surface.*

The concentration of caffeine in the aqueous phase of the w/o emulsion was 82 mmol L^{-1}: this corresponded to full coverage of the silica particles (as indicated by the plateau of the adsorption isotherm). Accordingly, silica was theoretically able of transporting ~ 20 mg caffeine per gram of silica (corresponding to 2.5 % of the full caffeine content). Skin absorption experiments have shown that 16 % of the full caffeine content actually permeated through skin after 24 h exposure to Pickering emulsion [12].

4 CONCLUSIONS

Both the Pickering and the classical o/w emulsions enhanced the skin absorption of two model drugs when compared to an oil solution. The surfactant-based o/w emulsion allowed retinol to pass easily through the *stratum corneum*. In contrast, the Pickering emulsion favoured reservoir formation of retinol inside the *stratum corneum*. Such behavior indicates many possible applications for cosmetic formulations or delivery of pharmaceutically active compounds. In the case of vitamins such as retinol, the target is obviously the viable skin layers [43, 44]. Similarly, hydrophobic ingredients such as organic sunscreens should not penetrate deep in the skin but reside in the upper layers of the *stratum corneum* to improve resistance to washing and/or abrasion. Such applications would seem to appropriate for Pickering emulsions.

Skin absorption of a hydrophilic model penetrant (caffeine) from a w/o Pickering emulsion significantly differed of that from emulsifier-stabilised emulsion because of the different surface properties of the water droplets within each formulation. The results of the physical chemistry experiments indicated two transport mechanisms. First, transfer of hydrophilic mater from water droplets to *stratum corneum* was promoted by stronger adhesion of Pickering emulsion droplets to the skin surface. Secondly, caffeine transport may be enhanced to some degree when adsorbed from the surface of the silica particles used as emulsion stabilisers (which penetrated half the thickness of the *stratum corneum*).

In addition, the present study showed that the sustained release from w/o Pickering emulsions occurred with limited penetration of the silica nanoparticles vehicle. The lack of permeation of the vehicle into the viable epidermis tends to suggest that such formulations may be relatively safe for topical applications as cosmetics and transdermal delivery.

5 REFERENCES

1. S.U. Pickering, Emulsions, *J. Chem. Soc.*, 1907, **91**, 2001–2021.

2. B.P. Binks, Particles as surfactants – similarities and differences, *Curr. Opin. Colloid Interface Sci.*, 2002, **7**, 21–41.

3. R. Aveyard, B.P. Binks and J.H. Clint, Emulsions stabilised solely by colloidal particles, *Adv. Colloid Interface Sci.*, 2003, **100-102**, 503–546.

4. B.P. Binks and T.S. Horozov, *Colloidal particles at liquid interfaces*. Cambridge University Press, Cambridge, 2006.

5. B.P. Binks and S.O. Lumsdon, Stability of oil-in-water emulsions stabilised by silica particles, *Phys. Chem. Chem. Phys.*, 1999, **1**, 3007–3016.

6. P.M. Kruglyakov and A.V. Nushtayeva, Phase inversion in emulsions stabilised by solid particles, *Adv. Colloid Interface Sci.*, 2004, **108-109**, 151–158.

7. J. Frelichowska, M.-A. Bolzinger and Y. Chevalier, Pickering emulsions with bare silica, *Colloids Surfaces A*, 2009, **343**, 70–74.

8. S. Arditty, V. Schmitt, F. Lequeux and F. Leal-Calderon, Interfacial properties in solid-stabilized emulsions, *Eur. Phys. J. E*, 2005, **44**, 381–393.

9. J. Frelichowska, M.-A. Bolzinger and Y. Chevalier, Effects of solid particle content on properties of o/w Pickering emulsions, *J. Colloid Interface Sci.*, 2010, **351**, 348–356.

10. C.A. Prestidge and S. Simovic, Nanoparticle encapsulation of emulsion droplets, *Int. J. Pharm.*, 2006, **324**, 92–100.

11. S. Simovic and C. A. Prestidge, Nanoparticle layers controlling drug release from emulsions, *Eur. J. Pharm. Biopharm.*, 2007, **67**, 39–47.

12. J. Frelichowska, M.-A. Bolzinger, J.-P. Valour, H. Mouaziz, J. Pelletier and Y. Chevalier, Pickering w/o emulsions: Drug release and topical delivery, *Int. J. Pharm.*, 2009, **368**, 7–15.

13. J. Frelichowska, M.-A. Bolzinger, J. Pelletier, J.-P. Valour and Y. Chevalier, Topical delivery of lipophilic drugs from o/w Pickering emulsions, *Int. J. Pharm.*, 2009, **371**, 56–63.

14. Wacker Chemie AG, *HDK H20 technical data sheet*, 2006.

15. H. Barthel, Surface interactions of dimethylsiloxy group-modified fumed silica, *Colloids Surfaces A*, 1995, **101**, 217–226.

16. P. Clarys, B. Gabard, R. Lambrecht, A. Barel, E. Bieli and S. Lüdi, There is no influence of a temperature rise on in vivo adsorption of UV filters into the stratum corneum, *J. Dermatol. Sci.*, 2001, **27**, 77–81.

17. H. J. Weigmann, J. Lademann, S. Schanzer, U. Lindemann, R. von Pelchrzim, H. Schaefer, W. Sterry and V. Shah, Correlation of the local distribution of topically applied substances inside the stratum corneum determined by tape-stripping to differences in bioavailability, *Skin Pharm. Physiol.*, 2001, **14**, 98–102.

18. M. M. Jiménez, J. Pelletier, M.-F. Bobin and M.-C. Martini, Influence of encapsulation on the in vitro percutaneous absorption of octyl methoxycinnamate, *Int. J. Pharm.*, 2004, **272**, 45–55.

19. P. Borm, D. Robbins, S. Haubold, T. Kuhlbusch, H. Fissan, K. Donaldson, R. Schins, V. Stone, W. Kreyling, J. Lademann, J. Krutmann, D. Warheit and E. Oberdorster, The potential risks of nanomaterials: a review carried out for ECETOC, *Particle Fibre Toxicology*, 2006, **3**, 11.

20. H. Schaefer, F. Watts, J. Brod and B. Illel, Follicular penetration, in *Prediction of percutaneous penetration: Methods, measurement, modelling*, ed. R.C. Scott, R.H. Guy and J. Hadgraft, IBC Technical Services, London, 1990, pp. 163–173.

21. A. Rolland, N. Wagner, A. Chatelus, B. Shroot and H. Schaefer, Site-specific drug delivery to pilosebaceous structures using polymeric microspheres, *Pharm. Res.*, 1993, **10**, 1738–1744.

22. M.-A. Bolzinger, S. Briançon, J. Pelletier and Y. Chevalier, Penetration of drugs through skin, a complex rate-controlling membrane, *Curr. Opin. Colloid Interface Sci.* 2012, **17**, 156–165.

23. P. Clément, C. Laugel and J.-P. Marty, In vitro release of caffeine from concentrated W/O emulsions: effect of formulation parameters, *Int. J. Pharm.*, 2000, **207**, 7–20.

24. K.A. Walters, Dermatological and transdermal formulations, in *Formulation strategies for modulating skin permeation*, Marcel Dekker, New York, 2002, Chapter 6.

25. M.-A. Bolzinger, S. Briançon and Y. Chevalier, Nanoparticles through the skin: managing conflicting results of inorganic and organic particles in cosmetics and pharmaceutics *WIREs Nanomed. Nanobiotechnol.* 2011, **3**, 463–478.

26. M.-A. Bolzinger, S. Briançon, J. Pelletier, H. Fessi and Y. Chevalier, Percutaneous release of caffeine from microemulsion, emulsion and gel dosage forms, *Eur. J. Pharm. Biopharm.*, 2008, **68**, 446–451.

27. G. Calderó, M.J. García-Celma, C. Solans, M. Plaza and R. Pons, Influence of composition variables on the molecular diffusion from highly concentrated water-in-oil emulsions (gel-emulsions), *Langmuir*, 1997, 13, 385–390.

28. G. Calderó, M.J. García-Celma, C. Solans, M.-J. Stébé, J.-C. Ravey, S. Rocca and R. Pons, Diffusion from hydrogenated and fluorinated gel-emulsion mixtures, *Langmuir*, 1998, 14, 1580–1585.

29. S. Rocca, S. Muller and M.-J. Stébé, Release of a model molecule from highly concentrated fluorinated reverse emulsions Influence of composition variables and temperature, *J. Control. Release*, 1999, **61**, 251–265.

30. J.G. Fokkens and C.J. de Blaey, Drug release from non-aqueous suspensions. II. The release of methylxanthines from paraffin suspensions, *Int. J. Pharm.*, 1984, **18**, 127–138.

31. S.A. Wissing and R.H. Muller, Solid lipid nanoparticles as carrier for sunscreens: in vitro release and in vivo skin penetration, *J. Control. Release*, 2002, **81**, 225–233.

32. A.W. Adamson, *Physical chemistry of surfaces*, 5th Edn, Wiley, New York, 1990, Chap X, 379–414.

33. T. Okubo, Surface tension of structured colloidal suspensions of polystyrene and silica spheres at the air-water interface, *J. Colloid Interface Sci.*, 1995, **171**, 55–62.

34. A. Mavon, H. Zahouani, D. Redoules, P. Agache, Y. Gall and Ph. Humbert, Sebum and stratum corneum lipids increase human skin surface free energy as determined from contact angle measurements: A study on two anatomical sites, *Colloids Surfaces B: Biointerfaces*, 1997, **8**, 147–155.

35. A. Mavon, D. Redoules, P. Humbert, P. Agache and Y. Gall, Changes in sebum levels and skin surface free energy components following skin surface washing, *Colloids Surfaces B: Biointerfaces*, 1998, **10**, 243–250.

36. R. Alvarez-Román, A. Naik, Y.N. Kalia, R.H. Guy and H. Fessi, Skin penetration and distribution of polymeric nanoparticles, *J. Control. Release*, 2004, **99**, 53–62.

37. R. Alvarez-Román, A. Naik, Y.N. Kalia, H. Fessi and R.H. Guy, Visualization of skin penetration using confocal laser scanning microscopy, *Eur. J. Pharm. Sci.*, 2004, **58**, 301–316.

38. J. Lademann, H. Weigmann, C. Rickmeyer, H. Barthelmes, H. Schaefer, G. Mueller and W. Sterry, Penetration of titanium dioxide microparticles in a sunscreen formulation into the horny layer and the follicular orifice, *Skin Pharmacol. Appl. Skin Physiol.*, 1999, **12**, 247–256.

39. F. Pflücker, V. Wendel, H. Hohenberg, E. Gärtner, T. Will, S. Pfeiffer, R. Wepf and H. Gers-Barlag, The human stratum corneum layer: An effective barrier against dermal uptake of different forms of topically applied micronised titanium dioxide, *Skin Pharmacol. Appl. Skin Physiol.*, 2001, **14(suppl 1)**, 92–97.

40. A. Mavon, C. Miquel, O. Lejeune, B. Payre and P. Moretto, In vitro percutaneous absorption and in vivo stratum corneum distribution of an organic and a mineral sunscreen, *Skin Pharmacol. Physiol.*, 2007, **20**, 10–20.

41. G. J. Nohynek, E. K. Dufour and M. S. Roberts, Nanotechnology, cosmetics and the skin: Is there a health risk? *Skin Pharmacol. Physiol.*, 2008, **21**, 136–149.

42. M. Korn, E. Killmann and J. Eisenlauer, Infrared and microcalorimetric studies of the adsorption of monofunctional ketones, esters, ethers, and alcohols at the silica/carbon tetrachloride interface, *J. Colloid Interface Sci.*, 1980, **76**, 7–18.

43. A. Vahlquist, J. B. Lee, G. Michaëlsson and O. Rollman, Vitamin A in human skin: II Concentrations of carotene, retinol and dehydroretinol in various components of normal skin, *J. Invest. Dermatol.*, 1982, **79**, 94–97.

44. G. J. Fisher and J. J. Voorhees, Molecular mechanisms of retinoid actions in skin, *FASEB J.*, 1996, **10**, 1002–1013.

12

PICKERING EMULSIONS STABILISED BY BIODEGRADABLE PARTICLES OFFER A DOUBLE LEVEL OF CONTROLLED DELIVERY OF HYDROPHOBIC DRUGS

F Laredj-Bourezg[1], M-A Bolzinger[1], J Pelletier[1], M-R Rovere[2], B Smatti[3], Y Chevalier[1]

[1]Laboratoire d'Automatique et de Génie des Procédés, F-69622, Villeurbanne, France, [2]Laboratoire des Substituts Cutanés, and [3]Centre commun de Quantimétrie, Université de Lyon, Villeurbanne, France, F-69622, Villeurbanne, France.

1 INTRODUCTION

Pickering emulsions are stabilised by solid particles in place of surfactants [1]. Their "surfactant-free" character makes them attractive for cosmetic and pharmaceutical applications because classical emulsions comprising surfactants often show adverse effects such as irritancy, hemolytic behavior, etc. [2]. Thus, biocompatible and biodegradable Pickering emulsions would provide an obvious benefit over current technology.

Pickering emulsions can be formulated using oils used in pharmaceutical applications and organic solid particles made from biodegradable materials. Since solid stabilising particles are necessarily smaller than emulsion droplets, solid nano-sized particles need to be selected in order to allow the fabrication of Pickering emulsions over a wide droplet size range. The purpose of the present research is the preparation of o/w Pickering emulsions stabilised by block copolymer particles. Solid particles can spontaneously adsorb at fluid interfaces forming either a dense monolayer of particles or a thick layer of aggregated solid particles that behaves as arigid stabilising layer acting against coalescence [3, 4]. Many types of solid particles (hydrophilic silica, hydrophobic silica, clay, barium sulfate, calcium carbonate, polystyrene, spores, etc.) have been used to stabilise such emulsions [1, 5, 6].

Biodegradable particles should decrease the risk of toxicity observed with common chemical surfactants and inorganic particles and are expected to create a barrier to diffusion that allows controlled release of drug substances incorporated either in the oily layer or inside the polymeric particles. Diblock copolymers made of polyethyleneglycol (PEG) and polylactic acid (PLA) have raised much interest because they are biocompatible and partly biodegradable [7 – 10]. The PLA block is made of biodegradable polyester, and the PEG block is a water-soluble polymer of low molar mass that is bioresorbable. Such block copolymers are ideal candidates for application of emulsions to pharmaceutical or cosmetic domains. Such particles are often called "block copolymer micelles", although they act more as particles than classical micelles. Indeed, classical micelles composed of water-soluble surfactants form spontaneously and are at equilibrium with a residual soluble fraction in water. In contrast, block copolymer micelles do not form spontaneously and the residual concentration of block copolymer in solution is extremely low.

The work reported in this chapter concerns the preparation of o/w Pickering emulsions stabilised with block copolymer particles by means of two different processes. The resulting emulsions were tested for skin penetration of a hydrophobic drug (*all-trans* retinol) and were compared to a surfactant based emulsion a block copolymer nanoparticle suspension. Confocal fluorescence microscopy was used to visualise the skin distribution of the fluorescent probe Nile Red in skin layers after exposure to either Pickering emulsions or block copolymer particles.

2 MATERIALS AND METHODS

D,L-Lactide (3,6-dimethyl-1,4-dioxane-2,5-dione), anhydrous toluene, poly(ethylene glycol) monomethyl ether (mPEG) of molar mass $5,000 \, g \, mol^{-1}$, stannous 2-ethylhexanoate (Sn(Oct)$_2$, *all-trans* retinol, Nile Red and phosphate buffer pH 7.4, were purchased from Sigma-Aldrich. Acetone was from Laurylab; dichloromethane was from Acros Organics; HPLC grade methanol was from Carlo Erba. Polysorbate 80 emulsifier (Montanox® 80) was purchased from Seppic (France), butyl hydroxyl toluene (BHT) from Clariant (France) and caprylic/capric triglyceride (medium chain triglyceride, MCT, Miglyol 812N, Sasol) were kind gifts from their suppliers.

Full-thickness pig skin (1.2 ± 0.05 mm) was used in skin absorption experiments. The skins of 3 donor animals were washed and excised, the subcutaneous fatty tissue was carefully removed with a scalpel and the skin pieces were stored flat at -20 °C until use. Deionised water of 18 MΩ.cm resistivity was used throughout the work.

Preparation of PLA-b-PEG Diblock Copolymer

D,L-Lactide was recrystallised in dry ethyl acetate. mPEG was dried by azeotropic distillation of its solution in anhydrous toluene under dry nitrogen atmosphere. The PLA-*b*-PEG diblock copolymer was synthesised by ring-opening polymerisation of D,L-lactide with mPEG as a macroinitiator and stannous 2-ethylhexanoate as a catalyst [8, 10 – 13]. This block copolymer was not soluble in water [8].

Preparation of the PLA-b-PEG Particles.

Particles were prepared using the spontaneous emulsification process [14]. This method was modified in order to increase the concentration of solid particles beyond the usual limits of the nanoprecipitation process. Thus the relative amounts of acetone and water were inverted, allowing higher concentrations of the final suspensions [16]. PLA-*b*-PEG was first dissolved in acetone and then introduced drop wise in water with gentle stirring. The acetone was then evaporated under reduced pressure. The particles had a mean diameter of 40-50 nm, as measured by dynamic light scattering (NanoZS® instrument, Malvern, UK).

Formulation Preparation and Characterisation

All formulations contained 0.5% retinol (Table 1). Three o/w emulsions were prepared: two Pickering Emulsions (PE) and a classical emulsion (CE). These were compared to an oily solution of retinol (SOL).

For the CE (o/w) formulation, Polysorbate 80 (HLB = 15.0) was used as the emulsifying agent. The amount of emulsifier was adjusted in order to obtain the same droplet size distribution as for the Pickering emulsion. The viscosity of the classical emulsion was similar to that of the Pickering emulsion to avoid the influence of viscosity on the skin absorption profile. The oil/aqueous phases ratio for the classical and Pickering emulsions was 50:50 (wt:wt).

Two Pickering emulsions were prepared. The first (PE1) involved the mechanical stirring of an aqueous suspension of particles and medium chain triglyceride. The second (PE2) involved spontaneous emulsification by dissolving PLA-*b*-PEG block copolymer and MCT in acetone before mixing with water. The acetone was then evaporated under reduced pressure. All the tested formulations (loaded with retinol) were prepared in the dark and were stored under nitrogen in order to preserve retinol integrity.

Table 1 *Composition of the six retinol formulations investigated (all values expressed as w/v%). Pickering emulsions (PE1 and PE2), classical emulsion (CE), PLA-b-PEG block copolymer particles (BCPP), Polysorbate 80 (PS80) solution and oil solution (SOL).*

Composition	PE1	PE2	CE	BCPP	PS80	SOL
all-trans Retinol	0.5	0.5	0.5	0.5	0.5	0.5
Medium Chain Triglycerides (MCT)	49.0	47.825	49.0			99.5
Butyl Hydroxy Toluene	0.5	0.5	0.5	0.5	0.5	
Block copolymer particles solution (7% w/v)	50.0			99.0		
PLA-*b*-PEG block copolymer		3.35				
Polysorbate 80			2.5		6	
Water		47.825	47.5		93	

Formulations containing Nile Red were prepared in the same way using 0.05% of Nile Red. Incorporation of retinol or Nile Red into the Pickering emulsions was performed by dissolving the substances in oil (methods 1 and 2) or acetone (block copolymer). An oily solution of retinol or Nile Red (SOL) was prepared by simply dissolving retinol or Nile Red in medium chain triglycerides.

^{1}H NMR spectra of PLA-*b*-PEG diblock copolymer were measured using a Bruker DRX 300 spectrometer operating at 300 MHz. Either deuterated chloroform (CDCl$_3$) or deuterated water (D$_2$O) were used as solvents. Chemical shifts were measured in ppm from tetramethylsilane. Emulsion droplet size distributions were measured by small angle light scattering using a MasterSizer® 2000 (Malvern, UK). The average size and polydispersity index of PLA-*b*-PEG particles dispersed in water were measured by means of dynamic light scattering using a NanoZS® instrument (Malvern, UK). The viscosity of emulsions was measured at 20°C using a Couette rheometer Rheomat R180 (Lamy, France) equipped with a mobile system N° 11 rotating at a shear rate of 200 s^{-1}. Emulsions of MCT-stabilised micelles were imaged at the Centre Technologique des Microstructures (CTμ) technical platform of the University of Lyon [15] using a transmission electron microscope (Philips CM120) working at 80 kV acceleration. The particle suspensions and emulsions were placed on formvar/carbon-coated copper TEM grids and dried under atmospheric conditions for TEM analysis.

Skin Distribution of Formulations

Confocal laser scanning microscopy (CLSM: Leica TCS-SP2 microscope) was used to observe the effects of different formulations on the distribution of the fluorescent dye Nile Red within (pig) skin samples. The skin samples were mounted in Franz diffusion cells and exposed to Nile Red-loaded formulations (Pickering emulsions, classical emulsion and particles) for 24 h, after which the skin samples were carefully removed and rinsed prior to embedding in optimum cutting temperature compound. The embedded samples there then frozen (-20°C) and transverse sections (5 μm thick) produced using a cryostat. The sections were hydrated in phosphate buffer saline, mounted and cover-slipped. The transverse sections were visualised using an excitation (laser diode) wavelength of 532 nm. The resulting emissions were recorded between 589-708 nm using a Leica 20X/0.70 HLX PL APO objective. Images were constructed in 3D using 20-25 *xy* confocal slices with a *z*-separation of 1 μm. Both classical reflection and fluorescence microscopy images of skin slices were recorded. The images from the two techniques were superimposed to obtain information on the distribution of Nile Red within the different skin layers.

In Vitro Skin Absorption Studies

Full-thickness pig skin (1.2 ± 0.05 mm) was obtained from young animals euthanised at the Laboratoire de Physiologie, University of Lyon, France. The skin was cleaned with tap water and close clipped to remove excess hair. The thickness of each skin piece was measured with a micrometer gauge (Mitutoyo). Skin integrity was confirmed by measurements of transepidermal water loss (TEWL) using a Tewameter TM300 (Courage and Khazaka). Skin samples with an average (triplicate) TEWL rate higher than $15 \text{ g m}^{-2} \text{ h}^{-1}$ were discarded (as per OECD, 2004). The skin was mounted in two-chamber glass diffusion cells with an effective penetration area was 2.54 cm^2 and a receptor chamber volume of 10 mL. The receiver solution was composed of buffer at pH 7.4 with 1.5% Brij$^{®}$98 and 0.5% BHT (butylhydroxytoluene; BHT). Freshly prepared test formulation (1 g) was spread uniformly on the skin surface. The study was carried out under occlusive conditions for 24 h in static Franz cells. All experiments were performed in the dark to avoid retinol degradation. At the end of the study the receptor fluid was removed and analysed by HPLC. Franz cells were dismantled; the skin surface was washed with 10 mL of receptor fluid; then skin stripping of the *stratum corneum* was performed in one step using acrylic adhesive (Loctite$^{®}$, super glue 3, Henkel, Germany). The viable epidermis and dermis were separated by heat treatment in water at 60°C for 45 s. The resulting sheets of epidermis and dermis were then cut into pieces with a scalpel, subject to solvent extraction (20 min sonication in cold methanol containing 0.5%w/v BHT) and filtered.

The collected samples were analysed for retinol using reverse-phase HPLC (Waters, St Quentin en Yvelines, France). The chromatography system comprised a Waters 717 injector, a Waters 600 pump, a reverse phase column XTerra$^{®}$MS C18 (3.9mm×150mm, 5μm) and a Waters 2996 photodiode array UV detector (325 nm). The mobile phase (85% methanol, 15% water) was maintained at a temperature of 30°C and a flow rate of 1.4 mL min^{-1}. The retention time of retinol was 6 min. Injection volume was 20 μL. The calibration curve was linear up to 40 μg mL^{-1} with a detection limit of 20 pg. The mean and standard error of the mean (S.E.) of $n = 6$ determinations were calculated. Statistical comparisons were made using the Student's *t*-test (two-sample assuming equal variances) and analysis of variance (ANOVA, single factor) with the level of significance at $p \leq 0.05$.

3 RESULTS AND DISCUSSION

NMR Characterisation of Particles

Samples of PLA-*b*-PEG diblock copolymer preparations were analysed by ^{1}H NMR to assess their physical state. The spectrum of the copolymer in CDCl$_3$ indicated whole polymer lines pertaining to the PLA and PEG blocks. The spectrum in CDCl$_3$ (Figure 1) was in agreement with the chemical structure of the PLA-*b*-PEG block copolymer solution [16]. In contrast, ^{1}H NMR spectrum of PLA-*b*-PEG micelles dispersed in D$_2$O showed sharp peaks for the PEG line at 3.6 ppm and the terminal methyl located at its chain end at 3.0 ppm (Figure 1). The protons of the PLA block could not be observed in the spectrum. The PLA lines were so broad that they were almost absent, indicating restricted mobility typical of the solid state of the corresponding PLA core of the micelles. Thus, the ^{1}H NMR spectra demonstrated that PLA-*b*-PEG micelles were solid particles containing a central hydrophobic core surrounded by a hydrophilic corona of PEG blocks swollen by water.

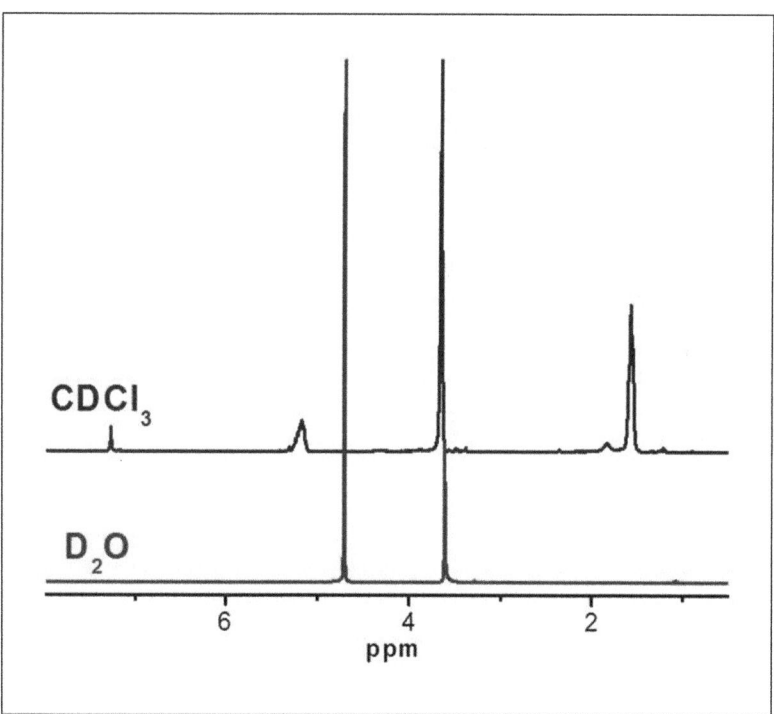

Figure 1 *^{1}H NMR spectra of PLA-b-PEG in D$_2$O and CDCl$_3$ solutions at room temperature indicating the different states of the PLA-b-PEG molecules in different media.*

Formulation Characterisation

The viscosity and droplet size of the three emulsions was the same, regardless of the preparation process (Table 2).

Pickering emulsions were observed using Transmission Electron Microscopy (TEM). Samples were dried prior to their observation by TEM, so that broken emulsions were observed. Images of the dried emulsions (PE1) showed clear areas corresponding to the oil droplets, which were surrounded by particles accumulated at the periphery of the broken oil droplets (Figure 2).

Skin Distribution of Retinol

Studies were performed using porcine skin which has been shown to be the most relevant animal model for human skin [18, 19]. Retinol was not detected in the receptor fluid which is in agreement with previous studies [21, 32]. Indeed, retinol has a high log*P* value (5.65) and consequently this lipophilic molecule accumulates preferentially in the *stratum corneum* from the tested formulations as illustrated in Figure 3. Therefore, only the retinol distribution within skin layers will be compared.

As indicated in Figures 3 and 4 and Table 3, retinol mainly remained inside the *stratum corneum* after 24 h exposure with a higher accumulation for PE2 (82%) and

Table 2 *Physicochemical characterisation (viscosity and mean droplet size) of Pickering emulsions (PE1 and PE2) and the corresponding classical emulsion (CE). All values expressed as mean ± S.E. of n = 3 samples.*

Parameter	PE1	PE2	CE
Viscosity at 20°C (mPa s)	14.5	14.7	14.5
Mean droplet size (μm)	2.7 ± 1.5	2.0 ± 0.4	2.7 ± 1.6

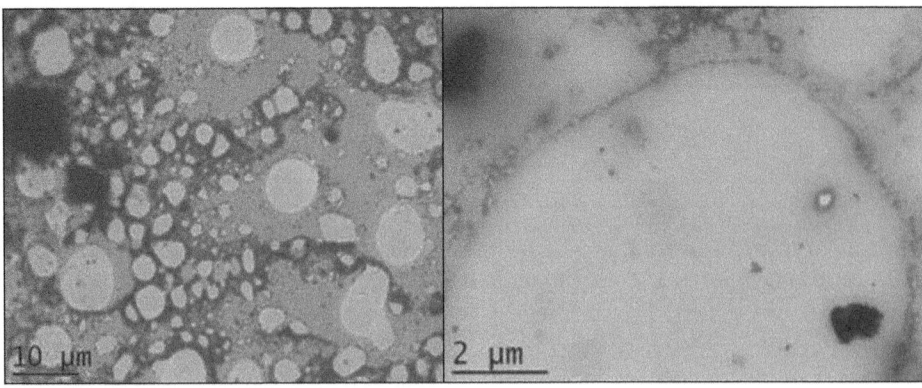

Figure 2 *Transmission electron microscopy (TEM) images of dried Pickering emulsion 1 (PE1) stabilised using block copolymer particles at two magnifications.*

PLA-*b*-PEG particles (87%). The classical emulsion and Polysorbate 80 preparation accelerated the transport through the *stratum corneum* and larger amounts could therefore reach the epidermis and dermis (CE: 31%; Polysorbate 80 solution: 34%).

The fraction of dose recovered from the *stratum corneum* was higher for the Pickering emulsions and the PLA-*b*-PEG particles when compared to CE and SOL and the total amount was much higher for the Pickering emulsions with a higher absorption observed for PE2 compared to PE1 (2-fold). Conversely, PLA-*b*-PEG particles allowed a relatively high accumulation of retinol in *stratum corneum* compared to the Polysorbate 80 preparation (19-fold), CE (90-fold).

Overall, there was no significant difference ($p < 0.05$) in retinol distribution between the surfactant based emulsion, the oily solution and the Pickering emulsion 1 (PE1).

Table 3 *Distribution of retinol recovered from different skin compartments after 24 h exposure to Pickering emulsions (PE1 and PE2), classical emulsion (CE), PLA-b-PEG block copolymer particles (BCPP), Polysorbate 80 (PS80) solution and oil solution (SOL). Values represent mean amount per surface area of skin.*

Skin Compartment	Amount of Retinol Recovered (ng cm^{-2})					
	PE1	PE2	CE	BCPP	PS80	SOL
Stratum corneum	590	1124	355	32180	1655	394
Epidermis	194	177	144	2819	758	104
Dermis	151	74	225	2129	1219	204
Total	935	1375	724	37128	3632	702

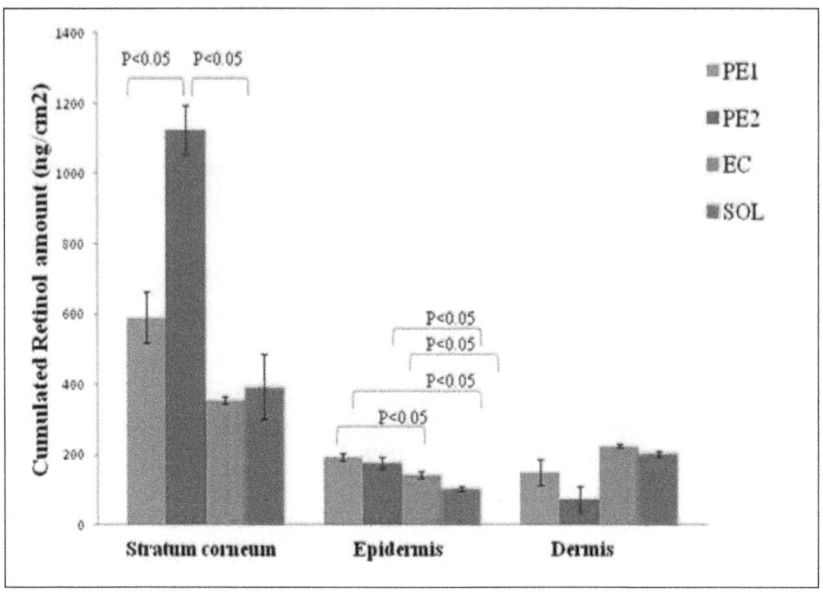

Figure 3 *Effects of formulation on skin distribution of retinol in excised pig skin after 24 h exposure to Pickering emulsions (PE1 and PE2), classical emulsion (CE) or retinol solution (SOL). All values are mean ± S.E. of n = 6.*

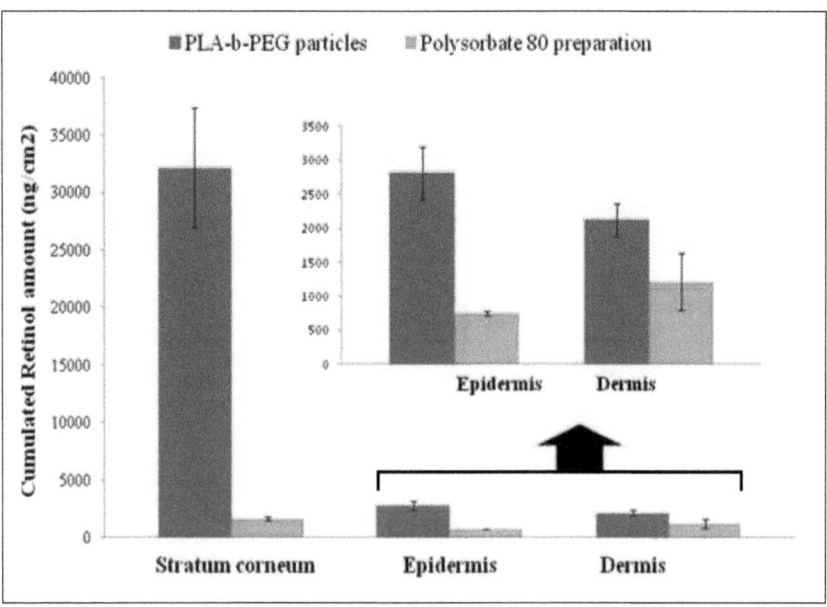

Figure 4 *Effects of formulation type on skin distribution of retinol in the excised pig skin after 24 h exposure to PLA-b-PEG particles or Polysorbate 80 solution. Mean ± S.E. of n = 6.*

Distribution of Nile Red in Skin Layers

Confocal fluorescence microscopy has recently become established as a technique to observe the drug distribution within the skin [23 – 26]. The relative fluorescence of Nile Red loaded emulsions was used to provide a semi-quantitative analysis of the distribution of the formulations within the epidermis, dermis and hair follicles. Unexposed pig skin was also evaluated using this technique: no background fluorescence was detected (data not shown).

After exposure of pig skin to the classical and Pickering emulsion 1, the dye-associated fluorescence was mostly located in the *stratum corneum* and epidermis (Figures 5A and 5B). The fluorescence extended into the dermis following application of PE2 (Figure 5C) and PLA-*b*-PEG particles (Figure 5D). Stronger fluorescence was observed in the dermis after exposure to PLA-*b*-PEG particles.

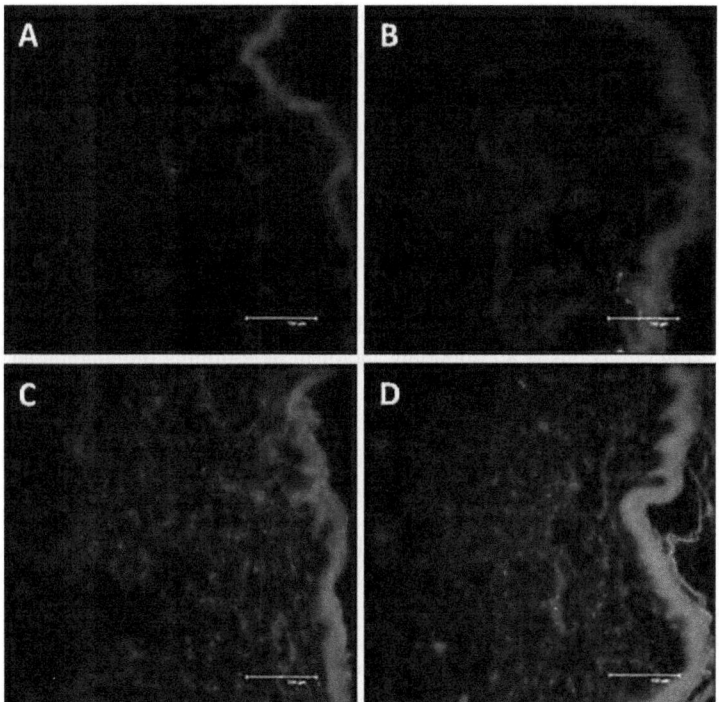

Figure 5 *Confocal fluorescence images revealing the penetration and distribution of Nile Red within pig skin after being exposed for 24 h to (A) Surfactant based emulsion, (B) Pickering emulsion 1, (C) Pickering emulsion 2 and (D) PLA-b-PEG particles. Each bar indicates 150 μm.*

4 DISCUSSION

It was important to select a model drug which would be fully encapsulated inside the droplets and particles of the experimental formulations. Retinol was chosen as a model lipophilic active ingredient as its highly lipophilic character ($\log P = 5.68$) guarantees the presence of retinol exclusively in the oil (dispersed) phase of the emulsion. It should be noted that the purpose of the study was to examine the relative effects of the different formulation rather than to develop a formulation that would provide a definite beneficial effect regarding the action of retinol [21, 32, 33].

The salient features of this study can be summarised as follows:

- Partitioning of retinol into the *stratum corneum* from Pickering emulsions was greater than from solution or classical emulsion. This effect has previously been observed using Pickering emulsions stabilised using silica [21].

- The PLA-*b*-PEG particles allowed a greater accumulation of retinol in the *stratum corneum*.

- Polysorbate 80 preparation showed higher accumulation of retinol in skin layers compared to Pickering emulsions. Previous studies have shown that perturbations in skin structure can occurred when non-ionic surfactants are applied to the skin [28] and others have demonstrated that surfactant solutions can significantly increase retinol absorption into skin when compared to emulsions [32].

- Pickering emulsion 2 showed higher accumulation of retinol in the *stratum corneum* compared to Pickering emulsion 1. Thus, the method used to prepare Pickering emulsions is an important factor in determining the amount and location of retinol in different skin layers.

Hydrophobic drugs are expected to be predominantly confined to the oil phase of Pickering emulsions, although it is likely that some proportion of a drug will become associated with the hydrophilic core of the stabilising particles and also residual particles that could coexist with the oil phase. Thus, it cannot be excluded that the block copolymers may also act as a vehicle for the transport of retinol deeper in the skin [28]. This would lead to a bi-modal delivery mechanism, with drug release from both the oil droplets and from PLA-*b*-PEG particles present in the emulsion.

The physicochemical characteristics of the formulation particles such as size, surface charge and hydrophobicity are obviously expected to influence their deposition and/or penetration into biological tissue. Although, the block copolymer particles were uncharged and had a hydrophilic corona made with PEG, they promoted a high accumulation of retinol in the *stratum corneum* compared to the Polysorbate 80 formulation. Previous studies using guinea pig skin have demonstrated that the shunt pathway (via hair follicles) is the predominant route of entry for Minoxidil-loaded particles of poly(caprolactone)-*block*-poly(ethylene glycol) with 2% of residual ethanol [22]. However, others have shown that the *stratum corneum* is the predominant route of entry for PLGA-microparticles loaded with rhodamine [34] and poly(caprolactone) nanoparticles loaded with the sunscreen octyl methoxycinnamate. Interestingly, nanoparticles achieved 3.4-fold increase in the concentration of octyl methoxycinnamate within the *stratum corneum* [29, 30]. A previous study in our laboratory using Pickering emulsions stabilised with hydrophobised fumed silica showed a similar behavior in retinol skin delivery [21] in that the nanometric silica particles were observed to penetrate up to half the thickness of the *stratum corneum* after 24 h exposure. The results of our present study indicate that the block copolymer micelles can effectively penetrate the skin through the *stratum corneum* and reach the epidermis in a high amount compared to the classical emulsion. This could be related to the high affinity of retinol for the medium chain triglyceride (MCT) present in the emulsion. Indeed, Teeranachaideekul *et al.* [23] observed that a Nile Red-loaded nanostructured lipid carrier (NLC) showed a higher penetration depth and intensity of dye for lower MCT content.

Confocal fluorescence microscopy clearly showed that Nile Red diffused more extensively through skin layers for block copolymer particles and Pickering emulsion 2 when compared to the classical emulsion. This confirmed the existence of residual particles containing the dye and permitting its skin penetration. Log$P$ is often used to predict the uptake into the *stratum corneum* [31]. It is worth noticing that Nile Red (Log$P$ = 3.44) is less hydrophobic than retinol (Log$P$ = 5.68), which indicates that the rate and extent of their delivery into the skin is disparate even when incorporated in the same formulations.

5 CONCLUSIONS

Pickering emulsions were designed using biodegradable particles as stabilisers. PLA-*b*-PEG micelles were found to be efficient stabilisers of high internal phase ratio emulsions which showed excellent stability on storage, even at high concentrations of medium chain triglycerides. Pickering emulsions and block copolymer particles proved to be very interesting carriers for transdermal delivery of hydrophobic drugs *in vitro*. Emulsions with stabilising block copolymer particles and/or residual particles acted as a vehicle for the transport of retinol deep into the skin. Dermal delivery of retinol was probably mediated by both the oil droplets and the PLA-*b*-PEG particles present in the emulsion. The penetration of retinol and Nile Red depended on the process by which the emulsions were prepared. Accordingly, judicious design of the preparation process may allow the development of future drug delivery systems which are able to control the delivery of drugs to specific layers of the skin while keeping the same ingredients in the formulation.

6 REFERENCES

1. R. Aveyard, B. P. Binks and J. H. Clint, Emulsions stabilized solely by solid colloidal particles. *Adv. Colloid Interface Sci.*, 2003, **100-102**, 503–546.

2. S. Simovic and C. A. Prestidge, Nanoparticles layers controlling drug release from emulsions, *Eur. J. Pharm. Biopharm.*, 2007, **67**, 39–47.

3. B. P. Binks and C. P. Whitby, Silica particle-stabilized emulsions of silicone oil and water: Aspects of emulsification, *Langmuir*, 2004, **20**, 1130–1137.

4. N. Yan and J. H. Masliyah, Adsorption and desorption of clay particles at the oil-water interface, *J. Colloid Interface Sci.*, 1994, **168**, 386–392.

5. B. P. Binks and S. O. Lumsdon, Influence of particle wettability on the type and stability of surfactant-free emulsions, *Langmuir*, 2000, **16**, 8622–8631.

6. B. P. Binks, J. H. Clint, G. Mackenzie, C. Simcock and C. P. Whitby, Naturally occurring spore particles at planar fluid interfaces and in emulsions, *Langmuir*, 2005, **21**, 8161–8167.

7. M. Chausson, A.-S. Fluchère, E. Landreau, Y. Aguni, Y. Chevalier, T. Hamaide, N. Abdul-Malak and I. Bonnet, Block copolymers of the type poly(caprolactone)-b-poly(ethylene oxide) for the preparation and stabilization of nanoemulsions, *Int. J. Pharm.*, 2008, **362**, 153–162.

8. A. S. Mikhail and C. Allen, Poly(ethylene glycol)-b-poly(epsilon-caprolactone) micelles containing chemically conjugated and physically entrapped docetaxel: synthesis, characterization, and the influence of the drug on micelle morphology, *Biomacromolecules*, 2010, **11**, 1273–1280.

9. G. S. Kwon and T. Okano, Polymeric micelles as new drug carriers, *Adv. Drug Deliv. Rev.*, 1996, **21**, 107–116.

10. C. Iojoiu, Th. Hamaide, V. Harabagiu and B.C. Simionescu, Modified poly(ε-caprolactone)s and their use for drug-encapsulating particles, *J. Polym. Sci. A: Polym. Chem.*, 2004, **42**, 689–700.

11. M. A. R. Meier, S. N. H. Aerts, B. B. P. Staal, M. Rasa and U. S. Schubert, PEO-b-PCL block copolymers: Synthesis, detailed characterization, and selected micellar drug encapsulation behavior, *Macromol. Rapid Commun.*, 2005, **26**, 1918–1924.

12. R. Li, X. Li, L. Xie, D. Ding, Y. Hu, X. Qian, L. Yu, Y. Ding, X. Jiang and B. Liu, Preparation and evaluation of PEG-PCL particles for local tetradrine delivery, *Int. J. Pharm.*, 2009, **379**, 158–166.

13. R. Diab, M. Hamoudeh, O. Boyron, A. Elaissari and H. Fessi, Microencapsulation of cytarabine using poly(ethylene glycol)-poly(caprolactone) diblock copolymers as surfactant agents, *Drug Dev. Ind. Pharm.*, 2010, **36**, 456–469.

14. H. Fessi, J.-P. Devissaguet, F. Puissieux and C. Thies, Procédé de dispersion de systèmes colloïdaux dispersibles d'une substance, sous forme de nanoparticules. Patents FR 2608988 (1986); EP 0275796 (1987); US 5118528 (1992).

15. http://microscopies.univ-lyon1.fr

16. F. Laredj-Bourezg, Y. Chevalier, O. Boyron and M.-A. Bolzinger, Emulsions stabilized with organic solid particles. *Colloids Surfaces A*, 2012, **413**, 252–259.

17. C. R. Heald, S. Stolnik, K. S. Kujawinski, C. De Matteis, M. C. Garnett, L. Illum, S. S. Davis, S.C . Purkiss, R. J. Barlow and P. R. Gellert, Poly(lactic acid)-poly(ethylene oxide) (PLA-PEG) particles: NMR studies of the central solidlike PLA core and the liquid PEG corona. *Langmuir*, 2002, **18**, 3669–3675.

18. R. L. Bronaugh, R. F. Stewart, E. R. Congdon and A. L. Giles Jr, Methods for *in vitro* percutaneous absorption studies, I. Comparison with *in vivo* results, *Toxicol. Appl. Pharmacol.*, 1982, **15**, 474–480.

19. G. Marti-Mestres, J. P. Mestres, J. Bres, S. Martin, J. Ramos and L. Vian. The "in vitro" percutaneous penetration of three antioxidant compounds, *Int. J. Pharm.*, 2007, **331**, 139–144.

20. S. El Hussein, P. Muret, M. Berard, S. Makki and P. Humbert, Assessment of principal parabens used in cosmetics after their passage through human epidermis–dermis layers (ex-vivo study), *Exp. Dermatol.*, 2007, **16**, 830–836.

21. J. Frelichowska, M.-A. Bolzinger, J. Pelletier, J.-P. Valour and Y. Chevalier, Topical delivery of lipophilic drugs from o/w Pickering emulsions, *Int. J. Pharm.*, 2009, **371**, 56–63.

22. J. Shim, H. S. Kang, W.-S. Park, S.-H. Han, J. Kim and I.-S. Chang, Transdermal delivery of mixnoxidil with block copolymer particles, *J. Control. Release*, 2004, **97**, 477–484.

23. V. Teeranachaideekul, P. Boonme, E. Barbosa Souto, R. H. Müller and V. B. Junyaprasert, Influence of oil content on physicochemical properties and skin distribution of Nile red-loaded NLC, *J. Control. Release*, 2008, **128**, 134–141.

24. H. Chen, X. Chang, D. Du, W. Liu, J. Liu, T. Weng, Y. Yang, H. Xu and X. Yang, Podophyllotoxin-loaded solid lipid particles for epidermal targeting, *J. Control. Release*, 2006, **110**, 296–306.

25. S. Lombardi Borgia, M. Regehly, R. Sivaramakrishnan, W. Mehnert, H.C. Korting, K. Danker, B. Röder, K.D. Kramer and M. Schäfer-Korting, Lipid particles for skin penetration enhancement–correlation to drug localization within the particle matrix as determined by fluorescence and parelectric spectroscopy. *J. Control. Release*, 2005, **110**, 151–163.

26. R. Alvarez-Román, A. Naik, Y.N. Kalia, H. Fessi and R. H. Guy, Visualization of skin penetration using confocal laser scanning microscopy. *Eur. J. Pharm. Biopharm.*, 2004, **58**, 301–316.

27. Nokhodchi, J. Shokri, A. Dashbolaghi, D. Hassan-Zadeh, T. Ghafourian and M. Barzegar-Jalali. The enhancement effect of surfactants on the penetration of lorazepam through rat skin. *Int. J. Pharm.*, 2003, **250**, 359–369.

28. M. J. Cappel and J. Kreuter, Effect of particles on transdermal drug delivery, 1991, **8**, 369–374.

29. R. Alvarez-Román, A. Naik, Y. N. Kalia, R. H. Guy and H. Fessi, Enhancement of topical delivery from biodegradable particles. *Pharm. Res.*, 2004, **21**, 1818–1825.

30. R. Alvarez-Román, G. Barré, R. H. Guy and H. Fessi, Biodegradable polymer nanocapsules containing a sunscreen agent: preparation and photoprotection. *Eur. J. Pharm. Biopharm.*, 2001, **52**, 191–195.

31. A. F. El-Kattan, C. S. Asbill, N. Kim and B. B. Michniak, The effects of terpene enhancers on the percutaneous permeation of drugs with different lipophilicities, *Int. J. Pharm.*, 2001, **215**, 229–240.

32. M. Förster, M.-A. Bolzinger, D. Ach, G. Montagnac and S. Briançon, Ingredients tracking of cosmetic formulations in the skin: A confocal Raman microscopy investigation. *Pharm. Res.*, 2011, **28**, 858–872.

33. G. J. Fisher, S. C. Datta, H. S. Talwar, Z.-Q. Wang, J. Varani, S. Kang and J. J. Voorhees, Molecular basis of sun-induced premature skin ageing and retinoid antagonism, *Nature*, 1996, **379**, 335–339.

34. E.G. de Jalón, M.J. Blanco-Prieto, P. Ygartua and S. Santoyo, PLGA microparticles: possible vehicles for topical drug delivery, *Int. J. Pharm.*, 2001, **226**, 181–184.

THE EFFECT OF HEAT ON DICLOFENAC PERMEATION THROUGH HUMAN SKIN

F Caserta[1, 2], D G Wood[2], M B Brown[1, 2] and W J McAuley[1]

[1]Department of Pharmacy, Centre for Topical Drug Delivery and Toxicology Research, University of Hertfordshire, Hatfield, United Kingdom.

1 INTRODUCTION

The application of thermal energy to enhance the percutaneous absorption of drugs offers an encouraging prospect for overcoming the skin barrier. External membranes such as the skin present a considerable barrier to the permeation of drugs, limiting the types of treatment that can be given via this route [1]. These barrier properties are mostly dependent on the *stratum corneum* (SC), the outermost layer of skin, which in most cases is known to be the rate limiting step for drug diffusion across the skin [2]. The idea of the SC as a permeability barrier has been represented as a two compartment model, which is often described as the 'bricks and mortar' model (Figure 1), with the corneocytes as the bricks and the lipid matrix as the mortar [3, 4]. The lipid matrix consists of long-chain ceramides, free fatty acids and cholesterol as their major classes [5, 6]. These lipids are organised in multiple lamellar structures between corneocytes and form a continuous medium through the SC [2]. Within these lamellae the lipids are highly organized, mostly in an orthorhombic arrangement [7] and the presence of this tightly packed orthorhombic structure is thought to have a key role in the skin's barrier properties [8-10].

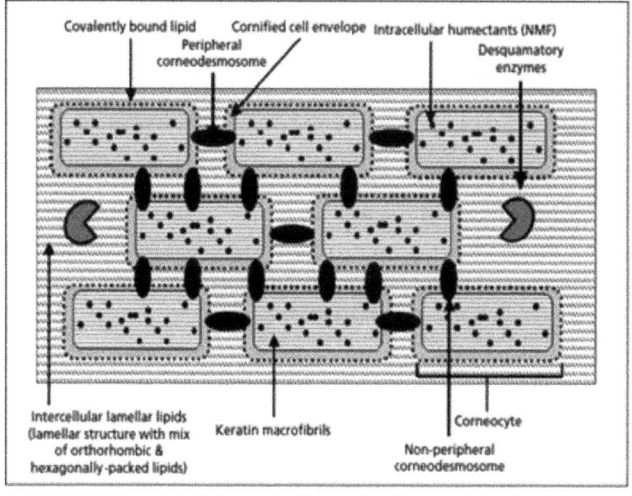

Figure 1 *Schematic 'bricks and mortar' representation of the structure of the SC [11].*

Currently, two different modalities are considered when using heat for optimising percutaneous absorption: one produces high temperatures causing thermal ablation of skin tissue whereas the other produces lower, physiologically tolerable temperatures. Although both strategies employ heat to improve the drug delivery via the skin, the mechanism behind the enhancing effect is different.

Thermal ablation devices comprise of micro-heater arrays, radio-frequency and lasers; such devices selectively remove the SC, without damaging the underlying tissues [12] by increasing the skin temperature (> 100 °C) for a short duration (< 1 s) of time. Following application of the ablation device, micropores of approximately 30 μm in diameter are formed [12] improving drug transport across the SC. Thermal ablation devices have shown to be effective in increasing the permeability of compounds ranging from low molecular weight to proteins and DNA [12-15] and to have acceptable safety profiles. However, clinical trials are still on-going [16].

The lower temperature approach typically uses transdermal patches employing an exothermic reaction such as heat produced by chemical reactions combining iron, carbon, sodium chloride, vermiculite and water (Figure 2).

Synera[TM] is the only licensed patch containing active drug available in the market using this strategy and it holds a eutectic drug reservoir of lidocaine and tetracaine. The heating system integrated into the patch is designed to increase the temperature of the skin from 32 °C (physiological skin surface temperature) to a maximum of 40 °C, with the aim of facilitating the topical delivery of the anaesthetics over the entire application of 30 minutes [18, 19]. A further example of a commercial therapeutic product employing the exothermic iron oxidation reaction is the Nurofen[®] back pain heat patch: it is a drug-free heat patch indicated for the relief of back pain and can generate a maximum temperature of approximately 45 °C for several hours at the delivery site [20]. Although a regulatory limit has not yet been set, such temperatures are thought to be physiologically tolerable and not to cause irreversible damage to the skin [21].

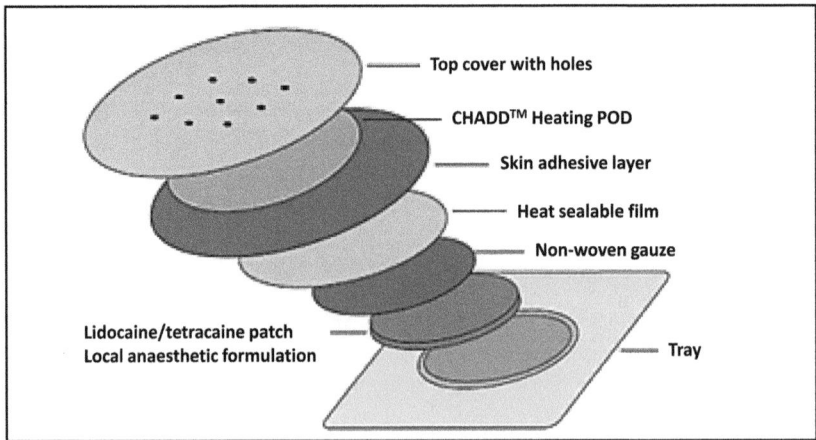

Figure 2 *Schematic representation of a heated lidocaine/tetracaine patch [17].*

The effect of elevated, physiologically-tolerable temperatures on solute permeation across skin has been previously investigated [22, 23] and recently there has been a renewed interest in the use of heat as a potential means of modifying the delivery profile of topically applied medicaments [20, 24-26]. The enhancing effect of this strategy is believed to rely on a combination of different mechanisms; increased drug diffusivity, both in the vehicle and in the skin [27], and (specifically for transdermal drug delivery) increased cutaneous blood flow. Changes in drug diffusivity can be explained by an increase in molecular motion of the permeant, described through the Stokes-Einstein equation [28]. The application of heat will therefore increase the rate of drug diffusion towards the skin/vehicle interface and subsequently through the SC. There are three possible pathways of permeation via the SC; intracellular, appendageal and/or intercellular routes. The latter is believed to be the most likely drug permeation route, with permeants passing via the lipid domains between corneocytes [29]. Differential scanning calorimetry studies, in combination with other techniques such as small-angle X-ray diffraction and Fourier Transform Infrared Spectroscopy, have shown that reversible alterations in the arrangement and state of the lipid bilayer are possible when increasing skin temperature [30, 31]. Four principal endothermic transitions (commonly referred to as T1-T4) have been identified at temperature intervals of 35-42, 65-75, 78-86 and 90-115 °C, respectively [30-33]. These have been associated with alterations of the SC lipids, lipids covalently linked to the corneocytes and proteins. Only the T1 phase transition is within the physiologically tolerable temperature range. This transition is thought to be a double step in the disordering of the lateral lipid packing, with a 'solid to fluid' change (at approximately 37 °C) followed by a transition from an orthorhombically-packed state, to a more penetrable hexagonal organization (at approximately 42 °C) which offers reduced resistance to drug diffusivity (Figure 3) in comparison to the orthorhombic state [30, 33, 34].

The effect of changes in the physical state of SC lipids on drug diffusion due to increased temperature (up to 45 °C) was confirmed by a previous *in vitro* study employing three model permeants of differing in lipophilicity [25]. The study assessed the change in diffusivity of the compounds as a function of temperature by employing an artificial membrane (non-rate limiting) to exclude the effect of heat on the SC lipids arrangement, thus enabling an estimation of the magnitude of the increased drug diffusivity in the vehicle directly related to the use of external heat. The former was found to account for 66, 68 and 81 % (mean values) of the enhanced permeation of the three permeants across the human epidermis. In contrast, the impact of heat on the skin's properties was responsible

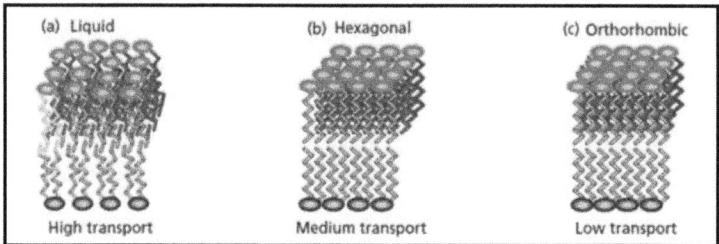

Figure 3 *Schematic representation of the positions of the alkyl chains in the liquid, hexagonal and orthorhombic arrangement SC. Adapted from [11].*

for only ≤ 34 % of the enhancement effect, demonstrating that the barrier properties of the SC lipids are only partially altered by these elevated temperatures.

A further effect of the application of heat is the vasodilatation of the cutaneous blood vessels. Vasodilation represents a cardiovascular response to a heat stress [35] and can cause drug delivery into the systemic circulation to increase [36] as result of an enhanced permeant clearance from the dermis. Although such an effect may be beneficial when attempting to design a system for transdermal delivery, it must be considered that the local concentrations of drugs in the skin or underlying deeper tissues could be potentially reduced.

The impact of dermal clearance on drug transport into deeper tissues has been assessed in several studies investigating the permeation of non-steroidal anti-inflammatory drugs (NSAIDs) via the skin into subcutaneous tissue [37-39]. They have shown that the vasoconstriction of subcutaneous blood vessels induced by the use of a vasoconstrictor [38] (i.e. phenylephrine) or a reduction in skin temperature [37] (20 – 25 °C) can cause the penetration of drugs to the joints to be enhanced. However, further studies are required to investigate the balance between the enhancement effect of temperature on skin drug transport and the impact of heat on the dermal clearance rate so that its effect for topical/regional drug delivery enhancement can be fully evaluated.

Overall, increasing the temperature of the skin within a physiologically tolerable range, could offer the chance of readily enhancing topical drug delivery, without requiring any costly or complicated devices (thermal ablation, ultrasound, iontophoresis, and microneedles) that potentially could limit usage and/or lead to poor patient compliance. Furthermore, as heat can increase the drug diffusion rate, its application could result in a faster onset of the action and therefore act as a better treatment of pain.

The aim of this present study was to determine the effect of temperature on the percutaneous permeation of the NSAID, diclofenac, prior to evaluating the feasibility of developing a novel thermophoretic system for delivery across the skin. Such delivery systems could offer therapeutic advantages for patients over diclofenac oral formulations, such as minimising side effects, by avoiding systemic exposure to the drug, and good patient compliance. However, realising these benefits relies on being able to deliver therapeutic concentrations of drug to the target site.

The application of heat is an interesting strategy to optimise the delivery of diclofenac across the skin as heat is known to be effective in the treatment of short-term relief of acute low back pain [40]. Thus, generating heat topically can potentially increase the drug permeation and at the same time may also optimise the treatment of pain.

2 MATERIALS AND METHODS

Diclofenac diethylamine and free acid was supplied by MedPharm Ltd. (Guildford, UK). Ammonium formate, phosphoric acid, absolute ethanol, PBS tablets, HPLC grade methanol and HPLC grade formic acid were purchased from Fisher UK.

The quantitative determination of diclofenac base was performed using HPLC. Chromatographic separation was achieved by employing a Gemini C18 reversed phase column (150 mm x 4.60 mm x 5 μm) (Phenomenex, UK). The mobile phase used for separation of both the free acid and salt form of diclofenac was 20 mM ammonium formate in methanol: 20 mM ammonium formate buffer (65%: 35% v/v) adjusted to pH 2.5 using formic acid with a Hanna pH 209 pH meter (Fisher Scientific, UK) . The analysis was carried out isocratically at room temperature with a UV detector wavelength of 280 nm, a 1.0 ml min^{-1} flow rate and an injection volume of 20 μl. This method was found to be suitable in accuracy, precision and linearity as prescribed by the International Conference on Harmonisation guidelines [41].

Frozen samples of excised human scrotal skin acquired from surgical procedures were obtained from donors with informed consent and appropriate ethical approval from the South East London Research ethics committee (ethics number, 10/H0807/51). The skin was stored at -20 °C until it was ready for use. After allowing the skin to defrost, subcutaneous fat was carefully removed using forceps and a scalpel. The full thickness scrotal skin was then used without any further preparation.

The statistical analysis software Minitab (version 15) was used to produce a 2 level full factorial experimental design, with the aim of identifying the effects of independent variables; such as temperature (32 °C and 45 °C), pH (3.5 and 7.5) and diclofenac in salt form (diethyl ammonium salt), on the response of drug flux. After running the factorial design, a series of experimental runs were generated which indicated the conditions to be employed for the *in vitro* diffusion studies such as skin temperature and pH of the donor solutions.

Six different scrotum skin donors were used during the diclofenac diffusion study; the variability between each donor was measured, by performing the centre point of the experimental design, using skin tissue from each of the donors. The given centre point was automatically calculated by the programme as the median; between the lowest and highest value for each factor.

Saturated solutions of the drug were prepared by adding diclofenac in excess (using both diethylammonium salt and free acid) to the vehicle and then adjusting the resulting pH value in accordance to the various runs generated from the factorial design. The suspension was then allowed to stir for 24 h at the experimental temperatures of 32 and 45 °C. The pH of the donor solution was re-checked before dosing Franz cells.

In vitro diclofenac diffusion studies were carried out using individually calibrated, unjacketed, horizontal Franz type diffusion cells with a receiver volume of approximately 3.0 mL and a surface area of about 1.0 cm^2. Full thickness scrotal skin was cut and placed between the donor and the receptor compartments of the Franz cells. The receptor compartments were filled with 20 % (v/v) ethanol in PBS and the cell was placed on a stirring plate immersed in a water-bath. A small magnetic bar was inserted into each receptor compartment to enable continuous stirring during the duration of the experiment. After allowing the scrotal skin to equilibrate with the receiver fluid for 0.5 h, 1 ml of saturated drug solution (containing undissolved/suspended solids) was then introduced into the donor chamber. The Franz cells were occluded and immersed in a water bath controlled at temperatures of 37, 44 °C and 50 °C for 34 h. Such temperatures were used to produce skin temperatures of 32, 39 and 45 °C respectively. These temperatures were monitored throughout the duration of the study using a digital thermometer, with a type K probe (Fisher Scientific, UK). At pre-determined time intervals, all of the receiver fluid was withdrawn and immediately replaced by an equivalent volume of receptor solution which had been equilibrated at the required temperature. The samples were then analysed by HPLC. The cumulative amount of drug, Q (μg cm^{-2}) per sample area was plotted against time (h) and the steady state flux (J_{ss}) was determined from the linear portion of the plot ($R^2 \geq 0.993$ over six points). As the Franz cells experiments indicated that diclofenac flux reached steady-state after 24 h, a further five samples were withdrawn (giving a total study duration of 34 hours) in order provide sufficient steady-state (linear) data to estimate J_{ss}.

A factorial ANOVA was performed in order to evaluate the interactions and the main effects of experimental input variables on diclofenac flux by using Minitab software (version 15). The Log$_{10}$ of the drug flux values was taken in order to transform the data from a non-normal distribution into a normal distribution. The chosen level of significance was $p \leq 0.05$.

3 RESULTS AND DISCUSSION

In this study, each of the factors, at two levels, were run on the factorial experimental design. One advantage of implementing such a statistical analysis is that it is known for its versatility in the design of various drug formulations. It is also a quick, efficient, reliable and precise method to verify the relationship between the factors and the response [42].

With regards to selecting the level for the parameters, the skin surface temperature 32 °C and a maximum temperature of 45 °C, which is thought to be physiologically tolerable by the skin [21] were used as low and high levels respectively, to identify whether application of exogenous heat had a significant effect on drug permeability.

As the main diffusional route for drugs across the SC is believed to be through the intercellular skin lipids [29], transport of the ionised drug through the skin is normally thought to be lower in comparison to its unionised form. In order to examine the significance of this effect, pH values were selected which covered the drug's ionisation range from 25 % to 100 % (pH = 3.5 and pH = 7.5, respectively). A pH of 2 (0 % ionisation) was not selected as this acidic condition could damage the skin's integrity and therefore affect the results from the Franz cell diffusion studies [43].

The diclofenac free acid is poorly soluble in water, thus the use of this drug in an organic salt form results in improved aqueous solubility and also facilitates the formation of ion-pairs which may assist partitioning into the skin and thus enhance percutaneous penetration [44]. Both diclofenac free acid and the diethylammonium salt were used in this study; half of experimental runs generated included the use of the diclofenac free acid and for the other half the diclofenac diethylammonium salt was used.

Analysis of the factorial experimental design enabled the assessment of the main effects and interactions between each independent variable on diclofenac flux. Temperature, when examined at its high level, was found to have a significant influence on the drug flux with an overall enhancement effect of about 8-fold ($p \leq 0.05$; Figure 4).

Additionally the modification of pH, from pH 3.5 to 7.5, had a significant effect on diclofenac permeation (overall enhancement effect approximately 12-fold) and thus has the potential to act as complementary strategy to further improve drug delivery across skin ($p \leq 0.05$) (Figure 5). Conversely, the use of diclofenac diethylammonium salt, independently as well as when combined with other factors, did not have a significant influence on the drug flux ($p \geq 0.05$).

The enhancement in diclofenac percutaneous absorption because of heat is consistent with that reported by other authors for other drug molecules [20, 25]. These authors attributed this enhancement to an alteration of the SC lipids, occurring at temperatures over T1 phase transition (35 - 42 °C) and to increased drug diffusivity in the formulation.

With regards to pH, considering the hydrophobic nature of SC, it is unexpected to find that the steady-state flux of diclofenac would significantly enhance with an increase of the percentage of the drug in ionised form (from pH 3.5 to 7.5). However it must be considered that, according to Fick's first law of diffusion, the steady-state flux of a drug (J), is the product of the permeability constant (K) and solubility in the vehicle (C_v). This implies that at a higher pH, the lower permeability of the ionized drug is more than compensated for, by the drugs' increased solubility. This is in agreement with the findings of previous studies, assessing the skin permeability of NSAIDs which reported that the steady-state flux across the skin is greater at a higher pH, when the fraction of ionized species is greater [45, 46].

The diclofenac anion is able to form ion-pairs in the presence of both organic and inorganic cations. Generally organic bases (cations) have been found to produce ion-pairs with partition coefficients that would be predicted to offer improved skin permeation [47]. Consequently, the use of diclofenac diethylamine salt could be a potential means to increase the percutaneous absorption of diclofenac across skin. A recent investigation comparing the differences in the *in vitro* diffusion of diclofenac free acid and seven of its salts (including diclofenac diethylamine salt) through rat skin showed that the percutaneous absorption significantly increased when changing from free acid to the salt form of the drug regardless of the types of counterions employed [48]. However the use of diclofenac diethylamine salt did not result in significantly improved drug permeation in this study when compared with other factors, such as temperature and pH, which showed a greater impact on drug flux.

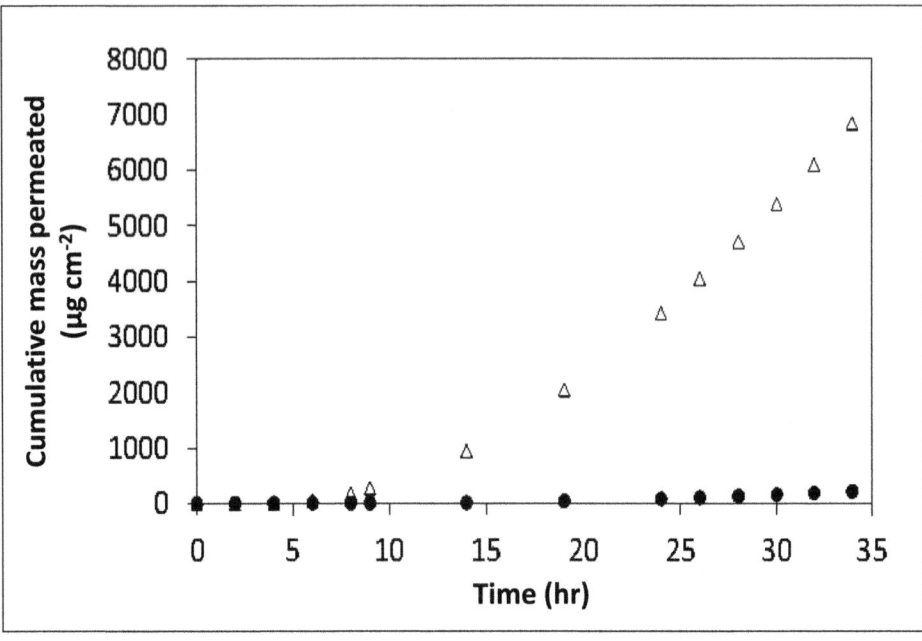

Figure 4 *Diclofenac free acid permeation at 32 °C (●) and 45 °C (Δ) at pH 7.5.*

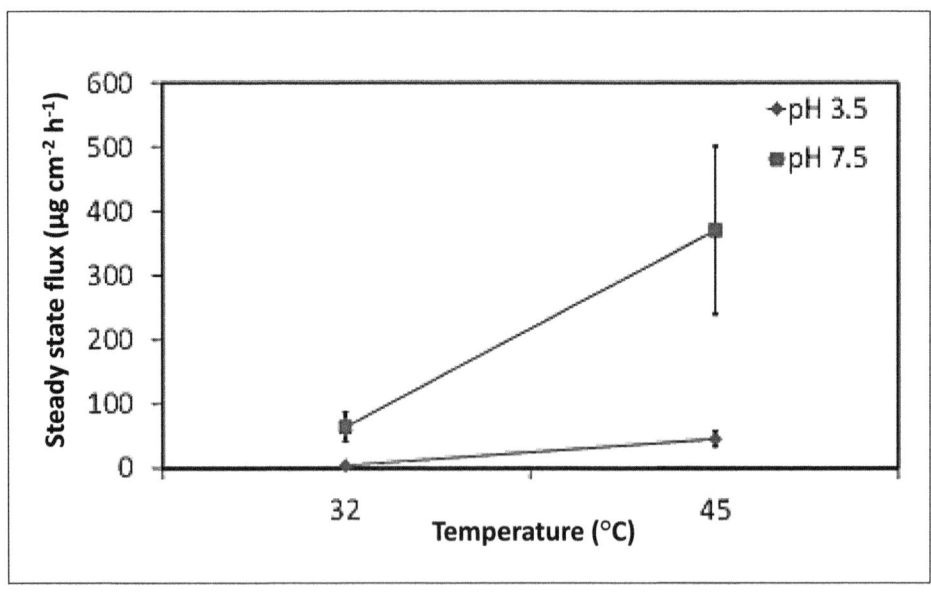

Figure 5 *Interaction plot showing the overall mean of diclofenac flux at 32 °C and 45 °C when increasing pH from 3.5 to 7.5 (mean ± SE, n=8).*

4 CONCLUSION

These diclofenac *in vitro* permeation studies have been performed as function of temperature. The analysis of the experimental data have shown that skin temperature, when increased from 32 °C to 45 °C, has a significant influence in enhancing diclofenac permeation across the skin barrier. Such an increase may therefore improve the therapeutic treatment of pain by increasing the diclofenac concentrations at the target site. To fully evaluate the suitability of heat for topical use, further work needs to be performed in order to identify an optimum level of intensity and duration of heat that would enhance the drug transport without compromising such delivery by increased dermal clearance.

5 REFERENCES

1. M. B. Brown, *et al.*, Transdermal drug delivery systems: Skin perturbation devices. *Methods in Molecular Biology*, 2008, **437**, 119–139.

2. J. A. Bouwstra, *et al.*, Structural Investigations of Human Stratum Corneum by Small-Angle X-Ray Scattering, *J Investig Dermatol*, 1991, **97**(6), 1005–1012.

3. A. Michaels, S. Chandrasekaran and J. Shaw, *Drug permeation through human skin: Theory and invitro experimental measurement,* AIChE Journal, 1975, **21**(5), 985–996.

4. P. M. Elias, Epidermal Lipids, Barrier Function, and Desquamation, *J Invest Dermatol*, 1983, **80**(1s), 44s–49s.

5. G. M. Gray and H. J. Yardley, Lipid compositions of cells isolated from pig, human, and rat epidermis, *Journal of Lipid Research*, 1975, **16**(6), 434–40.

6. G. M. Gray and R. J. White, Glycosphingolipids and Ceramides in Human and Pig Epidermis, *J Investig Dermatol*, 1978, **70**(6), 336–341.

7. J. Caussin, G. S. Gooris and J. A. Bouwstra, FTIR studies show lipophilic moisturizers to interact with stratum corneum lipids, rendering the more densely packed, *Biochimica et Biophysica Acta (BBA) - Biomembranes*, 2008, **1778**(6), 1517–1524.

8. G. S. K. Pilgram *et al.*, Aberrant lipid organization in stratum corneum of patients with atopic dermatitis and lamellar ichthyosis, *Journal of Investigative Dermatology*, 2001, **117**(3), 710–717.

9. J. A. Bouwstra and M. Ponec, The skin barrier in healthy and diseased state, *Biochimica et Biophysica Acta (BBA) - Biomembranes*, 2006, **1758**(12), 2080–2095.

10. F. Damien and M. Boncheva, The Extent of Orthorhombic Lipid Phases in the Stratum Corneum Determines the Barrier Efficiency of Human Skin In Vivo, *J Invest Dermatol*, 2009, **130**(2), 611–614.

11. A. V. Rawlings, Recent advances in skin 'barrier' research, *Journal of Pharmacy and Pharmacology*, 2010, **62**(6), 671–677.

12. J. Bramson, *et al.*, Enabling topical immunization via microporation: a novel method for pain-free and needle-free delivery of adenovirus-based vaccines, *Gene therapy*, 2003, **10**(3), 251–260.

13. W. R. Lee, *et al.*, Lasers and Microdermabrasion Enhance and Control Topical Delivery of Vitamin C, 2003, **121**(5), 1118–1125.

14. J. Birchall, *et al.*, Cutaneous gene expression of plasmid DNA in excised human skin following delivery via microchannels created by radio frequency ablation, *International Journal of Pharmaceutics*, 2006, **312**(1–2), 15–23.

15. W. R. Lee, *et al.*, Transdermal drug delivery enhanced and controlled by erbium:YAG laser: a comparative study of lipophilic and hydrophilic drugs, *Journal of Controlled Release*, 2001, **75**(1–2), 155–166.

16. A. Arora, M. R. Prausnitz and S. Mitragotri, Micro-scale devices for transdermal drug delivery, *International Journal of Pharmaceutics*, 2008, **364**(2), 227–236.

17. J. Sawyer, *et al.*, Heated lidocaine/tetracaine patch (Synera™, Rapydan™) compared with lidocaine/prilocaine cream (EMLA®) for topical anaesthesia before vascular access, *British Journal of Anaesthesia*, 2009, **102**(2), 210–215.

18. S. E. Curry and J. C. Finkel, Use of the Synera™ Patch for Local Anesthesia Before Vascular Access Procedures: A Randomized, Double Blind, Placebo Controlled Study, *Pain Medicine*, 2007, **8**(6), 497–502.

19. Available from: http://www.synera.com/PDFs/Synera_PI.pdf, [cited 03/05/2012].

20. D. G. Wood, M. B. Brown and S. A. Jones, Controlling barrier penetration via exothermic iron oxidation, *International Journal of Pharmaceutics*, 2010, **404**(1–2), 42–8.

21. A. R. Moritz and F. Henriques Jr, Studies of Thermal Injury: II. The Relative Importance of Time and Surface Temperature in the Causation of Cutaneous Burns, *The American Journal of Pathology*, 1947, **23**(5), 695.

22. I. H. Blank, R. J. Scheuplein and D. J. MacFarlane, Mechanism of percutaneous absorption. 3. The effect of temperature on the transport of non-electrolytes across the skin, *The Journal of Investigative Dermatology*, 1967, **49**(6), 582–9.

23. W. C. Fritsch and R. B. Stoughton, The effect of temperature and humidity on the penetration of C14 acetylsalicylic acid in excised human skin, *The Journal of Investigative Dermatology*, 1963, **41**, 307–11.

24. A. K. Jain, and R. Panchagnula, Effect of temperature on imipramine hydrochloride permeation: role of lipid bilayer arrangement and chemical composition of rat skin, *International Journal of Pharmaceutics*, 2003, **250**(1), 287–293.

25. F. Akomeah, *et al.*, Effect of heat on the percutaneous absorption and skin retention of three model penetrants, *European Journal of Pharmaceutical Sciences*, 2004, **21**(2–3), 337–345.

26. D. G. Wood, *et al.*, Characterization of Latent Heat-Releasing Phase Change Materials for Dermal Therapies, *The Journal of Physical Chemistry C*, 2011, **115**(16), 8369–8375.

27. D. G. Wood, M. B. Brown and S. A. Jones, Understanding heat facilitated drug transport across human epidermis, *European Journal of Pharmaceutics and Biopharmaceutics*, 2012, **81**(3), 642–649.

28. J. T. Edward, Molecular volumes and Stockes-Einstein equation, *Journal of Chemical Education*, 1970, **47**(4), 261–270.

29. J. Hadgraft, Skin deep, *European Journal of Pharmaceutics and Biopharmaceutics*, 2004, **58**(2), 291–299.

30. C. L. Gay, *et al.*, Characterization of low-temperature (i.e., < 65 degrees C) lipid transitions in human stratum corneum, *Journal of Investigative Dermatology*, 1994, **103**(2), 233–239.

31. P. A. Cornwell, *et al.*, Modes of action of terpene penetration enhancers in human skin; Differential scanning calorimetry, small-angle X-ray diffraction and enhancer uptake studies, *International Journal of Pharmaceutics*, 1996, **127**(1), 9–26.

32. B. F. Vanduzee, Thermal analysis of human stratum corneum, *Journal of Investigative Dermatology*, 1975, **65**(4), 404–408.

33. C. L. Silva, *et al.*, Study of human stratum corneum and extracted lipids by thermomicroscopy and DSC, *Chemistry and Physics of Lipids*, 2006, **140**(1–2), 36–47.

34. J. A. Bouwstra, *et al.*, Structure of human stratum corneum as a function of temperature and hydration: A wide-angle X-ray diffraction study, *International Journal of Pharmaceutics*, 1992, **84**(3), 205–216.

35. C. R. Wyss and L. B. Rowell, Lack of humanlike active vasodilation in skin of heat-stressed baboons, *Journal of Applied Physiology*, 1976, **41**(4), 528–531.

36. T. O. Klemsdal, K. Gjesdal and J. E. Bredesen, Heating and cooling of the nitroglycerin patch application area modify the plasma level of nitroglycerin, *European Journal of Clinical Pharmacology*, 1992, **43**(6), 625–628.

37. S. Sammeta and S. Murthy, "ChilDrive": A Technique of Combining Regional Cutaneous Hypothermia with Iontophoresis for the Delivery of Drugs to Synovial Fluid, *Pharmaceutical Research*, 2009, **26**(11), 2535–2540.

38. K. Higaki *et al.*, Enhancement of topical delivery of drugs via direct penetration by reducing blood flow rate in skin, *International Journal of Pharmaceutics*, 2005, **288**(2), 227–233.

39. N. Monteiro-Riviere, *et al.*, Topical Penetration of Piroxicam Is Dependent on the Distribution of the Local Cutaneous Vasculature, *Pharmaceutical Research*, 1993, **10**(9), 1326–1331.

40. G. Garra, *et al.*, Heat or Cold Packs for Neck and Back Strain: A Randomized Controlled Trial of Efficacy, *Academic Emergency Medicine*, 2010, **17**(5), 484–489.

41. ICH (1996) *Q2(R1): Validation of Analytical Procedures: Text and Methodology.* International Conference on Harmonisation of Technical Requirements for Registration of Pharmaceuticals for Human Use.

42. G. A. Lewis, D. Mathieu and R. T. L. Phan, *Pharmaceutical experimental design*, 1998, Informa Healthcare.

43. B. W. Barry, *Dermatological formulations: percutaneous absorption*, 1983, Marcel Dekker Inc.

44. M. Manconi, *et al.*, Ex vivo skin delivery of diclofenac by transcutol containing liposomes and suggested mechanism of vesicle-skin interaction, *European Journal of Pharmaceutics and Biopharmaceutics*, 2011, **78**(1), 27-35.

45. J. Hadgraft, J. d. Plessis and C. Goosen, The selection of non-steroidal anti-inflammatory agents for dermal delivery, *International Journal of Pharmaceutics*, 2000, **207**(1-2), 31–37.

46. V. Sarveiya, J. F. Templeton and H. A. E. Benson, Ion pairs of ibuprofen: increased membrane diffusion, *Journal of Pharmacy and Pharmacology*, 2004, **56**(6), 717–724.

47. A. Fini, *et al.*, Formation of ion-pairs in aqueous solutions of diclofenac salts, *International Journal of Pharmaceutics*, 1999, **187**(2), 163–173.

48. M. Wang and L. Fang, Percutaneous absorption of diclofenac acid and its salts from emulgel, *J Pharm Pharmacol*, 2008. **60**(4), 429–35.

14

TOPICAL MICRONEEDLE DRUG DELIVERY ENHANCED WITH MAGNETOPHORESIS

T W Prow[1], Y H Mohammed[2], A B Ansaldo[1] and H A E Benson[2]

[1]Dermatology Research Centre, The University of Queensland, School of Medicine, Translational Research Institute, Brisbane, Australia. [2]CHIRI Biosciences, School of Pharmacy, Curtin University, Perth WA, Australia.

1 INTRODUCTION

Proteins and peptides are coming of age as targeted biological therapies. A broad range of peptides have now been approved by the FDA and other regulatory agencies for use as therapeutics. Likewise, large proteins such as infliximab and botox are commonly used to treat skin conditions. The most widely used delivery approach for peptides and proteins is parenteral injection, largely due to oral instability and the lack of a suitable alternative. Intense research is being done to improve the short half-lives of these biological therapies and find suitable alternative delivery routes. Transdermal delivery holds promise as an improved delivery route because the skin has low levels of enzymatic activity and can help drugs avoid first-pass metabolism. Further, topical delivery has the potential for direct treatment of skin disease for both therapeutic and cosmetic applications. Access to therapeutic targets within skin is limited by the well-characterised skin barrier composed of the stratum corneum, viable epidermis, dermal-epidermal junction and dermis. Access to the systemic circulation is hindered by the more superficial skin barriers. Overcoming these barriers has implications particularly for the elderly and immune compromised, where the destruction of the skin barrier for drug delivery must be balanced with consideration for increased susceptibility for infection. Therefore, the optimal topical drug delivery enhancement technology would be minimally invasive and have a negligible effect on pathogen susceptibility.

For over two decades, microneedles have been the topic of intense research and development by commercial and academic laboratories. This sustained research activity has been propelled by significant investment based on the considerable commercial promise of minimally invasive transdermal drug and vaccine delivery. There are four major classes of microneedle delivery strategies: solid microneedles, coated-solid microneedles, dissolving microneedles and hollow microneedles [1]. Although microneedle delivery does improve drug delivery, combining this with other technologies holds promise to make a step change improvement in sustained topical drug delivery. Iontophoresis was combined with 430 µm long microneedles and was used to deliver antisense oligodeoxynucleotides in one of the earliest examples of a combinatorial approach [2]. The optimal configuration resulted in increased oligonucleotide delivery compared to microneedle treatment alone. This is accomplished by using microneedles to create fissures in the stratum corneum, viable epidermis, dermal-epidermal junction and the dermis. Drug delivery through these conduits is enhanced via iontophoresis, in this case. Several groups have explored a combination of iontophoresis and microneedling [3-8]. This chapter describes our study using magnetophoresis to enhance microneedle-based drug delivery.

Magnetophoresis, in the context of transdermal drug delivery, is the process of driving drugs into skin using magnetic fields. The skin is composed primarily of keratinocytes, fibroblasts and the endothelial cells that form the blood vessels. All of these cell types have been shown to be responsive to magnetic fields [9-11]. Magnetic field effects have been reported in a variety of skin related contexts including improving wound healing [12, 13], stimulating collagen synthesis[14] and improving chronic skin ulcers [15, 16]. Magnetophoresis with stationary permanent magnets alone, has been shown to enhance topical delivery of benzoic acid, lidocaine and tertbutaline sulphate [17, 18]. We have shown pulsed electromagnetic fields improve delivery of 5-aminolevulinic acid, naltrexone hydrochloride and 10 nm gold nanoparticles [19-21]. More recently, we have shown that a static magnetic field array can enhance skin permeation and hydration [22]. Carrying this work forward, we have explored the delivery enhancing capacity of a combined topical drug delivery approach using microneedles and magnetic field arrays.

2   METHODS

Full thickness human skin samples excised from female patients (30-50 year old) undergoing abdominoplasty were refrigerated immediately after surgery. Sampling was approved by the Human Research Ethics Committee of the Princess Alexandra Hospital (097/090) in compliance with the guidelines of the National Health and Medical Research Council of Australia. The following procedure was used to obtain epidermal sheets. The subcutaneous fat was removed by dissection, the full thickness skin then immersed in water at 60° C for 1 min, allowing the epidermis to be lifted off the dermis. The epidermis was placed onto aluminium foil, air-dried, then placed in a sealed bag and stored at –20° C until required.

Melanostatin peptide labelled with FITC was custom synthesised from Genscript USA Inc. Sodium fluorescein (NaF) was obtained from the Princess Alexandra Hospital pharmacy, RetinoFluor (Phebra).

Transepidermal water loss (TEWL) measurements were performed using an AquaFlux AF200 closed chamber evaporimeter (Biox Systems Ltd., London, UK). AquaFlux V6.2 software was used to analyse the data. After calibration, the TEWL probe was positioned in the centre of the skin sample and the handle maintained perpendicular to the surface of the skin. A slight pressure was applied to the skin through the probe and then the TEWL device activated. Data logging automatically ceased once the rate of change of the TEWL reading dropped below a specified amount or the test time had elapsed its specified duration. TEWL measurements were taken before and after microneedle treatment.

The configuration and geometries of the microneedle arrays were 700 µm microneedle length × 250 µm width × 50 µm thickness. There were 3 microneedles per 5 mm plate. The plates were assembled in banks of 2 with 2 mm spacing. A LaserPro S290 (M2 Lasers, Brisbane, QLD) laser milling machine was used to cut the 50µm thick 304 stainless steel sheet to the microneedle specifications. The cutting laser was a 20W source coupled to a cutting head with fibre optic cable with a resolution of 1000 dpi. Quality control was done on the microneedles before application and the observed configuration was 703.1 ± 16.1µm long and 257.8 ± 9.4µm wide with a tip at a 55° angle that was 250 µm long.

Microneedle Treatment: A piece of gauze was folded in half 4 times and then the skin sample placed on top. In-situ skin tension was approximately simulated by stretching the skin over the gauze and pinning in place. A pre-prepared microneedle impact applicator was then used to apply the microneedle to the skin at speed (approximately 3 m s^{-1}). This applicator consisted of a housing, a spring-loaded plunger and the banks of laser cut microneedles (two 3-needle banks gave a total of 6 microneedles). The banks were mounted into the plunger that was then drawn back, loading the spring, and held in place using a pin. Upon removal of the pin, the plunger accelerated towards the skin, driving the microneedles through the stratum corneum and into the viable epidermis and dermis.

A Franz cell was prepared by applying vacuum grease to the sealing surfaces of the receptor and donor. A stirring magnet was then added to the receptor. The skin sample was placed on the receptor with the skin side facing upwards. The donor was assembled on top and fastened in place using a collar. The receptor was then filled with saline. During this procedure, steps were taken to ensure no air bubbles remained in the receptor. A volume of 300µL Sodium Fluorescein (NaF) at 1 mg mL^{-1} in saline was pipetted onto the centre of the skin in the Franz cell.

A rubber o-ring (o.d. 10 mm and i.d. 7 mm) was placed onto the skin surface in the Sodium Fluorescein solution and the magnet (ETP 012) or non-magnetic control (NMC) was then laid on top. The magnetic and non-magnetic film was a gift from OBJ Ltd. (Perth, WA, Australia). The o-ring served to maintain the dye reservoir on the surface of the skin and prevent the magnet from occluding the pores. After dye/magnet treatment, vacuum grease was added to the top of the donor cell and a glass cover slip placed on this surface to seal the unit, thus minimising evaporation. The Franz cell was then placed into a circulating, heated water bath at a temperature of between 34° C and 35° C and covered in aluminium foil. The Franz cells were incubated in this bath for 15 minutes. The Franz cells were then disassembled and the skin samples placed back in the hydrating weight boat and covered with aluminium foil to prevent photobleaching.

Confocal laser scanning microscopy was used to measure fluorescent drug delivery. Either a VivaScope 1500 or 2500 was used with the 488 nm laser excitation source and a 550/88 nm band pass emission filter. The stage could move ± 8.0 mm in direction of x and y totalling a complete area of 16 mm^2. The confocal microscope imaged up to 9 frames per second with 1000 x 1000 pixels in an individual image. The laser power enabled imaging up to 200 µm. Reflectance confocal microscopy was used to identify the skin strata during these experiments. To prepare the samples for LSCM, a cut was made around the imprint made by the Franz cell. This reduced the size of the skin sample to allow easy and reproducible mounting into the LSCM stage. A small drop of immersion oil was added to the skin surface of the excised skin samples and massaged. A small compressive load was applied to the skin sample in the form of a custom-made sample cup to ensure a flat skin surface for imaging. A dermoscopic image was first taken. This gave a macroscopic image of the skin surface and aided in accurate targeting of the microneedle pores during LSCM imaging. The sample was then mounted into the LSCM and, using reflectance confocal microscopy (RCM), the skin surface was located. After adjusting the location so that a plane just above the skin surface was visible, the zero depth was set. The LSCM machine was then switched to the fluorescent mode with 488 nm excitation and 550 ± 88 nm emission band was recorded. A mosaic/z-stack was obtained consisting of ten 7 mm x 7 mm layers with a 20 µm step between each layer and each image was 0.5 x 0.5 mm^2 at 1000 x 1000 pixels. After completion, the LSCM machine was switched back to RCM mode and an identical stack acquired.

The image data names were encoded to blind the image analysis researcher. Once the analysis was complete, the codes were revealed and the data organised into groups. Each mosaic image was reduced in size to 10% and saved as a jpg file using Adobe Photoshop. Image J (NIH, Bethesda, MD) was then used to quantify fluorescence intensity and dye positive areas. Once a z-stack of mosaics was loaded into Image J, a threshold was applied at 100 - 255 to a 500 x 500 px (2.5 x 2.5 mm^2) area of interest. The "Analyse Particles" function was applied with a size range restriction of 500-infinity to exclude furrows and background. The output was "Positive Area" and "Integrated Intensity". These values were normalised to laser power by dividing the output with the laser power.

3   RESULTS

The average penetration area and average integrated density were calculated using Image J and were used to assess Melanostatin penetration. The average delivery signal of Melanostatin with microneedles and ETP012 was 1.48 times greater than that observed with microneedles and non-magnetic control material. This result was an average across the stratum corneum, viable epidermis and dermis. The area of penetration was larger and the measured intensity was greater in all three layers in ETP012 enhanced groups, supporting the hypothesis that magnets were capable of enhancing Melanostatin transport across the stratum corneum without microneedles. The combination of ETP012 and microneedles showed 7.12 times greater area of penetration when compared to non-magnetic control material. The average intensity of the penetration area represented by integrated density was also enhanced by 2.8 times supporting the hypothesis that combined penetration enhancers are capable of significant improvements in topical delivery.

Before any treatment the average TEWL was 11.3 ± 3.7 g m^{-2} h^{-1}. After microneedle treatment the TEWL increased to 17.9 ± 3.7 g m^{-2} h^{-1} and this was a significant $(p=0.0002)$ increase as determined by a Unpaired t-test with Welch's correction applied to these two groups. However, there were no significant differences between treatment groups analysed with a one way ANOVA (Kruskal-Wallis test) with a Dunns post-test. This is likely due to the small sample size of n=3 per group. Table 1 shows the mean TEWL values before and after microneedle treatment of each group, in addition to the average TEWL change. The changes in TEWL between the ETP012 and other microneedle groups were not significant, suggesting all microneedling was consistent in terms of TEWL.

Group	Before			After			Delta		
Saline	15.4	±	5.2	16.6	±	8.3	1.2	±	3.2
NaF	11.8	±	0.9	11.0	±	1.4	-0.8	±	0.7
NaF & MN	9.6	±	0.8	16.2	±	3.4	6.6	±	3.3
NaF & MN & ETP012	12.2	±	1.4	20.2	±	1.7	8.0	±	1.7
NaF & MN & NMC	9.7	±	1.1	17.4	±	3.0	7.7	±	4.0
NaF & NMC	9.3	±	0.7	9.6	±	0.5	0.3	±	0.3

Table 1 *TEWL rates before and after microneedle treatment. The treatment groups include treatment with fluorescein (NaF), microneedles (MN), ETP012 and the non-magnetic control (NMC). Delta is the difference between the before and after TEWL rates All values are mean ± standard deviation of n=3 replicates.*

One group was treated with the microneedle applicator without microneedles to control for any barrier disruption from the high velocity impact (Figure 1A). These data are marked "NaF & Empty applicator" and show a fluorescence pattern similar to NaF alone where there was no penetration of the dye beyond the stratum corneum. Dermoscopy imaging clearly showed microneedle pores in microneedle treated samples (MN, Figure 1B). RCM also confirmed the presence of microneedle pores (data not shown). LSCM revealed a fluorescence pattern that consisted of primarily furrow localisation in all NaF treated groups. The only cases where dye penetration occurred were those samples treated with microneedles (Figure 1B, bottom panel).

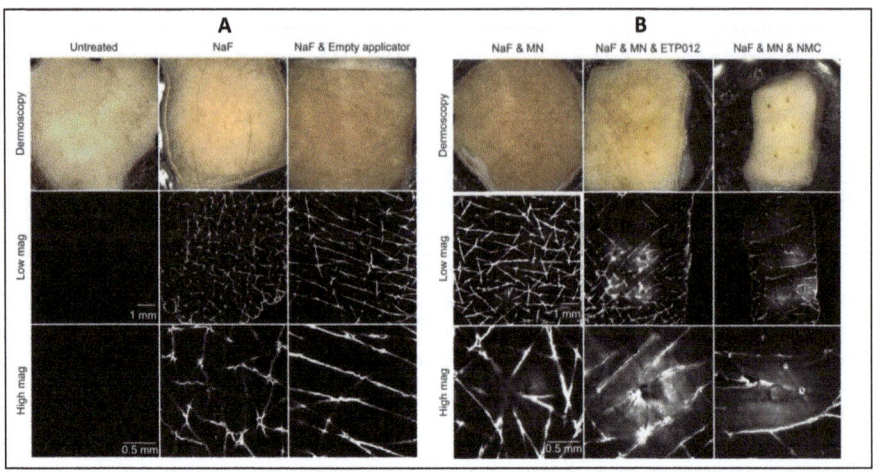

Figure 1 *Dermoscopy and confocal imaging of treated skin. Skin was treated with saline, sodium fluorescein (NaF), microneedles (MN), ETP012 or a non-magnetic control material (NMC). Dermoscopy images are shown in top row and LSCM images are shown at low and high magnification in the second and third rows. The images are separated into non-treated (A) and (B) skin exposed to the microneedle array.*

In all samples treated with microneedles, the dye diffused through the stratum corneum and epidermis, into the dermis. All of the fluorescence images shown in Figure 1 were acquired at a depth of 73 μm, which corresponds to the superficial dermis. Visually, the group with NaF & MN & EPT012 showed the highest levels of radial diffusion and depth of NaF penetration. However, the overall levels of fluorescence were quite variable from one microneedle pore to the next.

Image analysis was performed to identify the normalised area positive for NaF fluorescence. A threshold served to reduce the confounding influence of NaF positive furrow data and select highly fluorescent areas. The resulting data shown in Figure 2A and C illustrate that there was a clear increase in NaF positive area in skin treated with the combination of microneedles and ETP012. In Figure 2A, the NaF positive area for skin treated with microneedles and ETP012 remained above background levels (i.e. microneedles with the non-magnetic control (NMC)) to a depth of ~100 μm. This suggests that the ETP012 effects were limited to the stratum corneum, epidermis and uppermost dermis. However, one of the limitations of this study was that only *en face* imaging was used and in our experience, imaging signal rapidly declines at this depth (~100 μm) due to light scattering. This could be addressed in the future by processing the tissue for histological sections followed by LSCM. Figure 2C shows a summary of the NaF positive area data that sums the data from the entire z-stack. This comparison clearly indicates a positive effect of ETP012 over the non-magnetic control material. This said, the corresponding increase in variability likely prevented these groups from being statistically significant. In the future, additional replicates should clarify this issue.

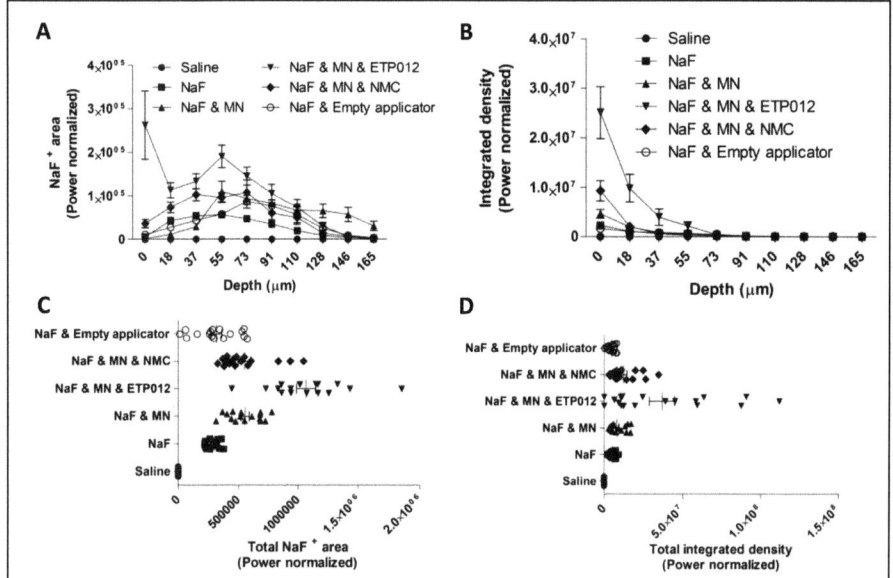

Figure 2 *Image analysis results from LSCM. Panel A shows the normalised size of the NaF positive (NaF+) areas at each depth. Panel B is a summary of that data wherein the values from Panel A were summed through the entire z-stack. Panel C represents the integrated density of the NaF+ areas at each depth analysed. The sum of those data is shown in Panel D.*

Integrated density is a measure of the intensity of the NaF signal and the data shown in Figure 2B clearly indicates that ETP012 treated skin had higher levels of fluorescence in the stratum corneum and epidermis than all other groups. These data suggest that there was enhanced influx of NaF in the top 37μm of skin. Figure 2D summarises the results from all groups and again shows that the ETP012 treatment clearly affected NaF fluorescence localisation, but also showed a corresponding increase in variability.

4 CONCLUSION

We observed no differences between the treatment groups with dermoscopy or RCM with the exception of the pores in those groups treated with microneedles. This illustrates the minimally invasive nature of magnetophoresis, as opposed to tape stripping, for example. LSCM revealed an interesting story where the ETP012 clearly had an enhancing effect on dye penetration and diffusion. This resulted in a clearly enhanced dye penetration kinetic profile for the ETP012 group versus all other controls. Our conclusion is that ETP012 treatment improved microneedle enhanced drug delivery. In future studies, *in vivo* studies with more replicates would help to translate this technology to the clinic.

5 ACKNOWLEDGEMENTS AND DECLARATIONS

The authors acknowledge the support of OBJ Ltd, in particular Mr Jeff Edwards and Dr Christopher Quirk, for the gift of magnetic and non-magnetic film used in this study. We are also grateful to the patients and surgeons who donated skin. YM acknowledges the financial support of an APA scholarship and Curtin University travel assistance. Heather Benson is a shareholder in OBJ Ltd.

6 REFERENCES

1.   Y. C. Kim, J.H. Park, and M. R. Prausnitz, Microneedles for drug and vaccine delivery, *Advanced drug delivery reviews*, 2012, **64**(14), 1547-1568.

2.   W. Lin, *et al.*, Transdermal delivery of antisense oligonucleotides with microprojection patch (Macroflux) technology, *Pharmaceutical research*, 2001, **18**(12): p. 1789-93.

3.   V. Vemulapalli, *et al.*, *In vivo* iontophoretic delivery of salmon calcitonin across microporated skin, *Journal of pharmaceutical sciences*, 2012, **101**(8), 2861-9.

4.   M. J. Garland, *et al.*, Dissolving polymeric microneedle arrays for electrically assisted transdermal drug delivery. *Journal of controlled release : official journal of the Controlled Release Society*, 2012. **159**(1), p. 52-9.

5.   C. S. Kolli, *et al.*, Microneedle assisted iontophoretic transdermal delivery of prochlorperazine edisylate. *Drug development and industrial pharmacy*, 2012, **38**(5): p. 571-6.

6.   H. Chen, *et al.*, Iontophoresis-driven penetration of nanovesicles through microneedle-induced skin microchannels for enhancing transdermal delivery of insulin. *Journal of controlled release : official journal of the Controlled Release Society*, 2009. **139**(1): p. 63-72.

7.   J. R. Wilson, L. J. Kehl, and S. Beiraghi, Enhanced topical anesthesia of 4% lidocaine with microneedle pretreatment and iontophoresis. *Northwest dentistry*, 2008, **87**(3), p. 40-1.

8.   X. M. Wu, H. Todo, and K. Sugibayashi, Enhancement of skin permeation of high molecular compounds by a combination of microneedle pretreatment and iontophoresis, *Journal of controlled release*, 2007, **118**(2), p. 189-95.

9.   C. A. Bassett, Fundamental and practical aspects of therapeutic uses of pulsed electromagnetic fields (PEMFs). *Critical reviews in biomedical engineering*, 1989, **17**(5), p. 451-529.

10.   C. A. Bassett, Beneficial effects of electromagnetic fields, *Journal of cellular biochemistry*, 1993. **51**(4), p. 387-93.

11. C. Polk and E. Postow, *Handbook of Biological Effects of Electromagnetic Fields*. 1996, CRC Press, Boca Raton.

12. M. Matic, *et al.*, Influence of different types of electromagnetic fields on skin reparatory processes in experimental animals, *Lasers in medical science*, 2009, **24**(3), p. 321-7.

13. M. S. Scardino, *et al.*, Evaluation of treatment with a pulsed electromagnetic field on wound healing, clinicopathologic variables, and central nervous system activity of dogs, *American journal of veterinary research*, 1998, **59**(9): p. 1177-81.

14. S. Ahmadian, S. R. Zarchi, and B. Bolouri, Effects of extremely-low-frequency pulsed electromagnetic fields on collagen synthesis in rat skin, *Biotechnology and applied biochemistry*, 2006, **43**(Pt 2), p. 71-5.

15. M. J. Callaghan, *et al.*, Pulsed electromagnetic fields accelerate normal and diabetic wound healing by increasing endogenous FGF-2 release, *Plastic and reconstructive surgery*, 2008, **121**(1), p. 130-41.

16. J. Milgram, *et al.*, The effect of short, high intensity magnetic field pulses on the healing of skin wounds in rats, *Bioelectromagnetics*, 2004. **25**(4), p. 271-7.

17. S. N. Murthy, S.M. Sammeta, and C. Bowers, Magnetophoresis for enhancing transdermal drug delivery: Mechanistic studies and patch design, *Journal of controlled release*, 2010, **148**(2), p. 197-203.

18. S. N. Murthy, Magnetophoresis: an approach to enhance transdermal drug diffusion, *Die Pharmazie*, 1999, **54**(5), p. 377-9.

19. G. Krishnan, *et al.*, Enhanced skin permeation of naltrexone by pulsed electromagnetic fields in human skin *in vitro*, *Journal of pharmaceutical sciences*, 2010, **99**(6), p. 2724-31.

20. S. Namjoshi, *et al.*, Liquid chromatography assay for 5-aminolevulinic acid: application to *in vitro* assessment of skin penetration via Dermaportation, *Journal of chromatography. B, Analytical technologies in the biomedical and life sciences*, 2007, **852**(1-2), p. 49-55.

21. S. Namjoshi, *et al.*, Enhanced transdermal delivery of a dipeptide by dermaportation, *Biopolymers*, 2008, **90**(5), p. 655-62.

22. H. A. Benson, *et al.*, Enhanced skin permeation and hydration by magnetic field array: preliminary *in-vitro* and *in-vivo* assessment, *The Journal of pharmacy and pharmacology*, 2010, **62**(6), p. 696-701.

15

IONTOPHORETIC DELIVERY OF GLYCOMIMETICS - A NEW APPROACH FOR THE TREATMENT OF INFLAMMATORY SKIN DISEASES

T Gratieri[1,2], B Wagner[3], D Kalaria[1], B Ernst[3], Y N Kalia[1]

[1]School of Pharmaceutical Sciences, University of Geneva & University of Lausanne, 30 Quai Ernest Ansermet, 1211 Geneva, Switzerland. [2]Faculdade de Ciências da Saúde, Universidade de Brasília. Campus Universitário Darcy Ribeiro, s/n, 70910-900, Brasília, DF, Brazil. [3]Institute of Molecular Pharmacy, University of Basel, Klingelbergstrasse 50, CH-4056 Basel, Switzerland.

1 INTRODUCTION

A key event in the inflammatory response is the recruitment of leukocytes from the microcirculation to the inflamed tissue through endothelial-dependent mechanisms that include leukocyte tethering and rolling, activation, firm adhesion and diapedesis to the interstitium [1, 2]. Leukocyte tethering and rolling along the endothelial surface is mediated by a family of carbohydrate-binding proteins (E-, P- and L-selectins). While L-selectin is constitutively expressed on the surface of leukocytes, expression of E- and P-selectin is upregulated on the endothelial surface during inflammation [3].

The cutaneous immune response is a normal physiological mechanism required for surveillance. However, the abnormal or excessive recruitment and influx of leukocytes is involved in the pathogenesis of several inflammatory diseases, such as psoriasis, eczematous disorders, atopic dermatitis or allergic contact dermatitis [4 – 8]. For example, psoriatic lesions reveal intense inflammatory infiltrate of the upper dermis, predominantly consisting of T-cells, which are activated by antigen-presenting cells and play a key role in the initiation and perpetuation of the disease, especially by influencing keratinocyte function leading to psoriasis-specific skin lesions [4]. Therefore, antagonism of the selectins and consequent inhibition of leukocyte trafficking between lymph nodes, blood and skin is considered to be a valuable approach for the treatment of these diseases [9].

Although there is evidence that administration of E- and P-selectin antagonists can significantly alleviate chronic inflammatory skin conditions [10 – 12], many compounds generated to-date have either not entered the clinic or failed in clinical trials [13, 14]. Among the drawbacks in targeting leukocyte migration into the skin with selectin antagonists are: (i) the considerable functional overlaps or redundancies of selectins; monospecific selectin antagonists are less effective (e.g., CDP-850) [13], (ii) poor binding affinities reflected by high IC_{50} values (e.g., Cylexin [1]) which reduce efficacy and significantly increase treatment costs and (iii) unfavorable pharmacokinetic properties such as high polarity which will prevent passive diffusion through the enterocyte layer in the small intestine (a prerequisite for oral availability) and facilitate rapid renal excretion [15, 16]. The topical administration of these molecules would be a non-invasive targeted alternative to systemic administration that could overcome most of the pharmacokinetic drawbacks.

The E-selectin antagonist CGP69669A (Figure 1) is a sialyl Lewis[x] (sLe[x])-glycomimetic [17 – 19]. Studies in tumor necrosis factor alpha (TNFα) stimulated mouse cremaster, which have induced expression of E-selectin and increased expression of P-selectin on endothelial cells, showed that treatment with CGP69669A markedly increased the velocity of rolling leukocytes if applied alone, and completely blocked E-selectin–dependent rolling when combined with anti–P-selectin antibody [17].

Although the potential of CGP69669A to block selectins has been established, its efficacy in the treatment of skin inflammatory disorders has not been tested due in large part to the challenges posed by its delivery. Topical administration could be used to target skin lesions and, as a consequence, reduce the amount of drug required for a therapeutic effect; furthermore, it may also be used to complement the administration of other therapeutic agents in multi-therapy approaches designed to overcome functional redundancy [15]. Such an approach might also reduce the risk of side effects as compared to (conventional) systemic immunosuppressive therapies, given that T cells would not be depleted, impaired in their general function or compromised with regard to their capacity for circulation into other organs.

In order to exert a pharmacologic effect, CGP69669A has to penetrate the stratum corneum (the skin's outermost layer and principal barrier to molecular transport). As it is a negatively charged molecule in aqueous solution under physiological conditions (log D at pH 7.4 - <-1.5), drug partitioning into the lipid-rich intercellular space in the stratum corneum, following topical application, will be limited and as a result it can be envisaged that it would display poor passive delivery. In contrast, its physicochemical properties suggest that CGP69669A is suitable for transdermal iontophoresis, which involves application of an electrical potential gradient and is used to enhance the delivery of hydrosoluble, ionised molecules through the skin through transport channels with more aqueous character [20]. Iontophoresis has been shown to significantly increase the

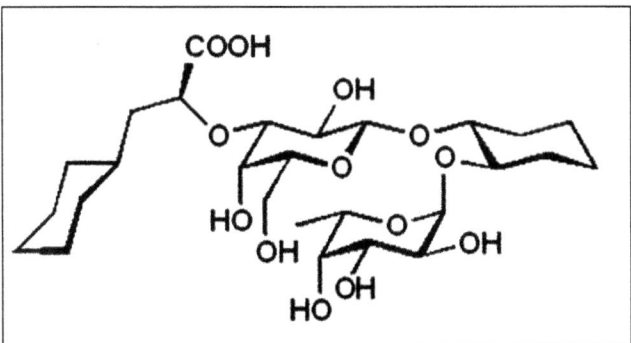

Figure 1 *Structure of the E-selectin antagonist CGP69669A (molecular weight = 600.64 Da).*

permeation of several ionised molecules (frequently by several orders of magnitude) [21–24].

Thus, the aim of the present work was (i) to evaluate the feasibility of delivering CGP69669A by transdermal iontophoresis and hence provide a first demonstration of its ability to administer "difficult-to-deliver" glycomimetics and (ii) to investigate the effect of experimental variables including applied current density, drug concentration, and formulation on iontophoretic transport rates in order to establish the optimal conditions for future *in vivo* studies.

2 MATERIALS AND METHODS

CGP69669A was synthesised in-house [19]. Rafinose, sodium hydroxide, silver wire and silver chloride were purchased from Sigma-Aldrich (Buchs, Switzerland). Sodium chloride, disodium hydrogenphosphate and potassium dihydrogenphosphate were purchased from Fluka (Buchs, Switzerland). PVC tubing (3 mm ID, 5 mm OD, 1 mm wall thickness) used to prepare salt bridge assemblies was obtained from Fisher Bioblock Scientific S.A. (Illkirch, France). All solutions were prepared using deionised reverse osmosis filtered water (resistivity $\geq$ 18 MΩ cm). All other chemicals were at least of analytical grade.

2.1 Skin source
Porcine ears were obtained from a local abattoir (CARRE; Rolle, Switzerland), the skin was excised (thickness 750 μm) with an air dermatome (Zimmer; Etupes, France), wrapped in aluminum foil and stored in polyethylene bags at −20 °C for a maximal period of 2 months. Human skin samples were collected immediately after abdominoplasty from the Department of Plastic, Aesthetic and Reconstructive Surgery, Geneva University Hospital (Geneva, Switzerland). After excision, the skin was cut into 10 cm x 10 cm pieces and the subcutaneous fatty tissue was removed from the skin specimen using a scalpel. The surface of each specimen was then cleaned with water and stored in the same manner as the porcine skin. The study was approved by the Central Committee for Ethics in Research (CER: 08-150 (NAC08-051); Geneva University Hospital).

2.2 Chromatography
CGP69669A was quantified by High Performance Anion Exchange Chromatography with Pulsed Amperometric Detection (HPAE-PAD). The ion chromatography system consisted of a GP50 gradient pump, ED50A electrochemical detector with a gold working electrode, AS50 autosampler, AS50TC thermal compartment, and Chromeleon® chromatography workstation (Dionex Corporation, Sunnyvale, CA, USA). A Dionex CarboPac^{TM} PA-200 anion-exchange column (3 x 250 mm) and guard column (3 x 50 mm) were used for analysis. The mobile phase was 100 mM NaOH solution. It was prepared in deionised water and purged with helium to minimise carbonate content. The flow rate was 0.50 ml/min and the injection volume was 10 μl. Detection was performed by using a quadruple potential waveform: +0.1 V from 0.0 to 0.4 s, -2.0 V from 0.41 to 0.42 s, +0.6V at 0.43 s and -0.1 V from 0.44 to 0.50 s, using the combination Ag/AgCl/pH reference electrode with the instrument set in the pH mode and with current integrated between 0.20 and 0.40 s for detection [25]. Rafinose was used as an internal standard.

The method was validated and showed good precision and accuracy both intra- and inter-day. CGP69669A concentrations of 1.0, 2.5, 5.0, 7.5, 10, 12.5, 15, 17.5, 20, 22.5 and 25 μg ml^{-1} in PBS buffer were used to prepare the calibration curves. 50 μl of the internal standard solution (Rafinose 55 μg ml^{-1}) was added to 1 ml of the CGP69669A sample. Calibration curves were fitted using least squares linear regression analysis ($Y = 0.0962$ x $+ 0.0436$; $r^2 = 0.997$). The specificity of the analytical method was verified with respect to endogenous compounds present in the skin. The results showed that there was no interference between the biological matrix and both the CGP69669A ($t_R = 3.0$ min) and internal standard peaks ($t_R = 3.6$ min). The limits of detection (LOD) and quantification (LOQ) for CGP69669A were 0.3 and 1.0 μg ml^{-1}, respectively.

Epidermal and dermal stabilities of CGP69669A were determined by placing 1 ml of 5 mg ml^{-1} CGP69669A solution (pH 7.5) in contact with epidermis (2 cm^2) and dermis (2 cm^2), respectively. Samples were collected every two hours for 6 h and were analysed by the HPAE-PAD method described above. The experiments were performed in triplicate.

The electrostability of a CGP69669A solution (pH 7.5; 5 mg ml^{-1}) was determined using a current density of 0.5 mA cm^{-2} applied for 6 h using salt bridges to ensure connectivity between the electrodes and the formulation. Samples were collected every two hours and were analysed by the HPAE-PAD method described above. The experiments were performed in triplicate.

Transdermal delivery by passive diffusion of CGP69669A was investigated across intact dermatomed porcine skin and tape-stripped human skin (15 tape-strips). The skin was clamped in two-compartment vertical diffusion cells (area 2.0 cm^2). The donor compartment was filled with 1 ml of a 5 mg ml^{-1} CGP69669A solution; the receptor compartment contained 12 ml of PBS buffer (pH 7.4) solution. During the experiment, samples (1 ml) were withdrawn from the receiver compartment hourly for 6 h and replaced with fresh buffer solution.

At the end of the permeation experiment, the diffusion cells were dismantled and the skin surface washed in running water to remove residual donor solution. The CGP69669A deposited in the skin was extracted by cutting the skin samples into small pieces and soaking them in 5 ml of PBS buffer. Samples were left overnight under constant stirring at ambient temperature and then filtered and analysed using the HPAE-PAD method described above. The extraction method was validated by spiking the skin samples with two different known amounts of CGP69669A in ethanol solution. After solvent evaporation, skin samples were subjected to the extraction procedure and the percentage recovery of CGP69669A was determined by calculating the ratio of the amount extracted from the skin samples to the amount added, determined by direct injection of the spiking solution in the absence of skin. The experiments were performed in triplicate. The recovery rates for the theoretical concentrations of 10 and 20 μg ml^{-1} were 95.9 ± 11.4 and 103.5 ± 16.4 %, respectively.

The experimental setup for assessing the electrically-mediated dermal absorption of CGP69669A was similar to that described in earlier studies [26]. Dermatomed porcine skin was clamped in two-compartment vertical diffusion cells (area 2.0 cm^2). The donor compartment was filled with either 1 ml of CGP69669A solution or 1 g of 2% hydroxyethyl cellulose gel containing 5 mg g^{-1} of CGP69669A and was connected to the cathode via a salt bridge assembly (3% agarose in 0.1 M NaCl). The receptor compartment housed the anode and was filled with 12 ml of PBS buffer 133 mM NaCl (pH 7.4) solution. During the experiment, samples (1 ml) were withdrawn from the receiver compartment hourly for 6 h and replaced with fresh buffer solution. Constant current was applied using Ag/AgCl electrodes connected to a power supply (Kepco® APH 1000 M, Flushing, NY). At the end of the permeation experiment (6 h), the drug was extracted from the skin and analysed as described above. Cumulative drug permeation was plotted as a function of time and the flux (J) was determined from the slope of the linear portion of the graph.

In order to investigate the effect of current on CGP69669A electrotransport kinetics, three different current densities (0.1, 0.3 and 0.5 mA cm^{-2}) were applied for 6 h using an aqueous solution of CGP69669A (5 mg ml^{-1}; pH 7.5) as the cathodal donor solution ($n \geq 4$ in all experiments).

The effect of drug concentration on CGP69669A delivery was studied by comparing CGP69669A iontophoresis at 0.5 mA cm^{-2} for 6 h from formulations containing 1, 3 and 5 mg/ml of CGP69669A in aqueous solution (pH = 7.5) ($n \geq 4$ in all experiments).

Iontophoretic delivery of CGP69669A across porcine skin and human skin samples was compared under the same experimental conditions – 5 mg ml^{-1} CGP69669A iontophoresed for 6 h at 0.5 mA cm^{-2} ($n \geq 4$ in both experiments).

Data were expressed as mean ± standard deviation (SD). Outliers determined using the Grubbs test were discarded. Results were evaluated statistically using either analysis of variance (ANOVA followed by Student Newman Keuls test) or Student's t-test. The level of significance was fixed at $\alpha = 0.05$.

3 RESULTS AND DISCUSSION

Prior to conducting the iontophoretic transport experiments, the solution stability of CGP69669A in contact with the epidermal and dermal skin surfaces and in the presence of current was evaluated. The amount of CGP69669A in solution after exposure to the skin samples for 6 h was 103.4 ± 1.7 and 98.1 ± 4.9 %, respectively, of that observed initially. After 6 h of current application, the amount of CGP69669A in solution was 96.5 ± 8.9 % of that at t = 0 h. Therefore, CGP69669A was considered to be stable under the conditions for the permeation experiments.

Passive permeation of CGP69669A across intact dermatomed porcine skin and tape-stripped human skin resulted in drug levels in the receiver compartment that were below the limit of detection (LOD 0.3 μg ml^{-1}). Furthermore, in both cases no CGP69669A was detected after skin extraction, implying that < 1.5 μg (considering the dilution of 5 ml for the extraction procedure) of CGP69669A was recovered from the skin. The results also showed that removal of the stratum corneum through tape stripping was unable to increase the passive permeation of CGP69669A. This was tentatively attributed to the ability of CGP69669A to interact with the epidermis through the formation of hydrogen bonds.

Cumulative CGP69669A permeation after 6 h of iontophoresis at 0.5 mA cm^{-2} was 160 ± 14 μg cm^{-2} and the steady state flux was 25 ± 6 μg cm^{-2} h. This confirmed that iontophoresis was extremely effective in enhancing the transport of molecules with such physicochemical properties – far more so than stratum corneum removal, which was not able to increase delivery. Given this observation, it is possible that stratum corneum ablation techniques might be less useful for these molecules.

The effect of current density on CGP69669A transport was investigated by determining cumulative permeation and steady state iontophoretic flux at 0.1, 0.3 and 0.5 mA cm^{-2}

(Figure 2). A linear relationship was observed when both cumulative CGP69669A permeation (70 ± 10, 114 ± 27 and 160 ± 14 µg cm^{-2}, respectively) and flux (11 ± 3, 19 ± 5 and 25 ± 6 µg cm^{-2} h^{-1}, respectively) were plotted against current density (r^2 = 0.999 in both cases). Statistical analysis (ANOVA; α=0.05) followed by Student Newman Keuls test) showed that there was a statistically significant difference between cumulative CGP69669A permeation at 0.1, 0.3 and 0.5 mA cm^{-2}. The extraction experiments showed that CGP69669A was retained within the membrane during iontophoresis. Similar amounts were recovered at 0.1 and 0.3 mA cm^{-2} (19 ± 4 and 23 ± 6 µg cm^{-2}, respectively); however, a further increase to 0.5 mA cm^{-2} did result in a statistically significant increase in skin deposition (50 ± 12 µg cm^{-2}).

These results were in contrast with those of an earlier study using cathodal iontophoresis [27] where increasing the current density did not produce an effect on the amount of dexamethasone sodium phosphate retained within the skin after 7 h of iontophoresis. It was suggested that the number of binding sites for anions was limited and readily saturated. The present study may indicate that such saturation was not achieved here or that, when using higher current density or concentration, interactions at other sites became possible.

Iontophoresis of CGP69669A solutions containing 1, 3 and 5 mg ml^{-1} for 6 h at 0.5 mA cm^{-2} also resulted in a linear and statistically significant increase (r^2 = 0.99; ANOVA (α=0.05) followed by Student Newman Keuls test) in both cumulative permeation (37 ± 13, 79 ± 23 and 160 ± 14 µg cm^{-2}, respectively) and the iontophoretic flux (7.1 ± 2.5, 14.4 ± 4.1 and 25.4 ± 6.4 µg cm^{-2} h^{-1}, respectively). Skin extraction also showed an increase in the amount of CGP69669A retained within the membrane at the highest drug concentration (Figure 3).

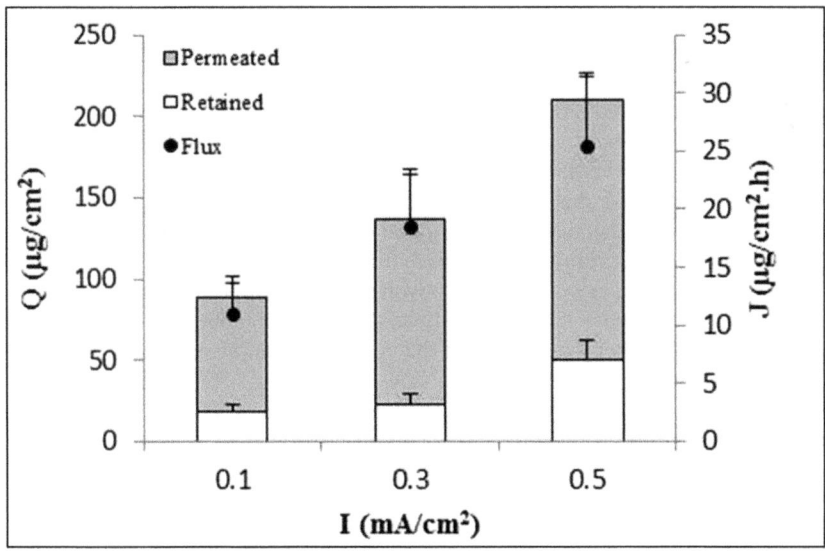

Figure 2 *Effect of current density (0.1, 0.3 and 0.5 mA $^{-2}$) on total delivery of CGP69669A (Q; sum of the amounts permeated across and retained within the skin) and steady state flux (J); transdermal iontophoresis for 6 h using a 5.0 mg ml^{-1} CGP69669A solution. (Mean ± SD; n ≥ 4).*

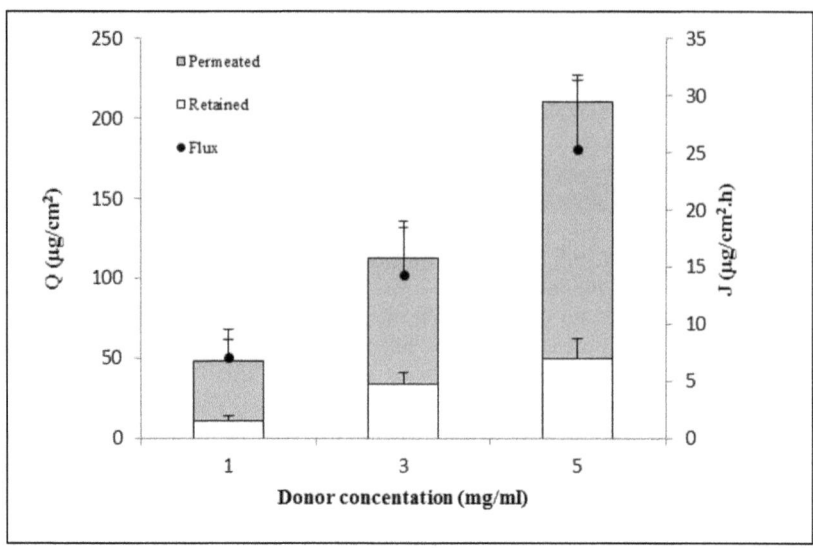

Figure 3 *Effect of donor concentration (1.0, 3.0 and 5.0 mg ml^{-1}) on total delivery of CGP69669A (Q; sum of the amounts permeated across and retained within the skin) and steady state flux (J); transdermal iontophoresis for 6 h at 0.5 mA cm^{-2}. (Mean ± SD; n ≥ 4).*

Porcine ear skin is routinely used since it is generally considered a good model for transdermal delivery across human skin [28, 29]; however, given the novelty of CGP69669A, it was important to investigate electrotransport across excised human skin and porcine ear skin under the same iontophoretic conditions (6 h current application at 0.5 mA cm^{-2} and using 5.0 mg ml^{-1} CGP69669A). Cumulative permeation (160 ± 14 and 146 ± 37 µg cm^{-2}) and steady state flux (25.4 ± 6.4 and 32.2 ± 6.7 µg cm^{-2} h^{-1}) across porcine and human skins were statistically equivalent (Student's t-test; α=0.05). The results confirmed that constant current iontophoresis was able to deliver CGP69669A across human skin and suggested that porcine ear skin was a valid model for studying the transdermal iontophoretic delivery of glycomimetics such as CGP69669A.

The feasibility of delivering CGP69669A from a 2% hydroxyethyl cellulose gel was also evaluated (Figure 4), since a semi-solid hydrogel formulation would represent a more patient-friendly formulation for topical application [30]. Iontophoretic permeation from solution was two-fold higher than that from the gel, which can be explained by the higher viscosity of the latter and as a consequence a slower molecular diffusion in the matrix [31]. However, the amount of CGP69669A accumulated in the skin was approximately three-fold higher using the gel formulation (163 ± 80 µg cm^{-2}), which would obviously be an advantage for local therapy. Considering the total amount of drug delivered (sum of the amounts permeated and retained in the skin) there was no statistically significant difference from the solution to the gel formulation; however, the selectivity for deposition over permeation was increased. Thus, even in a non-optimised form, ~10% of the applied amount of CGP69669A was delivered following iontophoresis using the gel: ~3% was detected in the receiver compartment and ~6.5 % was accumulated within the skin, demonstrating the feasibility of using such a formulation.

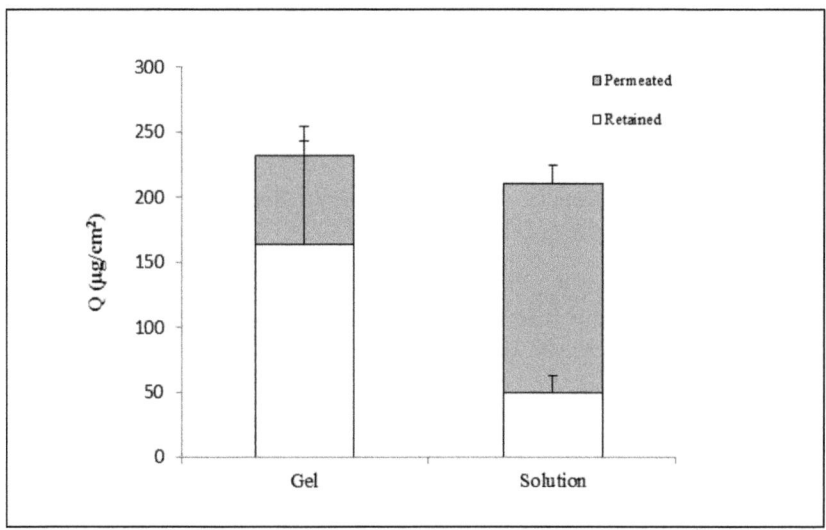

Figure 4 *Comparison of total delivery of CGP69669A (Q; sum of the amounts permeated across and retained within the skin) and steady state flux (J) after 6 h of transdermal iontophoresis at 0.5 mA/cm² from either solution (5.0 mg/ml) or 2% hydroxyethyl cellulose gel containing 5.0 mg/g of the drug (pH 7.5). (Mean ± SD; n ≥ 4).*

4 CONCLUSION

The results confirm that CGP69669A, a small anionic glycomimetic (and E-selectin antagonist), can be delivered non-invasively into and across intact skin by transdermal iontophoresis in a controlled and reproducible manner: the flux increases essentially linearly as a function of both the current density and CGP69669A concentration. The next step in the project is to optimise the CGP69669A formulation in preparation for *in vivo* studies. In the longer term, the iontophoretic transport of other glycomimetics with superior therapeutic potential for the treatment of inflammatory skin disorders will be investigated – either singly or in combination with other therapeutic agents that have different mechanisms of action. Experimental and computational studies will also be performed in order to understand the effect of glycomimetic molecular properties on electrotransport behavior.

5 REFERENCES

1. M. P. Schön and R. J. Ludwig, Lymphocyte trafficking to inflamed skin–molecular mechanisms and implications for therapeutic target molecules, *Expert Opin Ther Targets*, 2005, **9**, 225–243.

2. K. Ley, Molecular mechanisms of leukocyte recruitment in the inflammatory process, *Cardiovasc Res*, 1996, **32**, 733–742.

3. G. S. Kansas, Selectins and their ligands: current concepts and controversies, *Blood* 1996, **88**, 3259–3287.

4. D. Bock, S. Philipp and G. Wolff, Therapeutic potential of selectin antagonists in psoriasis, *Expert Opin. Investig. Drugs*, 2006, **15**, 963–979.

5. C. Robert and T. S. Kupper, Inflammatory skin diseases, T cells, and immune surveillance, *N. Engl. J. Med.*, 1999, **341**, 1817–1828.

6. A. Tanaka, H. Takahama, T. Kato *et al.*, Clonotypic analysis of T cells infiltrating the skin of patients with atopic dermatitis: Evidence for antigen-driven accumulation of T cells, *Hum Immunol.*, 1996, **48**, 107–113.

7. M. Vocanson, A. Hennino, A. Rozieres, G. Poyet and J. F. Nicolas, Effector and regulatory mechanisms in allergic contact dermatitis, *Allergy*, 2009, **64**, 1699–1714.

8. S. Vollmer, A. Menssen, J. C. Prinz, Dominant lesional T cell receptor rearrangements persist in relapsing psoriasis but are absent from nonlesional skin: evidence for a stable antigen-specific pathogenic T cell response in psoriasis vulgaris, *J. Invest. Dermatol.* 2001, **117**, 1296–1301.

9. M. P. Schön, Inhibitors of selectin functions in the treatment of inflammatory skin disorders, *Ther Clin Risk Manag.*, 2005, **1**, 201–208.

10. M. P. Schön, T. Krahn, M. Schon, *et al.*, Efomycine M, a new specific inhibitor of selectin, impairs leukocyte adhesion and alleviates cutaneous inflammation, *Nat. Med.*, 2002, **8**, 366–372.

11. M. Friedrich, D. Bock, S. Philipp *et al.*, Pan-selectin antagonism improves psoriasis manifestation in mice and man, *Arch. Dermatol. Res*, 2006, **297**, 345–351.

12. N. C. Kaneider, A. J. Leger and A. Kuliopulos, Therapeutic targeting of molecules involved in leukocyte-endothelial cell interactions, *FEBS J.*, 2006, **273**, 4416–4424.

13. M. Bhushan, T. O. Bleiker, A. E. Ballsdon, *et al.*, Anti-E-selectin is ineffective in the treatment of psoriasis: a randomized trial, *Br. J. Dermatol.*, 2002, **146**, 824–831.

14. M. Friedrich, S. Philipp, M. Hardtke *et al.*, Anti-L-selectin therapy is not effective in psoriasis: a randomized trial, *J. Invest. Dermatol.*, 2005, **124**, 232.

15. W. H. Boehncke, M. P. Schön, G. Giromolomi *et al.*, Leukocyte extravasation as a target for anti-inflammatory therapy – Which molecule to choose? *Exp. Dermatol.*, 2005, **14**, 70–80.

16. B. Ernst and J. L. Magnani, From carbohydrate leads to glycomimetic drugs, *Nat. Rev. Drug. Discov.*, 2009, **8**, 661–677.

17. K. E. Norman, G. P. Anderson, H. C. Kolb, K. Ley and B. Ernst, Sialyl Lewis(x) (sLe(x)) and an sLe(x) mimetic, CGP69669A, disrupt E-selectin-dependent leukocyte rolling *in vivo*, *Blood*, 1998, **91**, 475–483.

18. A. Titz, J. Patton and M. Smiesko *et al.*, Probing the carbohydrate recognition domain of E-selectin: the importance of the acid orientation in sLex mimetics, *Bioorg. Med. Chem.*, 2010, **18**, 19–27.

19. H. C. Kolb and B. Ernst, Development of tools for the design of selectin antagonists, *Chem. Eur. J.*, 1997, **3**, 1571–1578.

20. Y. N. Kalia and A. Naik, J. Garrison, R. H. Guy, Iontophoretic drug delivery, *Adv. Drug Deliv. Rev.*, 2004, **56**, 619–658.

21. L. P. Gangarosa Sr, A. Ozawa, M. Ohkido, Y. Shimomura, J. M. Hill, Iontophoresis for enhancing penetration of dermatologic and antiviral agents, *J. Dermatol.*, 1995, **22**, 865–875.

22. W. T. Zempsky, J. Sullivan, D. M. Paulson, S. B. Hoath, Evaluation of a low-dose lidocaine iontophoresis system for topical anesthesia in adults and children: A randomized, controlled trial, *Clin. Ther.*, 2004, **26**, 1110–1119.

23. M. L. Welch, W. J. Grabski, M. L. McCollough, *et al.*, 5-fluorouracil iontophoretic therapy for Bowen's disease, *J. Am. Acad. Dermatol.*, 1997, **36**, 956–958.

24. N. Abla, A. Naik, R. H. Guy, Y. N. Kalia, Topical iontophoresis of valaciclovir hydrochloride improves cutaneous aciclovir delivery, *Pharm. Res.*, 2006, **23**, 1842–1849.

25. R. D. Rocklin, A. P. Clarke and M. Weitzhandler, Improved long-term reproducibility for pulsed amperometric detection of carbohydrates via a new quadruple-potential waveform, *Anal. Chem.*, 1998, **70**, 1496–1501.

26. S. Dubey and Y. N. Kalia, Non-invasive iontophoretic delivery of enzymatically active ribonuclease A (13.6 kDa) across intact porcine and human skins, *J. Control Release* 2010, **145**, 203–209.

27. J. Cazares-Delgadillo, C. Balaguer-Fernandez, A. Calatayud-Pascual *et al.*, Transdermal iontophoresis of dexamethasone sodium phosphate *in vitro* and *in vivo*: Effect of experimental parameters and skin type on drug stability and transport kinetics, *Eur. J. Pharm. Biopharm.*, 2010, **75**, 173–178.

28. I. P. Dick and R. C. Scott, Pig ear skin as an in-vitro model for human skin permeability, *J. Pharm. Pharmacol.*, 1992, **44**, 640–645.

29. U. Jacobi, M. Kaiser, R. Toll *et al*, Porcine ear skin: an *in vitro* model for human skin, *Skin Res. Technol.*, 2007, **13**, 19–24,

30. A. K. Banga and Y. W. Chien, Hydrogel-based iontotherapeutic delivery devices for transdermal delivery of peptide/protein drugs, *Pharm. Res.*, 1993, **10**, 697–702.

31. J. F. Huang, K. C. Sung, O. Y. P. Hu, J. J. Wang, Y. H. Lin, J. Y. Fang, The effects of electrically assisted methods on transdermal delivery of nalbuphine benzoate and sebacoyl dinalbuphine ester from solutions and hydrogels, *Int. J. Pharm.* 2005, **297**, 162–171.

16

FOAM – A UNIQUE TOPICAL DRUG DELIVERY SYSTEM

D Tamarkin

Foamix Ltd., Weizmann Science Park, 2 Holzman Street, Rehovot 76704, Israel

1 INTRODUCTION

Topical therapy is one of the foundations of a dermatologist's therapeutic tools as it is used for symptomatic relief, control, or cure of the underlying disease. The vehicles used to deliver topical therapy can considerably influence drug performance. The vehicle can have direct effects on disease; it can affect the delivery of the active agent, its physical appearance and contribute to sensory properties which affect compliance. While semi-solid compositions such as creams, lotions, gels and ointments are commonly used by consumers, new forms are desirable in order to achieve better control of the application, improve skin absorption and maintain or bestow the beneficial properties of such topical products.

Foam is becoming a prominent delivery system for topical drugs. This platform provides an innovative, easy to apply, modern alternative to creams and ointments. A significant advantage of the foam formulation is that it spreads easily on large skin areas, is not greasy or oily and does not leave a greasy or oily film on the skin after application.

The use of foam in dermatology was first reported in 1977 by Woodward and Berry who studied the therapeutic benefit of Betamethasone benzoate, in a hydroalcoholic "quick-break" foam in comparison with a corresponding semi-solid dosage form [1]. The activity of the foam, as determined by a vasoconstriction test, was similar to the corresponding ointment and better than cream. In 1995, Deaffontio et. al. investigated the anti-inflammatory and analgesic profile of a topical foam formulation of ketoprofen lysine salt, which exhibited anti-inflammatory and analgesic effectiveness [2, 3]

A comprehensive review on foam drug delivery in dermatology was written by Carryn et. al. in 2003 [4]. Tamarkin et. al. published a broad review, titled "Emollient foam in topical drug delivery" in 2006 [5]; and more recently, in 2010, Steckel et. al. wrote a review on foam technology, entitles "Foams for pharmaceutical and cosmetic application" [6].

2 OVERVIEW OF THE MARKET: CURRENT FOAM TECHNOLOGIES

Currently, only a few prescription dermatological foam products are commercially available (Table 1). The development of new foams is led today by Stiefel and Foamix, which is engaged in the development of innovative foams in collaboration with several pharmaceutical companies.

To-date, all foams are collectively designated "Medicated Foams" by the European Pharmacopoeia; and the U.S. Pharmacopoeia simply lists "Foam Aerosol" as a sub-part of the Aerosol section [17].

3 THE ROSETTA STONE OF FOAM

Not all foams are the same. While in the past there were few types of medicated foam (i.e., aqueous, hydroethanolic and emulsion-based emollient foams) today there are several new classes of foam formulations under development which are distinct from each other by their composition and functionality. For example, there is petrolatum-based foam which is the foam equivalent of an ointment; a foam based on hydrophilic solvents such as PEG and propylene glycol, which is the foam version of a hydrophilic ointment; and an oil-based foam which corresponds to oil solutions or suspensions.

Table 1 *Current, commercially available dermatological foam products.*

Name	Active(s)	Vehicle	Indication	Ref
EpiFoam®	1% Hydrocortisone acetate 1% pramoxine hydrochloride	Aqueous foam	Inflammatory and pruritic manifestations of corticosteroid-responsive dermatoses	7
Olux®	0.05% Clobetasol propionate	Thermolabile steroid hydroethanolic foams (60% ethanol)		8
Luxiq®	0.12% betamethasone valerate			9
Evoclin®	1% clindamycin	Hydroethanolic foam	Acne	10-12
Olux-E®	0.05% clobetasol propionate	Emulsion aerosol foam	Corticosteroid-responsive dermatoses	13
Verdeso®	0.05% desonide			14
Soriluxt™	0.005% calcipotriene		Psoriasis	15
Fabior™	0.1% tazarotene		Acne	16

Foams that are based on potent solvents, such as dimethyl isosorbide and DMSO contribute to high solubility and enhanced transdermal drug delivery and hydroethanolic foams (containing high levels of ethanol) are suitable mostly for scalp treatment. There are also techniques to stabilise suspensions in foam and there are foams that contain high levels of saccharides and honey for wound & burn therapy.

Such versatile foam classes have been used to develop a large number of foam products, containing a variety of active ingredients, including antibiotic agents, antifungals, antiviral agents, immune-modulators, corticosteroids, steroid hormones, anti-acne agents, anti-psoriasis agents and skin barrier-building agents for the treatment of atopic dermatitis.

It is important for the pharmaceutical scientist, as well as the clinical practitioner, to understand the difference between the above classes of foam formulations and to be able to select the right type of formulation for a given clinical condition. The current review presents the "Rosetta Stone" of foam. It introduces the various types of foam technology platforms and suggests a functional "translation" to their respective traditional topical dosage forms. Table 2 lays out a series of foam classes, in correspondence with the current topical dosage forms, with a summary of the main features and attributes of each class of foams; and the following sections will provide further features of each of these classes.

4 WATER CONTAINING FOAMS

The initial generations of medicated foams included aqueous, hydroethanolic and the newer platform of emulsion-based foams, also termed "emollient" foams. Water-containing foams have several general advantages, *viz*:

- Usability: emollient foams spread easily onto the skin and absorb quickly.

- Stability: the pressurized aerosol container is an impermeable packaging system and so prevents contact of the formulation with air, light and contaminants during storage and also during the use period, unlike tubes which are exposed to the environment once they are first opened. Hence, drugs which are prone to oxidation or are photosensitive can have longer shelf life and in-use life when formulated in foam.

- Skin hydration and conditioning. The hydrophobic components of emulsion-based foam, such as petrolatum and liquid oils act to correct skin dryness through their emollient and humectant properties. They make the external layers of the skin (epidermis) softer and more pliable, thereby increasing the skin's hydration (water content) by reducing water evaporation.

The following sections will review the compositions and of the various water-containing foam platforms.

Table 2 *Classification of foam technology platforms, with equivalent traditional topical dosage form designations.*

Type	Class	Foam Formulation	Attributes	Traditional Equivalent
Aqueous	Aqueous Foam	Main ingredients are water, gelling agents and surfactants	Non-greasy.	Gel
	Hydroethanolic Foam	Main ingredients are ethanol and water	Serves to solubilise drugs and thus improve bioavailability. Adequate for oily skin areas. Does not require preservatives.	Solution, Tincture
	Hydrophilic Emulsion Foam	Oil in water emulsion	Emollient, skin conditioning vehicle. Can carry and stabilise both lipophilic and hydrophilic drugs.	Emulsion, Cream, Hydrophilic cream.
	Lipophilic Emulsion Foam	Water in oil emulsion	Favourable usability.	Emulsion, Cream, Lipophilic cream.
	Potent-Solvent Foam	Water and strong solvents	Serves to solubilise drugs thus improve bioavailability. Induces skin penetration suitable for transdermal drug delivery. Does not require preservatives.	Gel, solution.
	Suspension Foam	Suspended drug in a foam formulation	Emollient, skin conditioning vehicle. Can carry stable drug suspensions. Favourable usability.	Topical suspension
Non-aqueous	Ointment Foam	Single phase, main ingredient petrolatum	Occlusive, builds up skin barrier. Serves to keep drugs in prolonged contact with the skin. In the absence of water, protects water-sensitive drugs Does not require preservatives.	Ointment, White ointment, Hydrophobic ointment
	Hydrophilic Ointment Foam	Single phase, main ingredient PEG, propylene glycol, glycerin or other hydrophilic solvents	Greaseless ointment base. Humectant, moisturiser. Solubilise drugs thus improve bioavailability. Protects water-labile drugs. Does not require preservatives.	Polyethylene glycol ointment, Hydrophilic ointment
	Oil Foam	Single phase, main ingredient is liquid oil	Builds up skin barrier, lubricates the skin, keep drugs in prolonged contact with the skin. Protects water-labile drugs. Does not require preservatives.	Oil solution or suspension
	Saccharide Foam	Monosaccharides, disaccharides, honey.	- Hygroscopic, absorbs exudates. Antibacterial. Burn and wound applications.	-

Hydroethanolic Foam

These are the foam version of alcohol solutions and tinctures. Olux® Foam and Luxiq® Foam (Table 1) were the first commercially available dermatological products and have gained high acceptance by physicians and patients. These products are thermo-labile, consisting of ethanol (about 60%), water, propylene glycol, cetyl alcohol, stearyl alcohol, polysorbate 60, citric acid, and potassium citrate and a hydrocarbon propellant. These were followed by Extina® (ketoconazole foam 2%) and Evoclin® (clindamycin foam 1%). *In vitro* studies have demonstrated that drugs formulated in hydroethanolic foam exhibit delivery of the drugs at an increased rate compared with other vehicles. For example, an *in vitro* study demonstrated that the hydroethanolic foam vehicle delivered more clobetasol propionate through the skin (5.3%) than the comparator solution, cream and lotion vehicles (2.8%, 2.7%, 2.1% and 1.8% respectively) [11]. These findings suggest that components within the foam (probably the alcohol) act as penetration enhancers and alter the barrier properties of the outer stratum corneum, thus driving the delivered drug across the skin membrane via the intracellular route.

Alcohol also promotes fast drying or the skin and thereby attempts to address the sticky feeling left by many topical formulations after application. However, alcohol is a defatting agent and may cause skin to become dry and cracked. Due to this undesirable property, hydroethanolic foams have not been proposed for the treatment of atopic dermatitis, a childhood inflammatory skin disorder which involves dry, itchy skin and rashes on various body areas.

The incidence skin irritation (burning, itching and stinging) as shown for example in the package insert of Luxiq® Foam is high (54%) and this is probably attributable to the combination of high alcohol content and inclusion of surfactants. Moreover, the incidence of skin irritation caused by the respective foam vehicle was 75% (of which 27% was moderate-to-severe). Furthermore, since alcohol is irritant to mucosal surfaces, the label of these products state "Avoid getting the foam in or near your eyes, mouth, lips, or broken skin." In addition, the current hydroethanolic foams are thermolabile (temperature-sensitive) and their usage is hindered by the recommendation not to dispense them directly onto the hands, as the foam melts immediately upon contact with warm skin. Instead, the foam is to be dispensed onto a cool surface, and then a small amount of foam should be picked up with fingers and gently massage into the affected area [18].

Thus, while alcohol is useful in solubilising an active agent and enabling effective dermal penetration of drugs, the development of less irritable foam vehicles, which overcome the evident skin drying and irritation caused by alcohol plus surfactants is warranted. One of the means to achieve this goal is to add emollient oils to the foam composition. Such emollients provide skin conditioning effects, build up the skin barrier properties and reduce skin irritation. An example of such foam is Scytera™ (Promius Pharma, developed by Foamix) which contains 2% coal tar for the treatment of psoriasis. Scytera™ contains alcohol but it also contains emollients. This novel foam vehicle is versatile and may be used to treat psoriasis even in challenging areas of the body, such as the scalp, palms and soles [19 – 21].

An alternative way to overcome the usability limitations of the traditional hydroethanolic foams is to replace the surfactant by polymeric agents, resulting in foams which are thermally stable [22, 23]. The absence of surfactants in the formulation further decreases the irritability of such formulations.

Cream Foam (Emollient Foam)

The term emollient foam relates to foams that exert soothing and moisturising effects when applied to the skin. Emollient foams are water and oil emulsions and, as such, possess vehicle properties similar to traditional creams and lotions. The emulsions can be oil-in-water or inverted (water-in-oil) emulsions, which correspond to "hydrophilic creams" and "hydrophobic creams", respectively.

The oil components of the foam increase the water-holding capacity of the stratum corneum, thereby correcting skin dryness, wrinkles and scaling and provide symptomatic relief of dry skin conditions like psoriasis, and atopic dermatitis [24, 25]. Currently available emollient foams include Olux-E®, Verdeso®, Sorilux foam and Fabior™ foam (Table 1).

The primary components of emollient foams are water and oil, which are present in the formulation as an emulsion. The oil component can be selected from all pharmaceutically-acceptable oils, including mineral oil; triglycerides, such as plant oils and capric/caprylic triglyceride; fatty acid esters, e.g., isopropyl myristate, isopropyl palmitate and diisopropyl adipate; and silicone oils, which possess therapeutic benefits. Petrolatum, also termed "Vaseline™", is a less desirable hydrophobic component, due to its greasy nature [26]. Formulations that include high petrolatum concentrations leave a greasy and sticky feeling after application and occasionally stain cloths.

Foaming agents that are required to stabilise the emulsion and evolve foam with desirable texture include surfactants, polymers and foam adjuvants. The surfactants should be carefully selected. Ionic surfactants are effective as foaming agents but are irritants and so non-ionic surfactants are preferred, especially when the target area of treatment is inflamed or infected skin.

A gelling agent is a useful component for the creation of a foam with desirable texture and spreading properties. Some gelling agents also possess film-forming properties, which serve to maintain drugs at the site of application.

Another group of components that contribute to the stability and sensory properties of the foam are "foam adjuvants" which are selected from the variety of fatty alcohols and fatty acids [27, 28]. Optionally, polar solvents such as glycerol, propylene glycol, hexylene glycol, dimethyl isosorbide and DMSO are added to the foam composition in order to increase the solubility of the active agents and to enhance skin penetration [29].

The propellant can be a standard hydrocarbon propellant (mix of butane, propane and isobutene) or a fluorocarbon gas.

A pharmaceutical emollient foam product may consist of a single drug or a combination of drugs, which can be dissolved in the water phase or the hydrophobic phase of the carrier composition. Yet, in certain cases, when the drug is not fully soluble in either the water or oil phase of the composition, it can still be dispersed in the emulsion. Examples of drugs that have been successfully incorporated in emollient foam formulations include antibiotics, antifungals, antivirals, corticosteroids, non-steroidal anti-inflammatory agents, retinoids, keratolytic agents, immunomodulators, anesthetic drugs, anti-allergic agents and anti-proliferative drugs [30 – 36].

Emollient foams possess several advantages, when compared with hydroethanolic foams:

- Breakability: the emollient foam is thermally stable. Unlike hydroethanolic foams, it does not readily collapse upon exposure to body temperature. Sheer-force breakability of the foam is preferable since it allows comfortable application and directed administration to the target area.

- Skin hydration and skin barrier function: the oil components of the foam build up the desirable skin barrier function thereby improving the condition of damaged skin.

- Decreased irritability: due to the lack of alcohol and improvement in skin barrier function, skin irritability is reduced or eliminated.

- Usability: foam provides significant usability advantages. When foam is released, it expands and allows easy spreading on the target area and is absorbed into the skin without any extensive rubbing. This feature is particularly relevant for the treatment of large skin surfaces.

The following examples demonstrate the value of the above mentioned advantages.

Example 1: Betamethasone Valerate Emollient Foam.

An emollient foam composition, containing 0.12% of betamethasone valerate, was developed with the aim of treating patients with psoriasis and atopic dermatitis. The formulation comprised oils and non-ionic surfactants in order to minimise skin irritation. A Phase II, randomized, blinded, right-left comparison within-patient clinical trial was performed with 30 patients with mild to moderate psoriasis. Two similar plaque areas of psoriasis (i.e. both knees or both elbows) were selected for treatment on each patient. Patients received six weeks of administration of the foam on one side and a commercially available betamethasone valerate 0.12% cream (Betnovate; GSK) on the other side.

In terms of efficacy, both treatments were equally effective in the treatment of the psoriatic lesions. After three weeks of treatment, there was a statistically-significant improvement from baseline in all parameters, including thickness (42-43% improvement), redness (36-44%), scaling (49-56%), itch (77-78%) and global score (42-44%). These clinical improvements persisted following an additional three weeks of treatment (Figure 1). Patients rated the foam as better than the cream in skin absorption, oily residue, shiny look, stickiness and odour (Figure 2). No drug-related adverse effects were recorded.

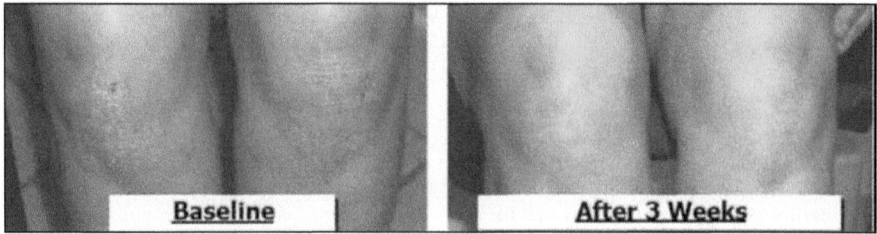

Figure 1 *Betamethasone Emollient Foam - clinical improvement of psoriasis lesions following three weeks of treatment. Dry, scaly areas of patellar skin can be observed prior to onset of treatment ("Baseline").*

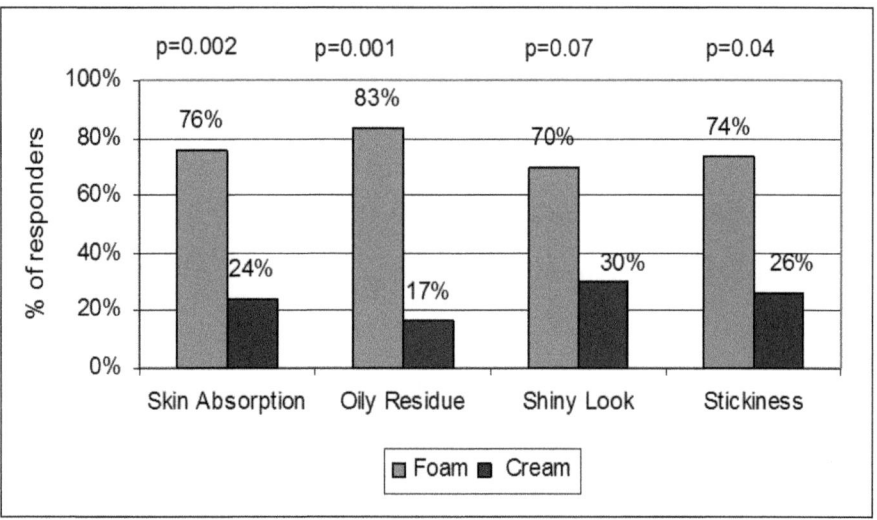

Figure 2 *Usability preference – Patients rated the foam as better than the cream in skin absorption, oily residue, shiny look, stickiness and odour.*

In conclusion, the emollient foam offered an attractive alternative to mid-potency steroid creams. As such, it is more likely that psoriasis patients would use their medication as frequently as prescribed and thus gain the desirable therapeutic benefits.

Example 2: *Metronidazole 1% Emollient Foam*

Metronidazole, the leading topical drug for rosacea, is currently available in gel, cream and lotion at 0.75% and 1% concentrations. Since the solubility threshold of metronidazole in water is relatively low ($\leq$ 0.75%) 1% metronidazole is not expected to fully dissolve in an aqueous vehicle.

Emollient foam compositions (comprising emollient oils and non-ionic surfactants) were designed with the aim of dissolving 1% of metronidazole. Surprisingly, the foam solubilised up to 2% (Figure 3).

An *in vitro* skin penetration study was conducted using excised human skin, aiming to compare the penetration profile of 1% metronidazole emollient foam. Two foam compositions were tested – one with 2.5% propylene glycol as a penetration enhancer and the other without propylene glycol (MZPG and MZ, respectively), in comparison with a commercial 1% metronidazole cream, "Noritate®" (Dermik).

As shown in Figure 4, the total cutaneous penetration of metronidazole following 16-hour exposure was 2-3 fold higher for the two foams than Noritate. Propylene glycol significantly increased the delivery of metronidazole through the skin. Thus, the enhanced solubility of the drug in the emollient foam appear to be useful in enhancing the effectiveness of topical metronidazole.

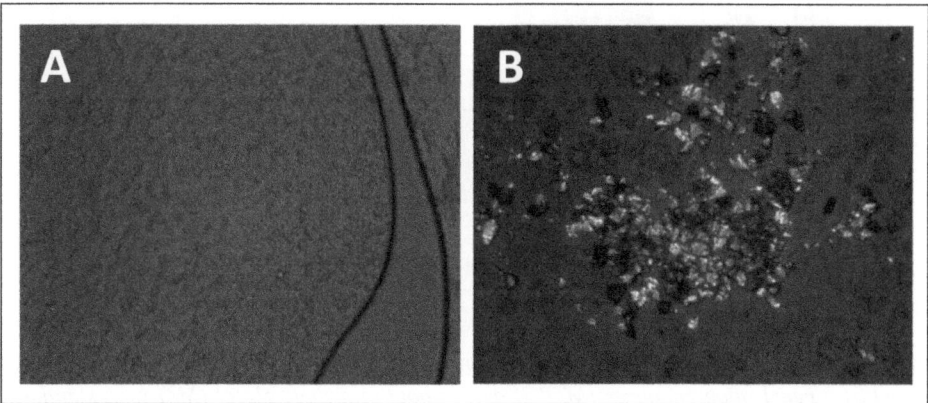

Figure 3 *Photomicrographs of Metronidazole 1% emollient foam (A) versus Noritate® Cream (B). Note the absence of crystals in the 1% emollient foam compared to the presence of Metronidazole crystals in Noritate.*

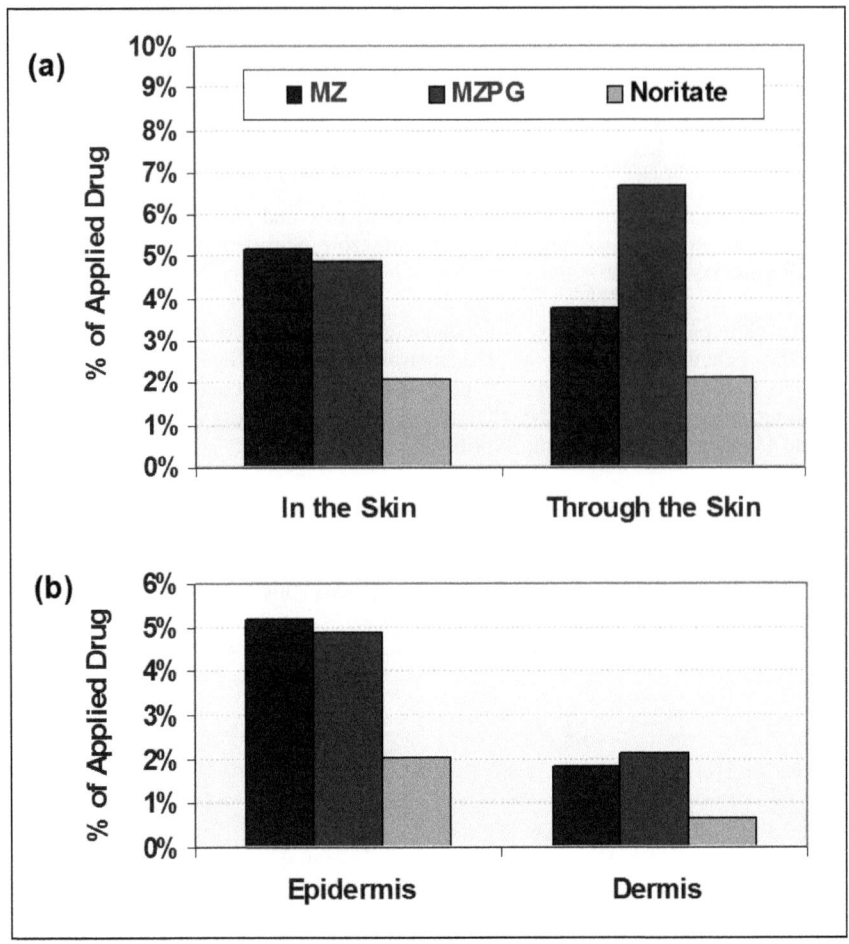

Figure 4 *Skin penetration profile of Metronidazole 1% emollient foam with 2.5% propylene glycol as a penetration enhancer (MZPG) and without propylene glycol (MZ) vs. a standard cream (Noritate). (a) Enhanced intradermal delivery and controllable transdermal delivery by both foams; and further, induction of transdermal delivery by propylene glycol. (b) Relative delivery of metronidazole to the two skin layers.*

Potent Solvent Foam

A new platform of foam formulation is intended to promote transdermal delivery of drugs via the addition of high concentrations of skin penetration enhancers. Following the recent FDA approval of products containing up to 40% DMSO, an aqueous foam comprising 40% DMSO was developed by Foamix as a carrier for non-steroidal inflammatory drugs that are intended to treat osteoarthritis as well as other drugs that can be administered transdermally [37].

Nano-Emulsion Foam

The use of nano-emulsion formulations is at the forefront of topical drug development innovation. Nano-emulaions can improve the solubility and ultimate bioavailability of drugs, when applied topically. New nano-emulsion foam formulations are under development by Foamix.

5 WATER-FREE FOAMS

The creation of foam formulations without water is counterintuitive. It is known in the art that foams can easily be formulated based on high amounts of water in combination with surface active agents, foam adjuvants and polymeric agents. As described in the literature, hydrophobic excipients, such as petrolatum, oils and hydrophilic solvents can have a de-foaming effect which makes the formulation of foams based on such solvents "challenging". To overcome this, substantial levels of surfactants that act as foaming agents have been used in the past. However, surface active agents are known to be irritating, especially ionic surface active agents, and repeated application to the skin or mucosa in high concentrations can damage the integrity of the skin and cause dryness and irritation. Therefore, there is a well-defined need to develop water-free foams which contain only minute quantities of surfactants or, ideally, no surfactants at all. Water-free foams have several potential advantages:

- Stability: the first and foremost advantage is that water-free foams are perfect vehicles for water-sensitive drugs. Many drugs, including corticosteroids, steroid hormones, immunomodulators and antibiotics and any drugs that contain ester groups, tend to degrade in the presence of water, so a vehicle that does not contain water is preferred. Moreover, the pressurised aerosol container is an impermeable packaging system and it prevents contact of the formulation with ambient moisture even during the use period, unlike tubes, which are exposed to the environment once they are first opened.

- Self-preservation: microorganisms require water to grow and reproduce. A water-free foam formulation retards the growth of bacteria, moulds and fungi during storage and, as mentioned above, the entry of moisture into the aerosol pressurised can is prevented during the use period of the drug, so water-free foams do not require preservatives.

- Usability: today's water-free topical formulations are primarily ointments. These are characteristically thick and greasy and require extensive rubbing for efficient topical application. In contrast, foams are structurally soft and their application is facile. They spread easily onto the skin and absorb quickly.

- Skin hydration and conditioning: hydrophobic excipients such as petrolatum and liquid oils act to correct skin dryness and ameliorate inflammation through their emollient and humectant properties. They make the external layers of the skin (epidermis) softer and more pliable, thereby increasing the skin's hydration (water content) by reducing water evaporation. Hydrophilic excipients, such as polyethylene glycol, propylene glycol and glycerin are hygroscopic; they attract ambient water and retain skin moisture. Water-free foams are rich with such emollients and humectants, so they maximise effects relating to skin hydration and conditioning.

While water-free foams are currently unavailable commercially, several products based on such foams are currently under development, as will be detailed in the following sections.

Ointment Foam – Petrolatum Based Foam

The Ointment Foam, developed by Foamix, is equivalent to traditional petrolatum-based ointments [38]. Petrolatum is occlusive and by creating a hydrophobic skin barrier it lowers transepidermal water loss (TEWL) which is thought to hydrate the upper layers of the stratum corneum.

Petrolatum-based foam formulations are not obvious to make, especially due to the high viscosity of the hydrocarbon; however foams that contain up to 90% petrolatum are currently under development. The foaming agents include small amounts of foam adjuvants and non-ionic surfactants. The propellant is typically hydrocarbon propellant.

Due to the unique texture of the foam, it instantly liquefies and spreads easily onto the skin upon application and so no extensive rubbing is required. Thus, the benefit of petrolatum's occlusive shield is retained without the thick texture and greasy feel of traditional ointments. This usability feature is especially valuable in the treatment of infants and children who suffer from dry skin conditions like atopic dermatitis. In such cases, the effect of the drug is accompanied by the synergistic effects of skin barrier restoration, lubrication and protective properties of the vehicle.

Examples of drugs that can benefit from this type of formulation include corticosteroids, which are typically applied to large areas of dry, inflamed and damaged skin, anti-infective agents (antibacterial, antifungal and antiviral drugs) and immunomodulators (such as pimecrolimus and tacrolimus) which treat atopic dermatitis. An illustrative example is a unique petrolatum – zinc oxide foam (petrolatum and natural oils 91% and zinc oxide 15%). Petrolatum is approved by FDA as an OTC active ingredient that helps treat and prevent nappy (diaper) dermatitis, seals out wetness and temporarily protects and helps relieve chapped or cracked skin, minor cuts, scrapes and burns.

Likewise, zinc oxide, the active ingredient in many nappy dermatitis products, is a skin protectant. It provides protection by forming a protective barrier on skin, preventing wetness and other irritants from reaching the skin underneath. Unlike traditional petrolatum-based pastes for diaper dermatitis which are very thick and hard to apply to the baby's sensitive skin, the petrolatum – zinc oxide foam is easy to apply, and still provides the same protective and healing effects. This synergistic composition can be further enhanced by the addition of an antimycotic agent (such as miconazole, ketoconazole, clotrimazole and nystatin) to eradicate the yeast infection; and/or a corticosteroid (e.g., hydrocortisone or betamethasone), to heal the inflammation.

Oil Foam

The Foamix Oil Foam is the foam version of traditional oil-based solutions and suspensions [39]. Oil foam is one of the most promising foam platforms for use in dermatology and can utilise a broad range of pharmaceutical liquid oils, including mineral oil, plant oils (e.g., olive oil, soybean oil and castor oil), emollients (e.g., isopropyl myristate, isopropyl palmitate, diisopropyl adipate, isostearic acid and oleyl alcohol) and silicone oils.

Despite the fact that oils are generally known as de-foaming agents and their incorporation in foam formulations is challenging, it has been possible to formulate unique foamable compositions that contain more than 90% oils, very small amounts of foaming agents and no water. The foaming agents can include surfactants with low hydrophilic-lipophilic balance (HLB), foam adjuvants (fatty acids and fatty alcohols), waxes and polymers. In cases where the drug to be included in the vehicle is incompatible with surfactants, it is possible to make foam compositions without surfactants at all [40 – 42]. The most suitable propellants for such foams are hydrocarbon propellants.

Oil foams have a very soft and airy texture, spread effortlessly on the target surface and quickly absorb into the skin leaving no greasy residue. In fact, the oil foams are deemed to be sufficiently cosmetically elegant that they can be used for the treatment of facial conditions such as rosacea and acne.

Oil foams are most suitable to accommodate unstable drugs. For example, it has been used as a vehicle for calcipotriene and calcitriene, two vitamin D3 analogs that treat psoriasis and atopic dermatitis, resulting in stable drugs with more than two years of shelf life (Table 1). The most advanced oil foam product in current development is Minocycline Foam (1% and 4%) [43, 44]. Minocycline is known to be a very unstable active agent and is degraded by a wide range of commonly used pharmaceutical excipients such as hydrophilic solvents (e.g. water, glycerin, sodium PCA, propylene glycol and polyethylene glycols), polymers (xanthan gum, poloxamers, carbomers and methocel) and surfactants (polysorbates, sorbitan esters, polyoxyalkyl esters and lanolin-based surfactants).

So, the technical challenge is to attain a stable foam in the absence of common yet incompatible ingredients. A series of development efforts resulted in a water-free, alcohol-free and surfactant-free formulation which contained more than 80% liquid oils using foaming agents such as fatty alcohols, fatty acids and waxes. The general advantages of oil-only foams are considered in the following example.

Example 3: Minocycline Oil Foam.

Both Minocycline Foam 1% and 4% show high stability. They remain within the designated specifications following 12 months' storage at 40°C and over 24 months at 25°C.

In-vitro studies demonstrated that Minocycline Foam 1% and 4% inhibited the growth of *Streptococcus pyogenes*, *Pseudomonas aeruginosa*, *Staphylococcus aureus*, a methicillin-resistant strain of *Staphylococcus aureus* (MRSA), and *Propionbacterium acnes*, the causative microorganism in acne.

Irradiation of the skin by UV radiation is known to decrease cell viability and total antioxidant capacity, while increasing the levels of inflammation (pro-inflammatory cytokines secretion) and epidermal cell apoptosis. An exploratory study has indicated beneficial effects of Minocycline Foam on cell viability and apoptosis of skin cells (Figure 5): apoptosis activation was significantly decreased by Minocycline Foam in a dose-dependent manner.

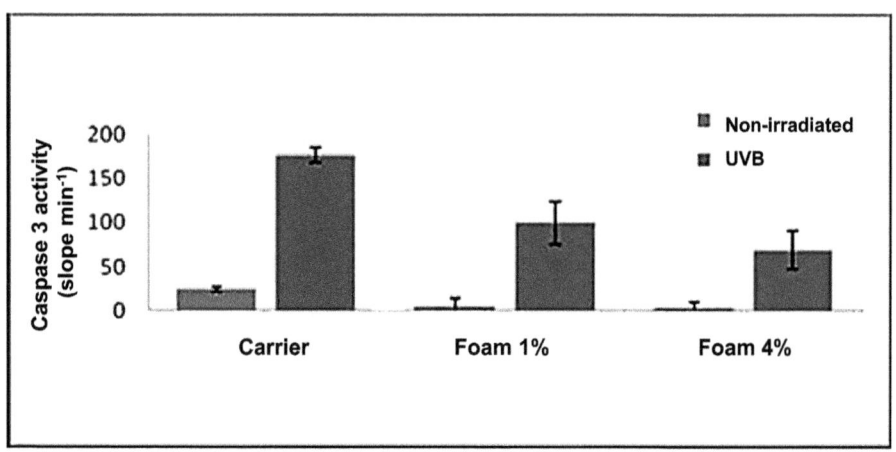

Figure 5 *Effect of Minocycline Foam 1% and 4% on apoptosis in skin organ culture after UVB radiation: treatment prior to irradiation results in more than 50% inhibition of apoptosis, as measured by caspase 3 activity.*

An additional experiment evaluated the same affects when the skin was treated <u>after</u> UV damage was induced. As shown in Figure 6, Minocycline Foam 4% decreased epidermal cell apoptosis by 60%, as measured by caspase 3 activity.

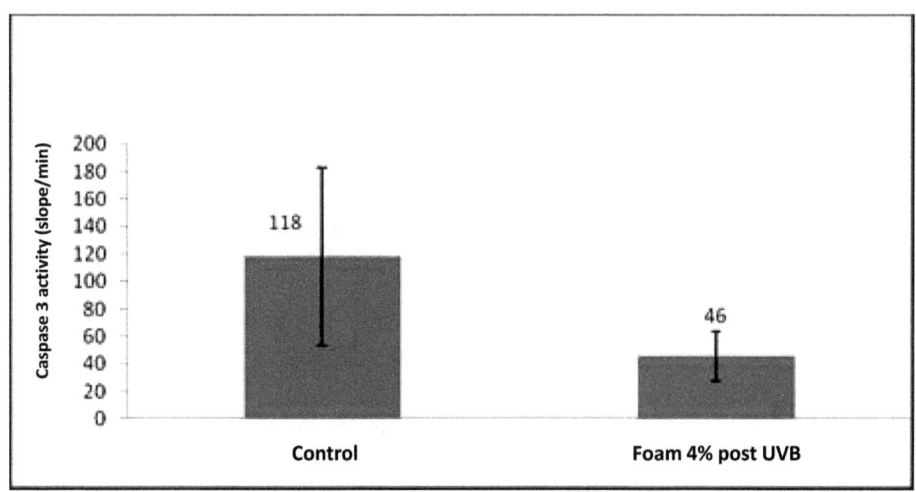

Figure 6 *Effect of Minocycline Foam 4% on apoptosis in skin organ culture after UVB radiation: treatment results in 60% inhibition of apoptosis, as measured by caspase 3 activity.*

The transdermal penetration of minocycline was tested using an *in vitro* Franz cell diffusion system containing porcine ear skin. Approximately 500 mg of product was placed in each cell and receptor chambers were sampled at baseline and 3, 6, 9 and 24 hours following application. After 24 hours the amounts of minocycline in the stratum corneum and viable skin were analysed. The study demonstrated that the drug was delivered exclusively into the skin (Table 3), with the mean amount of Minocycline in the skin following 24 hours of exposure being 9.5 μg cm^{-2} for the 1% formulation and 43 μg cm^{-2} for the 4% formulation.

Table 3 *In vitro skin delivery of Minocycline from 1% and 4% foam formulations: High amounts were recovered from the stratum corneum and viable epidermis, but no measurable transdermal penetration (into the receptor chamber). All values are mean ± standard deviation of n=5 (1% foam) or n=6 (4% foam).*

Compartment	Minocycline Penetration (μg cm^{-2})	
	1% Foam	**4% Foam**
Stratum Corneum 1	7.77 ± 4.32	33.63 ± 20.41
Stratum Corneum 2	0.93 ± 0.77	7.49 ± 8.67
Total Stratum Corneum	*8.70 ± 4.97*	*41.12 ± 16.89*
Viable Skin	0.79 ± 0.19	2.00 ± 0.81
Total Intradermal Delivery	*9.49 ± 4.99*	*43.12*
Receiving Compartment (Transdermal Delivery)	0.00	0.00

The weight of skin at the delivery area was about 100 mg, which implies that the concentration of minocycline in the skin following 24 hours of exposure was approximately 168 µg g^{-1} and 760 µg g^{-1} skin for the 1% and 4% formulation, respectively. This exceeds the dose required to treat bacterial skin infections. No transdermal absorption of minocycline was observed indicating that Minocycline Foam should not involve minimal systemic absorption (and thus reduce or eliminate any systemic adverse effects).

A randomised double blind dose-ranging Phase II clinical study was designed to assess the efficacy, safety and tolerability of two strengths of the Minocycline Foam in paediatric patients with impetigo. The study enrolled 32 patients (age 2 to 15) with at least two impetigo lesions.

Patients received the foam twice daily for 7 days and were examined again on day 14. Robust efficacy was demonstrated in both 1% and 4% strengths. Clinical success was defined as the absence of treated lesions, or treated lesions had become dry without crusts with or without erythema compared to baseline, or had improved (defined as a decline in the size of the affected area, number of lesions or both) such that no further antimicrobial therapy was required.

Notably, about 80% of the patients in both groups cured or improved and met the efficacy success criterion after 3 days of treatment. Clinical response at the end of the treatment was 92% and 100% respectively for the low or high doses; and all patients (100%) showed success on day 14 (Table 4 and Figure 7).

Eight patients were infected at baseline by methicillin-resistant *Staphylococcus aureus* (MRSA) and in all of them the infection was eradicated on Day 7. Minocycline Foam was well-tolerated and no drug related side effects were recorder in any of the patients throughout the study. Questionnaires, filled by the patients' caregivers revealed high satisfaction from treatment.

Table 4 *Efficacy of Minocycline Foam (1% and 4% formulations) in cohort of paediatric patients. Data expressed as success rate (defined in text, above) at days 3, 7 (End of Treatment) and 14 (Follow Up).*

Study Day	Formulation		
	Minocycline 1%	Minocycline 4%	All
Day 3	81.3%	78.6%	80.0%
Day 7 (EOT)	92.3%	100.0%	95.8%
Day 14 (FU)	100.0%	100.0%	100.0%

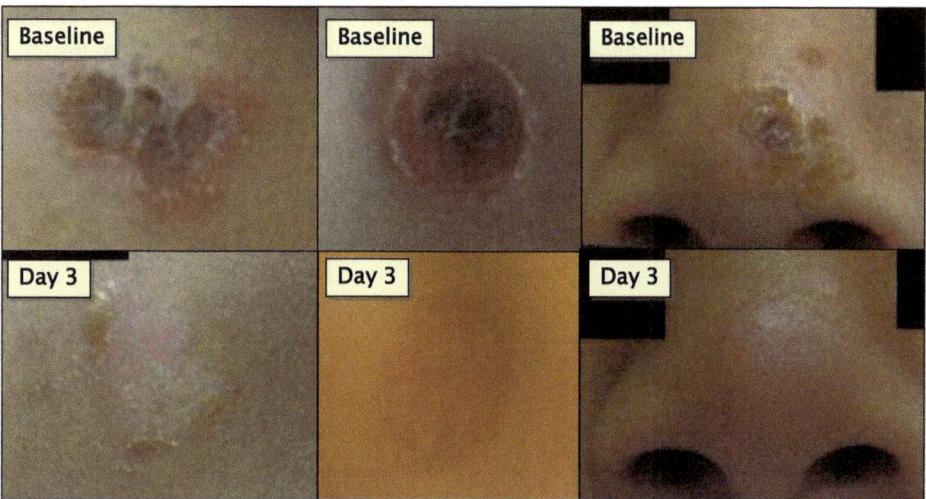

Figure 7 *Photographic documentation of the effect of Minocycline Foam in paediatric patients with impetigo, demonstrating visible improvement or clearance of lesions within 3 to 7 days of treatment.*

Hydrophilic Waterless Foams

Hydrophilic, waterless foams are the equivalent of traditional hydrophilic ointments and may contain up to 98% of a polar hydrophilic solvent. The solvent may be selected from polyols (organic solvents that contain at least two hydroxy groups in their molecular structure) or polyethylene glycols (PEGs). Examples of polyols include diols (such as propylene glycol, butanediol and diethylene glycol) and triols (such as glycerin) whereas the PEGs are primarily selected from low-molecular weight liquids such as PEG 200, PEG 400, PEG 600 and PEG 1000. However, mixes of liquid PEGS with higher PEGS like PEG 4000, PEG 6000 and PEG 8000 are feasible providing the viscosity, prior to filling of the composition into aerosol canisters, is less than ~10,000 cP.

The addition of secondary polar solvents, such as dimethyl isosorbide, transcutol, DMSO, and α-hydroxy acids, such as lactic acid and glycolic acid, is sometimes warranted in order to enhance solubilisation and skin permeation of the drug [45]. These properties enable increased permeability across the skin, resulting in an enhanced therapeutic effect. Foaming agents include up to 5% surfactants and small amounts of polymers with hydrocarbon or fluorocarbon propellants.

Polyols, PEGs and other polar solvents have a great affinity for water, and as such, exhibit hygroscopic properties. Microorganisms require water to grow and reproduce. Thus, the high concentration of these polar solvents can inhibit the growth of bacteria and fungi. Consequently, waterless hydrophilic foams do not require preservatives in their composition and their application onto infected skin surfaces can be used as topical treatment of superficial infections. It is further possible to add an anti-infective (antibacterial or antifungal) drug, resulting in a synergistic effect and consequently higher treatment success [46 – 47].

Saccharide Foam

A saccharide foam containing up to 90% monosaccharides, disaccharides and honey has been developed for the treatment of wounds and burns [48]. It is hygroscopic and so has anti-infective attributes and also absorbs exudates.

6 CONCLUSIONS

Creams and ointments have been used historically in skin care and dermatology. Foams offer an innovative and more convenient means of topical drug delivery. The continuing development of versatile foam technology platforms will facilitate achieving new topical products, including valuable drugs for the treatment of dermatological disease.

Foam offers many advantages over traditional formulations, including enhanced usability and compliance, improved clinical safety, increased tolerability and efficacy, stability and targeted drug delivery.

Are the current foam platforms the final word in topical drug delivery? The answer is probably "no": new generations of foam are currently being developed and so the foam story is just beginning.

7 REFERENCES

1. R. Woodford and B. W. Barry, Bioavailability and activity of topical corticosteroids from a novel drug delivery system, the aerosol quick-break foam, *J Pharm Sci.*, 1977, **66**, 99–103.

2. M. Parrini, P. Cabitza, A. Arrigo and M. Vanasia, Efficacy and Tolerability of Ketoprofen Lysine Salt Foam for Topical Use in the Treatment of Traumatic Pathologies of the Locomotor Apparatus, *Clin Ter.*, 1992, **141**, 199–204.

3. L. Daffonchio, A. Bestetti, G. Clavenna, G. Fedele, M. P. Ferrari and C. Omini, Effects of a New Foam Formulation of Ketoprofen Lysine Salt in Experimental Models of Inflammation and Hyperalgesia. *Arzneimittelforschung*, 1995, **45**, 590–4.

4. C. H. Purdon, J. M. Haigh, C. Surber and E. W. Smith, Foam Drug Delivery in Dermatology: Beyond the Scalp, *American Journal of Drug Delivery*, 2003, **1**(1), 71–75.

5. D. Tamarkin, D. Friedman and A. Shemer, Emollient foam in topical drug delivery, *Expert Opin Drug Deliv.*, 2006, **3**, 799–807.

6. A. Arzhavitina and H. Steckel, Foams for Pharmaceutical and Cosmetic Application, *Int J Pharm.*, 2010, **394**, 1–17.

7. See EpiFoam Prescribing Information: http://www.medapharma.us/products/pi/Epifoam_PI.pdf.

8. Olux Foam Prescribing Information: http://www.stiefel.com/content/dam/stiefel/globals/documents/pdf/US_Olux_Foam.pdf .

9. Prescribing Information: http://www.stiefel.com/content/dam/stiefel/globals/documents/pdf/US_Luxiq.pdf.

10. Luxiq Foam Prescribing Information: http://www.stiefel.com/content/dam/stiefel/globals/documents/pdf/US_Evoclin_Foam.pdf.

11. X. Huang, H. Tanojo, J. Lenn, C. H. Deng and L. Krochmal, A Novel Foam Vehicle for Delivery of Topical Corticosteroids, *J Am Acad Dermatol.*, 2005, **53**(Suppl 1), S26–38.

12. A. R. Shalita, J. A. Myers, L. Krochmal and A. Yaroshinsky: The Safety and Efficacy of Clindamycin Phosphate Foam 1% Versus Clindamycin Phosphate Topical Gel 1% for The Treatment of Acne Vulgaris. *J Drugs Dermatol.* 2005, **4**, 48-56.

13. Olux-E Prescribing Information: http://www.stiefel.com/content/dam/stiefel/globals/documents/pdf/US_Olux-e_Foam.pdf.

14. Verdeso Foam Prescribing Information: http://www.stiefel.com/content/dam/stiefel/globals/documents/pdf/US_Verdeso_Foam.pdf.

15. Sorilux Foam Prescribing Information: http://www.stiefel.com/content/dam/stiefel/globals/documents/pdf/US_Sorilux_Foam.pdf.

16. Fabior Foam Prescribing Information: http://www.stiefel.com/content/dam/stiefel/globals/documents/pdf/US_Fabior.pdf.

17. European Pharmacopoeia 05, 2005, pp. 604–605.

18. See for example the Use Instructions of Evoclin Foam http://www.accessdata.fda.gov/drugsatfda_docs/label/2012/050801s015lbl.pdf.

19. A. J. Frankel, J. A. Zeichner and J. Q. Del Rosso, Coal Tar 2% Foam in Combination with a Superpotent Corticosteroid Foam for Plaque Psoriasis: Case Report and Clinical Implications, *J Clin Aesthet Dermatol.*, 2010, **3**, 42–45.

20. J. A. Zeichner: Use of Topical Coal Tar Foam for the Treatment of Psoriasis in Difficult-to-treat Areas, *J Clin Aesthet Dermatol.*, 2010, **3**, 37–40.

21. D. Tamarkin *et. al.* (Foamix), US Pat Appl. No. 20090180970: Foamable Composition Combining A Polar Solvent And A Hydrophobic Carrier.

22. D. Tamarkin *et. al.* (Foamix), WO 2011/013008: Non Surface Active Agent Non Polymeric Agent Hydro-Alcoholic Foamable Compositions, Breakable Foams and Their Uses.

23. D. Tamarkin *et. al.* (Foamix), WO 2011/013009: Non Surfactant Hydro-Alcoholic Foamable Compositions, Breakable Foams and Their Uses.

24. I. Nola, K. Kostovic, L. Kotrulja and L. Lugovic: The use of emollients as sophisticated therapy in dermatology, *Acta Dermatovenerol Croat.*, 2003, **11**, 80–87.

25. M Loden and AB Aco Hud: Role of Topical Emollients and Moisturizers in the Treatment of Dry Skin Barrier Disorders. *Am J Clin Dermatol.*, 2003, **4**, 771–788.

26. D.S. Morrison, *Petrolatum a Useful Classic Cosmetics & Toiletries*, 1996, **111**, 59–69.

27. D. Tamarkin *et. al.* (Foamix), WO 2006/0140984: Cosmetic and Pharmaceutical Foam

28. D. Tamarkin *et. al.* (Foamix), WO 2005/0186147: Cosmetic and Pharmaceutical Foam With Solid Matter

29. D. Tamarkin *et. al.* (Foamix), US Patent No. 7,700,076: Penetrating pharmaceutical foam.

30. D. Tamarkin *et. al.* (Foamix), WO 2005/0186142: Kit and Composition of Imidazole with Enhanced Bioavailability.

31. D. Tamarkin *et. al.* (Foamix), WO 2005/0232869: Nonsteroidal Immunomodulating Kit and Composition and Uses Thereof.

32. D. Tamarkin *et. al.* (Foamix), WO 2005/0205086: Retinoid Immunomodulating Kit and Composition and Uses Thereof.

33. D. Tamarkin *et. al.* (Foamix), WO 2006/0018937: Steroid Kit and Foamable Composition and Uses Thereof.

34. D. Tamarkin *et. al.* (Foamix), WO 2005/0271596: Vasoactive Kit and Composition and Uses Thereof

35. D. Tamarkin *et. al.* (Foamix), US Patent No. 8,119,150: Non-flammable Insecticide Composition and Uses Thereof.

36. D. Tamarkin *et. al.* (Foamix), US Patent No. 8,119,109: Foamable Compositions, Kits and Methods for Hyperhidrosis.

37. D. Tamarkin *et. al.* (Foamix), WO 2010/125470: Foamable Vehicles and Pharmaceutical Compositions Comprising Aprotic Polar Solvents and Uses Thereof.

38. D. Tamarkin *et. al.* (Foamix), WO 2008/0260655: Substantially Non-Aqueous Foamable Petrolatum Based Pharmaceutical and Cosmetic Compositions and Their Uses.

39. D. Tamarkin *et. al.* (Foamix), US Patent No. 7,820,145: Oleaginous Pharmaceutical and Cosmetic Foam.

40. D. Tamarkin *et. al.* (Foamix), WO 2011/039637: Surfactant-Free Water-Free Foamable Compositions, Breakable Foams and Their Uses.

41. D. Tamarkin *et. al.* (Foamix), WO 2011/138678: Compositions, Gels And Foams with Rheology Modulators and Uses Thereof.

42. D. Tamarkin *et. al.* (Foamix), WO 2011/064631: Surfactant-Free, Water-Free, Foamable Compositions and Breakable Foams and Their Uses.

43. Acknowledgement: The development of Foamix's Minocycline Foam was supported by the Bird Foundation (www.birdf.com).

44. D. Tamarkin *et. al.* (Foamix), WO 2011/039638: Topical Tetracycline Compositions

45. D. Tamarkin *et. al.* (Foamix), US Patent No. 8,114,385: Oleaginous Pharmaceutical and Cosmetic Foam

46. D. Tamarkin *et. al.* (Foamix, US Patent No. 7,704,518: Foamable Vehicle and Pharmaceutical Compositions Thereof.

47. D. Tamarkin *et. al.* (Foamix), WO2008/0299220: Hydrophilic Non-Aqueous Pharmaceutical Carriers and Compositions and Uses.

48. D. Tamarkin *et. al.* (Foamix), US Patent No. 7,645,803 Saccharide Foamable Compositions.

17

TOPICAL BIOEQUIVALENCE: A COMPREHENSIVE APPROACH USING MULTIPLE SURROGATE METHODS

P A Lehman, S G Raney, and T J Franz

Dermal and Transdermal Research, PRACS Institute, Fargo, North Dakota, United States of America

1   INTRODUCTION

At the present time, establishing the bioequivalence (BE) of topical drug products is not a simple process: in contrast to oral medications, costly and time-consuming clinical trials are required for most topical products. In the United States there are provisions within the regulations that allow other methods (i.e. surrogates for clinical trials) to be used to establish BE. The stated preference of the Food and Drug Administration (FDA) is for pharmacodynamic effect studies, but *in vivo* animal studies as well as *in vitro* studies are potentially acceptable. Although there is currently only one approved surrogate test within the topical products area (the "skin blanching" or "vasoconstrictor assay" for glucocorticoids [1]), FDA activity over the past twenty five years through sponsored scientific meetings and research, as well as in-house research and initiatives, has led to the development of several alternative test methods for topical products [2 – 12].

It now appears possible, through the use of multiple appropriately selected surrogate tests, to establish the BE of any topical drug product without the need for clinical trials. The strength of this combinatorial approach lies in the fact that, while all tests have potential limitations, they don't all have the same limitation, and the information from one test can complement another. Table 1 illustrates a partial list of different tests that could be combined to make a comprehensive assessment of a drug's relative bioavailability from test and reference product lots. For example, the *in vitro* permeation test (IVPT) is one such alternative to a clinical endpoint study that has been correlated with (and predictive of) human *in vivo* bioavailability, thereby supporting its use as a valid surrogate test to establish bioavailability (BA) or BE [13]. The *in vitro – in vivo* correlation of the IVPT model has been conclusively demonstrated in respect to systemic delivery [14], but there is some debate over the ability of this test to reliably detect small changes in local delivery to various skin compartments; the critical target for topical BE [5]. This potential shortcoming of the IVPT can be compensated for by the use of a complementary second or third surrogate test that directly or indirectly measures drug concentration within the skin itself. Pharmacokinetic methods such as microdialysis and skin stripping, or pharmacodynamic methods such as the vasoconstrictor (VC) assay, cumulative irritation

Table 1 *List of well-known pharmacokinetic and pharmacodynamic methods which have potential application as bioequivalence (BE) surrogates.*

Pharmacokinetic (PK)	Pharmacodynamic (PD)
In vitro release test	Vasoconstrictor assay
In vitro permeation test	Transepidermal water loss
In vivo stratum corneum tape stripping	21-day cumulative irritation test
In vivo microdialysis	

assay, and measurement of transepidermal water loss (TEWL) are among the candidate surrogate tests that can be selected as part of such a comprehensive analysis, depending on the drug product being evaluated.

The objective of the present study was to assess the validity of this comprehensive approach using three hydrocortisone cream products as a model test system. Two standard (1%) hydrocortisone cream products were selected (Hytone 1% and Cortizone-10), as well as one higher concentration (2.5%) cream product (Hytone 2.5%), representing both potentially equivalent and potentially inequivalent comparators based upon concentration and/or manufacturer. Products were first evaluated using the *in vitro* release test (IVRT), which focused on the product formulation and its relative ability to release the drug through a porous, non-biological membrane. The greatest limitation of this method is recognised to be its lack of biological relevance and its inability to consistently correlate with *in vivo* bioavailability data. As compensatory measures, the IVPT, tape stripping, and VC tests were also employed. These tests focus on the biologically relevant partitioning of the drug from the topical product into the stratum corneum, and/or into and through the viable layers of the skin. The goal was to evaluate whether these different surrogate tests could individually differentiate between different aspects of product performance, and thereby collectively illustrate the utility of a comprehensive approach to the determination of BE between topical product lots.

2 MATERIALS AND METHODS

Three hydrocortisone products (Cortizone-10, Hytone 1% and Hytone 2.5%) were obtained from local pharmacies and were used prior to their expiration date. (Hytone products are no longer marketed in the US.) Neat (USP grade) hydrocortisone (HC) and betamethasone diprionate were obtained as a gift from the Lemmon Company (Sellersville, PA). Methanol and acetonitrile were HPLC grade and purchased from Fisher Scientific (Fair Lawn, NJ). ^3H_2O (5 mCi ml^{-1}) was obtained from New England Nuclear Products (Boston, MA). Oleth-20 was a gift from Croda Inc. (New York, NY). All other chemicals and reagents were analytical grade or better. Mixed cellulose filter membranes (HAWP 024 12) were obtained from the Millipore Corporation, Bedford, MA. Transpore™ tape (#1527-1) was obtained from 3M Health Care, St. Paul, MN.

The *in vitro* release rate of hydrocortisone from the three test products was measured using mixed cellulose acetate/nitrate filter membranes (HAWP 024 12) mounted on 0.8 cm^2 jacketed Franz diffusion cells (PermeGear Inc., Hellertown, PA). The receptor chamber was filled with phosphate-buffered isotonic saline (pH 7.3-7.4) containing 0.5% polyethylene glycol 20 oleyl ether (Oleth-20, Volpo-20), a non-ionic surfactant used to increase HC solubility and ensure sink conditions in the receptor solution [15]. The cells were then placed in the diffusion apparatus in which the receptor cell was maintained at 32 ± 1° C and stirred magnetically at 600 rpm. Each product (2 cm^3) was applied onto the membrane using a syringe after which the dosing chamber was occluded with polyethylene film (Saran Wrap™). At 1, 2, 4, 6, and 8 hours, samples (1 ml) of receptor chamber solution were collected and replaced with fresh solution. Samples were stored at -20° C until assayed for HC by high pressure liquid chromatography (HPLC). Test products were compared side-by-side on three cells each and the mean ± standard error calculated. The cumulative amount of HC recovered in the receptor solution at each sample time was determined and the release rate (μg cm^{-2} h$^{-\frac{1}{2}}$) calculated by linear regression on the square root of time. The statistical significance of differences in the release rates between products was assessed using the Student's t-test.

Cryopreserved human cadaver skin, dermatomed to 0.5 mm and without obvious signs of skin disease or damage, was obtained from a local skin bank. It was stored in a water-impermeable plastic bag at $-70°$ C until the day of an experiment. The frozen specimen was thawed prior to use by placing the sealed bag in warm (37 ° C) water, following which the skin surface was rinsed in tap water to remove any blood or other extraneous material. Skin from a single donor was cut into multiple smaller sections to fit on 0.8 cm^2 Franz diffusion cells. The dermal chamber (receptor compartment) was filled with phosphate-buffered isotonic saline, pH 7.3-7.4 and stirred magnetically at 600 rpm. The epidermal chamber (donor compartment) was left open to ambient laboratory conditions (approximately 40-60% relative humidity, 21-23° C) and skin surface temperature maintained at 32 ± 1° C by circulating temperature-controlled water through the jacket surrounding the dermal chamber.

To assure the integrity of each skin section, its permeability to ^{3}H$_2$O was determined before application of the hydrocortisone products. Following a brief (0.5 to 1 hour) equilibration period, ~150 μl ^{3}H$_2$O (specific activity 0.5 μCi ml^{-1}) was layered across the top of the skin section so that the entire exposed surface was covered. After 5 minutes the aqueous layer was removed and the surface of the skin carefully blotted dry. Twenty-five minutes later the saline receptor solution was removed and analysed for radioactive content by liquid scintillation counting. Only skin specimens in which the absorption of ^{3}H$_2$O was less than 0.75 μl (approximately 0.5% applied dose) were utilised in this study.

Following the ^{3}H$_2$O integrity test, the receptor solution was replaced with a 1:10 dilution of phosphate buffered saline (reduced salt concentration improved the hydrocortisone recovery during sample processing). Subsequently each test formulation was applied to three to four replicate sections from multiple donors at a target dose of 10 mg cm^{-2}. At 4, 8, 12, 24, 30, 36 and 48 hours the receptor solution was removed in its entirety and replaced with fresh receptor solution. Samples were stored at -20° C for subsequent hydrocortisone analysis by HPLC.

The data were used to determine the rate of absorption profile as well as calculate total absorption for the entire 48-hour period. The rate of absorption (μg cm^{-2} hr^{-1}) was determined for each sampling interval by dividing the amount absorbed in that interval by the length (hours) of the interval and correcting for area. Total absorption (μg cm^{-2}) for the experimental period (48 hours) was calculated from the sum of drug content found in all seven receptor samples corrected for area. The mean ± standard error for both the rate of absorption and total absorption was calculated by averaging the data from all skin sections per product.

Stratum corneum tape stripping studies were conducted on twenty four adult human subjects between 18 to 40 years of age. These studies were conducted with Institutional Review Board approval and written informed consent was obtained from all subjects. Subjects were free of all dermatologic disease and had not used oral retinoids or glucocorticoids in the previous 3 months, nor had the skin of the ventral forearms been treated with any topical medications in the previous month. On both the right and left volar forearms, four rectangular (2 cm x 5 cm) areas were outlined with ink. The long axis of the rectangle was positioned perpendicular to the long axis of the arm and each test site was spaced 1.5 cm from adjoining sites. Test products were applied at a dose of 19 ± 0.1 mg (1.9 mg cm^{-2}) using a positive displacement pipette and evenly spread over the entire area with a glass rod. Five minutes following application of product, either a non-occlusive screen guard or an occlusive polyethylene film was placed over the subject's forearm.

In these studies four different exposure conditions were evaluated: 1) two hours non-occluded, 2) eight hours non-occluded. 3) eight hours occluded and 4) sixteen hours occluded. Six subjects were used for each exposure condition. At the end of the specified exposure duration, the test sites were tape stripped twenty two times using 1 inch wide Transpore™ tape, the width being slightly larger than the width of the rectangular application area. No wash procedure was utilised prior to stripping. Each strip was applied, rubbed several times to ensure good contact and then removed in one continuous and rapid motion. The first two strips were pooled as representing non-absorbed surface dose. The next 20 strips were pooled into four sets (5 strips each) and placed into 20 ml screw-cap glass vials. To each vial was added 10 ml methanol containing 20 ng ml^{-1} betamethasone diproprionate as an internal standard. Vials were then horizontally shaken (150 RPM) at room temperature for two days, following which 1.8 ml of extraction solution was taken and dried by vacuum centrifugation, then re-dissolved in 100 μl methanol. The recovery efficiency from spiked tape samples put through the same procedures was >90%. Hydrocortisone was quantified by HPLC.

A vasoconstrictor study to compare the potency of the two 1% hydrocortisone products was conducted in six adult subjects. This study was conducted with Institutional Review Board Approval and written informed consent was obtained from all subjects. To duplicate 5 cm^2 sites on the right and left volar forearms, 19 ± 0.1 mg (3.8 mg cm^{-2}) of product was applied using a positive displacement pipette and evenly spread over the entire area with a glass rod. An additional site on both arms was left as a non-treatment control. Five minutes following dosing an occlusive polyethylene film was placed over the subject's forearms and held in place with an elastic cloth sleeve. After 16 hours the occlusive cover was removed and the skin allowed to dry prior to assessing the extent of vasoconstriction. Two readings were taken, one at 17 hours and another at 20 hours. Vasoconstriction was measured using a Chroma Meter (Model # CR-300, Minolta Corp, Ramsey, NJ) set for automatic averaging of 3 readings. Using the "LAB" colour scale, the "a" value (representative of skin redness) was recorded for each site. The change in "a" value between pre-dosing and the 17- and 20-hour readings was calculated for each site. Readings were summed across all subjects by product and the mean and standard error determined.

Rate of release study receptor solutions were assayed directly without concentration. Receptor solution samples from the in vitro skin permeation study were concentrated by evaporating a 4 ml aliquot by vacuum centrifugation (Speed-Vac, Savant) and then redissolved in 100 μl methanol (recovery efficiency from spiked samples was >98%). Hydrocortisone was quantified by high pressure liquid chromatography on a Hewlett-Packard model 1090 M series II system with a diode-array-detector using a reverse-phase Absorbosphere HC C18 (3μm, 4.6 x 100 mm, Altech, Deerfield, IL) column maintained at 40° C. The solvent system consisted of a linear gradient flow at 0.5 ml min^{-1} starting at 75/25 (water/acetonitrile) changing to 25/75 (water/acetonitrile) over 5 minutes followed by a re-equilibration to starting conditions over the next 4 minutes. Twenty microliters of sample were injected and eluting peaks were monitored at 245 nm (4 nm bandwidth) referenced to 450 nm (80 nm bandwidth). Hydrocortisone concentration was calculated from external standard curves prepared with neat hydrocortisone which were found to be linear from 2-200 μg ml^{-1}. The lower limit of detection was 0.3 μg ml^{-1}.

3 RESULTS

The results obtained by IVRT are presented in Figure 1. There was no difference in the rate of HC release from the two 1% HC products: 84.1 ± 10.6 versus 87.1 ± 11.2 μg cm^{-2} h$^{\frac{1}{2}}$ for Cortizone-10 and Hytone, respectively. The rate of release from the 2.5% product was approximately 1.8-times greater (154.1 ± 37.7 μg cm^{-2} h$^{\frac{1}{2}}$) than either lower strength product ($p < 0.05$).

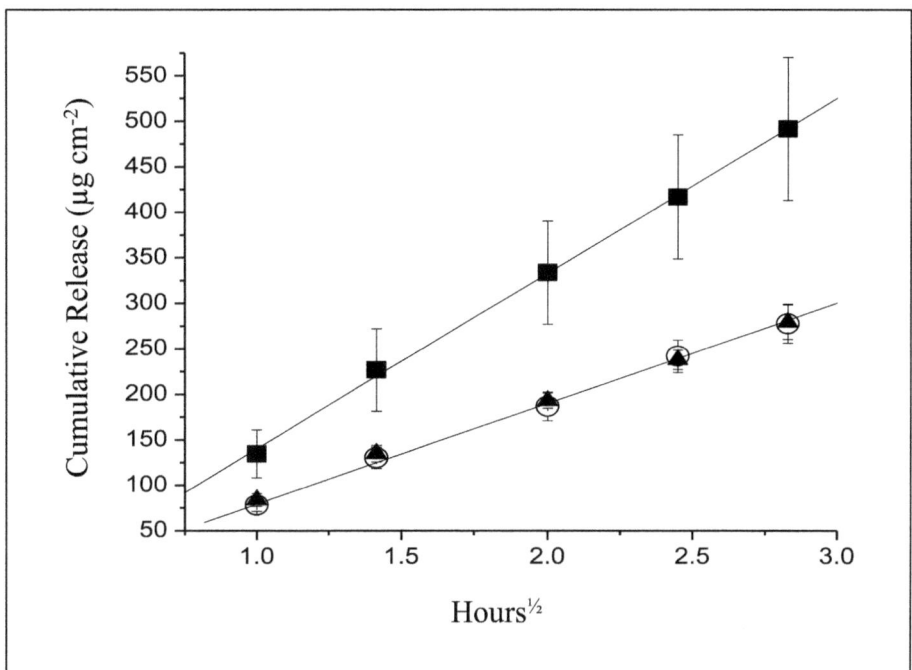

Figure 1 *In vitro release test - average rate of HC release through a mixed cellulose filter membrane after application of Cortizone-10 (▲), Hytone 1% (○) or Hytone 2/5% (■). Data expressed as mean ± SEM.*

Data showing the rate of absorption of HC from the three test products is presented in Figure 2. Hytone 1% and 2.5% yielded absorption profiles that were essentially indistinguishable, despite their 2.5-fold difference in concentration. However, both products generated a substantially greater HC flux than that measured from Cortizone-10. Between 6 – 30 hours the flux was > 60 ng cm^{-2} h^{-1} for both Hytone products, but no greater than 10 ng cm^{-2} h^{-1} for Cortizone-10. Total absorption over the 48-hour period of measurement is given in Table 2 and shows an order of magnitude difference between Cortizone-10 and the two Hytone products, a difference that was highly statistically significant (p<0.001).

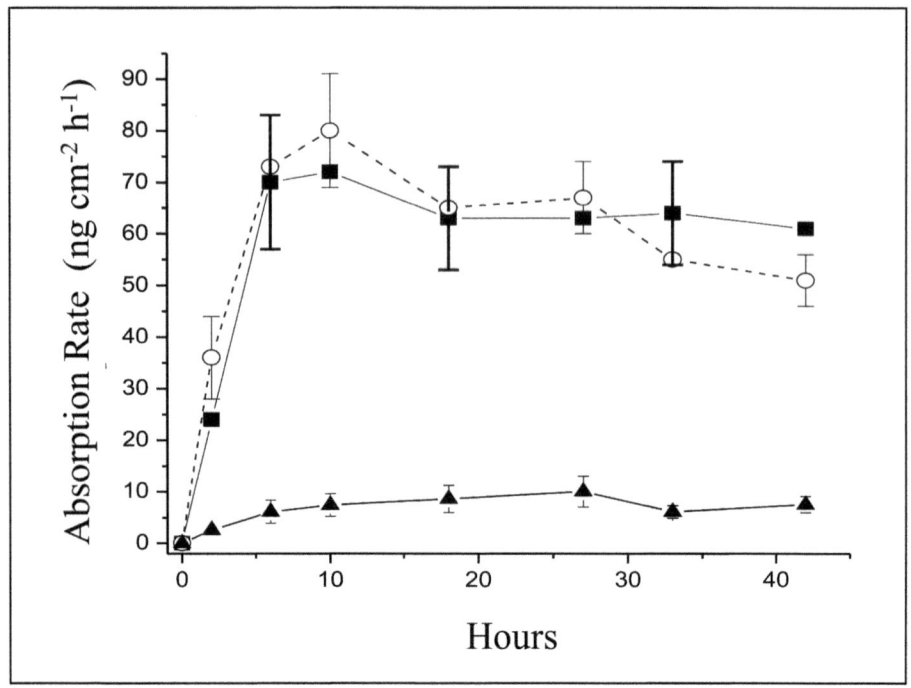

Figure 2 *IVPT: Average rate of HC absorption through human skin after application of Cortizone-10 (▲), Hytone 1% (○) and Hytone 2.5% (■). Data expressed as mean ± SEM.*

Table 2 *Summary of in vitro skin permeation test results. All values are mean ± SEM.*
†Number of skin sections/ number of donors; Hytone 1% and 2.5% were tested side-by-side using skin from the same four donors.

Product	Total Absorption $\mu g\ cm^{-2}$ (48 h)	Range	Peak Flux $ng\ cm^{-2}\ h^{-1}$	$N^\dagger$
Cortizone-10	0.28 ± 0.06	$0.09 - 0.59$	10.0 ± 3.0	9/3
Hytone 1%	2.24 ± 0.26	$0.49 - 3.68$	80.1 ± 11.2	16/4
Hytone 2.5%	2.21 ± 0.35	$0.94 - 4.75$	72.2 ± 11.6	16/4

Since Cortizone-10 and Hytone 1% were found to differ widely in HC bioavailability as determined by IVPT, these two products were carried forward into the next phase of the study to determine if the other two biologically relevant surrogate tests (tape stripping and vasoconstrictor (VC) assay) could also confirm this difference. The results obtained from non-occluded two hour and eight hour applications of both products are presented in Figure 3. No difference in drug distribution through the stratum corneum was observed after 2 hours exposure, but greater HC levels were detected with Hytone after 8 hours. The amounts of HC recovered in the first twelve tape strips were similar. However, there were significant (p<0.05) differences between the two test conditions in the amounts of HC recovered from the last ten strips (representative of the deeper portion of the stratum corneum). In our experience with Transpore tape, 25-35 strips are required to completely strip the stratum corneum in most subjects.

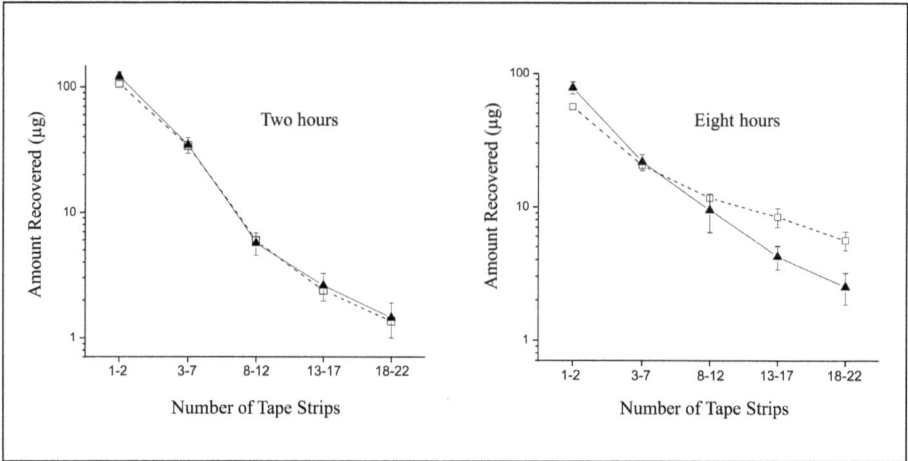

Figure 3 *Recovery of HC from tape strips following a two hour and eight hour non-occluded application of Cortizone-10 (▲) and Hytone 1% (□). All values are mean ± SEM.*

Two additional tape strip studies were conducted to serve as a "bridge" to the VC studies which were to follow. Since HC is a low potency glucocorticoid, its ability to induce vasoconstriction is limited and occlusion is necessary to measure it in the majority of subjects. As a result it was necessary to verify that prolonged hydration (8 – 16 hours) did not alter conditions at the surface of the skin such that differences between the two test products were greatly reduced or eliminated. The results are presented in Figure 4 and confirm that under occlusive conditions HC levels within the horny layer remain greater for Hytone than Cortizone-10.

An important aspect of the data presented in Figure 4 is the extent to which they support the results obtained with the IVPT. The permeation data showed that the maximum difference in HC flux between Cortizone-10 and Hytone was seen in the second receptor sample taken at eight hours, and the difference was maintained over the next forty hours

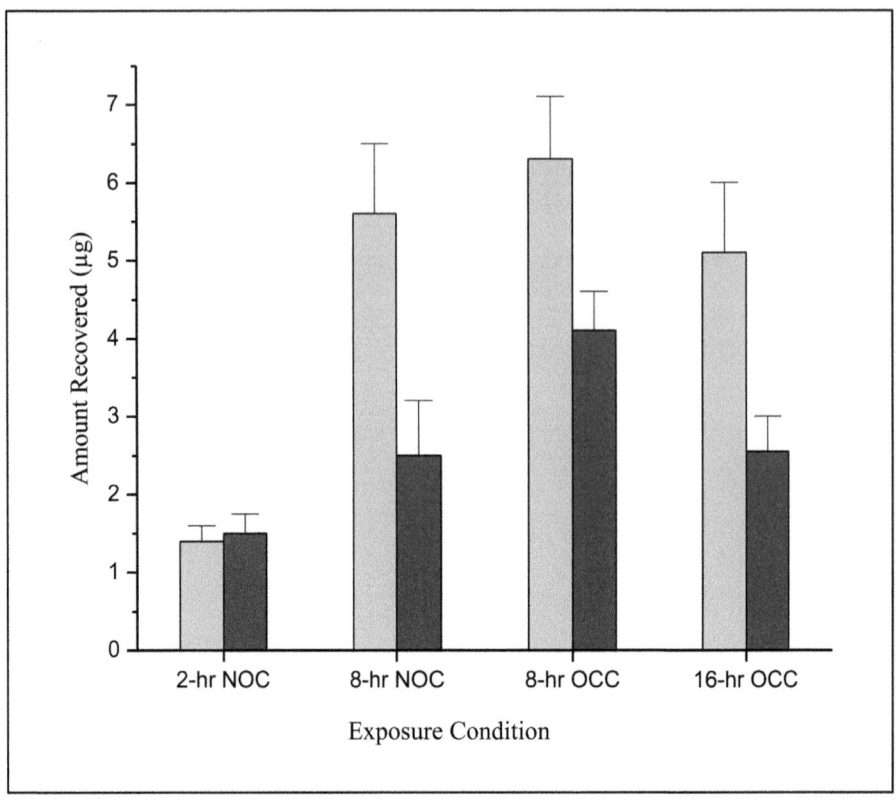

Figure 4 *Total amount of HC recovered in the last five tape strips (#18-22) following Cortizone-10 (■) and Hytone 1% (□) application under four different dosing regimens. OCC = occluded, NOC = non-occluded. Error bars represent SEM.*

(Figure 2). Tape stripping also showed a difference between the products at eight as well as sixteen hours. Furthermore, tape stripping showed an increasing level of HC in the stratum corneum at eight hours versus two hours and this parallels the increasing flux seen with both products at eight hours versus the earlier sample taken at four hours. Unfortunately, no sample was taken at two hours in the permeation experiment that could be directly compared to the 2-hour tape stripping data.

The results of the vasoconstriction (VC) assay are presented in Table 3. Hytone was found to be the more potent product in that it induced a greater vasoconstrictor response than that seen with Cortizone-10. At the 17-hour reading, the difference between products fell just short of statistical significance, but at 20 hours a statistically significant difference was observed ($p < 0.05$). These results are in agreement with those obtained by the IVPT and skin stripping.

Table 3 *Summary of Vasoconstrictor Assay. Values represent the difference in "a" value between pre-dose readings and 17- or 20-hour post-dose measurements (Mean ± SEM).*

	17-Hour	20-Hour
Cortizone-10	1.82 ± 0.42	1.72 ± 0.37
Hytone	2.56 ± 0.45	2.82 ± 0.51
p value	0.067	0.015

4 DISCUSSION

A consistent picture of agreement was demonstrated between the three surrogate tests in identifying the inequivalence of two 1.0% hydrocortisone (HC) products. When examined within the context of percutaneous absorption theory, the results are not surprising. The most likely explanation for the difference between products is due to a greater distribution of drug between vehicle and stratum corneum from one versus the other. This difference in partitioning would result not only in a greater HC level within the stratum corneum but, because of the greater driving force it creates, would also lead to greater percutaneous absorption. Both of these expected consequences were detected in this study. A greater HC content within the stratum corneum was found *in vivo* through tape stripping and a greater flux through the skin was found using the *in vitro* permeation test (IVPT).

Although these two surrogate tests do not directly assess HC drug levels within the living layers of the skin where its site of action must reside, both predict that there will be a difference in the drug levels resulting from application of the two 1.0% HC products. This, in fact, was confirmed using the VC assay. A difference in the response of the superficial blood vessels of the dermis can only occur as a result of a difference in HC concentration within the dermis. Given the complete harmony between the results obtained by three surrogate tests of divergent nature (2 PK vs 1 PD, 2 *in vivo* vs 1 *in vitro*), it is irrefutably clear that the relative bioavailability of HC is greater from one test product than the other: though the two products are pharmaceutically equivalent, they are not bioequivalent.

A positive correlation between the VC assay and tape stripping has been seen before with both betamethasone dipropionate and triamcinolone acetonide [4, 10]. Likewise, positive correlation between the VC assay and IVPT has also been observed before [13]. A further positive correlation between IVPT and the VC assay was observed in this study, beyond that for Cortizone-10 versus Hytone 1.0%. The permeation test found no difference in bioavailability between Hytone 1.0% and 2.5% and this is in agreement with the results reported by Stoughton and Wullich who found that there was also no difference in vasoconstrictor potency [16].

The results obtained using the *in vitro* release test (IVRT) were completely at odds with the results obtained by the three biologically relevant surrogate tests. IVRT found no difference between the two 1.0% HC products, but did find a lower rate of release from both versus the 2.5% product. Given the considerable difference in composition of the two 1.0% products (Table 4), it is surprising that no difference was detected.

The current status of IVRT is somewhat ambiguous in that it was originally conceived as a test analogous to the *in vitro* dissolution test for solid oral dosage forms and, therefore, would be useful in establishing BE. In 1997 the FDA issued a guidance defining limited conditions in which a manufacturer could make changes to one of its own already approved products and use IVRT data to demonstrate "sameness" of the new and old formulations, avoiding the need to conduct an *in vivo* BE test [7]. Thus, it was implicitly recognised as a BE surrogate in a limited number of precisely specified situations. However, in the same document it stated that, because of the absence of sufficient data demonstrating good *in vitro/in vivo* correlation, IVRT was not yet considered to be a valid BE surrogate and not to be used "for comparing different formulations across manufacturers". This latter position may be changing since, most recently, the FDA issued a draft guidance proposing a very short path to approval of generic acyclovir ointment 5.0% in which IVRT plays a pivotal role (along with other physicochemical tests) and the need for a clinical trial is eliminated [17].

Table 4 *Declared excipient listing of the three products tested in this study (Cotizone-10, Hytone 1% and Hytone 2.5%).*

Cortizone-10	Hytone 1% & 2.5%
Aloe Vera Gel	Cetyl Alcohol
Aluminum Sulfate	Cholesterol and related sterols
Calcium Acetate	Glyceryl Monostearate SE
Cetearyl Alcohol	Isopropyl Myristate
Glycerin	Polyoxyl 40 Stearate
Light Mineral Oil	Polysorbate 60
Methylparaben	Propylene Glycol
Potato Dextrin	Sorbic Acid
Propylparaben	Sorbitan Monostearate
Sodium C12-15 Alcohols Sulfate	
Sodium Lauryl Sulfate	
Water	Water
White Petrolatum	
White Wax	

While the present data reveal a situation in which the results obtained using IVRT are completely erroneous as they pertain to HC dermal bioavailability from three specific products, it does not negate the potential utility of the test for other actives in other products. It seems likely that changes within the matrix of some formulations, as detected by IVRT, might translate directly to changes in bioavailability. Demonstration of good *in vitro-in vivo* correlation and a thorough validation of the test method for those situations would be necessary to advocate its use for establishing or supporting BE.

5 CONCLUSION

The use of multiple surrogate methods, rather than clinical trials, to satisfy regulatory requirements for the approval of generic topical drug products is proposed. The validity of this approach has been demonstrated by showing that the data obtained from three different surrogate methods (IVPT, tape stripping, VC assay) were all in agreement in finding the relative bioavailability of two 1% HC products to be significantly different. A fourth test, IVRT, failed to find a difference between the two products. This work illustrates the value of a comprehensive approach to the assessment of BE, where the collective weight of evidence from the multi-dimensional analysis using appropriate and relevant surrogate tests would provide the most accurate information regarding the equivalence (or lack thereof) of two topical products.

6 REFERENCES

1. Food, Drug and Cosmetic Act and implementing regulations: 21 CFR § 320 and 21 CFR § 320.24.

2. Food and Drug Administration, Guidance for Industry: Topical Dermatologic Corticosteroids: *In Vivo* Bioequivalence. Center for Drug Evaluation and Research, Rockville, MD, 1995. http://www.fda.gov/downloads/Drugs /GuidanceComplianceRegulatoryInformation/Guidances/ucm070234.pdf.

3. J. Skelly, V. Shah, H. Maibach, R. Guy, R. Wester, G. Flynn and A. Yacobi, FDA and AAPS report of the workshop on principles and practices of *in vitro* percutaneous penetration studies: relevance to bioavailability and bioequivalence, *Pharm. Res*, 1987, **4**, 265–267.

4. L. Pershing, B. Silver, G. Krueger, V. Shah and J. Skelley, Feasibility of measuring the bioavailability of topical betamethasone dipropionate in commercial formulations using drug content in skin and skin blanching assay, *Pharm. Res.*, 1992, **9**, 45–51.

5. V. Shah, C. Behl, G. Flynn, W. Higuchi and H. Schaefer, Principles and criteria in the development and optimization of topical therapeutic products, *Pharm. Res.*, 1992 **9**, 1107–1111.

6. L. K. Pershing, L. D. Lambert, V. P. Shah and S. Y. Lam Variability and correlation of chromameter and tape-stripping methods with the visual skin blanching assay in the quantitative assessment of topical 0.05% betamethasone dipropionate bioavailability in humans, *Int. J. Pharm.*, 1992, **86**, 201–210.

7. Food and Drug Administration, Guidance for Industry: Nonsterile Semisolid Dosage Forms, Scale-up and Post-Approval Changes: Chemistry, Manufacturing, and Controls; *In Vitro* Release Testing and *In Vivo* Bioequivalence Documentation. Center for Drug Evaluation and Research, Rockville, MD, 1997. http://www.fda.gov/downloads/Drugs/GuidanceComplianceRegulatoryInformation/Gui dances/UCM070930.pdf

8. V. Shah, G. Flynn, A. Yacobi, H. Maibach, C. Bon, N. Fleischer, T. Franz, S. Kaplan, J. Kawamoto, L. Lesko, J. Marty, L. Pershing, H. Schaefer, J. Sequeira, S. Shrivastava,

J. Wilkin and R. Williams, Bioequivalence of topical dermatological dosage forms-methods of evaluation of bioequivalence., *Pharm. Res.*, 1998, **15**, 167–171.

9. G. Flynn, V. Shah, S. Tenjarla, M. Corbo, D. DeMagistris, T. Geldman, T. Franz, D. Miran, D. Pearce, J. Sequeira, J. Swarbrick, J. Wang, A. Yacobi and J. Zatz, Assessment of value and applications of *in vitro* testing of topical dermatological drug products, *Pharm. Res.*, 1999, **16**, 1325–1330.

10. L. Pershing, S. Bakhtian, C. E. Poncelet, J. L. Corlett and V. P. Shah, Comparison of skin stripping, *in vitro* release, and skin blanching response methods to measure dose response and similarity of triamcinolone acetonide cream strengths from two manufactured sources, *J. Pharm. Sci.*, 2002, **91**, 1312–1323.

11. C. Herkenne, I. Alberti, A. Naik, Y. Kalia, F. Mathy, V. Preat and R. Guy, In vivo methods for the assessment of topical drug bioavailability, *Pharm. Res.*, 2008, **25**, 87–103.

12. B. N'Dri-Stempfer, W. Navidi, R. Guy and A. Bunge, Improved bioequivalence assessment of topical dermatological drug products using dermatopharmacokinetics, *Pharm. Res.*, 2009, **26**, 316–328.

13. T. Franz, P. Lehman and S. Raney, Use of excised human skin to assess the bioequivalence of topical products, *Skin Pharmacol. Physiol.*, 2009, **22**, 276–286.

14. P. Lehman, S. Raney and T. Franz, Percutaneous absorption in man: in vitro-in vivo correlation, *Skin Pharmacol. Physiol.*, 2011, **24**, 224–230.

15. R. L. Bronaugh and R. F. Stewart, Methods for in vitro percutaneous absorption studies III. Hydrophobic compounds, *J. Pharm. Sci.*, 1984, **73**, 1255–1258.

16. R. B. Stoughton and K. Wullich, 1989, The same glucocorticoid in brand-name products. Does increasing the concentration result in greater topical biologic activity?, *Arch. Dermatol.*, 1989, **125**, 1509–1511.

17. Food and Drug Administration, Draft Guidance on Acyclovir, March, 2012. http://www.fda.gov/Drugs/GuidanceComplianceRegulatoryInformation/Guidances/ucm075207.htm

SECTION III: COSMECEUTICALS

R P Chilcott

Department of Pharmacy, University of Hertfordshire, College Lane Campus, Hatfield, United Kingdom.

Cosmetics have been part of human culture for thousands of years but have only recently started to incorporate substances that have been specifically designed to achieve a particular function through rationale design and scientific evidence [1 – 3]. Clearly, marketing is a major driver for the development of cosmetic products containing "active" compounds (cosmeceuticals). However, the identification and formulation of such bespoke molecules represents a significant development in cosmetic technology and has a number of parallels with topical therapies such as appropriate formulation, consumer safety and demonstration of "efficacy". Indeed, there is considerable transfer of knowledge between the cosmetic and pharmaceutical sectors which has both advanced our understanding of the skin and led to many innovative techniques for measuring skin function. However, the development of new cosmeceuticals and products is potentially fraught with difficulty for two main reasons: regulatory interpretation and demonstration of safety/efficacy.

The US Food and Drug Administration (FDA) defines cosmetics as "articles intended to be … applied to the human body ... for cleansing, beautifying, promoting attractiveness, or altering the appearance" [4]. In contrast, the FDA defines a drug as "articles intended for use in the diagnosis, cure, mitigation, treatment, or prevention of disease" and "articles (other than food) intended to affect the structure or any function of the body of man or other animals" [5]. Therefore, a cosmetic ingredient which causes a measureable physiological response could be considered by regulatory agencies to be a drug and thus subject to extensive controls which would preclude or at the very least severely limit the availability of certain cosmetic products. The distinction between a pharmacologically active substance and an active cosmetic ingredient is becoming increasingly blurred as new compounds are developed with increasingly sophisticated mechanisms of action. Chapter 19 of this book provides a brief historical perspective on the development of cosmeceutics and sets out an interesting intellectual case that the fundamental need for cosmetic products arises from exposure to sunlight. This thought-provoking chapter establishes that many active ingredients in contemporary cosmetic formulations model the intended outcome of pharmacologically active substances (such as retinoids) but via subtly different mechanisms which currently obviate their regulatory classification as drugs.

The theme of clarifying the differences between pharmaceutics and cosmeceutics is continued in Chapter 20, where the origins of definitions and descriptions currently utilised by regulatory authorities are reviewed and questioned where appropriate. The author points out that the use of a relatively antiquated regulatory process (whilst sufficiently generic and having evolved to some degree to incorporate advances in technology) may require modernisation to protect the consumer, regulator and cosmetic companies against unsubstantiated, inappropriate or confusing information. For example, whilst the category of 'cosmeceutics' has been used for some time, it is not currently recognised by any regulatory agency, despite calls for such classifications [6]. Furthermore, the development of new formulations and excipients arguably necessitates a fundamental reappraisal of the properties, characteristics and definitions associated with cosmetic products based on a functional approach.

The need to demonstrate cosmetic safety without recourse to animal experimentation has stimulated great interest in alternative methods [7–9]. Specifically, legislation recently introduced within Europe [10] now prohibits the use of animal research for the development of cosmetic products. Whilst this change has been met with widespread approval (and for good reason), it has become increasingly difficult for companies to demonstrate the effectiveness and safety of its products using the range of *in vivo* methods available to pharmaceutical scientists. However, many *in vitro* approaches fail to accurately model the unique interactions of cosmetic products with human skin under "in use" conditions. Chapter 21 reports the use of an *in vitro* human skin diffusion cell system to address issues relating to the systemic absorption and toxicity of aluminium from antiperspirant products. The use of diffusion cells is a long-established method for measuring the percutaneous absorption of substances across the skin [11] and can be readily applied to the risk assessment of cosmetics.

Work reported in Chapter 22 presents a good example of the rational design of cosmeceutics based on existing therapeutic entities. In this case, drugs traditionally employed for the treatment of progeria [12] are considered as candidates for anti-aging formulations. This work also illustrates similarities in the development of cosmetics and topical therapies in that dermal absorption of the active ingredients from the vehicle (formulation) must be subject to appropriate measurement to ensure delivery of a biologically relevant dose with an appropriate margin of safety.

Overall, the chapters in this section challenge existing dogma concerning the classification and use of cosmeceutics and provide examples of how cosmetic formulations can be assessed for safety and efficacy without recourse to animal studies.

REFERENCES

1. F. S. Brandt, A. Cazzaniga and M. Hann, Cosmeceuticals: current trends and market analysis, *Semin Cutan Med Surg.*, 2011, **30**(3), 141–3.

2. D. Li, Z. Wu, N. Martini and J. Wen, Advanced carrier systems in cosmetics and cosmeceuticals: a review, *J Cosmet Sci.*, 2011, **62**(6), 549–63.

3. K. A. Nolan and E. S. Marmur, Over-the-counter topical skincare products: a review of the literature, *J Drugs Dermatol.*, 2012, **11**(2), 220–4.

4. Federal Food, Drug, and Cosmetic Act. Chapter II, Section 201, Paragraph i.

5. Federal Food, Drug, and Cosmetic Act. Chapter II, Section 201, Paragraph g.

6. B. J. Vermeer, B. A. Gilchrest, Vermeer, S. L. Friedel, LLB, Cosmeceuticals: A Proposal for Rational Definition, Evaluation, and Regulation, *Arch Dermatol.*, 1996, **132**(3), 337–340.

7. T. Hartung, S. Bremer, S. Casati, S. Coecke, R. Corvi, S. Fortaner, L. Gribaldo, M. Halder, S. Hoffmann, A. J. Roi, P. Prieto, E. Sabbioni, L. Scott, A. Worth and V. Zuang, A modular approach to the ECVAM principles on test validity, *Altern. Lab. Anim.*, 2004, **32**, 467–472.

8. Y. Ohno, The validation and regulatory acceptance of alternative methods in Japan, Altern Lab Anim., 2004, **32 Suppl 1B**, 643–55.

9. M. Pauwels and V. Rogiers, Safety evaluation of cosmetics in the EU. Reality and challenges for the toxicologist, *Toxicol Lett.*, 2004, **151**(1), 7–17.

10. Council Directive 76/768/EEC of 27 July 1976 on the approximation of the laws of the Member States relating to cosmetic products. OJEC L262 of 1976-09-27, pp. 169–200. ref: Cosmetic Directive 76/768/EC.

11. R. U. Pendlington, *In vitro* percutaneous absorption measurements, in *Principles and practice of skin toxicology*, ed. R. P. Chilcott and S. Price, John Wiley & Sons Ltd, Chichester, 2008, chapter 8, pp. 129–148.

12. R. L. Pollex and R. A. Hegele, Hutchinson-Gilford progeria syndrome, *Clin Genet.*, 2004, **66**(5), 375–81.

19

MAKING SENSE OF THE COSMECEUTICALS CONCEPT

D Fernandes

The Renaissance Body Science Institute, 183 Bree Street, Cape Town 8001, South Africa

1 INTRODUCTION

It is well-known that conventional cosmetics should have no physiological effect on the skin and should make no changes. Theoretically they should only moisturise the stratum corneum and in that way make wrinkles and fine lines less apparent. They should have no effects deeper than that level. Cosmetics may legally advertise their action as:

- Reducing the appearance of wrinkles.
- Making the surface appear smoother.
- Making blemishes less apparent.
- Allowing the skin to look more radiant.

On the other hand, 'cosmeceutics' may be described as:

- Products that address photo-damage and the chemical changes induced by sunlight.
- Products that enhance the production of collagens and elastin to tighten the skin.
- Products that lighten skin blemishes.

We cannot ignore the fact that the consumer wants wrinkles to actually be shallower than before so that the skin *is* smoother - not merely *seem* smoother. They also want blemishes to disappear as much as possible and finally they want the healthiest skin imaginable. Furthermore, the consumer wants this from a cosmetic and not from a medicine (the eventual costs would be prohibitive). The consumer is slowly becoming aware of active ingredients and they want these effects from cosmetics. As it turns out, we can achieve all of this with the special ingredients that we already have available on the market. With the laws as they are in the developed world, manufacturers of products that could really be described as cosmeceutics, cannot legally claim to make changes in skin and can only advertise as simple cosmetics. Not surprisingly, unscrupulous companies on the other hand have laid claim to being cosmeceutics when they have no functional effect at all. These pseudo-cosmeceutics contain glycolic acid generally. The term "cosmeceutic" should be reserved for products that actually make demonstrable changes to skin. They should particularly address photoaging because photodamage causes thinner skin, loss of collagen and elastin and pigmentation blemishes.

Optimal formulation is vitally important to ensure that ingredients applied to the surface of the skin will get through the lipid bilayers of the stratum corneum in order to have effects on the keratinocytes, melanocytes, Langerhans cells and fibroblasts. We have to ensure with our formulations that the active ingredients do not pass right through the epidermis and get absorbed into the bloodstream and produce other generalised effects.

Fundamentally, real cosmeceutics are arguably products that address the concatenation of chemical changes that occur when we go out into sunlight. That is why we need to understand photodamage before we design a real cosmeceutic.

2 OVERVIEW OF PHOTODAMAGE

We all go out into the sun and within seconds chemical changes start happening as the energy of various light waves start interacting with chromophores in our skin. Photo-damage occurs not only because of UVA and UVB rays but also from visible light such as blue, which penetrates deeper than UV light. It seems that as the light becomes redder, less damage occurs but then one starts to see the effects of temperature changes from infra-red rays. Most of our visible photo-damage occurs from wave lengths shorter than blue. The thicker the epidermis, the "softer" the effects of UV-B rays shorter than 300 nm [1]. Melanin further softens the effects of UVB and is also provides the main endogenous protection from UVA. The more melanin one has in the stratum spinosum, the greater the protection. If the melanin remains active up to the stratum corneum, then we get maximal protection from UV rays.

In the stratum corneum we also have other chromophores for UVB as well as UVA. The stratum corneum cells normally have retinyl palmitate which absorbs both UVB as well as UVA and, depending on its concentration, can be a powerful natural sunscreen [2]. We have many other chromophores in our skin such as trans-urocanic acid and carotenoids. Trans-urocanic acid is isomerised to cis-urocanic acid which actually aggravates the free radical challenge [3]. Beneath the epidermis, Collagen fibres in the dermis scatter light rays.

Photo-damage is indicated by the development of a tan. When keratinocyte DNA is damaged by free radicals, nitric oxide is released and stimulates melanogenesis [4]. We know that blue light [5] and even green light [6] damages DNA because these light waves can induce a tan – however, the damage is significantly less than UV light.

We normally have vitamin A and antioxidants in our skin but after a relatively short exposure to light, the vitamin A in our skin absorbs the energy of UV light in the higher UVB and lower UVA range and consequently becomes inactive. After about 20 minutes the vitamin A levels in skin are only about 20% of the normal levels [7]. The normal skin antioxidants are also damaged by UV light. Vitamin A in physiological doses inhibits the conversion of pre-matrix-proteinases to active collagenases (MMPs) etc. [8] and when we are deprived of vitamin A then the MMPs are released and destroy collagen and elastin. We also lose the control that vitamin A normally exerts on cellular growth, differentiation and maturation and, as a result, mutations develop that could eventually cause wrinkles, pigmented marks or skin cancer [9].

Vitamin C is an essential antioxidant with vitamin E for cellular membrane stability and both of them are destroyed by light exposure. Vitamin C is also essential for the production of healthy collagen [10].

Niacin (niacinamide) plays a very important part in photo-protection [11] and maintenance of healthy skin [12]. A major advantage of niacin is that it reduces deposition of melanin.

In summary, light depletes vitamin A (particularly as retinyl palmitate), vitamin C, E and other antioxidants and therefore these molecules need to be replaced by every cosmeceutic aiming to treat photodamage. Fortunately, topical vitamin A (as retinyl palmitate), B3, C and E etc. have been shown to have physiologic benefits for skin. Studies indicate that despite all the advances in cosmetic chemistry, vitamin A remains the only complete age-controlling physiologic ingredient. While aging may be extremely complex, at this stage the dominant cosmeceutic agent has to be vitamin A either as retinyl palmitate, retinol, retinaldehyde or the medical form retinoic acid.

3 COSMECEUTIC HISTORY AND BIOCHEMISTRY OF VITAMIN A

The era of cosmeceuticals was heralded in by the use of retinoic acid by Albert Kligmann in the early 1980's [13] which was considered the very first ingredient that could reduce photo-damage and intrinsic aging. A virtual media fire-storm ensued as people celebrated the first step in controlling skin aging. The general impression was that retinoic acid was the only effective form of vitamin A and its pre-cursors were considered ineffective even though the very first time in history that aged skin had been improved was from retinyl palmitate.

Greater scrutiny of the international literature reveals that in the 1950's Reiss had already described that senile wrinkles on the legs could be reduced by using retinyl palmitate, (synthesised for the first time just a few years before), the pre-cursor of retinoic acid [14]. Reiss' work was ignored maybe because at that time the perceived wisdom was that aging was inevitable, irreversible and natural. At that stage photo-aging was not considered a disease but was generally seen as a natural progression that happened as years passed. Today we know that wrinkles and sun-blemishes are the first stage of photo-damage that ultimately ends up in the formation of solar keratoses and skin cancer which are diseases. The concept that wrinkles are the earliest manifestation of a disease is still not widely accepted even today. Once we recognise the continuum of wrinkles progressing to cancer, then evidence shows that as early as 1959 Stuttgen had recognised that vitamin A (as retinoic acid; Figure 1) could treat photoaging [15]: he showed that topical applications of retinoic acid onto solar keratoses could restore normal skin. However, this was written in German and at that time the term photoaging did not exist, so again valuable information was ignored.

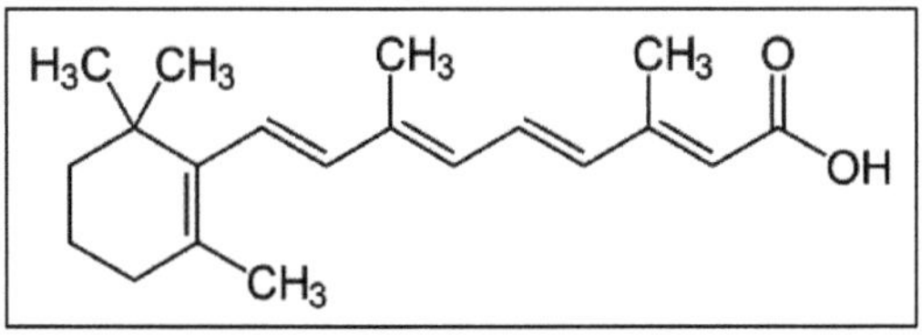

Figure 1 *Molecular structure of retinoic acid.*

Jarrett [16] recognised that the normal metabolism of vitamin A indicated that every form of vitamin A should have the same effects because they are all naturally converted at physiological levels to retinoic acid. We know that we eat most of our vitamin A as retinyl palmitate, absorb it and transfer it in the blood as retinol and then store it in the liver (and in keratinocytes) as retinyl palmitate which is stable. Then, as required by our physiology, we reversibly de-esterify it to retinol [17] and then further, also reversibly, to retinyl aldehyde and finally but irreversibly to retinoic acid. This happens virtually every minute of every day in our lifetime. From Jarrett's point of view, retinyl acetate was far more convenient to use than retinoic acid. Once we understand the physiology of vitamin A we must realise that by supplying retinyl palmitate we actually use the skin cells as factories to make retinoic acid in physiological ratios. It seems that the ratio of retinyl palmitate to retinol to retinal to retinoic acid is 91:3:3:3 and this ratio is tightly controlled [18]. Retinol, retinal and retinoic acid are relatively unstable and generally only found in very low doses in the human cell. This effectively means that retinyl palmitate in adequate doses will have the same effects as retinoic acid. If we compare international units then we can get an indication of the necessary concentrations of retinyl palmitate that produces the same effects as retinoic acid. That is the most important clue in making effective cosmeceutics delivering the power of vitamin A. 1000 i.u. of retinyl palmitate will produce the same changes as 1000 i.u. of retinoic acid. The masses of these ingredients will be vastly different because retinyl palmitate is basically retinol attached to a very large palmitate moiety and that has to be cleaved off before one has retinol and the retinol then has to be oxidised to retinal and further oxidised to retinoic acid. This all happens in the intracellular reticulum and retinoic acid is delivered to the nucleus to act on the DNA [19].

While we have shown that retinyl palmitate is converted to retinoic acid, the simple presence of retinyl palmitate in a cosmetic product will not give it a cosmeceutical effect. It has to be at an appropriate doses that will give physiological effects. Because retinyl palmitate has a much heavier side chain, and because once absorbed it still has to be metabolised into retinoic acid, we need far greater concentrations of it in a cosmetic product to get the same I.U. activity as retinoic acid. In fact, we need a greater concentration of retinyl palmitate in creams than is usually found in order to achieve the desirable cosmeceutical effect.

Clinically, we know that creams containing less than 200 i.u. g^{-1} of retinyl palmitate will not give sufficient changes to warrant a product as being a cosmeceutic [20]. In general, most creams containing retinyl palmitate are less than 10% of an effective dose. Above 200 i.u. g^{-1}, one sees increasing signs typical of the changes of retinoic acid and in fact in order to get cosmeceutic effects, one has to accept the possibility of inducing a transient retinoid reaction. The typical changes in photoaged skin induced by retinoic acid or retinyl palmitate etc. require a daily dose above 1000 i.u. per day in order to have the following effects:

1) An increase in the rate of mitosis in keratinocyte stem cells of the basal layer (growth layer). Stem cells ensure that the growth of the epidermis is maintained. Some cells remain as stem cells others become, under the influence of vitamin A, differentiated keratinocytes which is why the skin becomes thicker. Not only does the skin get thicker; it also heals faster because the cells are growing faster (in fact just the normal growth rate) [21]. Vitamin A stimulates stem cell division and at the same time promotes differentiation into stem cells or into keratinocytes that will grow into the stratum spinosum and eventually mature with the production of keratins into corneocytes of the stratum corneum that is the real destiny of a differentiated keratinocytes.

2) Stimulated epidermal growth results in an improvement in the function of the horny layer (stratum corneum) which becomes more resistant to environmental pollution [22].

3) Melanin in keratinocytes becomes more evenly distributed and by some currently unexplained mechanism reduces excessive abnormal pigmentation [23]. It is possible that vitamin A's influence on keratinocytes inhibits the production of melanin by melanocytes and so reduces it down to the normal rate for the natural colour of the skin [24].

4) The production of sebum is decreased in oily skin. This can cause temporary dryness when one starts using vitamin A cosmeceutics. Many people make the mistake of thinking that the vitamin A cream does not suit their skin because they find it drying. However, once natural moisturisers are produced in greater quantities, skin hydration returns to normal.

5) Vitamin A stimulates CD 44 trans-membrane receptors [25] which stimulate increased production of hyaluronic acid which then normalises the natural skin hydration system [26]. It increases the secretion of natural moisturising factors into the inter-cellular space by the fibroblast cells of the dermis. That allows the skin to retain more water with some puffing out of the wrinkles. These natural moisturising factors filter up into the interstitial fluid of the epidermis. Glycosaminoglycans are some of the chemicals created by the fibroblast to help retain moisture.

6) Vitamin A supports and potentiates the Langerhans cells of the skin and thereby improves detection of errant cancer-inducing mutations [27] as well as stimulating general skin immunity [28].

7) Vitamin A influences the DNA of fibroblast cells, the most important cell in the dermis, particularly the genes for the production of collagen: healthier collagen is formed and unhealthy collagen is removed by enzyme activity [29].

8) The blood supply to the deeper layers of the skin is improved, which means that nutrition of the skin is improved and the skin also takes on a healthier colour [30].

4 DISADVANTAGES OF TOPICAL RETINOIDS

The greatest problem with using topical retinoids, and probably the main reason why they are not in adequate concentration in most cosmetics, is that the skin can become irritated, pink, dry and acne may be exacerbated. This is known as the retinoid reaction [31]. It is often mistaken for an allergic dermatitis but this is a transient phenomenon that is really a chemical irritation. We need cellular receptors to transfer retinoids from the intercellular space into the cell. If we don't have sufficient cellular receptors then the retinoids accumulate in the interstitial fluid and cause a chemical irritation. Retinoid receptors are removed by UVA exposure and can only be built up by increased levels of vitamin A (retinoic acid). In a way one is in a "catch 22" situation because one can only build up retinoid receptors with the very same molecule that cannot easily get into the cell.

The best way to deal with the retinoid reaction is to try and avoid it by introducing vitamin A in gradually increasing amounts to the skin so that the vitamin A can stimulate the production of more retinoid receptors at about the same rate as one increases the exposure dose. In some people this happens rapidly, whereas in others it occurs rather slowly so the rate of increase has to be determined for each individual. It is interesting to note that almost all children are born with abundant retinoid receptors and maintain them because they are not exposed to much to sunlight. As they grow up, their retinoid receptors are gradually diminished as they experience photo-damage. Experience has shown that very young children can tolerate higher doses of topical vitamin A in general than teenagers.

One way to avoid a retinoid reaction is to use retinoids infrequently in the beginning. They could be used only every third of fourth day in the beginning and then gradually increased to alternate days and then to once a day and finally to twice a day. The problem with this is that it generally does not easily fall into the way that people like to treat their skin.

Another less common but more effective way to avoid a retinoid reaction is for the manufacturer to make products in a gradually stepped up concentrations of vitamin A so that the client can start on a very low dose of vitamin A that is unlikely to cause a retinoid reaction and skin gradually builds up more vitamin A receptors. People experience dry skin if they do not have sufficient retinoid receptors and the skin becomes more comfortable when sufficient retinoid receptors have developed. When one reaches this stage, one can move up to a stronger concentration and by doing this over successive stages, one can eventually build up sufficient retinoid receptors to accommodate the highest permissible concentrations of topical vitamin A. This seems the best way to successfully get people to use topical vitamin A and to be able to use levels of vitamin A that induce dramatic changes to photoaged skin. The changes seen from using cosmetic forms of vitamin A (e.g. retinyl palmitate) could match the changes seen from retinoic acid.

5 VITAMIN C (ASCORBIC ACID) AND OTHER ANTIOXIDANTS

Vitamin C is probably one of the most common cosmeceutic agents because, like vitamin E, it is relatively easy to use. Ascorbic acid (Figure 2) is an essential co-factor in the production of normal collagen and if one intends to smoothen out skin then we have to employ ascorbic acid. Unfortunately, ascorbic acid is very unstable and easily becomes oxidised within a few weeks. Also, because of its hydrophilic nature, it cannot easily cross the lipid bilayers and has to be used in high concentrations and rather acidic preparations in order to be absorbed [32]. It furthermore poses the problem that it cannot be taken through the lipid cell wall because there are no ascorbic acid receptors. Usually it passes through cell walls mimicking a sugar as de-hydro-ascorbic acid.

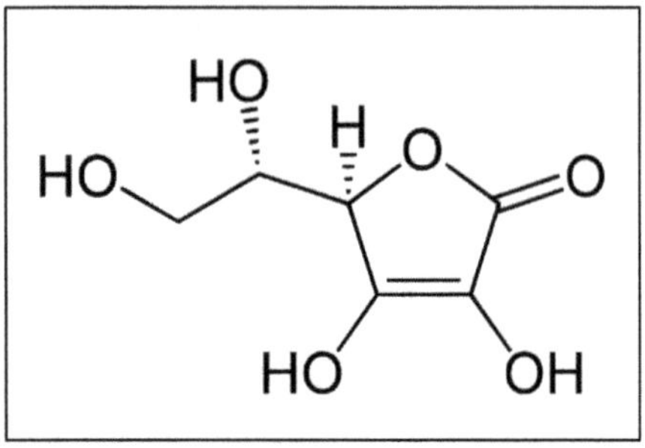

Figure 2 *Molecular structure of vitamin C (ascorbic acid).*

The magnesium and sodium phosphate salts of ascorbic acid are more effective [33] because while they are water-soluble, they are more easily absorbed and can easily enter into keratinocytes. However, their main problem in cosmeceutics is that they decay fairly rapidly with a shelf life of only about six to twelve months. Ascorbyl palmitates are more stable and are absorbed into skin. At this stage the ideal seem to be ascorbyl tetra-isopalmitate which is easily absorbed through the skin and into cells where the palmitic moiety is cleaved off and the ascorbic acid can exert its effects in the production of collagen as well as stimulating DNA expression [34].

One of the mistakes that has led the concept of Cosmeceutics awry is the tendency of marketing departments to concentrate on single antioxidants. Various 'fashionable' antioxidants have each been hailed as the solution to aging. All antioxidants are important but should be used in conjunction with as many others as possible. Vitamin E is a popular ingredient but it is extremely unstable in contrast to antioxidants contained in green tea and others like Rooibos tea (indigenous to South Africa). Co-enzyme Q10, resveratrol, carnosic acid and a whole list of other effective antioxidants are all worthwhile ingredients to be included into cosmeceuticals. The list is too great to mention in a short chapter like this.

6 PEPTIDES

Peptides are an extremely complex and exciting group of cosmeceutic-type ingredients and are found today in many skin and hair-care products on the shelves in stores. Peptides provide an opportunity to arrest the natural degradation that occurs from photoaging as well as chrono-ageing. Some peptides are similar to those normally found in the body and are bio-equivalent. Others are synthetic and their effects have been demonstrated by research. Peptides are agglomerations of some of the 20 amino acids that we use in our body and the permutations are enormous if we consider only the peptides that have less than 10 amino acids (oligopeptides). There are literally hundreds of thousands of possible oligopeptides but not all will have any function. Polypeptides contain 10 to 100 amino acids. Again,not every peptide has a function in the body. The amino acid sequence and the arrangement of the chains determine the function of the peptide.

Peptides for cosmetics can generally be classed as follows:
1. Signal peptides: messengers between cells such as those which stimulate the formation of collagen and elastin (e.g. Matrixyl™, copper tri-peptides). Many of these peptides are similar to the terminal ends of normal collagen fibres [35]. Matrixyl was introduced about late 1998 and gained popularity when it was shown in 2003 to give effects similar to retinoic acid [36]. Retinoic acid no doubt drives its effects through the production of peptides. At this stage Matrixyl remains one of the important peptides for rejuvenating skin.
2. Neurotransmitter-affecting peptides active on nerves. For example, peptides that inhibit the stimulation of muscle fibres (Argireline, Leuphasil, Violox, Syn-ake etc). These can be absorbed in such a low concentration through the skin and still be effective on muscles relatively far away. They can express biological activity in concentrations of one part per billion.
3. Carrier peptides like the copper peptides that convey copper into the skin.

Peptides offer a wide range of benefits if they are well formulated with penetrant enhancers and/or as used with penetrant enhancing techniques. The benefits will depend on the type of peptide and the concentration. If they are well formulated then the difference should become obvious to the user within several weeks. Many commercially available peptides approach or exceed the molecular weight of 500 which is above the generally accepted threshold for dermal absorption or are strongly hydrophilic (and so have difficulty partitioning into the skin). In such instances, we need to adopt methods to reduce or eliminate skin barrier function in order to enhance delivery of such products into the epidermis and dermis.

7 ENHANCING DERMAL ABSORPTION OF COSMETIC INGREDIENTS

A simple method to enhance penetration is micro-needling through the stratum corneum (the barrier to penetration). Prefabricated micro-needle patches [37] are impractical whereas cosmetic instruments such as the Environ Cosmetic Roll-Cit™ or the Environ Focus-Cit™ allow the consumer to achieve the best results at home. The introduction of microscopic holes in the stratum corneum allow larger molecules such as peptides to traverse the skin barrier layer and become available to living cells below. Potentially this system can enhance penetration as much as 100 times more than conventional topical application – but this does require an intensive treatment of the skin for about 10 minutes.

That is why it is essential that the needles do not pierce the skin too deeply otherwise treatment may cause pain or discomfort and inhibit a thorough treatment. Instruments that penetrate only the stratum corneum are comfortable and no pricking sensation is experienced at all.

By using an appropriate iontophoretic charge, one can deliver a charged peptide into the skin. The peptide charge is dependent on the net charge of the various amino acids in a peptide. This method can enhance dermal penetration up to 4 times greater than simply applying the product onto the skin. An important advantage of this technique is that it can get relatively large molecules through skin. The main disadvantage is that iontophoretic delivery generally has to be performed by a professional skin care therapist.

Mitragotri and associates introduced an even more dramatic advancement in penetration techniques in 1996 when they proposed that low frequency sonophoresis (LFS; not to be confused with conventional high frequency ultrasound) could be used to enhance penetration by up to 40 times more than topical skin application [38]. The combination of iontophoresis and LFS is the most effective method yet described to enhance the penetration of active topical products and can deliver large peptides such as insulin through skin.

8 SUMMARY

Cosmeceutic describes a type of cosmetic product that legally should not exist. While many companies claim to market cosmeceutics without any valid reason, we have to realise that there are cosmetics that do make changes to skin. We already know of products that tighten skin and smoothen out fine wrinkles. They refine not only the surface of the skin but make cellular changes that can be detected with the microscope. Pigment marks may be lightened because of effects on the keratinocyte as well as the melanocytes. These are changes that could best be described as due to a cosmeceutic. It is doubtful that any official recognition will ever be made to accommodate effective cosmeceutics. They will remain as simple cosmetics and the term cosmeceutic will have to be reserved for products like retinoic acid products that have been tested thoroughly for their effects on photoaging. Dermatologists and other specialists will need to examine the pre-requisites for what really constitutes a cosmeceutic and what differentiates a cosmeceutic from a standard cosmetic so that they may best advise their patients.

9 REFERENCES

1.   R. R. Anderson and J. A. Parrish, The optics of human skin, *J Invest Dermatol*, 1981, **77**(1), 13–9.

2.   C. Antille *et al.*, Vitamin A exerts a photoprotective action in skin by absorbing ultraviolet B radiation, *J Invest Dermatol*, 2003, **121**(5), 1163–7.

3.   F. De Fine Olivarius, *et al.*, Urocanic acid isomers: relation to body site, pigmentation, stratum corneum thickness and photosensitivity, *Arch Dermatol Res,* 1997, **289**(9), 501–5.

4.   C. Romero-Graillet *et al.*, Nitric oxide produced by ultraviolet-irradiated keratinocytes stimulates melanogenesis, *J Clin Invest*, 1997, **99**(4), 635–42.

5.   J. G. Peak and M. J. Peak, Comparison of initial yields of DNA-to-protein crosslinks and single-strand breaks induced in cultured human cells by far- and near-ultraviolet light, blue light and X-rays, *Mutation Research*, 1991, **246**(1), p. 187–91.

6.   J. G. Peak and M. J. Peak, Induction of slowly developing alkali-labile sites in human P3 cell DNA by UVA and blue- and green-light photons: action spectrum, *Photochem Photobiol*, 1995, **61**(5), 484–7.

7.   E. H. Cluver and W. M. Politzer, The pathology of sun trauma, *S Afr Med J*, 1965, **39**(41), 1051–3.

8.   R. E. Watson *et al.*, Retinoic acid receptor alpha expression and cutaneous ageing, Mech Ageing Dev, 2004, **125**(7), 465–73.

9.   P. P. Fu *et al.*, Physiological role of retinyl palmitate in the skin, *Vitam Horm*, 2007. **75**, 223–56.

10.   T. L. Duarte, M. S. Cooke and G. D. Jones, Gene expression profiling reveals new protective roles for vitamin C in human skin cells, *Free Radic Biol Med*, 2009, **46**(1), 78–87.

11.   H. T. Kang, H. I. Lee and E. S. Hwang, Nicotinamide extends replicative lifespan of human cells, *Aging Cell*, 2006, **5**(5), 423–36.

12.   C. A. Benavente, M. K. Jacobson and E. L. Jacobson, NAD in skin: therapeutic approaches for niacin, *Curr Pharm Des*, 2009, **15**(1), 29–38.

13.   L. H. Kligman, Photoaging. Manifestations, prevention, and treatment, *Dermatol Clin*, 1986, **4**(3), 517–28.

14.   F. Reiss and R. M. Campbell, The effect of topical application of vitamin A with special reference to the senile skin, *Dermatologica*, 1954, **108**(2), 121–8.

15.   G. Stuttgen, Zur Lokalbehandlung von Keratosen mit Vitamin A-Säure, *Dermatologica*, 1962, **124**, 65–80.

16.   A. Jarrett, R. Wrench and B. Mahmoud, The effects of retinyl acetate on epidermal proliferation and differentiation. I. Induced enzyme reactions in the epidermis, *Clin Exp Dermatol*, 1978, **3**(2), 173–88.

17.   D Sklan, Vitamin A in Human Nutrition, *Prog Food Nutrition Sci*, 1987, **11**(1), 39–55.

18.   R. L. Sedjo *et al.*, Circulating endogenous retinoic acid concentrations among participants enrolled in a randomized placebo-controlled clinical trial of retinyl palmitate, *Cancer Epidemiol Biomarkers Prev*, 2004, **13**(11 Pt 1), 1687–92.

19.   S. B. Kurlandsky *et al.*, Biological activity of all-trans retinol requires metabolic conversion to all-trans retinoic acid and is mediated through activation of nuclear retinoid receptors in human keratinocytes, *J Biol Chem*, 1994, **269**(52), 32821–7.

20.   R. E. Watson *et al.*, Repair of photoaged dermal matrix by topical application of a cosmetic 'antiageing' product, *Br J Dermatol*, 2008, **158**(3), 472–7.

21.   P. M. Elias, Epidermal effects of retinoids: supramolecular observations and clinical implications, *J Am Acad Dermatol*, 1986, **15**(4 Pt 2), 797–809.

22.   V. Goffin *et al.*, Topical retinol and the stratum corneum response to an environmental threat, *Skin Pharmacol*, 1997, **10**(2), 85–9.

23.   J. P. Ortonne, Retinoid therapy of pigmentary disorders, *Dermatol Ther*, 2006. **19**(5), 280–8.

24.   G. E. Costin and V. J. Hearing, Human skin pigmentation: melanocytes modulate skin color in response to stress, *FASEB J*, 2007, **21**(4), 976–94.

25.   E. Calikoglu *et al.*, UVA and UVB decrease the expression of CD44 and hyaluronate in mouse epidermis, which is counteracted by topical retinoids, *Photochem Photobiol*, 2006, **82**(5), 1342–1347.

26.   O. Sorg *et al.*, Proposed mechanisms of action for retinoid derivatives in the treatment of skin aging, *J Cosmet Dermatol*, 2005, **4**(4), 237–44.

27.   K. J. Dunlop, G. M. Halliday and R. S. Barnetson, All-trans retinoic acid induces functional maturation of epidermal Langerhans cells and protects their accessory function from ultraviolet radiation, *Exp Dermatol*, 1994, **3**(5), 204–11.

28.   L. Meunier *et al.*, Retinoic acid upregulates human Langerhans cell antigen presentation and surface expression of HLA-DR and CD11c, a beta 2 integrin critically involved in T-cell activation, *J Invest Dermatol*, 1994, **103**(6), 775–9.

29.   R. Hein, H. M., P. K. Muller, O. Braun-Falco, and T. Krieg, Effect of vitamin A acid and its derivatives on collagen production and chemotactic response of fibroblasts, *Br J Dermatol*, 1984, 111–37.

30.   M. T. Goldfarb, C. N. Ellis and J. J. Voorhees, Topical tretinoin: its use in daily practice to reverse photoageing, *Br J Dermatol*, 1990, **122 Suppl 35**, 87–91.

31.   B. H. Kim, Y. S. Lee and K. S. Kang, The mechanism of retinol-induced irritation and its application to anti-irritant development, *Toxicol Lett*, 2003, **146**(1), 65–73.

32.   S. R. Pinnell *et al.*, Topical L-ascorbic acid: percutaneous absorption studies, *Dermatol Surg*, 2001, **27**(2), 137–42.

33.   A. Kawada *et al.*, A new approach to the evaluation of whitening effect of a cosmetic using computer analysis of video-captured image, *J Dermatol Sci*, 2002, **29**(1), 10–8.

34.   P. M. Maia Campos *et al.*, In vitro antioxidant and in vivo photoprotective effects of an association of bioflavonoids with liposoluble vitamins, *Photochem Photobiol*, 2006, **82**(3), 683–8.

35.   A. Perrin *et al.*, Stimulating effect of collagen-like peptide on the extracellular matrix of human skin: histological studies, *Int J Tissue React*, 2004, **26**(3-4), 97–104.

36.   L. R. Robinson *et al.*, Topical palmitoyl pentapeptide provides improvement in photoaged human facial skin, *Int J Cosmet Sci*, 2005, **27**(3), 155–60.

37.   S. Henry *et al.*, Microfabricated microneedles: a novel approach to transdermal drug delivery, *J Pharm Sci*, 1998, **87**(8), 922–5.

38.   S. Mitragotri, D. Blankschtein and R. Langer, Transdermal drug delivery using low-frequency sonophoresis, *Pharm Res*, 1996, **13**(3), 411–20.

20

DISCUSSING COSMETICS THROUGH A FUNCTIONAL SCOPE – A POTENTIAL CONTRIBUTION TO FACILITATE DECISIONS AND PROCEDURES

L M Rodrigues

Universidade Lusófona - CBIOS (Research Center for Health Sciences & Technologies) and Universidade de Lisboa - Pharmacol Sciences Dept. (Lab Experimental Physiology - Faculty of Pharmacy) Lisboa, Portugal.

1 INTRODUCTION

According to the last published report by Cosmetics Europe, The Personal Care Association, previously known as Colipa [1], the European Union (EU) cosmetic market represents over €66 billion, making it by far the most important consumer market in the world. It is responsible for an estimated employment (direct and indirect) of 1.7 million people, consisting of more than 4,000 manufacturers of which more than two thirds are small medium enterprises (SMEs).

The well-known European Union Directive 76/768/EEC [2] continues to play a key role in the present regulatory panorama, even in view of the new Cosmetic Products Regulation 1223/2009 [3] replacing the Cosmetics Directive. With this regulation, Europe is declaring its reinforcement of social responsibility and product safety, even though most of the provisions of this new regulation will be only applicable as from 11 July 2013. The ban and the strict regime which aims at phasing out animal testing was not modified.

The so called "cosmetic's legislation" embraces many related legislations in the EU, from the protection of experimental animals to misleading consumer information [3]. The central idea that cosmetic products must be safe for consumer use is systematically reinforced. In the EU, consumers generally consider cosmetics as "safe"; this is a net result of the huge investment (by industry in particular) in ingredients research and safety assurance to produce better and safer finished products [4 – 6]. However, because legislation is still unclear regarding many aspects, this leaves sufficient room for less than objective approaches and interpretations. This lack of clarity starts within the definition of "cosmetic" and continues with the concept of "safety" and many other aspects that involve efficacy and cosmetic claims.

From the consumer's perspective, these products are definitely associated with "well-being" and "health". This is (partially) due to very aggressive industrial marketing which always tries to create preferences and meet consumer's expectations by labelling its products with extraordinary (often therapeutic) properties. Such expectations are systematically cultivated by complex claims apparently supported by evidence-like data, in an ostensibly science-based environment.

Consumer associations, whether public or private, regularly try to draw the public's attention to some of these aspects, pressuring for better and more objective information. The "quality seals" are recent tools which provide additional choice-argument to the final consumer. Every market includes products with different quality standards aiming at different consumer targets. Nevertheless, basic aspects of cosmetic product safety and efficacy must be equally addressed not only by the industry but also by the regulator, regardless of which member state is involved. We are aware of the controversial "hot topics" which frequently involve the Product Information File (PIFs) [3], especially when some attributes are mentioned. The present paper reflects the author's opinion on some topics relevant to cosmetic safety in order to develop practical interpretations of current guidelines and legislation. It is inspired by the EU market experience, good practice procedures and a functional perspective on cosmetics. The aim is to facilitate processes and decisions in order to contribute to a more competitive industry.

2 THE COSMETIC PARADIGM IN THE EU

The central idea supporting a totally different regulatory approach to cosmetics when compared with medicines is the absolute non-application of the risk-benefit relationship. This means that when placed in contact with the epidermis, hair system, nails, lips and external genital organs the intention is not to heal but rather to *".... exclusively or mainly to clean[ing] them, perfume[ing] them, change[ing] their appearance, protect[ing] them, keep[ing] them in good condition or correct[ing] body odours"* [3].

Cosmetics in the EU are extensively regulated by vertical legislation (dealing with different categories of products) and horizontal legislation (dealing with specific related aspects) accumulated over the last 40 – 50 years [4]. This also means that precautions have been systematically adopted regarding the development and selection of new substances (ingredients) and their respective risk assessment. Currently, the European Scientific Committee on Consumer Safety (SCCS) is responsible for the inclusion or admission decision of a cosmetic substance. This is a multidisciplinary panel of scientists and is part of the European Health & Consumer Directorate General (DG SANCO) which carries out full safety evaluations. Thus, there are enough reasons to consider cosmetics as safe but this situation should not be perceived as "inert". Indeed, modern science has clearly demonstrated that the interaction of any substance with biological systems always causes the modification of local cellular events until the balance known as homeostasis is fully re-established [7, 8]. This also applies to cosmetic products acting on living systems such as the human epidermis and mucosa.

What ought we to conclude about cosmetics? We recognise that they are often formulated with the same pharmaceutical technology as that used in medicines and may include complex (Galenical) mixtures and even substances acting as carriers and penetration enhancers. On the other hand, should we expect the same impact on human skin from a simple solution (e.g. eau de cologne) as from a complex shampoo or a day cream? To keep up with the scientific developments from biopharmaceutical and biomedical research applied to dermatology and cosmetology, we probably need to discuss cosmetics under a different conceptual frame, rather than asking for new definitions.

The EU definition of a "cosmetic product" is based upon the main functional aspects intended for these types of products which is essential in differentiating them from drugs and medical products [9, 10, 11]. Emerging technologies and advances in knowledge constantly raise new questions and doubts that need to be addressed. The following sections address some of these aspects.

3   CLARIFYING COSMETIC INTENTION TO IDENTIFY SAFETY PROCEDURES

As stated above, cosmetics are intended to clean, perfume, change appearance or protect. Furthermore, the definition states that cosmetics are also destined "to keep [the various external parts of the human body] in good condition", and "to correct body odours". Physical boundaries are fixed; *viz.*, the epidermis, hair system, nails, lips and external genital organs, with the focus on "intention". Other aspects such as the complexity and technical sophistication of formulations are disregarded, although use restrictions and target populations are considered.   Therefore, how should we proceed in order to determine the adequate safety requirements for each cosmetic product, considering the impressive variety of presentations and ingredients for each one?   Should we treat all the cosmetic products the same way or should we consider defining specific requirements for each specific type of product?

Cleansing products are probably the most important representatives of cosmetics in terms of consumption.   Normal daily activities promote the adhesion of many different components to the surface of the skin, hair and nails: water alone is not sufficient to eliminate them. Lipids are a major component of this material and generally carry a positive charge which facilitates adhesion to keratins [12]. Another factor which will promote accumulation of exogenous compounds on the skin surface are sebaceous secretions which, when accumulated, also promotes growth of saprophytic flora on the skin's surface which in turn may elicit a positive feed-back effect on sebaceous and sudoriparous secretions [12 – 14]. This, in turn, may affect desquamation (contributing to cellular debris on the skin surface) and the appearance of non-aqueous components of sweat.   Topically applied cosmetics, whether decorative or protective, may potentially interact with these biological systems. To remove skin surface contaminants, tensioactive substances (detergents) are crucial as they promote (1) humectation e.g. reduction of water surface tension which moistens the fat, (2) solubleness, separating lighter fat fractions that mechanically float inside micelles, (3) emulsification which allows for the dispersion and separation of the fat droplets that are then easily eliminated and (4) detergency, determining a modification of superficial charges on the skin's surface (and dirt) that facilitates the dissolution or emulsification of those materials and their subsequent elimination [12, 14].

Cleansing formulations are mainly solutions and emulsions designed to act mostly on keratinic surfaces (stratum corneum, hair and nails) to which dirt consistently adheres through electrostatic attraction.   This is due to the anionic character of the surface, and hydrophobic attraction between protein's hydrophobic residues and fat components. Interaction with the underlying viable epidermis is not expected, but some precautions are recommended when contact occurs with more sensitive areas (mucosa). The increase in use of cosmetics for the eye area has determined a need for special cleansing products and it is an exception, especially as the development of upper eyelid dermatitis syndrome is complex and often associated (although not really demonstrated) with eye makeup [15]. Another concern regards the application of these products to the skin and scalp of newborns, infants and children. The immature nature of the "barrier" in this context may indicate that some additional precautions are needed [16].

Fragrances are an essential component of many products and are considered by some as *"the realm"* of cosmetics. Apart from being the principal component of perfumes, eau de toilet and colognes, fragrances may also contribute to many other cosmetic products, ranging from a minor role in some forms (e.g. eye make-up and nail products) to a major component when it influences consumer choice [17]. Fragrances may be of "natural" (plant or animal) origin or synthetic, but in either case they are thought to be responsible as the most common cause of cosmetic reactions from irritant or allergic contact dermatitis [17 – 19]. The current legislation, in the case of perfume and aromatic compositions, demands the description of the name and code number of the composition and the identity of the supplier as part of the Cosmetic Product Safety report (Annex I) and several fragrances are included in the restriction list (*ANNEX II, List of Substances Prohibited in Cosmetic Products)*[3]. Fragrances are also associated with vasomotor rhinitis, a condition involving significant discomfort (watering eyes, sneezing, a runny congested nose, and headaches) arising from contact with fragrances from perfume, air fresheners and detergents, etc. However, we are still far from a full understanding of this matter [17].

Decorative cosmetics, which aim to change personal appearance, include a wide variety of forms which include powder, foundation paste, make-up bases, rouge or "blush" and lipstick, as well as a large diversity of nail and hairdressing colouring/non-colouring preparations. It is a very complex area and sensitive to fashion tendencies. In particular, the use of colorants, pigments, oxidation tints and dyes, and hairdressing chemicals in Europe is strictly controlled by restriction lists of current legislation in light of their sensitising potential. Even so, a significant number of skin reactions are attributed to facial make-up products including lipsticks, rouge, make-up bases and facial powders [18 – 21].

Special attention is given to hair preparations, colouring or non-colouring formulations and straighteners. These involve the use of several chemicals that, unlike shampoos, generally have a longer contact time with the scalp which sunsequently increases their irritancy and sensitising potential [22, 23]. Hair dyes are still associated with reactions from mild irritancy to acute vascular scalp eruption [24].

Nail preparations are another challenge, both for the industry which continuously pursues more environmentally friendly components and for dermatologists still dealing with few but often complex situations. These preparations include hardeners, plasticisers, polishers and removers. More recently, prosthetic (sculptured) polymerised nails have become popular and are sometimes photobonded *in situ*. Associated reactions vary from irritancy (sometimes beyond the fingers), to allergy, paronychia, onycholysis and onychia [25]. The prolonged use of sculptured nails is reported to cause mechanical damage, with occlusion of the entire nail plate being an obvious factor [25, 26].

The term "protection" usually means to protect underlying tissue from contact stimulae (mechanical for instance) or against potential aggressors such as air exposure or light (UV radiation). This effect is normally obtained with hydro-miscible emulsions which leave a flexible film on the skin surface when the containing water evaporates, although miscible powders are also popular [27]. Some formulations include ingredients which are intended to reinforce or to re-establish the epidermal barrier [28 – 30]. These normally interact exclusively with the most external epidermal layers. However, depending upon the pharmaceutical form (mostly creams, aerosols and lotions), some chemicals used to absorb UV radiation may be present in relatively high concentrations [31, 32] and so may interact with the deeper skin layers, potentially resulting in photosensitisation [33, 34].

Body odours result from the action of bacteria upon sterile apocrine secretions, producing a characteristic, individualised odour that varies with physical condition, emotional state and diet [35]. Eccrine sweat is also sterile and odourless upon secretion [35, 36]. As in the previous case with apocrine secretions, odour may develop following bacterial metabolism but is not as "strong". Apparently the presence of hair greatly increases the odour since it acts as a collecting site for secretions, bacteria and debris [35].

In order to "correct body odours" cosmetic formulations are designed to mask, remove or decrease these odours and, additionally, may also prevent their development [36, 37]. It is a very popular class of products which may be used from very young ages (pre-adolescent and teenagers) and is available in a wide variety of forms which are regarded as safe. In fact, most of the chemicals used in deodorant and in antiperspirant formulations are not regarded as sensitisers, and most of the associated allergic reactions are due rather to other components such as fragrances [38]. Nevertheless, irritant reactions to aluminum salts are common, mostly due to environmental heat, moisture and inflammation caused by depilation/shaving the axilla; restrictions on their use have been reviewed in the current legislation [3].

Finally, it is very difficult to analyse objectively the statement "to keep ... in good condition" and so we shall not try to guess the legislator's objectives in order to justify it.

This brief review on the functional definition of a cosmetic product, according to the current EU legislation, enables us to address this wide variety of cosmetic products, no matter what the type and level of technical sophistication involved and to become aware of the adequate safety and efficacy requirements (Table 1). Many cosmetic products are basically designed and presented according to cosmetic intention. These products do not determine or require any particular demonstration regarding their safety or efficacy and we should consider them as **Regular (R)** products. On the other hand, products that require specific indications for their use such as safety warnings or which involve particular efficacy claims, require specific care, use restrictions and special safety and efficacy proof, should be considered as **Special Requirements (SR)** products (Table 1).

As shown in the table, cleansing products include classical formulations for body and hair which do not imply any specific conditions concerning their purpose and conditions of use. However, if they are being used for specific conditions (acneic skin, baby skin) or if they include special properties (photoprotection) or are to come into contact with mucosa (oral, vaginal), then specific safety and efficacy information should be provided. With regard to the issue of fragrance inclusion, only baby products mainly or exclusively intended to perfume should require more information. Decorative cosmetics for skin, lips and nails do not need special requirements if not intended to address special claims (for example, photoprotection). Hair colouring or non-colouring preparations include many chemicals with irritancy and sensitising potential, meaning that specific information regarding safety and efficacy is justified. The same comments may be applied to occlusive, long-lasting prosthetic nails. The protecting product's genre, like cleansing, also involves classical formulations for skin, lip mucosa, and hair which do not imply any specific conditions

Table 1 *Functional organization of cosmetics, according to intention as defined in current EU legislation regardless of presentation or form. As indicated, special (safety and/or efficacy) claims going beyond the regular (**R**) intention definition determine special (safety and/or efficacy) requirements (**SR**).*

Cosmetic function	Typical formulations	Specific safety requirements	Specific efficacy requirements	Class of requirements
Cleansing	Facial, body and hair cleansing and/or conditioning.	No	No	**R**
	As above but with special claims. Eye, mucosal, dental or baby cleansing and conditioning.	Yes	Yes	**SR**
Perfuming	Body and oral Perfuming.	No	No	**R**
	Baby perfuming.	Yes	Yes	**SR**
Changing appearance (decoration)	Makeup bases, rouge or "blush" for skin and lips. Nail products and nail liners.	No	No	**R**
	As above, but with special claims. Hair preparations and dyes.	Yes	Yes	**SR**
Protecting	Body and facial formulation to protect from environment. Lip protection (no UV).	No	No	**R**
	As above, but with special claims. Formulations with UV protection (skin, mucosa, and hair). Tan promoters.	Yes	Yes	**SR**
Correcting body odors	Deodorant.	No	No	**R**
	Intimate deodorant. Antiperspirants.	Yes	Yes	**SR**

when considering the expected intention and conditions of use. However, when involving special claims (such as photoprotection which includes sun-protection products), special information should be required. The same comment applies to the body odour correctors which include formulations for application to intimate areas.

This classification strategy intends to simplify procedures both for industry and competent authorities, taking into account the interest of all parties including the consumer. Similar views have been adopted by other legislative bodies [39, 40], and so demonstrates its practicability. Nevertheless, the present proposal aims to go a little further by organising the requirement for a rationale decision through a functional perspective. It is an objective approach that facilitates the classification of products and is especially important in achieving more relevant information regarding safety and other related claims, as discussed in the next section.

4 (BIOLOGICAL) EFFICACY SUBSTANTIATION OF COSMETICS

Over the last 40 – 50 years, industrial developments have made remarkable progress in creating new molecules and formulations, technology and methods to explore human skin *in vivo* and the biological impact of topical interventions. Still, industry marketing naturally tends to stretch boundaries in order to differentiate their products from others. Complexity and originality, from the product presentation to its claim, are always present and therapeutic intentions are frequently suggested, if not directly addressed, giving the consumer impression of a "science-based proof-of-effect". Clear limits between cosmetics, drugs and medical products are sometimes difficult to establish objectively and overlapping cannot be avoided, despite efforts made by regulators. Nevertheless, whilst the nature of the claim determines the classification as a cosmetic or as a drug in the USA [41], European legislators have only recently decided to address cosmetic claims substantiation [2 – 4, 42]. However, a very discrete posture was adopted on this matter which basically explains the current market situation where misleading and often meaningless claims do not help to differentiate between "the good, the bad and the ugly". Current EU legislation specifically recognises the consumer's right to be protected from misleading claims concerning efficacy and other characteristics of cosmetic products and is committed to defining common criteria in relation to specific claims for cosmetic products, which includes guidelines applied in all member states [43]. Formally, the legislation requires the proof of the effect claimed for the cosmetic product [44]. However, controversial situations still arise.

Scientific and technological advances have changed our view about skin physiology and pathophysiology, which necessarily extends to the biological effects of cosmetics. More than ever before, we now have a plethora of sophisticated means that are used to identify and publish much evidence of cosmetic activity involving specific ingredients, formulations, or both. This has revived discussion about so-called "functional cosmetics", also "clinically correct cosmetics" [45] or, especially in the US, the controversial "cosmeceuticals". This last designation, dated from 1961, attributed to a founding member of the US Society of Cosmetic Chemists [46], has been progressively applied to the so-called "physiologically active cosmetics" and adopted by clinical dermatologists [47 – 49]. Consequently, it has rapidly promoted a fast growing market, although allegedly undefined, and unregulated [49 – 51]. In fact, EU authorities, like the FDA (the United States Food and Drug Administration), still do not recognise such a category [51].

Thus, in order to deal objectively with this efficacy issue, should we require a new definition of a cosmetic product? Alternatively, should we admit a claim only if supported by peer-reviewed, double-blinded, statistically significant experimental studies?

We must recall that, regardless of the evidence accumulated about the physiological impact of cosmetics on human skin, such effects must occur in the absence of pharmacological, immunological or metabolic activities [45]. There is no point discussing the meaning of this wording which is reasonably comprehensible albeit in some cases potentially controversial; for what really is a pharmacological action? However, because claims are essential for the cosmetic industry, the type and variety of claims have reached a high level of complexity. For this reason a clear-cut separation through scientific reasoning is recommended, justifying many studies and classifications [52 – 54]. So, why not classify these claims according to their physiological (meaning "functional") significance?

Some of these claims are based on one or more properties or characteristics which are directly related to one or more measurable variables with physiological meaning (Type I). Other claims are based on one or more properties or characteristics only partially related to one or more measurable variables with physiological meaning (Type II). Finally, there are complex claims which bear no relation to physiologically measurable variables (Type III; Table 2).

This systematisation immediately suggests an appropriate approach to substantiate the claim. "In use" controlled tests, biometrical tests and *in vitro* tests may be employed to substantiate efficacy, with varying levels of sophistication and rigour. In any case, this classification will easily allow for the identification of those variables with physiological relevance, which would interest all parties (*viz.*, the industry, the authority and ultimately the consumer). Table 3 illustrates with some examples the technical substantiation approach in the present proposal.

No doubt, marketers will continue to create new properties to differentiate their products from others, but these proposals will have to be comprehensible and demonstrable thus complying with a Type I or II classification. Moreover, the objective demonstration of these effects will avoid fantasy designations such as "physiocosmetics" and "cosmeceuticals". Studies developed to support these claims must be suitably designed and their methods carefully chosen with rigorous data analysis and rational conclusions presented. This will mean that different levels of scientific quality control will be expected. However, the utilisation of this systematic strategy will definitively help to promote product differentiation.

5 SAFETY-RELATED ATTRIBUTES AND HUMAN CLINICAL TESTING

Assuming the cosmetic product is safe, the current EU legislation chooses the term "safety evaluation", avoiding the technically adequate "risk assessment" procedure, whilst paradoxically banning any animal testing. This raises the question of how to reconcile consumer health with this new reality [4], and leave room for speculative arguments concerning the interest of human clinical testing.

Human clinical trials involving finished cosmetic products are apparently justified in reinforcing consumer safety [5, 55], but reliable published data on the frequency, type and application of these tests is practically non-existent. They are intended to demonstrate the innocuous nature of the formulation and usually involve exaggerated application conditions such as under patch (occlusive, semi-occlusive) for variable periods of time or near in-use conditions. An example of the latter are open tests [56, 57] emerging from classical dermatotoxicological methodologies previously developed for other purposes [55 – 57]. We must emphasise that these trials (exclusively applied to finished products) are not meant to investigate nor confirm the risk potential and so cannot replace the appropriate safety testing needed for the PIF, which means that they don't exempt the preclinical information regarding individual ingredients.

Data from human clinical testing is not requested by EU legislation. In fact, clinical safety studies are totally ignored in the present regulatory context. Despite this, big manufacturers regularly carry out this type of testing often for marketing purposes. Not infrequently, authorities request this information as part of the PIF and that clearly modifies the context. Apart from the obvious financial implications involved, one should discuss objectively the significance of this testing and, if relevant, discuss conditions in

Type I Claim	Type II Claim	Type III Claim
Hydrating, cleansing, protection, "barrier", reaffirming, "mate", bright / luminous, kitocine regulator.	Anti-age / anti-old, rejuvenating, anti-seborrhea, anti-acne, repair, optimiser.	Nutrient, repulping, volume sculpt, detoxifying, regenerator, "second skin", "healthy", "natural", "saciator", "vivificator".

Table 2: Tentative, systematic proposal for the most frequently cosmetic claims used in current consumer advertisements. According to this functional view, Type I claims are directly related to one or more measurable variables with physiological meaning, Type II claims are only partially related to variables with physiological meaning, while Type III bear no relation to physiologically measurable variables.

	Type I Claim		Type II Claim	Type III Claim	
Hydration	Epidermal capacitance / conductance, TEWL dynamical tests (POST), controlled in-use tests, other.	**"Anti-acne"**	Sebometry/seb-U-tape, controlled in-use tests, other.	**Detoxifier/Purifier**	?
"Barrier"	TEWL, imaging techniques, controlled in-use tests, other.	**Repairing (functional)**	TEWL, microcirculation, pH, challenge tests, controlled in-use tests, other.	**Revitaliser**	?
Photoprotection	Colorimetry, erythema / melanin, controlled in-use tests, other.	**"Anti-age"**	"Barrier" function, skin relief, skin thickness, biomechanics, controlled in-use tests, other.	**Healthy**	?

Table 3: Regarding the systemization proposed in Table 2, it is clear that different levels of analysis with the same purpose may be chosen to substantiate the type I and Type II claims.

which it should be considered. Table 4 summarises the main characteristics of well-known attributes widely used in these contexts.

Under this view, data from human volunteers may fall into two major categories. The first we may call **Compatibility Studies**, which are designed to confirm the safety of use of the finished product and are related to clinical references and methodologies. The other category comprise **In Use Studies**, designed to demonstrate some specific aspects related to the product's utilisation under normal conditions as well as other consumption aspects (including consumer acceptability).

Compatibility studies include well-known tests such as the Primary Skin Irritation and the Accumulated Skin Irritation tests which are based on classical methods developed from animal models [58 – 60] to assess the irritation and sensitisation on human skin [60 – 62] and may be additionally performed under ultraviolet light exposure to assess photoirritation and photosensitisation 63 – 66]. Again, these tests correspond to free modifications of the original tests proposed to assess irritants and allergens, meaning that their use and application is far from harmonised. The type and number of volunteers varies enormously, being almost exclusively based on non-representative subpopulations of participants. Methodologies range from simple open tests to single or repeated patch tests (PT or HRIPT) and vary in duration of exposure. Data processing rarely includes adequate

Table 4 *Most common cosmetic safety related attributes, type and class of human tests frequently chosen and corresponding remarks (PT: patch test; HRIPT : human repeated insult patch test).*

Attribute	Test	Categories	Remarks
"Clinically tested" or "Dermatologically tested"	PT / HRIPT Open test In use	Compatibility clinical study or In use controlled study	– Dubious, cannot be extrapolated to the general population – Does not add any new information related to the product's safety profile
"Hypoallergenic"	HRIPT	Compatibility clinical study	– Creates false expectations since the allegation is totally meaningless
"Ophthalmologically tested"	Open test or In use	Compatibility clinical study	– Dubious, cannot be extrapolated to the general population
"Non-burning / irritating to eyes"	In use	In use controlled study	– Dubious, cannot be extrapolated to the general population
"Non comedogenic" or "non acnegenic"	In use	In use controlled study	– Dubious, cannot be extrapolated to the general population
"Clinically tested for sensitive skin"	PT/HRIPT	Compatibility clinical study or In use controlled study	– Extremely dubious; sensitive skin involves very different skin conditions and sensitivities. Extrapolation to all types of complaints included in this designation is not possible.
"Clinically tested for children"	PT/HRIPT	Compatibility clinical study or In use controlled study	– Dubious; it raises several ethical related concerns; it cannot be extrapolated to the general population

comparative statistics. Only clinical evaluation criteria seem to be unanimously accepted, generally adopting systems from clinical dermatology (e.g. The Standardization Group of the European Society of Contact Dermatitis). In the end, whilst performed under a "clinical" banner, such volunteer studies are not true clinical trials according to international guidelines. In practical terms, these studies are almost exclusively used to justify marketing claims of "clinically/dermatologically tested" or "hypoallergenic" despite being inadequate in scientific terms.

The picture is a little different when considering the In Use tests. Sometimes they are referred to as a "clinically (or dermatologically) controlled application test" and are intended to identify the consumer's perception of a certain product when repeatedly applied according to the normal (recommended) conditions of use. Safety-related aspects resulting from cumulative action or application near sensitive areas (eyes) may be easily identified, while cosmetic acceptability and efficacy perception may also be assessed. A recently published retrospective analysis [5] involving around 50,000 volunteers involved in Patch (43,000) or In Use tests (6,400) revealed that patch testing evoked no reaction in around 88% of the test products and only 1 out of 43,000 individuals revealed an allergic response. In Use tests revealed 11.5% of reactions with a very low percentage of clinically visible reaction where the allergy was totally absent. Thus, the relatively small sample sizes used in the "clinical assessment" of cosmetics are unlikely to provide substantive evidence for broad claims relating to irritancy or sensitisation. A final consideration is that some of these studies are designed to target specific applications such as "for sensitive skin" or "for children" that should require more stringent and controlled test conditions.

So, is there any added value resulting from these tests?

Technically, the Compatibility Studies are clearly redundant since potential toxicity or irritancy should be systematically eliminated by pre-clinical analysis. However, this cannot rule out any non-representative, sporadic, random, individual reactions. In Use Studies, may provide further information with commercial potential but may not contribute much additional technical information. So in fact these tests generally add nothing to the safety requirements, although they may help to reinforce some aspects (use conditions, vigilance) as discussed earlier. From the manufacturers' point of view, the cost – effectiveness of such studies must be pondered. From the authority's perspective, the attribute approach must be technically evaluated in terms of the information relevant for consumer safety and for objective decision-making. For the consumer, no added value is foreseen. On the contrary, these attributes create a diversity of misleading expectations and false confidence.

6 CONCLUSIONS

Any intervention in a biological system will alter local homeostasis; this is a fundamental concept in human physiology. Such may be the case of a cosmetic product when brought contact with human skin. Alterations differ in accordance with the object and nature of the intervention, meaning that they may vary from superficial and transient effects through to long-term structural changes. This sets the basis for this functional perspective of cosmetics and eliminates discussion about dubious designations. Such a functional approach would eliminate the need to produce new legislation or classifications. In fact, it would complement and clarify existing aspects of cosmetic safety and efficacy claims. In summary:

- EU cosmetic legislation is guided by intended use and the experience has proved that this is an acceptable criteria since definitions are clear and objective for most of the defined intentions. Meaningless statements such as *"keeping ... in good condition"* should be eliminated.

- The functional view based on the cosmetic intention simplifies the formal approach, both for the manufacturer and the authority, as illustrated in Table I. Every cosmetic intention may demand Regular or Special Requirements depending on the alleged level of sophistication.

- The functional approach extends to proof of effect which itself must have a physiological meaning (Table 2).

- In every situation, all claims and attributes must have an objective meaning and avoid misconceptions (especially if false) as happens with dubious allegations (Table 4). Authorities must consensually agree which type of allegations should be allowed and define the respective validation conditions.

Finally it is noteworthy to underline that the current definitions in the EU legislation (as in the FDCA act of 1938) still include the central concepts defined in the original documents. This displays remarkable resilience even after such obvious progress in science and technology. Knowledge helps to eliminate old boundaries and gaps, but it cannot avoid new limits and overlapping areas. In science as in life, there are no absolutes.

7 REFERENCES

1. COLIPA Annual Report 2010.

2. Council Directive 76/768/EEC of 27 July 1976, On the approximation of the laws of the member states relating to cosmetic products, OJEC L262/169 (and respective amendments).

3. EU Regulation 1223/2009 of the European Parliament and of the Council on cosmetic products (recast), Brussels, 30 November 2009.

4. M. Pauwels and V. Rogiers, Human health safety evaluation of cosmetics in the EU : a legally imposed challenge to science, *Toxicol. and Appl, Pharmacology*, 2010, **243**, 260–274.

5. N. Gerlach, M. Wiebusch, U. Heinrich, H. Tronnier, Side effects of Cosmetics on human skin, *IFSCC Magazine*, 2011, **4**, 281–7.

6. European Consumer Exposure To Cosmetic Products, a Framework For Conducting Population Exposure Assessments, *Food and Chemical Toxicology*, 2007, **45**, 2097–2108; www.elsevier.com/locate/foodchemtox.

7. S. J. Cooper, From Claude Bernard to Walter Cannon, Emergence of the concept of homeostasis, *Appetite*, 2008, **51**, 419–427.

8. B. Lemmer, Discoveries of rhythms in human biological functions: A historical review, *Chronobiology International*, 2009, **26**(6), 1019–1068.

9. Directive 2001/83/EEC of the European Parliament and of the Council on the Community code relating to medicinal products for human use, O.J.E.C, L311/67 28/11/2001.

10. Council Directive 90/385/EEC of June 1990, On the approximation of the laws of the Member States relating to active implantable medical devices, O.J.E.C, L189. 20/07/1990 P. 0017.0036.

11. Council Directive 93/42 EEC of 14 June 1993 concerning medical devices, O.J.E.C. L169. 12/07/1993 P. 0001.0043.

12. A. R. Latven, Fundamentals and comprative actions of cleansing creams, *Am Perfum.*, 1058, **72**, 29–30.

13. P. Somasundaran and P. Purohit, Polymer/surfactant interactions and nanostructures: current development for cleansing, release, and deposition of actives, *J Cosmet Sci.*, 2011, **62**(2), 251–8

14. K. Subramanyan, Role of mild cleansing in the management of patient skin, *Dermatol Ther.*, 2004, **17 Suppl 1**, 26–34.

15. S. P. Modjtahedi, J. R. Toro, P. Engasser and H. Maibach, Cosmetic Reactions, in *Dermatotoxicology* ed. H. Zhai and H. I. Maibach, CRC Press, 6th edn, 2004.

16. W. L. Epstein, Contact type delayed hypersensitivity in infants and children: induction of rhussensitivity, *Pediatrics*, 1961, **27**, 51–53.

17. Z. D. Draelos, Fragrances, dermatitis and vasomotor rhinitis, in *Cosmetics and dermatological problems and solutions - a Problem Based Approach*, Informa Healthcare Edit., London, 3rd edn, 2011, pp. 137–140.

18. R. M. Adams and H. Maibach, A five years study of cosmetic reactions, *J Am Acad Dermatol.*, 1985, **13**, 1062–1069

19. L. Naldi, The epidemiology of fragrance allergy: questions and needs, *Dermatology*, 2002, **205**(1), 89–97.

20. S. J. Gilpin, X. Hui and H. I. Maibach, Volatility of fragrance chemicals: patch testing implications, *Dermatitis*, 2009, **20**(4), 200–7.

21. M. M. Chowdhury, Allergic contact dermatitis from prime yellow carnauba wax and coathylene in mascara, *Contact Dermatitis*, 2002, **81**, 80–83.

22. E. W. Brauer, Cosmetics for the Dermatologist, in *Clinical Dermatology*, ed. D. J. Dennis and J. McGuire, Harper and Row, Philadelphia, 1984, vol 4.

23. F. J. Storrs, Permanent wave contact dermatitis: contact allergy to glyceryl monothioglycolate, *J Am Acad Dermatol.*, 1984, **11**, 74–85.

24. M. J. Cruz, V. De Vooght, X. Muñoz, P. H. Hoet, F. Morell, B. Nemery, J. A. Vanoirbeek, Assessment of the sensitization potential of persulfate salts used for bleaching hair, *Contact Dermatitis*, 2009, **60**(2), 85–90.

25. R. Baran, Pathology induced by the application of cosmetics to the nail, in *Principles of Cosmetics to the Dermatologist*, ed. P. Frost and S. N. Horwitz, CV Mosby, St Louis, 1982

26. A. A. Fisher and R. Baran, Adverse reactions to acrylate sculpture nails with particular reference to prolonged paresthesia, *Am J Contact Dermatitis*, 1991, **2**, 38–42.

27. J. Zhang, E. W. Smith and C. Surber, Galenical principles in skin protection, *Curr Probl Dermatol.*, 2007, **34**, 11–18

28. A. zur Mühlen, A. Klotz, S. Weimans, M. Veeger, B. Thörner, B. Diener and M. Hermann, Using skin models to assess the effects of a protection cream on skin barrier function, *Skin Pharmacol Physiol*, 2004, **17**(4), 167–75

29. J. Levin and S. B. Momin, How Much Do We Really Know About Our Favorite Cosmeceutical Ingredients?, *J Clin Aesthet Dermatol*, 2010, **3**(2), 22–41.

30. J. A. Segre, Epidermal barrier formation and recovery in skin disorders, *J Clin Invest*, 2006, **116**(5), 1150–1158.

31. M. A. Pathak, Sunscreens: topical and systemic approaches for protection of human skin against harmful effects of solar radiation, *J Am Acad Dermatol*, 1982, (3), 285–312.

32. M. Berwick, The good, the bad, and the ugly of sunscreens, *Clin Pharmacol Ther*, 2011, **89**(1), 31–3

33. N. Cook and S. Freeman, Report on 19 cases of photoallergic contact dermatitis to sunscreens seen at the skin and cancer foundation, *Australas. J Dermatol.*, 2001, **42**(4), 257–259

34. E. Linos, E. Keiser, T. Fu, G. Colditz, S. Chen and J. Y. Tang, Hat, shade, long sleeves, or sunscreen? Rethinking US sun protection messages based on their relative effectiveness, *Cancer Causes Control*, 2011, **22**(7), 1067–71

35. J. N. Labows, K. Z. J. McGinley and A. M. Klingman, Axillarey oddor - current status in *Principles of Cosmetics for the Dermatologist*, ed. P. Frost P and S. N. Horwitz, CV Mosby, St Louis, 1982.

36. S. Plechner, Antiperspirants and Deodorants, in *Cosmetics: science and technology*, ed. M. S. Balsan, S. D. Gershon, M. M. Rieger, E. Sagarin and S. J. Strianse, Wiley Interscience, NY, 1972, vol. 2.

37. G. E. Piérard, P. Elsner, R. Marks, P. Masson, M. Paye, and the EEMCO Group, EEMCO guidance for the efficacy assessment of antiperspirants and deodorants, *Skin Pharmacol Appl Skin Physiol*, 2003, **16**(5), 324–42.

38. S. D. Soileau, Dermal penetration of calcium salts and calcionis cutis in *Toxicology of the skin*, ed. A. W. Hayes, J. A. Thomas, D. E. Gardner and H. I. Maibach, London, Taylor and Francis, 2001.

39. BRASIL. Agência Nacional de Vigilância Sanitária – Resolução-RDC 79, de 31 de agosto de 2000. Rules and procedures to register personal hygiene, cosmetics and parfumes (portuguese). DO RFB, DF, 31 ago. 2000, n. 169-E, Seção 1, p. 34.

40. ANVISA Guide for the safety evaluation of cosmetic products, 2003, (available at www.anvisa.gov.br)

41. Federal Food, Drug, and Cosmetic Act (FDCA), 1938

42. V. Rogiers, Efficacy claims of cosmetics in Europe must be scientifically substantiated from 1997 on, *Skin Research and Technology*, 1995, **1**, 44–46

43. EU Regulation 1223/2009 of the European Parliament and of the Council on cosmetic products (recast), Brussels, 30 November 2009, pp. 51–52.

44. EU Regulation 1223/2009 of the European Parliament and of the Council on cosmetic products (recast), Brussels, 30 November 2009, Art.11.

45. P. Morganti, S. Paglialunga, EU borderline cosmetic products review of current regulatory status, *Clinics in Dermatology*, 2008, **26**, 392–397.

46. R. Reed, The definition of "cosmeceutical", *J Soc Cosmet Chem*, 1962, **13**, 103–6.

47. A. M. Kligman, Why cosmeceuticals?, *Cosm & Toil*, 1993, **108**, 37–40.

48. A. M. Kligman, Cosmeceuticals as a third category, *Cosm &Toil*, 1995, **113**, 33–8.

49. A. M. Kligman, Cosmeceuticals: a broad-Spectrum Category between Cosmetics and Drugs, in *Cosmeceuticals and Active Cosmetics*, ed. H. I. Maibach and P. Elsevier, Taylor & Francis, Boca Raton, 2005.

50. Z. D. Draelos, Cosmeceuticals: undefined, unclassified, and unregulated, *Clinics in Dermatology*, 2009, **27**, 431–434.

51. A. M. Newburger, Cosmeceuticals: myths and misconceptions, *Clinics in Dermatology*, 2009, **27**, 446–452.

52. R. Smithles, Substantiating performance claims, *Cosmet Toilet*, 1984, **99**, 79–84.

53. J. Weichers, Mind over matter: cosmetic claim substantiation issues facing the future, *Cosmet Toilet*, 2005, **120**, 57–64.

54. M. Paye and A. Barel, Introduction to the "proof of claims", in *Handbook of cosmetic science and technology*, ed. M. Paye, A. Barel and H. Maibach, CRC Press, New York, 2006.

55. I. Lauerma and H. I. Maibach, Provocative tests in Dermatology, in *Provocation in Clinical Practice*, ed. S. L. Spector, Marcel Dekker, NY, 1995.

56. T. Fischer and H. I. Maibach, Improved but not perfect patch testing, *Am J Contact Dermatitis*, 1990, **1**, 73–90.

57. M. Hannuksela and H. Salo, The repeated open application patch test (ROAT), *Contact Dermatitis*, 1986, **14**, 221–227.

58. J. H. Draize, G. Woodard and H. Calvery, Methods for the study of irritation and toxicity of substances applied topically to the skin and mucous membrane, *J. Pharmacol. Exp. Ther.*,1944, **82**, 377–390.

59. J. H. Kay, and J. C. Calandra, Interpretation of eye irritation tests, *J. Soc. Cosmet. Chem.*,1962, **13**, 281–289.

60. J. H. Draize, Appraisal of the safety of chemicals. Foods, drugs and cosmetics. pp. 46–49, OECD 404, 1965.

61. H. I. Maibach and W. L. Epstein, Predictive patch testing for sensitization and irritation, *Am. Perf. Cosm.*,1965, **80**, 55–56.

62. A. M. Klingman and W. M. Wooding, A method for the measurement and evaluation of irritants of human skin, *J. Invest Dermatol.*,1967, **49**, 78–94.

63. H. Stork, Photo allergy and photosensitivity, *Arch. Dermatol.*, 1965, **91**, 469–482.

64. M. A. Pathak, Photobiology of Melanin Pigmentations, *J. Am. Acad. Dermatol.*, 1983, **9**, 724–733.

65. K. H. Kaidbuy and A. M. Klingman, Identification of topical photosensitizing agents in humans, *J. Invest. Derm.*, 1978, **70**, 149–152.

66. K. H. Kadbey and A. M. Klingman, Photo maximization test for identifying photoallergic contact sensitizers, *Contact Dermatitis*, 1980, **6**, 161–9.

21

EFFECT OF OCCLUSION ON THE PERCUTANEOUS ABSORPTION OF ALUMINIUM FROM ANTIPERSPIRANT PRODUCTS

T Mistry[1], K Staff[2], K Anjum[1], J D Owen[1], J Stair[1], S C Wilkinson[3], G P Moss[4]

[1]School of Pharmacy, University of Hertfordshire, Hatfield, United Kingdom. [2]School of Pharmacy and Medical Sciences, University of South Australia, Adelaide, Australia. [3]Medical Toxicology Centre, Wolfson Unit, Medical School, University of Newcastle-upon-Tyne, United Kingdom. [4]The School of Pharmacy, Keele University, Keele, United Kingdom.

1 INTRODUCTION

Despite the widespread, and generally very safe, use of aluminium-containing antiperspirants there is a paucity of studies examining, in vitro or in vivo, the percutaneous absorption of aluminium and even fewer that examine this phenomena from consumer formulations. Aluminium has been considered to be a neurotoxicant, with suggestions that it may play a role in brain disorders, including Alzheimer's disease [1]. However, the United Kingdom's Alzhemier's Society currently suggest that, while there might be "circumstantial evidence linking this metal with Alzhemier's disease … no causal relationship has yet been proved" [2]. They further comment that, as evidence for other causes of Alzheimer's disease continues to grow, "a possible link with aluminium seems increasingly unlikely" (Alzheimer's Society, 2013) [2].

Aluminium absorbed from antiperspirants has also been suggested as potentially presenting a risk of breast cancer due to the prolonged residence time and lack of product wash-off [3]. However, studies of 800 women found no link between antiperspirant use and breast cancer rates, finding no correlations with depilation (shaving with a razor), antiperspirant use, or both [4]. Similar findings were reported by Fakri et al., who studied 100 women and reported no correlation with antiperspirant use [5].

There are very few studies that examine the percutaneous absorption of aluminium. For example, Flarend et al. examined the in vivo absorption of aluminium (as [26]Al) in two volunteers following a single application of 15.5 mg aluminium (as chlorohydrate) [6]. The results of this study showed that, while aluminium was absorbed topically, that the amount absorbed was 3.6 µg. This is well below the limit proposed by the Agency for Toxic Substance and Disease Registry, whose toxicological profile for aluminium suggested a minimum risk level of oral aluminium exposure of 1mg kg^{-1} day^{-1} [7]. While this is not an ideal comparison to make with percutaneous absorption, it suggests that dermal bioavailability of aluminium is low compared to other routes.

Recently, Pineau *et al.* measured aluminium absorption into and across normal and stripped human skin in occluded (stripped skin) and non-occluded (normal skin) *in vitro* experiments. They examined aerosol, emulsion roll-on and stick antiperspirant formulations and analysed absorption over a 24-hour period using atomic absorption spectrometry. For normal skin, they found little difference between the amount of aluminium that diffused into and across the skin for each formulation tested. For the stripped skin experiments they found that a greater amount of aluminium was absorbed from the stick formulation. However, they found a negligible amount of aluminium had passed across the skin into the receptor fluid and that there was an increased concentration of aluminium in the stratum corneum [8].

Therefore, the aim of this study was to investigate the *in vitro* absorption of aluminium from commercially available antiperspirant products. In particular, the effect of occlusion was examined. While such experiments are widely established and provide important information for many fields, including risk assessment and drug delivery, they can provide preliminary information only and it is in this context that this study has been conducted.

2 MATERIALS AND METHODS

Nitric acid 70%, redistilled 99.999%, aluminium standard calibration solution 1000 ppm (Fluka Ltd.), nitric acid 4M, citric acid analytical reagent grade, sodium hydroxide pellets, methanol, ethyl acetate and Presept 2.5 g disinfection tablets (Johnson and Johnson Ltd., UK) were all obtained from Sigma-Aldrich Ltd. (Poole, UK) unless otherwise specified. The following antiperspirant products were purchased from Boots The Chemist (St. Pancras, London, UK): Sure for Men Absolute Protection Roll-on (Batch number 92591LWB); Nivea for Men Cool Kick Roll-on (93338710); Dove Original Anti-Transpirant Roll-on (92684LWC); Driclor (663L). All products were used as supplied. Franz diffusion cells (including donor compartment lids and glass sidearm caps) were manufactured and supplied by Soham Scientific Ltd. (Soham, UK). All other glassware and equipment was supplied by Fisher Scientific Ltd. (Loughborough, UK).

Full ethical approval was granted to obtain human tissue from Hammersmith Hospital (10/H0807/51; London, UK) and the University of Hertfordshire Ethics Committee (School of Pharmacy). Scrotal skin was obtained from surgery and immediately frozen until required. The skin was stored in plastic containers at -30°C for a maximum of four months prior to use. Skin samples were defrosted immediately prior to use in this study and prepared as described previously [9]. Subcutaneous fat was removed and the remaining full-thickness skin was cut into sections approximately 2x2 cm, sufficiently large to cover the receptor phase of 10 mL (receptor chamber volume) Franz diffusion cells. All glassware used was acid washed with 70% nitric acid and rinsed with purified water for 24 hours prior to use to minimise surface metal adherence. Furthermore, additional skin samples (n = 3) were digested in order to determine the background concentration of aluminium [10]. These skin samples were weighed and digested with 5 ml high-purity nitric acid 70% overnight at room temperature. The samples were then diluted 10:1 with deionised water, centrifuged (Thermo Scientific, UK, series CL31) for 45 minutes at 8000 RPM and filtered using 0.22μm syringe filter.

A small magnetic stirring bar was added to each receptor chamber compartment which was then filled with citric acid buffer (pH 5.5) equilibrated at 37°C. Skin samples were weighed and placed above the receptor compartment of the Franz cells. The donor compartment was then placed onto the top of the skin surface and the cell was then clamped securely together. The assembled Franz cells were placed on a weighing scale and, using a plastic pipette, 0.5g of each antiperspirant product was added to the donor compartment, which was then sealed with Parafilm® for occluded experiments or left uncovered for non-occluded experiments. The cells were placed in the water bath at 37°C and stirred continuously at 400 rpm [11, 12].

Sampling was carried out 1, 2, 4, 6 and 24 hours after the beginning of each experiment, whereupon the entire receptor phase was removed and replaced with fresh citric acid buffer (maintained at 37°C from the water bath). Sampling intervals were chosen based on previous studies and the aqueous solubility of the aluminium salts, which suggested that 'sink' conditions would be maintained throughout the experiment [13]. Samples were collected in glass vials, pre-filled with 10 mL of 2% nitric acid diluent to allow sufficient sample volume for ICP-OES analysis [14]. The vials were then covered with Parafilm®. All vials were covered and refrigerated until analysed. Further, at the end of the experiments skin samples were gently but thoroughly cleaned to remove excess product and the skin was digested as described above.

Aluminium determinations were carried out on a Varian 710 ICP-OES (Varian / Agilent Ltd, Stockport, UK) axial spectrometer. The wavelengths used, as well as the instrument parameters are summarised in Table 1. Calibration standards were made from Al 1000 ppm standard at concentrations of 0.01 ppm, 0.05 ppm, 0.1 ppm, 0.5 ppm and 1.0 ppm. The blank used was 2% nitric acid. Each sample was measured in replicate three times by the ICP-OES instrument. The limits of detection (LOD) for each wavelength were calculated from a total of 20 replicate values of 2% nitric acid blank, and are as follows: 396.152nm (0.002ppm), 237.312nm (0.009ppm) and 308.215nm (0.010ppm).

3 RESULTS AND DISCUSSION

Human scrotal skin was used in this experiment. While clearly preferential to other mammalian tissue or artificial membranes, scrotal skin is not widely associated with percutaneous absorption experiments. Human scrotal skin was considered to be a suitable tissue with which to conduct such experiments as it is widely considered to have a thinner stratum corneum and epidermis, and hence may exhibit higher permeability to topical

Table 1 *ICP-OES parameters used in this study. Aluminium detection wavelengths were 396.152, 237.312 and 308.215 nm.*

Parameter	Value
Power (kW)	1.20
Plasma flow (L/min)	15.00
Auxiliary flow (L/min)	1.50
Nebulizer pressure (kPa)	200

chemicals than skin from other anatomical sites used more frequently in *in vitro* studies, whilst maintaining the stratum corneum integrity that is required for such studies. In addition, the penetration of metals occurs mainly through the skin's appendages such as the hair follicles and eccrine sweat glands, both of which are present in axillary and scrotal skin [15, 16]. Thus, aluminium penetration across scrotal skin can be considered maximal in toxicological terms. It is anticipated therefore that scrotal skin would provide a reasonable "worst case scenario" for modelling absorption of aluminium through axilla skin.

The concentration of Al found in the test products, along with the amount of residual aluminium found in the scrotal skin samples (0.443 ± 0.0692 µg g^{-1}; n=3) was used to assess the percentage absorption of aluminium from each test product (Table 2). The residual level detected our method is in accordance with expected aluminium levels in human tissue providing confidence in the extraction and detection methods used in this communication [17].

The skin absorption profiles (Figure 1) and corresponding diffusion cell recovery data (Table 2) demonstrate that overall, very low percentages of the applied dose permeate across the skin membrane and into the receptor phase after 24 hours. The highest percentage absorption was observed for the Driclor product (0.022 and 3.437%, for occluded and unoccluded experiments, respectively). It should also be commented that the non-occluded experiment examining the Driclor product, while statistically not different, appears to be the only result that shows an average absorbance above 0.05%. Further statistical examination of the flux (Table 2) for each product (Kruskal Wallis test, $p = 0.05$) indicated that there was significant variation ($p = 0.037$) between all the products but no significant difference between any two products.

In comparing occluded and non-occluded experiments, no significant differences (Independent t-test, $p = 0.05$) were found in three experiments. The only significant difference was observed for the Nivea product ($p > 0.05$). In comparing the effects of different products (Kruskal Wallis test, $p = 0.05$) no significant differences were observed for occluded experiments ($p = 0.238$) whereas, for non-occluded experiments, a significant difference was observed ($p = 0.034$) possibly due to the comparatively high aluminium associated with the non-occluded Driclor experiment. While occlusion generally increases permeability, it does not do so in all cases [18, 19]. This lack of a significant trend may be due to the low overall concentrations of aluminium found at the conclusion of the experiment, but may also indicate the variability of absorption which is consistent with findings in the wider field. This is an important consideration given the semi-occluded, or often variable occlusion, associated with the site of application – the axillae.

Table 2 *Aluminium absorption (delivered from antiperspirant formulations) across human skin in vitro under occluded (**O**) or unoccluded (**U**) conditions, expressed as percentage of the applied dose and total amount recovered from the receptor phase after 24 hours. All values presented as mean ± standard deviation of n≥3 replicates.* [A] *Active ingredients were aluminium chloride hexahydrate (ACH) or aluminium chlorohydrate (AC).* [B] *Concentration of aluminium [Al] in each product.* [C] *Percentage of applied dose penetrated at 24 hours.* [D] *Total amount of aluminium penetrated at 24 hours*

Product	Active Ingredient[A]	[Al][B] (mg mL^{-1})	% Dose[C]		Amount[D] (μg)		Flux (μg cm^{-2} hr^{-1})	
			O	U	O	U	O	U
Driclor	ACH	24.6 ± 1.2	0.022 ± 0.003	3.437 ± 0.923	0.22 ± 0.030	34.4 ± 9.260	0.007	0.378
Nivea for men		35.4 ± 2.3	0.015 ± 0.001	0.050 ± 0.005	0.15 ± 0.010	0.50 ± 0.050	0.017	0.024
Sure for men	AC	37.6 ± 1.0	0.009 ± 0.001	0.008 ± 0.001	0.09 ± 0.10	0.08 ± 0.010	0.007	0.008
Dove Original		38.2 ± 1.6	0.012 ± 0.003	0.015 ± 0.001	0.13 ± 0.030	0.15 ± 0.01	0.011	0.014

Table 3 *Aluminium absorption (delivered from antiperspirant formulations) through human skin in vitro, expressed as recovery of Al from within skin after 24 hours. All values are mean ± SD of n ≥ 6). Asterisks (8) indicate significant p-values (t-test; α = 0.05).*

Product	Amount in skin (mg g^{-1})		P
	Occluded	**Unoccluded**	
Driclor	1.06 ± 0.48	0.93 ± 0.30	0.487
Nivea for men	0.98 ± 0.38	0.17 ± 0.07	0.000*
Sure for men	0.89 ± 0.60	0.39 ± 0.14	0.085
Dove Original	0.51 ± 0.56	1.42 ± 0.74	0.029*

As an *in vitro* indication of the total aluminium absorbed into and across the skin, the skin content of aluminium was assessed as a whole; that is, tape strips of the *stratum corneum* were not taken and assessed independently from the remaining epidermal and dermal layers. In general, the amount of aluminium found in the skin samples was substantially higher than that found in the receptor compartment of the diffusion cell (Table 3). It is apparent that, in general, the amount of aluminium found in the skin after 24 hours is similar, with only the non-occluded experiment for Dove Original showing a high degree of variance. Again, no clear trends were found between occluded and non-occluded experiments, with the only significant differences ($p = 0.05$) between such being shown for the two emulsion products tested (Nivea and Dove).

Very low levels of aluminium ($0.09 - 0.22$ µg g^{-1} for occluded conditions and $0.08 - 34.4$ µg g^{-1} for unoccluded experiments) were measured in the "systemic" compartment. In contrast, substantially more aluminium ($0.51 - 1.06$ mg and $0.17 - 1.42$ mg for occluded and unoccluded conditions, respectively) were found in the skin digests after the diffusion study. In some cases, particularly the Driclor (non-occluded) receptor concentration and the Dove Original skin digest sample (non-occluded) the range of results and variability may be influenced by two single high readings, one in each experiment. No data points from successfully completed experiments were removed from this study and this may be reflected in the variability of the data. However, presentation of all the data including those experiments that showed the highest amounts of aluminium, is an important consideration in the context of this assessment. It must be stressed however, that since the skin digests include both the viable skin and stratum corneum, no conclusion can be drawn about the local concentration of aluminium in the viable skin layers and hence there is insufficient information to comment on the potential local toxic effects of aluminium in the skin.

With regard to toxicity and the safe limits of aluminium exposure (1mg kg^{-1} day^{-1}) proposed by the Agency for Toxic Substances and Disease Registry [7], the results of this preliminary study show that the total aluminium absorbed from the range of antiperspirant roll-on products examined is well below this limit, although such limits clearly should consider the total internal dose of aluminium by all routes and should be considered within the constraints of the experimental protocols employed in this study. This compares well with previous findings particularly in the case of occluded experiments [6, 8]. Comparison with Flarend's results [6] might also suggest that the higher levels of aluminium found within the skin will permeate slowly across the skin in the systemic circulation, possibly reducing further issues over the total amount of aluminium that is available following the application of antiperspirant products. Pineau's recent study [8] suggests that the uptake of aluminium into and across normal skin from antiperspirant products is minimal; however, they further recommended that manufacturers of antiperspirant products should, in the light of the stripped skin experiments, proceed with caution in product development.

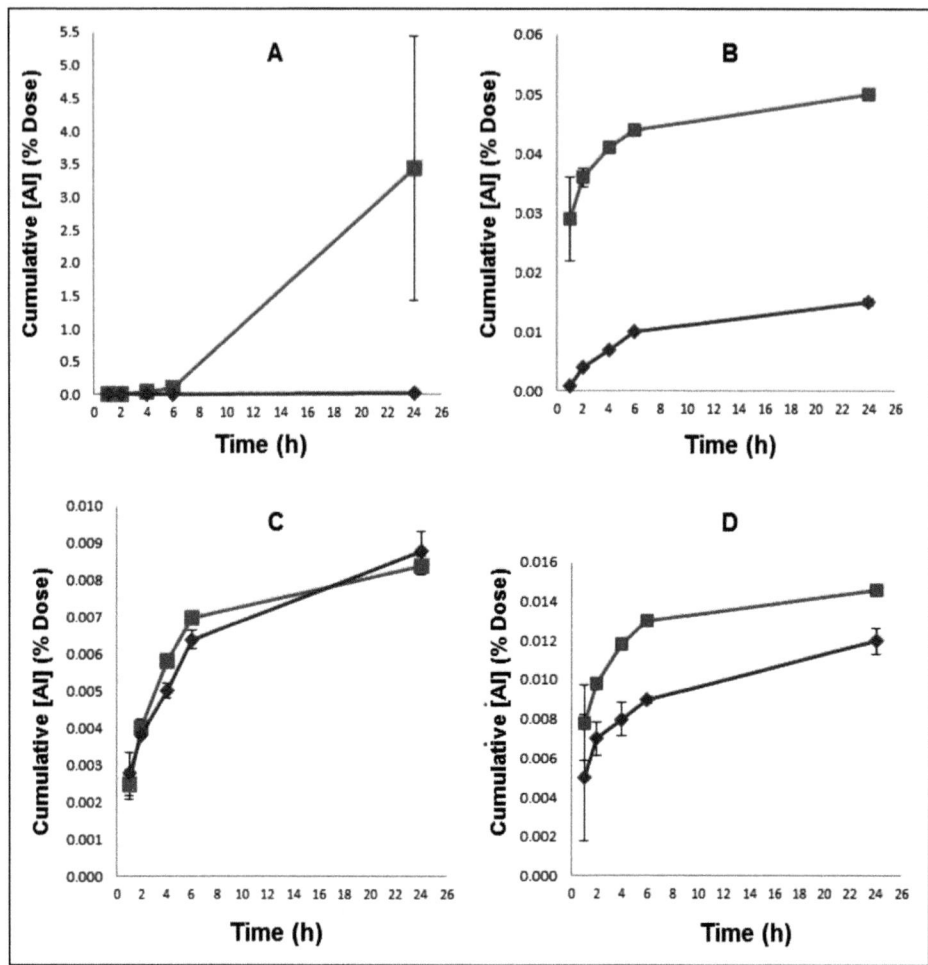

Figure 1 *Cumulative absorption (as percentage of the applied dose) of aluminium across human skin in vitro for (A) Driclor, (B) Nivea, (C) Sure and (D) Dove antiperspirant products. All values are mean ± SD of n ≥ 6. Studies were performed under occluded (■) or unoccluded (♦) conditions.*

However, it is clear that such results must be considered within the context of the obvious limitations of *in vitro* experiments, compared to the *in vivo* situation, although it might also be mentioned that such limitations also include the absence of biological mechanisms of clearance. Formulation type should also be considered, and the results of this study cannot readily be extrapolated to other antiperspirant types, such as aerosols and sticks.

4 ACKNOWLEDGEMENTS

The authors would like to thank the School of Pharmacy, University of Hertfordshire, UK, for funding this project.

5 REFERENCES

1. R. Yokel, S. Rhineheimer, R. Brauer, P. Sharma, D. Elmore and P.J. McNamara, Aluminium bioavailability from drinking water is very low and is not appreciably influenced by stomach contents or water hardness, *Toxicology*, 2001, **161**, 93–101.

2. Alzheimer's Society. 2013. Am I at risk of developing dementia? Available at: http://alzheimers.org.uk/site/scripts/documents_info.php?documentID=102. [accessed August 2013].

3. P.D. Darbre, Underarm cosmetics and breast cancer, *Journal of Applied Toxicology*, 2003, **23**, 89–95.

4. D.K. Mirick, S. Davis, D.B. Thomas, Antiperspirant use and the risk of breast cancer, *Journal of the National Cancer Institute*, 2002, **94**, 1578–1580.

5. S. Fakri, A. Al-Azzawi, N. Al-Tawil. Antiperspirant use as a risk factor for breast cancer in Iraq. *Eastern Mediterranean Health Journal*, 2006, **12**, 478–482.

6. R. Flarend, T. Bin, D. Elmore, S.L. Hem. A preliminary study of the dermal absorption of aluminium from antiperspirants using aluminium-26, *Food and Chemical Toxicology*, 2001, **39**, 163–168.

7. ATSDR, United States. 2008. Department of Health and Human Services, *Toxicological profile for aluminium,* Agency for Toxic Substances and Disease Registry, Atlanta.

8. A. Pineau. O.Guillard, B. Fauconneau, F. Favreau, M-H.Marty, A. Gaudin, C.M. Vincent, A. Marrauld, J-P. Marty, In vitro study of percutaneous absorption of aluminium from antiperspirants through human skin in the Franz^{TM} diffusion cell, *Journal of Inorganic Biochemistry*, 2012, **110**, 21–26.

9. T.J. Franz, Percutaneous absorption – relevance of in vitro data, *Journal of Investigative Dermatology*, 1975, **64**, 190–195.

10. K. Uysal, Y. Emre, E. Kose, The determination of heavy metal accumulation ratios in muscle, skin and gills of some migratory fish species by inductively-coupled plasma-optical emission spectrometry (ICP-OES) in Beymelek Lagoon (Antalya / Turkey). *Microchemical Journal*, 2008, **90**, 67–70.

11. F. Akomeah, T. Nazir, G. Martin, M. Brown. Effect of heat on the percutaneous absorption and ski retention of three model penetrants, *European Journal of Pharmaceutical Sciences*, 2004, **21**, 337–345.

12. F. Larese, A. Gianpietro, M. Venier, G. Maina., N. Renzi, In vitro percutaneous absorption of metal compounds, *Toxicology Letters*, 2007, **170**, 49–56.

13. G.P. Moss, D.R. Gullick, W.J. Pugh, Multiple-application dosing: effect on the percutaneous absorption of aluminium from commercially available products, *Journal of Pharmacy and Pharmacology*, 2005, **57S**, 104.

14. T. Burden, J. Powell, R. Thompson. Optimal accuracy, precision and sensitivity of inductively-coupled plasma optical-emission spectrometry – bioanalysis of aluminium, *Journal of Analytical Atomic Spectrometry*, 1994, **10**, 259 – 266.

15. J.J. Hostynek. Factors determining percutaneous metal absorption. *Food and Chemical Toxicology*, 2003, **41**, 327–345.

16. K. Staff, M.B. Brown, R.P. Chilcott, R.C. Hider, X.L. Kong, S.A. Jones. Ga(III) complexes – The effect of metal coordination on potential systemic absorption after topical exposure, *Toxicology Letters*, 2011, **202**, 155–160.

17. N.D. Priest. The biological behaviour and bioavailability of aluminium in man, with special reference to studies employing aluminium-26 as a tracer: review and study update, *Journal of Environmental Monitoring*, 2004, **6**, 375–403

18. A.L. Stinchomb, F. Pirot, G.D. Touraille, A.L. Bunge, R.H. Guy. Chemical uptake into human stratum corneum in vivo from volatile and non-volatile solvents, *Pharmaceutical Research*, 1999, **16**, 1288–1293.

19. L. Taylor, R.S. Lee, M. Long, A.V. Rawlings, J. Tubek, L. Whitehead., G.P. Moss. Effect of occlusion on the percutaneous penetration of linoleic acid and glycerol, *International Journal of Pharmaceutics*, 2002, **249**, 157–164.

22

RECENT ADVANCES IN TOPICAL APPLICATIONS FOR A NEW ANTI-AGING DRUG

B Cantecor[1,4], M P Savelli[1], G Marti-Mestres[2], V Bonniol[3,4], M A Mostefa Side Larbi[1,3,4] and P Piccerelle[1,5]

[1]Biopharmacy Laboratory, Faculty of Pharmacy, Aix-Marseille University, France. [2]Faculty of Pharmacy, University of Montpellier I, France. [3]Faculty of Medicine, Aix-Marseille University, France. [4]Prenyl Bio, Marseille, France. [5]IMBE, Aix-Marseille University, France.

1 INTRODUCTION

The human life span has more than doubled over the last two centuries. With this increase in life expectancy, a significant segment of the aging population will seek to improve their quality of life including the ability to preserve a more youthful appearance for as long as possible [1]. Skin is notably sensitive to aging because it is affected by both intrinsic (genetically determined) and extrinsic (sun radiation, stress, pollution, etc.) processes. In response to consumer demand, numerous laboratories have developed new topical anti-aging formulas labeled as "innovative" but containing common active ingredients (alpha-hydroxy acids, antioxidants, depigmenting agents, etc.) which represent nothing revolutionary because they have been used for many years to reduce signs of ageing [2-4].

Progeria, also known as Hutchinson-Gilford Progeria Syndrome (HGPS), is one of 7,000 identified rare diseases [5]. Progeria is characterised by premature and accelerated aging. The first signs of the disease appear in the first 12-18 months of life and are characterised by a break in the growth curve. The average age of death is 13.5 years, often caused by myocardial infarction or stroke. Progeria children have a characteristic phenotype including very thin skin, superficial veins, pinched nose, alopecia and microretrognathia (abnormal development of the face). Progeria children exhibit severe generalised atherosclerosis and suffer from a significant reduction in bone density and lipoatrophy [5, 6]. The mutation responsible for progeria, found in the *LMNA* gene encoding lamin A and C, was identified independently in 2003 by French and US researchers [7, 8]. Three different genetic diseases resulting in accelerated aging, i.e., HGPS, restrictive dermopathy and mandibuloacral dysplasia, have been shown to exhibit the same pathophysiological mechanism [9]. Mutation of either the *LMNA* gene or the gene encoding the protease involved in the cleavage of the last 15 C-terminal amino acids of the lamin A precursor (FACE1/ZMPSTE24) leads to the presence of farnesylated progerin, which is anchored to membranes through its farnesyl anchor. The phenomenon results in the accumulation of farnesylated progerin in the nuclear lamina, inducing characteristic changes in nuclear shape and size and resulting in a decrease of "soluble" mature lamin A in the remaining nucleoplasm. This abnormality in the composition of the nuclear matrix causes several disorders in nuclear genome activity that trigger cell aging, including DNA repair and

RNA transcription and maturation. Without *LMNA* mutations, progerin is also produced by cells in aged subjects, due to age-related dysfunction of mRNA splicing machinery [10]. Finally, the expression of progerin by stem cells derived from adult tissues belonging to either epidermal [11] or mesenchymal lineages [12] leads to rapid exhaustion of the stem cell pool: a further pro-aging event.

Research on progeria provides an alternative source of active ingredients. The combination of two drugs, a statin and an aminobisphosphonate, decreases the accumulation and/or persistence of prenylated (farnesyl or a geranylgeranylated) nuclear proteins that are responsible for cell aging in progeria patients. The same combination also corrects several disorders exhibited in a mouse model reproducing human progeria and increases the life span of these mice [13]. Among the available statins and aminobisphosphonates on the market, we selected a combination of sodium pravastatin (PRA) and sodium alendronate (ALN) on the basis of screening results obtained with cultured fibroblast cells [14]. This drug pair improved the mitotic index of cells significantly and is widely available.

The purpose of the present investigation was to study the transdermal diffusion of the drug pair from a topical vehicle to assess its skin bioavailability for future applications. The active ingredients must penetrate the stratum corneum (SC) and diffuse in sufficient concentrations to the intended target in the skin over a consistent time course [15]. It is well known that choice of vehicle strongly influences skin absorption [16, 17] and so the *in vitro* study was performed with Franz static diffusion cells [18] using an oil-in-water (O/W) emulsion vehicle directly therein, fulfilling the necessary criteria for collecting relevant cosmetic information. Dermal kinetics were determined using porcine skin following the guidelines of the Scientific Committee on Consumer Safety (SCCS) [19] and the Organisation for Economic Co-operation and Development Guideline TG 428 [20]. Pig (ear) skin resembles human skin morphologically and functionally, and drug permeation rates appear to be similar in pig and human skin [21].

2 MATERIALS AND METHODS

Test product formulation and characterisation

The following chemical products were used for donor formulation: purified water (10.0 ± 5.0 μS, Elix Advantage 3, Millipore, France), sodium alendronate trihydrate (Ph. Eur. Quality; Cadila Pharmaceutical Limited, India), sodium pravastatin (Ph. Eur. Quality; Hisun, China), isononyl isononanoate (Seppic, France), caprylic/capric triglyceride (Evonik Goldschmidt GmbH, Germany), glycerin (Spiga Nord, Italy), arachidyl alcohol and behenyl alcohol and arachidyl glucoside (Seppic, France), PEG-100 stearate and glyceryl stearate (Seppic, France), polyacrylate-13 and polyisobutene and polysorbate 20 (Seppic, France), glycol palmitate (Seppic, France), caprylyl glycol (Schülke, France), ethylhexylglycerin (Schülke, France), tocopheryl acetate (DSM, Switzerland), dye (Sensient Cosmetic Technologies, France), methyisothiazolinone (Schülke, France) and sodium hydroxide (Cooper, France).

Differential Scanning Calorimetry (DSC 131, Setaram Instrumentation, France) was used to characterise the potential reactivity of active ingredients. Samples were prepared by mixing the drugs (50/50, w/w) and heating the mixtures from 25 °C to 200 °C with a ramp of 10° C min^{-1}. The samples were sealed in 30 µl aluminum pans and the experiments were performed in ambient air.

Preliminary results obtained on young and old fibroblast cultures have demonstrated a synergistic effect for PRA and ALN on fibroblast mitotic index [14]. This mitosis-activating effect was induced by 1 µM of each drug, i.e., 4.465 x 10^{-5} % (w/w) of PRA and 3.251 x 10^{-5} % (w/w) of ALN. We therefore chose to study a mixture with equimolar equivalents of PRA and ALN. As the most effective dose on fibroblasts was very low (1 µM), the starting dose of PRA was set at a theoretical concentration of 0.5 % (w/w) in the donor formulation, and ALN was set at a theoretical amount of 0.375 % (w/w). The formulation was an O/W emulsion prepared by stepwise addition of the inner oily phase to the aqueous phase at 85 °C. The addition of oil was performed under rotor stator agitation at 1700 rpm (Turbotest, VMI Rayneri, France). An Eclipse E600 optical microscope (Nikon) and a PC running the Lucia G version 4.8 software image analyser (Roucaire, France) were used to determine the mean globule size. Viscosities were determined using a rotating viscometer (Brookfield DV-II+ pro, USA). The measurements were conducted at 25 °C with a No. 4 spindle at 6 rpm. Sample pH was directly measured without dilution with a Cyberscan pH 510 pH meter (Eutech Instrument, Netherlands) at 25 °C. Emulsion density was measured at 20 °C using a pycnometer (Erichsen, France) with a 50 mL sample cell.

To visualise emulsion stability, a water-soluble blue colorant (methylene blue) was added to the formulation which was then centrifuged at 4,400 rpm for 30 minutes at 25 °C (using a 5702 RH Centrifuge, Eppendorf, Germany). Sample homogeneity was visually inspected. The system was considered destabilised when phase separation or coalescence was observed. Emulsion stability was assessed by accelerated aging of the prepared emulsion by measuring physicochemical properties (viscosity, pH, and globules size) during storage of the emulsion in a glass container under shelf-life testing conditions at room temperature for 6 months and under accelerated conditions of 4 and 40 °C for 6 months or under accelerated conditions of 50 °C for 3 months (vacuum drying oven, VD23/53/115 WTB Binder, Germany).

Measurement of transdermal delivery

Static Franz glass diffusion cells (Legallais, France) were used for this investigation. These cells consist of donor and receptor chambers between which a piece of whole porcine skin was positioned. The receptor chamber fluid was physiological serum (Versol NaCl 0.9 %, Aguettant, France) and 0.1 M sodium citrate (Sigma, France). After an initial visual inspection of the skin, a TM210 tewameter (Courage-Khazaka, Germany) was used to determine transepidermal water loss (TEWL), which reflects skin integrity [22]. The rate of percutaneous penetration and TEWL was greatly increased when the barrier function of the SC was compromised. If the barrier was disrupted, the sample was not used in the study. Skin thickness was measured using a dial thickness gauge (0.01-10 mm; Mitutoyo Ehy 331, Japan).

Skin permeability studies were performed using methods described elsewhere [17, 23, 24]. Porcine ears (three different donors) were obtained from freshly killed animals at a local slaughterhouse (Pézenas, France). After cleaning with cold tap water, full-thickness, non-dermatomed skin (approximately 0.9-1.1 mm) was removed with a scalpel from the cartilage of the dorsal pinnae. Intact skin discs with an internal diameter of 3 cm were sealed in plastic bags and stored for no more than 6 weeks at -20 °C [25]. The skin samples were mounted inside modified Franz diffusion cells [18] in such a way that the dermal side of the skin was exposed to the receptor fluid. The diffusion area was 0.95 cm². The precise volume of the acceptor compartment (approximately 9 mL) was measured for each cell and was included in the calculations. The continuously stirred receptor medium (physiologic serum Versol NaCl 0.9 % solution to preserve skin conditions) was maintained at 37 ± 1 °C by water circulation (Polystat CC1, Huber). Under these conditions, the skin disk temperature was 32 ± 1 °C, which corresponds to the skin surface temperature *in vivo* (measured with Thermocouple thermometer, ecoScan).

Skin absorption experiments were started by the topical application of 30.1 - 57.5 mg of test product to the surface of the skin through the donor compartment after which the donor chambers were covered with Parafilm® to minimise evaporation.

After an exposure time of 24 h, the diffusion cells were dismounted, and the skin surface was washed with 10 mL of 0.1 M sodium citrate, which also served to flush the Parafilm® covering the donor compartment. These 10 mL were re-used to wash (by passage) the walls of the donor compartment.

Tape stripping was used to separate the superficial layers of the SC [26]. The SC of the treated area was removed by 15 successive tape-stripping steps using D-Squam™ ($\varphi = 14$ mm; Monaderm, Monaco) with a constant pressure (225 g cm^{-2}) for 5 s. After eliminating the SC from skin samples via tape stripping, the viable epidermis and dermis samples were chopped into small pieces, and the test substances were extracted with 5 mL of 0.1 M sodium citrate solution for 24 h under constant shaking.

A total of eight Franz cells were used for the study. During the Franz test, all cells were subject to application of the test "preparation" with the exception of Cell 1, which was used as a blank cell with no product application, and Cell 8, which received a placebo (a preparation without active substances; vehicle control).

Sample analysis

All reagents used for chromatography were analytical reagent grade: sodium citrate (Sigma, France), sodium borate decahydrate (Sigma, France), 9-fluorenylmethyl chloroformate (FMOC) (Fluka, France), orthophosphoric acid 85 % (Sigma, France), acetonitrile (Carlo Erba, France), dichloromethane (Carlo Erba, France), methanol (Carlo Erba, France), glacial acetic acid (VWR, France), and triethylamine (Sigma, France). Ultrapure water was obtained from a Direct Q5 water purification system (Millipore, France).

Determination of drug concentrations (PRA and ALN) was determined by two methods according to the analytical substrate (solid or liquid). Solid samples included SC tape strips, epidermis and dermis. Liquid samples included receptor chamber fluid and diffusion cell washing solutions. Solid substrates were extracted with 5 mL of 0.1 M sodium citrate solution under magnetic stirring for 24 hours. Liquid samples were analysed following filtration (0.45 μm pore, 33 mm diameter Millipore PVDF filters; Merck Millipore, France).

Drug analysis was performed by high-performance liquid chromatography (HPLC) using an AN-HPLC-001 and AN-HPLC-004 high-performance liquid chromatography system equipped with a quaternary pump, automatic injector, column oven and an Agilent 1100 or 1200 series detector (Agilent Technologies, France) using a diode array or equivalent and Agilent Chemstation software (Agilent Technologies, France).

Separation of PRA was performed using a C_{18} column (Agilent Eclipse XDB, 250 × 4.6 mm, 5 μm) at 25 °C and detected at 238 nm with a retention time of 15 min using a 90/10 phase A/methanol mobile phase at a flow rate of 1.5 mL min^{-1}. The analysis time was 40 min, and 10 μL injection volumes were used. Mobile phase A was prepared using a mixture of 450 mL of methanol, 1 mL of acetic acid, 1 mL of triethylamine and 550 mL of water in 1,000 mL flask. The mobile phase was stirred magnetically for a few minutes and filtered through a 0.45 μm membrane filter. The diluent was obtained using a 45/55 (v/v) mixture of methanol and water. PRA standards were prepared by dilution of a PRA stock solution.

Separation of ALN was performed by using a PLRP-S column (Polymer Lab, 150 × 4.6 mm, 5 μm), at 25 °C. ALN was detected at 266 nm with a retention time of 5 min using a 75/5/20 phase A/methanol/acetonitrile mobile phase at a flow rate of 1.0 mL min^{-1}. Analysis times were 30 min for the assay and 15 min for the standards, and 20 μL injections were used. Mobile phase A consisted of a diluent/sodium phosphate solution adjusted to pH 8 with orthophosphoric acid. The diluent was a 0.1 M sodium citrate solution. Sodium phosphate buffer was obtained by adjusting a 0.1 M sodium phosphate solution to pH 8 with orthophosphoric acid. The tetraborate solution was 0.1 M. A derivatisation solution was prepared by dissolving FMOC in acetonitrile to obtain a 0.1 % w/v solution. The derivatisation step consisted of adding 1 mL of a standard or sample solution, 1 mL of tetraborate solution and 0.8 mL of derivatisation solution to 10 mL samples tubes with caps. The tubes were mixed by vortex agitation for 30 seconds and allowed to stand for 30 minutes at ambient temperature. Next, 5 mL of dichloromethane was added, and the tubes were vortexed for 30 seconds. The tubes were then centrifuged for 10 minutes at 2,000 rpm. The supernatants were transferred to vials after filtration.

The amounts of each drug were quantified by reference to the calibration standards and converted to percentage of applied dose. The "unabsorbed dose" was defined as the amount of drug in the donor compartment and tape strips. "Absorbable dose" was represented by the quantities in the epidermis and dermis. The amount of test compound in the receptor fluid was defined as the "absorbed dose".

3 RESULTS AND DISCUSSION

The physicochemical properties of PRA and ALN are listed in Table 1. It is well known that physicochemical criteria influence transdermal bioavailability. Ideal transdermal candidates are characterised by an aqueous solubility greater than 1 mg mL^{-1}, a lipophilicity (log octanol-water partition coefficient) between 1 and 3, a molecular weight less than 500 Daltons (Da), a melting point below 200 °C and saturated aqueous solution pH between 5 and 9 [27-30].

Table 1 *Formulae and physical characteristics of sodium pravastatin (PRA) and sodium alendronate (ALN).* *Values from experimental data.* **P = octanol/water partition coefficient.* ****K_p (m s^{-1}) = permeability coefficient; K_p values were calculated from Potts and Guy's equation [16].*

Parameter	Test Compound	
	sodium pravastatin (PRA)	sodium alendronate (ALN)
CAS number	81131-70-6	121268-17-5
Formula		
Molecular weight (Da)	446.5	325.1
Melting point (°C)	171.2 - 173.0	257.0 - 262.5
H-Bond Donors	3	8
H-Bond Acceptors	7	11
Aqueous solubility (mg mL^{-1})	> 300 [31]	40 [32]
pH of saturated aqueous solution*	7.1	4.0
log P** (pH 7)	-0.23 [33]	-2.6 [34]
Log K_p***	-5.63	-6.57
pKa	4.7 [35]	1.78 / 2.60 / 6.73 / 11.51 / 12.44 [36]

It is difficult to predict the amplitude of skin penetration of the studied drugs from the overall data because only some of the physicochemical parameters are favorable. Molecular weight is one of the main determinants in skin delivery [27, 29]: both drugs are sufficiently small to pass through the stratum corneum. PRA also has a low melting point, suggesting lower intermolecular forces and a higher propensity to cross the skin barrier [37, 38]. Nevertheless, the marked hydrophilicity (high water solubility, log P < 0) may hinder transfer and diffusion into the SC (transdermal drug delivery) and therefore be a limiting factor. Drug mobility is also a function of hydrogen bonding groups and is inversely proportional to the diffusion coefficient [39, 40]. Molecules with more than four hydrogen bonding groups have very low diffusivity [38, 41] and so it was expected that the two active ingredients would have very low diffusivities across the stratum corneum. The diffusion of permeants is also affected by the concentrations of ionised forms generated at different pH values (see later).

The differential scanning calorimetry (DSC) curve for PRA displays two endothermic peaks at 113 °C and 151 °C (Figure 1) which correspond to (a) coordinated and crystal water and (b) melting point, respectively. This result is in good agreement with previous PRA DSC data [42].

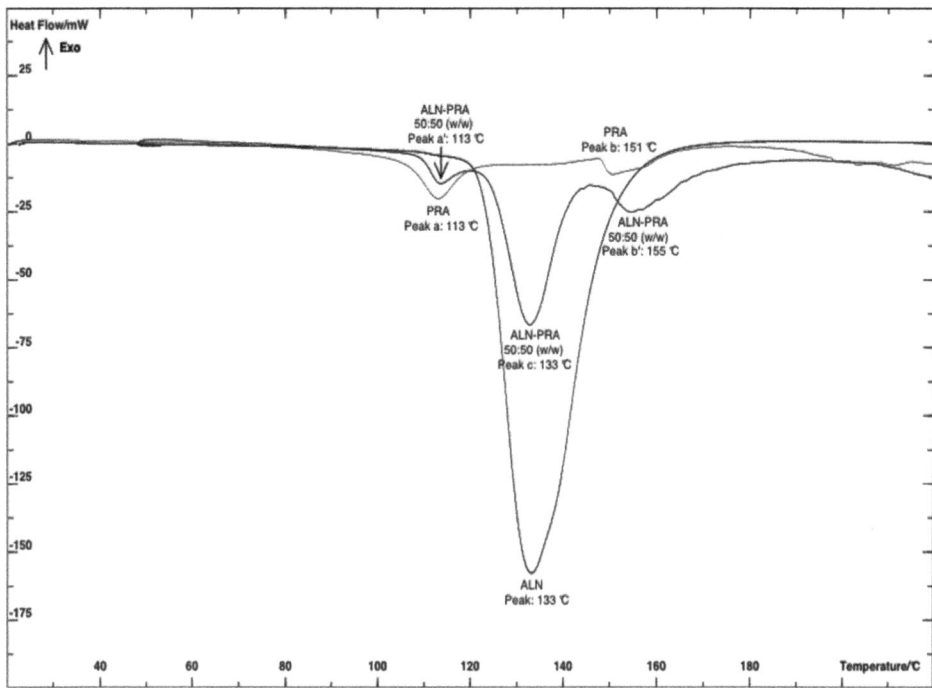

Figure 1 *Differential scanning calorimetry curves (DSC) for sodium pravastatin and sodium alendronate.*

The DSC curve for ALN indicates an endothermic peak at 133 °C which corresponds to coordinated and crystal water. The melting point of ALN was not reached because we did not study the behavior of the two drugs beyond the PRA melting point (approximately 172 °C). The DSC curve of a 50/50 mixture of the two drugs presents the same peaks as those for each separate active ingredient: the active ingredients behave identically to the single components. Thus, we concluded that there was no interaction between the two active ingredients.

Both PRA and ALN are salts of weak acids and strong bases. Dissolution in water leads to dissociation of Na^+ and the ionised (weak acid) form according to each pKa and solution pH (Table 1). The pKa value of pravastatin is 4.7 and at a pH equal to the pKa, 50 % of the salt exists as the molecular species and 50 % is ionised (Figure 2). Below pH 3, only the molecular species exists; above pH 7, only the ionised species exists. The pKa of PRA (4.7) is close to cutaneous pH (approximately 5.5) [1]; near this value, small changes in pH drastically alter the ionic/molecular species ratio. The pH of the donor formulation (pH 6) results in a mixture of 5 % of molecular species coexisting with 95 % ionised species. Therefore, pH is a critical parameter for the absorption of PRA.

Alendronate (ALN) dissociation as a function of pH is much more complex. Alendronate has five pKa values (Table 1) that corresponding to six functional groups that can be ionised: five H^+ donors and one amino group H^+ acceptor [36]. As illustrated in Figure 3, in the pH range of interest (5-7), there are principally two ionised species: H_3L^- and $H_2L_2^-$. At pH 6, 84 % of the molecules are in the H_3L^- form, and 16 % exist in the $H_2L_2^-$ form.

The ionised species of a drug has a lower permeability coefficient than its respective molecular species. Thus, the free acid or free base should be preferentially used to improve permeation [43]. Pugh and coworkers found that charge and MW are equally important predictors of diffusion [44]. Accordingly, PRA and ALN have pKa values that are not conducive to the efficient spread of drugs through the skin.

The donor formulation was an O/W emulsion with a light, non-oily texture that was easy to apply and left no residue after application. These are characteristics that influence consumer acceptance and patient compliance for a cosmeceutical product [45]. The physicochemical characteristics of the O/W emulsion were the following: viscosity, 40,000 mPa s (6 rpm; spindle No. 4; 25 °C); pH, 6 (25 °C); density, 0.99 (20 °C); and a globule size < 5 μm. These results are in agreement with data typically reported for this type of formulation. To determine the exact amounts of the active ingredients in the emulsion, the preparation used for this study was analysed by HPLC/UV. The donor formulation contained 0.55 % (w/w) PRA and 0.39 % (w/w) ALN. These concentrations were selected for the permeability study.

From the perspective of transdermal delivery, the formulation does not meet the required criteria. The concentration of PRA and ALN are very low, the drugs are highly hydrolysable and easily ionised, and the vehicle is essentially aqueous. Indeed, the SC (constituting the major natural skin barrier) is predominantly lipoidal in nature, implying high lipid solubility for maximal output into the SC. However, water is the most natural penetration enhancer. Water increases the hydration of the SC and so increases the transdermal diffusivity of a variety of chemicals [46]. Moreover, high aqueous solubility is essential for maximal output into the viable epidermis. The ideal penetrant must therefore have two opposing prerequisites (lipophilicity and hydrophilicity) to penetrate the different skin layers effectively [47].

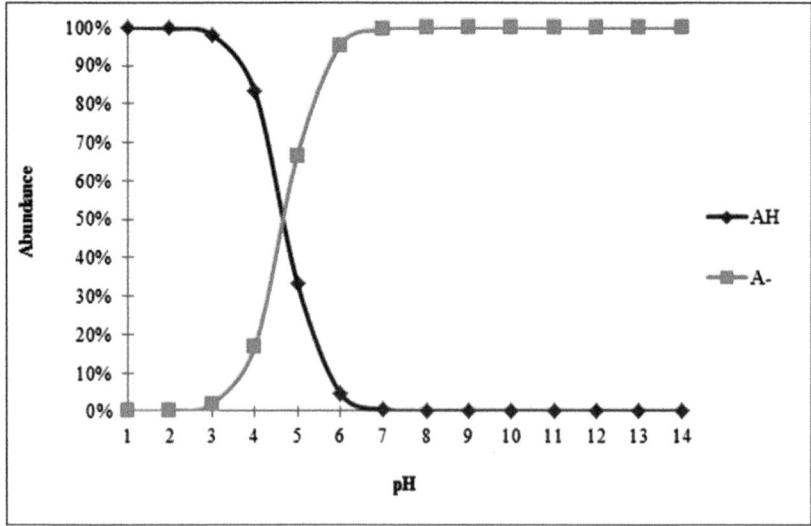

Figure 2 *Protonation distribution diagram showing the relative abundances of PRA species [unionised ("AH") and ionised ("A^{-1}")] at equilibrium as a function of pH.*

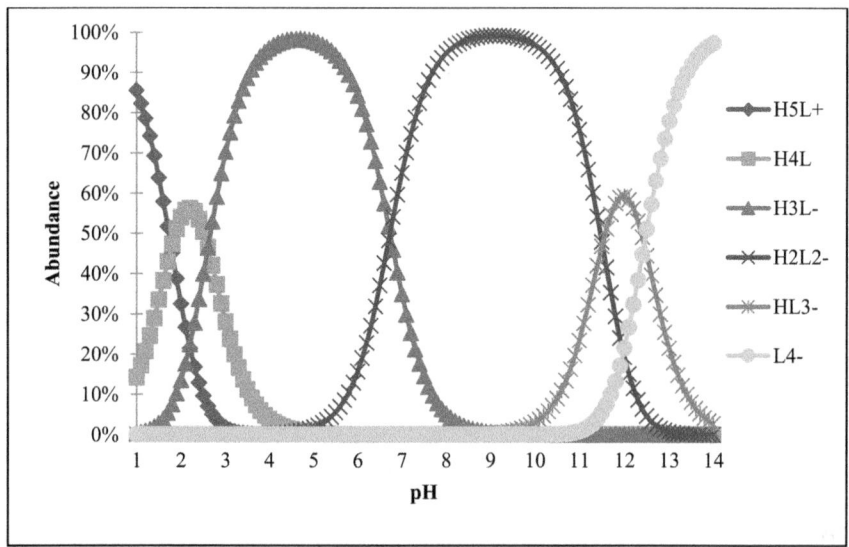

Figure 3 *Protonation distribution diagram showing the relative abundances of ALN species [unionised; "H4L" or ionised; "H5L⁺", "H3L⁻", "H2L²⁻" and HL³⁻] vs. pH at equilibrium.*

Viscosity and pH of the test formulation were stable at room temperature over the duration of the test period (6 months; Figure 4). The donor formulation was stable at 40 °C over the entire 6 months and at 50 °C for 3 months. No instability was observed after the centrifuge test (30 min at 4,400 rpm at 25 °C).

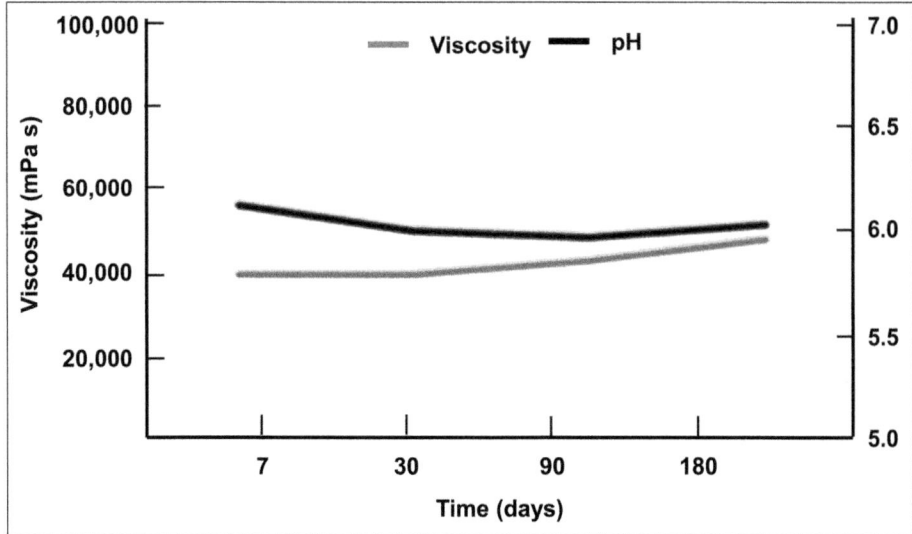

Figure 4 *Viscosity and pH of test formulation as a function of time.*

Table 2 *Transepidermal water loss (TEWL) values obtained from skin within each diffusion cell before application of test formulation. Values presented as average ± standard deviation.*

	Cell 1	Cell 2	Cell 3	Cell 4	Cell 5	Cell 6	Cell 7	Cell 8
TEWL $(g\ m^{-2}\ h^{-1})$	14 ± 1.2	3.6 ± 0.9	11.9 ± 4.2	13.9 ± 4.1	6.8 ± 2.8	4.0 ± 3.2	11.9 ± 2.5	11.5 ± 1.2

After an initial visual observation of the skin pieces, TEWL rates were recorded from the skin surface in each diffusion cell (Table 2). The rate of percutaneous penetration and TEWL would be greatly increased if the barrier function of the SC was compromised. Thus, TEWL was measured before starting the experiment. All TEWL values were less than $15\ g\ m^{-2}\ h^{-1}$, indicating that the pig-skin samples used in the subsequent test were structurally viable.

Initial dermal absorption studies were performed with test emulsions containing 0.5 % w/w PRA and 0.375 % w/w ALN. These studies showed that the amounts of active ingredients that pass through the skin after 1, 3, 8 and 24 hours were low and are therefore difficult to measure: many samples contained amounts of drug that were below the limit of detection for the analytical method. This low transdermal flux can be attributed to the finite dose of product deposited on the skin surface. Indeed, the concentration used was well below the saturated concentration usually used for this type of experiment (approximately 500 times lower for PRA and approximately 100 times lower for ALN). Therefore, we opted for a single determination of percutaneous absorption after 24 hours rather than a kinetic study (involving multiple samples of receptor chamber fluid). Skin absorption after 24 h allowed us to verify whether these low concentrations of drugs penetrated the skin layers in significant quantities.

The Scientific Committee on Consumer Safety requires that the mass balance recovery must be in the range of 85-115 % of the applied dose [19]. The total recovery of PRA exceeded 85 % (Table 3 and Figure 5). The absorbed dose (equivalent to PRA found in the receptor chamber fluid) was 12.1 %, with the amount of PRA found in viable skin (6.6 %) being taken as the potentially absorbed dose. The amount retained by the SC (34.7 %) was not considered to be dermally absorbed and thus did not contribute to the systemic dose. Thus overall, the percutaneous penetration of PRA was low and the majority of the drug remained on the surface or in the SC. These results are in agreement with the predicted transdermal delivery described earlier.

The total ALN recovered was much less than 85 % (Table 3 and Figure 6). More than 40 % of the ALN was unaccounted for and under these conditions it is difficult to determine the percutaneous penetration of ALN. Nevertheless, the results show that ALN was found in viable skin and receptor fluid, so ALN penetrates the various skin layers. The reliability of the analytical method was verified by checking the specificity and repeatability parameters. Moreover, all low-concentration dosages were tested with reference to a range of standards.

Table 3 *Quantities of sodium pravastatin (PRA) and sodium alendronate (ALN) recovered after 24 hours, expressed as percentage of applied dose. Values represent average ± standard deviation.*

Drug Recovery (% Applied Dose)	Recovery of formulation	Stratum corneum	Epidermis + dermis	Receptor fluid	TOTAL
PRA	35.1 ± 8.9	34.7 ± 9.0	6.6 ± 1.6	12.1 ± 8.2	88.5 ± 27.7
ALN	30.9 ± 11.9	14.9 ± 2.7	4.9 ± 1.1	5.9 ± 4.5	56.6 ± 20.2

The entire residual formulation added to the donor compartment at the end of 24 h was recovered at the end of the study. Given the apparent loss of ALN in the dose recovery, we hypothesised that some ALN evaporated or degraded under the experimental conditions. To test this hypothesis, we added a defined quantity of donor formulation to three cups, sealed them with Parafilm® and placed them in a thermostatic chamber at 32 °C for 24 h. The cups and Parafilms® were rinsed with extraction solvent after 24 hours and the rinse samples were analysed. The Parafilm® did not contain ALN, so ALN did not evaporate. In addition, the total amount of ALN added to the cup was recovered, so ALN does not degrade under the test conditions after 24 hours.

Collectively, these data suggest that the unrecovered dose may have permeated the skin but was undetected due to strong interactions of the drugs with molecules present in the skin. For example, it is known that bisphosphonates such as ALN strongly chelate multivalent ions [48] and so ALN may have formed insoluble Ca^{2+} complexes in the skin. Moreover, bisphosphonates are known to bind plasma proteins such as albumin [49, 50]. Thus, there is a strong possibility that ALN binds skin proteins such as keratin, collagen and elastin. However, we did not explore this issue further in this study. Due to its high water solubility, ALN is not likely to be metabolised in the body [49, 50].

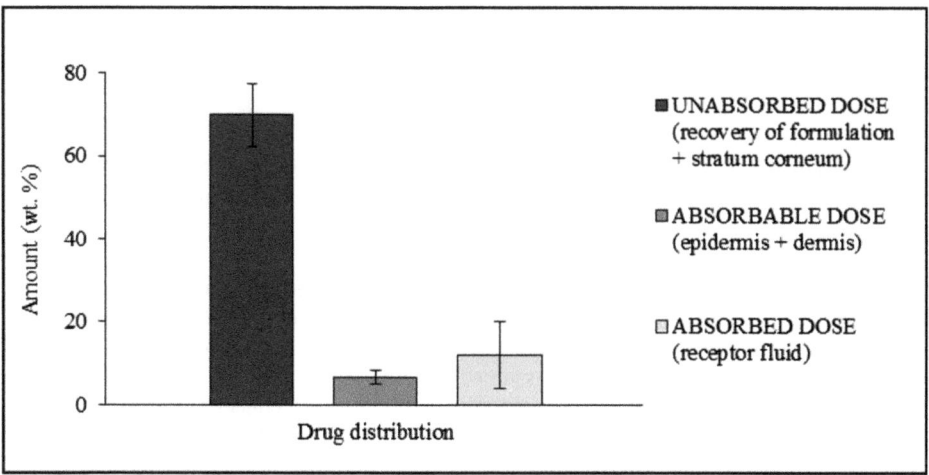

Figure 5 *Skin distribution of sodium pravastatin (PRA) after 24 hours, expressed as percentage of applied dose. Values represent average ± standard deviation.*

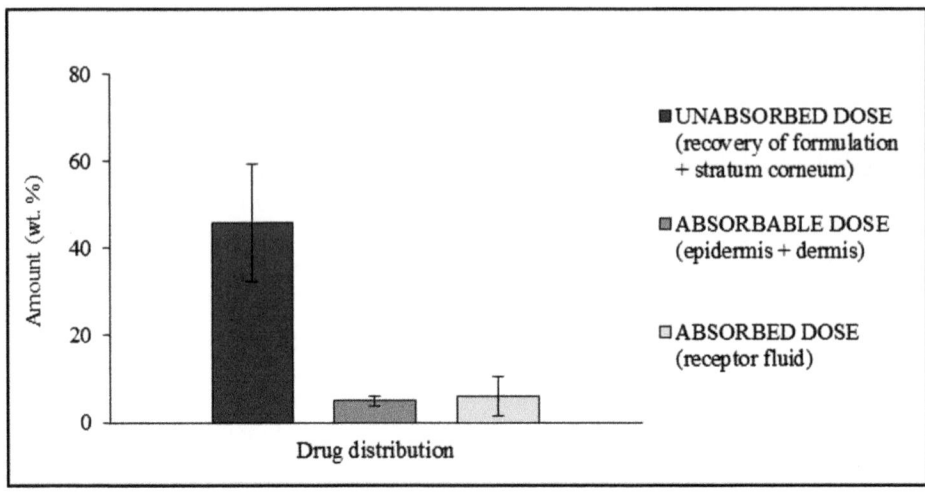

Figure 6 *Skin distribution of sodium alendronate (ALN) after 24 hours, expressed as percentage of applied dose. Values represent average ± standard deviation.*

In order to perform a safety assessment, the Systemic Exposure Dosage (SED) was estimated using Equation 1 which takes into account the skin surface intended for treatment with the product as well as the frequency of product application and its retention factor [19, 20, 51].

$$SED = \frac{\left[(A \times 1000) \times \left[\frac{C}{100} \right] \times \left[\frac{DA}{100} \right] \right]}{BW} \tag{1}$$

Where SED is the Systemic Exposure Dose (expressed as mg kg^{-1} day^{-1}), A (g day^{-1}) is the amount of finished product applied (generally assumed to be 800 mg for a single facial application), C is the concentration (%) of the active ingredient, DA is the dermal absorption (expressed as percentage of applied dose) and BW is body weight (default = 60 kg).

The experimental value for DA was derived from the *in vitro* permeability study and represents the amount of active ingredient recovered in the epidermis, dermis and receptor fluid. The value of DA used for the calculation was based on the average (18.7) plus two standard deviations (17.6) to give an assumed dose of 36.3 % (Table 4) to incorporate a margin of safety [19]. Given the poor recovery of ALN, its concentration in the final product was considered to be entirely absorbed (DA = 100 %) for the SED calculation [19].

Table 4 *Dermal absorption of sodium pravastatin (PRA) and sodium alendronate (ALN) expressed sum of {absorbed and absorbable} percentage of applied dose derived from (I) experimental measurements and (II) assumed for the purpose of calculating the SED (systemic exposure dose; Equation 1).*

Drug	Dermal Absorption (DA; % applied Dose)		Calculated SED (mg kg^{-1} day^{-1})
	(I) Experimental	*(II) Assumed*	
PRA	18.7 ± 8.8	36.3	0.027
ALN	10.8 ± 4.6	100	0.52

It is generally accepted that the Margin of Safety (MoS) of a substance can be calculated by dividing its lowest No Observable Adverse Effect Level (NOAEL) value by its calculated SED. The selection of endpoints for NOAELs is difficult but critical because these numbers are often used by risk assessors. Selecting the most appropriate endpoint is fundamental, but only two relevant studies have been published for PRA and only one for ALN. Table 5 therefore provides only an estimate of the MoS; these conclusions should be modified with an extended analysis of the literature on repeated exposure tests for PRA and ALN.

Based on studies in dogs using the oral route of administration, the lowest NOAEL was reported to be 25 mg kg^{-1} day^{-1} for PRA [52 – 54] and 2 mg kg^{-1} day^{-1} for ALN [37]. Factoring in a safety margin of 10 to cover inter-species differences and an additional factor of 10 to cover inter-individual differences, the acceptable MoS should be >100 [51].

In the future, the amount of ALN in the formulation can be decreased, and the MoS can be increased to obtain an acceptable safety margin for ALN. It is also possible to improve the test performance to obtain a usable DA value that does not consider the starting ALN concentration and thereby increases the MoS.

4 CONCLUSIONS

A synergistic combination of two anti-aging active ingredients (PRA and ALN) was incorporated into an oil-in-water emulsion with the required cosmetic characteristics for topical application. An *in vitro* permeability study using this cosmeceutical vehicle

Table 5 *Calculated margins of safety (MoS) for sodium pravastatin (PRA) and sodium alendronate (ALN) derived from experimental measurements of dermal absorption and reported no observable adverse effect level (NOAEL).*

Drug	NOAEL (mg kg^{-1} day^{-1})	Species and route of administration	MoS
PRA	25	Dog, oral [52-54]	926
ALN	2	Dog, oral [32]	3.8

demonstrated that both molecules are able to diffuse through the skin barrier into the dermis in small amounts despite their high hydrophilicity and strong ionisation.

These results provide evidence to support clinical studies demonstrating the *in vivo* efficiency of the PRA/ALN combination to ameliorate the signs of skin aging, although the concentrations of PRA and ALN in the product must first be optimised. To satisfy to required safety criteria for cosmetics (i.e., an MoS > 100) the ALN concentration should be reduced and the relative quantity of the two drugs should be adjusted to account for their different skin permeation properties.

This study provides preliminary data for the development of a new anti-aging transdermal delivery system.

5 ACKNOWLEDGEMENTS

We wish to thank the slaughterhouse (Pézenas, France) for kindly donating the pig ears. The authors are grateful to David Bergé-Lefranc for his expertise in physical chemistry. Finally, we thank Pierre Cau for his assistance in laminopathy.

6 REFERENCES

1. M. A. Farage, K. W. Miller and H. I. Maibach, Degenerative changes in aging skin, in *Textbook of Aging Skin*, ed. M. A. Farage, K. W. Miller and H. I. Maibach, 2010, Springer, Berlin, pp. 25–35.

2. J. K. Rivers, The role of cosmeceuticals in antianging therapy, *Skin Therapy Lett*, 2008, **13**(8), 5–9.

3. M. Amer and M. Maged, Cosmeceuticals versus pharmaceuticals, *Clin Dermatol*, 2009, **27**(5), 428–30.

4. F. S. Brandt, A. Cazzaniga and M. Hann, Cosmeceuticals: current trends and market analysis, *Semin Cutan Med Surg*, 2011, **30**(3), 141–3.

5. R. C. Hennekam, Hutchinson-Gilford progeria syndrome: review of the phenotype, *Am J Med Genet A*, 2006, **140**(23), 2603–24.

6. M. A. Merideth *et al.*, Phenotype and course of Hutchinson-Gilford progeria syndrome, *N Engl J Med*, 2008, **358**(6), 592–604.

7. A. De Sandre-Giovannoli *et al.*, Lamin a truncation in Hutchinson-Gilford progeria, *Science*, 2003, **300**(5628), 2055.

8. M. Eriksson *et al.*, Recurrent de novo point mutations in lamin A cause Hutchinson-Gilford progeria syndrome, *Nature*, 2003, **423**(6937), 293–8.

9. C. L. Navarro, P. Cau and N. Levy, Molecular bases of progeroid syndromes, *Hum Mol Genet*, 2006, **15 Spec No 2**, R151–61.

10. P. Scaffidi and T. Misteli, *Science*, 2006, **312**, 1059–1063.

11. J. Espada *et al.*, Nuclear envelope defects cause stem cell dysfunction in premature-aging mice, *J Cell Biol*, 2008, **181**(1), 27–35.

12. P. Scaffidi and T. Misteli, *Nat Cell Biol*, 2008, **10**, 452–459.

13. I Varela *et al.*, Combined treatment with statins and aminobisphosphonates extends longevity in a mouse model of human premature aging, *Nat Med*, 2008, **14**(7), 767–72.

14. LIMP, *Effet de 4 compositions cosmétiques sur la division de ficroblastes humains agés et jeunes sur 72H*, 2008.

15. L. E. Millikan, Cosmetology, cosmetics, cosmeceuticals: definitions and regulations, *Clin Dermatol*, 2001, **19**(4), 371–4.

16. R. O. Potts and R.H. Guy, Predicting skin permeability, *Pharm Res*, 1992, **9**(5), 663–9.

17. L. Duracher *et al.*, The influence of alcohol, propylene glycol and 1,2-pentanediol on the permeability of hydrophilic model drug through excised pig skin, *Int J Pharm*, 2009, **374**(1–2), 39–45.

18. T. J. Franz, Percutaneous absorption. Relevance of in vitro data, *J. Invest. Dermatol*, 1975, **64**, 190–195.

19. SCCS, Basic criteria for the in vitro assessment of dermal absorption of cosmetic ingredients, in *SCCS/1358/10*, 2010, European Commission, Brussels.

20. OECD, Guideline for the Testing of Chemicals. Skin absorption: In vitro method., in *Draft Guideline 428*, 2000, OECD, Belgium.

21. G. A. Simon and H. I. Maibach, The pig as an experimental animal model of percutaneous permeation in man: qualitative and quantitative observations--an overview, *Skin Pharmacol Appl Skin Physiol*, 2000, **13**(5), 229–34.

22. A. Nangia *et al.*, In vitro measurement of transepidermal water loss: a rapid alternative to tritiated water permeation for assessing skin barrier function, *Int J Pharm*, 1998, **170**, 33–40.

23. C. Fernandez *et al.*, LC analysis of benzophenone-3 in pigskin and in saline solution: application to determination of in vitro skin penetration, *J Pharm Biomed Anal*, 2000, **22**(2), 393–402.

24. L. Duracher *et al.*, Irradiation of skin and contrasting effects on absorption of hydrophilic and lipophilic compounds, *Photochem Photobiol*, 2009, **85**(6), 1459–67.

25. E. Kurul and S. Hekimoglu, Skin permeation of two different benzophenone derivatives from various vehicles, *Int J Cosmet Sci*, 2001, **23**(4), 211–8.

26. P. Clarys *et al.*, There is no influence of a temperature rise on in vivo adsorption of UV filters into the stratum corneum, *J Dermatol Sci*, 2001, **27**(2), 77–81.

27. J. D. Bos and M. M. Meinardi, The 500 Dalton rule for the skin penetration of chemical compounds and drugs, *Exp Dermatol*, 2000, **9**(3), 165–9.

28. A. Naik, Y. N. Kalia and R. H. Guy, Transdermal drug delivery: overcoming the skin's barrier function, *Pharm Sci Technolo Today*, 2000, **3**(9), 318–326.

29. B. M. Magnusson *et al.*, Molecular size as the main determinant of solute maximum flux across the skin, *J Invest Dermatol*, 2004, **122**(4), 993–9.

30. M. Schneider *et al.*, Nanoparticles and their interactions with the dermal barrier, *Dermatoendocrinol*, 2009, **1**(4), 197–206.

31. B. M. S. Company, Princeton, New Jersey, 08543, USA.

32. EMEA, *Scientific discussion - Fosavance,* 2005.

33. H. N. Joshi, M. G. Fakes and A. T. Serajuddin, Differentiation of 3-Hydroxy-3methylglutaryl-coenzyme A Reductase Inhibitors by their Relative Lipophilicity, *Pharm Phamacol Commun*, 1999, **5**, 269–271.

34. EDQM, *Safety data sheet sodium alendronate*, 2008.

35. I. Tamai *et al.*, Proton-cotransport of pravastatin across intestinal brush-border membrane, *Pharm Res*, 1995, **12**(11), 1727–32.

36. M. Meloun *et al.*, Thermodynamic dissociation constants of alendronate and ibandronate by regression analysis of potentiometric data, *J Chem Eng*, 2011, **56**, 3848–3854.

37. J. Hadgraft and W. J. Pugh, The selection and design of topical and transdermal agents: a review, *J Investig Dermatol Symp Proc*, 1998, **3**(2), 131–5.

38. A. Alikhan, S. Farahmand and H. I. Maibach, Correlating percutaneous absorption with physicochemical parameters in vivo in man: agricultural, steroid, and other organic compounds, *J Appl Toxicol*, 2009, **29**, 590–596.

39. M. S. Roberts, W. J. Pugh and J. Hadgraft, Epidermal permeability: penetrant structure relationships. 2. The effect of H-bonding groups in penetrants on their diffusion through the stratum corneum, *Int J Pharm*, 1996, **132**, 23–32.

40. J. du Plessis *et al.*, Physico-chemical determinants of dermal drug delivery: effects of the number and substitution pattern of polar groups, *Eur J Pharm Sci*, 2002, **16**(3), 107–12.

41. M. S. Roberts, S. E. Cross and M. A. Pellet, Skin transport, in *Dermatological and transdermal Formulations*, ed. A. W. Walters, 2002, Marcel Dekker, New York.

42. Y. Garg and K. Pathak, Design and in vitro performance evaluation of purified microparticles of pravastatin sodium for intestinal delivery, *Pharm Sci Tech*, 2011. **12**(2), 673–82.

43. M. E. Lane *et al.*, Passive Skin Permeation Enhancement, in Topical and Transdermal Drug Delivery: Principles and Practice, ed. H. A. E. Benson and A. C. Watkinson, 2012, Wiley, pp. 23–42.

44. W. J. Pugh, I. T. Degim and J. Hadgraft, Epidermal permeability-penetrant structure relationships: 4, QSAR of permeant diffusion across human stratum corneum in terms of molecular weight, H-bonding and electronic charge, *Int J Pharm*, 2000, **197**(1–2), 203–11.

45. D. Karadzovska *et al.*, Predicting skin permeability from complex vehicles, *Adv Drug Deliv Rev*, 2012.

46. M. S. Robert and M. Walker, Water: the most natural penetration enhancer., in Pharmaceutical skin penetration enhancement, ed. K. Walter and J. Hadgraft, 1993, New York, pp. 1–30.

47. J. W. Wiechers, The barrier function of the skin in relation to percutaneous absorption of drugs, *Pharm Weekbl Sci*, 1989, **11**(6), 185–98.

48. A. G. Porras and B. J. Gertz, Pharmacokinetics and Pharmacodynamics of Alendronate Development, in *Applications of Pharmacokinetic Principles in Drug Development*, ed. R. Krishna, 2003, pp. 464–465.

49. J. H. Lin, Bisphosphonates: a review of their pharmacokinetic properties, *Bone*, 1996, **18**(2), 75–85.

50. F. Karamustafa and N. Celebi, Bisphosphonates and Alendronate, *J Pharm Sci*, 2006, **31**, 31–42.

51. SCCP, Opinion for basic criteria for the in vitro assessment of dermal absorption of cosmetic ingredients, in *SCCP/0970/06*, 2006, European Commission: Brussels.

52. S. Manabe, *et al.*, Preliminary dose finding study for subacute toxicological study of pravastatin sodium in monkeys, *J Toxicol Sci*, 1989, **14 Suppl 1**, 41–55.

53. S. Manabe *et al.*, Subacute toxicological study in monkeys treated orally with pravastatin sodium for 5 weeks, *J Toxicol Sci*, 1989, **14 Suppl 1**, 57–83.

54. C. Tarumi *et al.*, Long term oral administration study of pravastatin sodium to beagles for 104 weeks, *J Toxicol Sci*, 1989, **14 Suppl 1**, 85–101.

SECTION IV: REGULATORY AND TOXICOLOGY

R P Chilcott

Department of Pharmacy, University of Hertfordshire, College Lane Campus, Hatfield, United Kingdom.

Regulatory guidelines are generally written with great attention to detail in order to promote an unambiguous understanding of requirements. However, the necessarily comprehensive and rigorous nature of such documentation can paradoxically lead to misinterpretation and, consequently, submissions which may not meet the expectations of regulatory organisations. Chapter 24 provides some insight and clarification on topical product development from a regulatory perspective and focuses on hydro-alcoholic gel formulations. Whilst written in a personal capacity (and thus not an official interpretation of US Food and Drug Administration policy), the chapter provides useful guidance on best practice and an insight into the regulatory expectations for topical products. Of interest are case studies for issues specific to topical products, such as the effect of showering on systemic absorption, person-to-person transfer of medicaments and interactions with other commonly applied products such as sunscreens.

Current guidelines for the assessment of chemical toxicity have evolved from many decades' experience and, in some circumstances, have developed into internationally harmonised standards [1]. However, tests involving animals are constantly being questioned on the basis of ethical concerns relating to their adequacy and relevance. Consequently, a great deal of effort has been expended on identifying and validating alternative models to classify substances according to their irritant or corrosive effects on the skin [2]. Whilst the judicious use of *in vitro* test systems can provide a reasonable indication of irritation potential, those based on cell cultures have difficulty dealing with poorly soluble compounds and lack an important component: a fully developed and patent skin barrier layer (stratum corneum). In order to circumvent such technical obstacles, a hybrid system comprising excised human skin and cell culture is described in Chapter 25. Although the study can at best be considered preliminary, the initial results provide proof-of-principle for a hybrid test system and clearly merits further investigation.

The need to constantly challenge the adequacy of established guidelines through evidence-based studies is necessary for maintaining or improving the relevance of regulations or legislation primarily designed to protect human health. The study presented in chapter 26 provides evidence which both challenges the accuracy of an *in vitro* release test (IVRT) for predicting bioavailability and provides further evidence that human skin diffusion cell methodology may be acceptable for regulatory approval of generic products. Skin diffusion cells have been in use for some 60 years and have demonstrated reliability in a series of independent studies by indicating good *in vitro* – *in vivo* correlations. Moreover, there have been several inter-laboratory studies which have validated or characterised specific aspects of diffusion cell methodology [3, 4]. However, it is tempting to speculate that the fundamental issue preventing the international adoption of skin diffusion cell methodology for regulatory submissions is the lack of rigorous validation process, as has been the case for alternative methods developed within the past two decades [5].

REFERENCES

1. A. Wooley, Regulatory dermatotoxicology and international guidelines, in *Principles and practice of skin toxicology*, ed R. P. Chilcott and S. Price, John Wiley & Sons Ltd, Chichester, 2008, chapter 19, pp. 333–346.

2. P. Jones, *In vitro* alternatives for irritation and corrosion assessment, in *Principles and practice of skin toxicology*, ed R. P. Chilcott and S. Price, John Wiley & Sons Ltd, Chichester, 2008, chapter 19, pp. 333–346.

3. R. P. Chilcott, N. Barai, A. E. Beezer, S. I. Brain, M. B. Brown, A. L. Bunge, S. E. Burgess, S. Cross, C. H. Dalton, M. Dias, A. Farinha, B. C. Finnin, S. J. Gallagher, D. M. Green, H. Gunt, R. L. Gwyther, C. M. Heard, C. A. Jarvis, F. Kamiyama, G. B. Kasting, E. E. Ley, S. T. Lim, G. S. McNaughton, A. Morris, M. H. Nazemi, M. A. Pellett, J. Du Plessis, Y. S. Quan, S. L. Raghavan, M. Roberts, W. Romonchuk, C. S. Roper, D. Schenk, L. Simonsen, A. Simpson, B. D. Traversa, L. Trottet, A. Watkinson, S. C. Wilkinson, F. M. Williams, A. Yamamoto and J. Hadgraft, Inter- and intralaboratory variation of in vitro diffusion cell measurements: an international multicenter study using quasi-standardized methods and materials, *J Pharm Sci.*, 2005, **94**(3), 632–8.

4. J. J. van de Sandt, J. A. van Burgsteden, S. Cage, P. L. Carmichael, I. Dick, S. Kenyon, G. Korinth, F. Larese, J. C. Limasset, W. J. Maas, L. Montomoli, J. B. Nielsen, J. P, Payan, E. Robinson, P. Sartorelli, K. H. Schaller, S. C. Wilkinson and F. M. Williams, In vitro predictions of skin absorption of caffeine, testosterone, and benzoic acid: a multi-centre comparison study, *Regul Toxicol Pharmacol.*, 2004, **39**(3), 271–81.

5. M. Bouvier d'Yvoire, S. Bremer, S. Casati, M. Ceridono, S. Coecke, R. Corvi, C. Eskes, L. Gribaldo, C. Griesinger, H. Knaut, J. P. Linge, A. Roi and V. Zuang, ECVAM and new technologies for toxicity testing, *Adv Exp Med Biol.*, 2012, **745**, 154–80, DOI: 10.1007/978-1-4614-3055-1_10.

24

SYSTEMIC TOPICAL DRUG DELIVERY: US REGULATORY PERSPECTIVES ON PRODUCT DEVELOPMENT, BIOPHARMACEUTICALS AND CLINICAL PHARMACOLOGY

T Ghosh

Office of New Drugs and Quality Assessment, Center for Drug Evaluation and Research, Food and Drug Administration, Silver Spring, MD, United States of America.

1 INTRODUCTION

Topical Drug Delivery which encompass both dermal (local) and transdermal (systemic) products is one of the most exciting and challenging areas of pharmaceutical research. Traditionally, topical drug delivery systems denote formulations that are intended to deliver drugs locally rather than systemically. These products differ from transdermal delivery systems, which are designed for drugs to penetrate through the skin and provide systemic action. Whether it is topical or transdermal, effectiveness of the formulation depends on the release of the active drug from its dosage form to the skin surface. Once the drug is presented to the skin surface, either it stays mostly on the skin surface or penetrates through the stratum corneum to other cutaneous layers all the way up to systemic circulation to manifest its pharmacological action depending upon the location of the disease to be treated.

While topical dermal systems are mostly used to treat superficial and local skin diseases, with the right combination of high permeability of a drug substance and right choice of drug product formulation, a topical product can be formulated specifically for systemic delivery of the drug molecules. Few such systemic topical products are currently on the US market. They are known as "Systemic Topical Dosage Form (STDF)" though their mode of action is similar to transdermal patches which deliver the drug molecules across the skin to the systemic circulation. This chapter will deal with these STDF products, all of which are currently available as hydro-alcoholic gel based systems (Table 1).

The transdermal route of administration avoids the discomfort and other problems associated with injections, as well as avoiding first pass hepatic metabolism associated with oral administration. However, the limitations of occlusive patch formulations currently available have problems related to skin adhesion, skin rash and irritation (resulting from high concentrations of drug and solvents or excipients in an occlusive patch).

Table 1: *US Approved Topical Gels for Systemic Use. API; active pharmaceutical ingredient. NDA #; New drug application number.*

Trade Name	API	Dosage Form	Sponsor	NDA #	Indication	Usage
Gelnique	Oxybutynin chloride	10% Gel	Watson	22204	Treatment of overactive bladder	Apply contents of one sachet (1 gram unit dose (1.14 mL) 100 mg/g oxybutynin chloride) of GELNIQUE once daily
Anturol	Oxybutynin	3% Gel	Arrow	202513	Treatment of overactive bladder	Apply three pumps of ANTUROL (84 mg) once daily
Divigel	Estradiol	0.1% Gel	Upsher Smith	22038	Treatment of moderate to severe vasomotor symptoms associated with menopause	Apply one unit dose packet (0.25, 0.5, or 1.0 mg Estradiol) each day
Elestrin	Estradiol	0.06% Gel	Azur Pharma	21813	Treatment of moderate to severe vasomotor symptoms associated with menopause	Daily administration of 0.87 or 1.7 g of ELESTRIN with a metered dose pump which delivers 0.87 g containing 0.52 mg of estradiol.
Estrogel	Estradiol	0.06% Gel	Ascend	21166	Treatment of moderate to severe vasomotor symptoms associated with menopause	Daily administration of 1.25-g doses containing 0.75 mg of estradiol
Testim	Testosterone	1% Gel	Auxilium	21454	Testosterone replacement therapy in adult males for conditions associated with a deficiency or absence of endogenous testosterone	Daily administration of 5 g/10 g of gel (5g/one tube containing 50 mg of testosterone)
AndroGel	Testosterone	1% Gel	Abbott	21015	Testosterone replacement therapy in males for conditions associated with a deficiency or absence of endogenous testosterone	5 g for adult males, applied topically once daily 2 x 75 g pumps (each pump dispenses 60 metered 1.25 g doses) 2.5 g packet or 5 g packet
AndroGel	Testosterone	1.62% Gel	Abbott	22309	Testosterone replacement therapy in males for conditions associated with a deficiency or absence of endogenous testosterone	Starting dose of AndroGel 1.62% is 40.5 mg of testosterone (2 pump actuations), applied topically once daily in the morning
FORTESTA™	Testosterone	2 % Gel	Endo	21463	Testosterone replacement therapy in males for conditions associated with a deficiency or absence of endogenous testosterone	Starting dose of FORTESTA is 40 mg of testosterone (4 pump actuations) applied topically once daily in the morning

Most of the STDF products currently available are in the gel formulation which achieve required plasma concentrations (mid to high range) but have not been associated with skin irritation or rashes and without the problem of maintenance of adhesion. Moreover, these products provide an easier, more controllable method for individualising drug therapy when compared to currently marketed transdermal patches. In essence, STDF products achieve the advantages of transdermal delivery of drug molecules without the drawbacks associated with the occlusive patch products. However, these STDF products have their inherent problems of being greasy, with higher potential of person-to-person transfer.

2 PRODUCT DEVELOPMENT

Though these topical gels are for systemic drug delivery, most of the development, manufacturing and testing steps for these products are similar to the steps in developing traditional topical products. The product development of a topical formulation generally includes the following stages:

- Selection of drug and/or appropriate salt form.
- Selection of the dosage form (e.g., cream, lotion, ointment, aerosol, solution, etc.).
- Preparation of prototype formulations and testing their cosmetic/aesthetic qualities.
- Development of analytical methods to assay drug in the formulations and skin layers.
- Evaluation of *in vitro* and *in vivo* skin penetration.
- Evaluation of cutaneous toxicity (i.e., sensitisation, photo-sensitisation, irritation, etc.).
- Microbial and preservative testing of selected formulations.
- Phase I, II, and III clinical studies.
- Scale-up activities including development of appropriate *in vitro* testing and specifications.

The design of any topical delivery system usually requires the formulator to make a choice of the dosage form. The choice of dosage form will depend on several factors, including the physicochemical characteristics of the drug, stability of the drug in the formulation, available manufacturing equipment and cost constraints. Topical formulations can be liquids, semisolids with multiple phases containing special internal structures, nanoparticles, vesicles (liposomes), microspheres/ microcapsules, etc. They are available in the dosage forms of solutions, suspensions, lotions, creams, ointments, gels, foams, sprays, wipes, sponges, swabs, patches etc.

Currently all of the approved systemic topical products are in the hydro-alcoholic gel formulation. Therefore, this chapter will focus on the development of systemic topical hydro-alcoholic gel products.

Gels are semisolids consisting of a suspension of small inorganic particles or large organic macromolecules that are interpenetrated with liquid. Such systems usually contain water-based or alcohol-based cosolvent systems with a thickening agent based on cellulose derivatives, polysaccharide polymers, or acrylate polymers. The nature of the solvent determines whether the gel is a hydrogel (water based) or an organogel (nonaqueous solvent). In addition to the drug, topical formulations contain a variety of excipients. These are either functional excipients that help to stabilise the drug/formulation or cosmetic excipients that improve esthetic properties of the formulation and aid in application to the skin. Some examples of excipients include surfactants, solvents, oils, waxes, thickeners, emulsifiers, suspending agents, gelling agents, anti-oxidants, preservatives, etc. Many are mixtures containing polymeric materials with internal structures.

The viscosity of gels decreases upon application of shear (such as squeezing from a tube) or with an increase in temperature. Gels may exhibit Newtonian, plastic, or pseudoplastic behavior, depending on the nature of the gelling agent.

Overall Product Development of the systemic gels, like any other dosage form, focuses primarily on (i) drug substance, (ii) drug product, (iii) manufacturing process, (iv) container closure systems and (v) evaluation of topical products:

i.   *Drug Substance*

For the drug substance, also known as active pharmaceutical ingredient (API), the applicant needs to address the following questions (which may not be all inclusive) before embarking on the drug product development:

- Is the API a new molecular entity (NME)? If so have the pharmacology/toxicology properties been properly evaluated?

- What is the nomenclature, molecular structure, molecular formula, and molecular weight of the API? How was the API structure elucidated and characterised?

- What are the physicochemical properties? For example, physical description, pKa, polymorphism, aqueous solubility (as function of pH), hygroscopicity, melting points, and partition coefficient.

- How do the manufacturing processes and controls ensure consistent production of the drug substance?

- What are the drug substance specifications? Generally API specifications include, but not limited to, description, identification, melting range, specific rotation, loss on drying, assays, impurities, and residual solvents,

- What drug substance stability studies support the retest or expiration date and storage conditions for the drug substance?

ii.   Drug Product

Once the API has been properly characterised, the applicant needs to focus on the type of dosage form they want to pursue as the final product. Many drugs are insoluble and are suspended in the formulation. Therefore, the solubility and particle size of the drug have implications on the clinical efficacy of the product. Solubility experiments need to be performed in the formulation development stage. Depending on the efficacy of the product, solubility may have to be modified with the use of solubilisers or penetration enhancers.

In this case, it is assumed that based on target product profile as well as desired pharmacokinetic profile of the API, a gel product has already been chosen as the to be marketed pharmaceutical dosage form. In light of this, the applicant needs to address the following areas (may not be all inclusive):

- Will it follow the New Drug (NDA) or Generic (ANDA) route?
    - If it is for an NDA, will it be first topical formulation for the API?
    - If it is for an ANDA, do the differences between the proposed formulation and the reference listed topical drug product (RLD) present potential concerns with respect to therapeutic equivalence?

- Which properties or physical chemical characteristics of the drug substance affect drug product development, manufacture, or performance?

- What are the components and composition of the final product?

- What are the function(s) of each excipient?

- Does any excipient exceed the allowable ICH limit for this route of administration?

- What evidence supports compatibility between the excipients and the drug substance?

- What are the specifications for the inactive ingredients and are they suitable for their intended function?

- What critical quality attributes (CQAs) should the drug product possess? How was the drug product designed to have these attributes? Specifically, maintenance of a desired viscosity and pH in the hydro-alcoholic gel keeping the API dissolved in the formulation.

- Is the formulation as close as possible to the reference listed drug (RLD) in terms of appearance, consistency and other CQAs?

- Is the *in-vitro* absorption profile of the proposed and the RLD products similar?

- Is the formulation optimised?

- Is the Quality by Design (QbD) approach applied?

- What are the drug product specifications at release as well as throughout stability to assure batch to batch uniformity and quality product performance throughout the shelf-life?

- What are the justifications for the proposed acceptance criteria?

- What are the specifications for stability studies, including justification of acceptance criteria that differ from the drug product release specification?

- What drug product stability studies support the proposed shelf life and storage conditions?

- What is the post-approval stability protocol?

iii.   Manufacturing Process Development

Following the development of the desired product, the applicant needs to focus on the manufacturing process development for scale up as well as production of commercial batches. The following issues, which may not be all inclusive, are important to be addressed during the manufacturing process development:

- Why was the manufacturing process selected for this drug product?

- How are the manufacturing steps (unit operations) related to the drug product quality?

- How were the critical process parameters identified, monitored, and/or controlled? Is any real time release testing (RTRT) in place?

- What is the scale-up experience with the unit operations in this process?

- Is the scale-up successful?

iv.   Container Closure System

Generally, topical gel products are available in tubes, sachets and pumps for single-use or multiple-use purposes. In finalising the Container Closure System, among others, it is important for the applicant to address the following:

- What specific container closure attributes are necessary to ensure product performance?

- Is that adequately addressed?

v.   Evaluation of Topical Products

Procedures and acceptable criteria for testing topically applied drug products can be divided into those that assess general product quality attributes and those that assess product performance. The product quality attributes include the following: description, identification, assay (strength), impurities, physicochemical properties, uniformity of dosage units, water content, pH, apparent viscosity, microbial limits, antimicrobial preservative content, antioxidant content, sterility (if applicable) and other tests that may be product specific. Product performance testing assesses drug release and other attributes that affect drug release from the finished dosage form.

3 PRODUCT QUALITY TESTS

Universal tests (see ICH Guidance Q6A—Specifications: Test Procedures and Acceptance Criteria for New Drug Substances and New Drug Products: Chemical Substances, available at www.ich.org) are listed below and are applicable to all topically applied drug products:

- Description: A qualitative description of the drug product including Color, Odor, and Appearance should be provided. The acceptance criteria should include the final acceptable appearance of the finished dosage form and packaging. The description should specify the content or the label claim of the article. This is part of the manufacturer's specification for the drug product.

- Identification: Identification tests should establish the identity of the drug or drugs present in the product and should discriminate between compounds of closely related structures that are likely to be present. Identity tests should be specific for the drug substance(s).

- Assay and Content Uniformity: Assay and content uniformity should be performed as an integral part of the testing of topical products. Content uniformity measurements can provide vital information on flocculation and particle size distribution of the drug in the formulation. The assay provides information on the stability of the drug in the formulation. A specific and stability-indicating test should be used to determine the strength (content) of the drug product.

- Impurities: Process impurities, synthetic by-products, impurities associated with the polymers (e.g., residual monomers), residual solvents, heavy metals, and other inorganic and organic impurities may be present in the drug substance and excipients used in the manufacture of the drug product and should be assessed and controlled. Impurities arising from the degradation of the drug substance and those arising during the manufacturing process of the drug product should also be assessed and controlled.

In addition to the universal tests listed above, the following specific tests should be considered on a case-by-case basis.

- Uniformity of Dosage Units: This test is applicable for dosage forms packaged in single-unit containers.

- Water Content: A test for water content should be included when appropriate. This test is generally formulation dependent.

- Microbial Limits: Microbial examination of nonsterile drug products is performed when appropriate. Acceptance criteria for nonsterile pharmaceutical products based on total aerobic microbial count (TAMC) and total combined yeasts and molds count (TYMC).

- Antimicrobial Preservative Content: Acceptance criteria for antimicrobial preservative content in multidose products should be established.

- Antioxidant Content: If antioxidants are present in the drug product, tests of their content should be established unless oxidative degradation can be detected by another test method such as impurity testing. Acceptance criteria for antioxidant content should be established. They should be based on the levels of antioxidant necessary to maintain the product's stability at all stages throughout its proposed usage and shelf life.

- pH: The skin has a pH of 4 – 6, and many topical products are designed to be in that pH range. The pH of the product can influence not only the solubility of the drug in the formulation, but may also affect its potential to cause skin irritation. Changes in pH throughout the shelf life of the product may also be indicative of stability problems and should be carefully monitored. When applicable, topically applied drug products should be tested for pH at the time of batch release and at designated stability time points for batch-to-batch monitoring. Because some topically applied drug products contain very limited quantities of water or aqueous phase, pH measurements may not always be warranted. This test is generally formulation dependent and is part of the manufacturer's specification for the drug product.

- Particle Size Distribution: An increase or decrease in the particle size of the drug in a formulation can affect its in vitro release and subsequently its bioavailability. Particle size and/or particle size distribution of the drug may change with time and should be monitored throughout the shelf life of the product. Such measurements should be an integral part of the scale-up process, and changes in the particle size from batch to batch should be monitored. The particle size distribution of the active drug substance(s) in topically applied drug products is usually determined and controlled at the formulation development stage. However, topically applied drug products should be examined for evidence of particle size alteration (i.e., appearance of particles, changes in particle form, size, shape, habit, or aggregation) of the active drug substance that may occur during the course of product processing and storage. Such examinations should be conducted at the time of batch release and at designated stability test time points for batch-to-batch monitoring because changes that are visually (macro- and microscopically) observable would likely compromise the integrity and/or performance of the drug product. These types of testing are generally formulation dependent and are part of the manufacturer specification for the drug product.

- Apparent Viscosity: Viscosity measures the flow characteristics of a topical formulation. Changes in viscosity of the product are indicative of changes in the stability or effectiveness of the product. The apparent viscosity of semisolid drug products should be tested at the time of batch release and initially at designated stability test time-points to set specifications for batch-to-batch and shelf life monitoring. Apparent viscosity specifications based on data obtained during product development and shelf life testing should be established for batch release and throughout their proposed shelf life. Consideration should be given to specifying thixotropic behavior (both loss of viscosity and recovery time) from both the quality control and clinical perspectives. While the viscoelastic performance of these products may change over time, it must be recognised that a single specification is what is approved. However, a more stringent in-process (or in-house) release critera may be implemented to assure that performance stays within the approved specification over shelf life.

- Uniformity in Containers: Topically applied semisolid drug products may show physical separation during manufacturing processes and during their shelf life. To ensure the integrity of the drug product, it is essential to evaluate the uniformity of the finished product at the time of batch release and throughout its assigned shelf life.

4 BIOPHARMACEUTICS

In addition to product performance tests identified above, other specific tests may be required. These include (i) *in vitro* release testing (to assess the release of a drug from the product formulation) and (ii) *in vitro* skin permeation testing.

i. *In Vitro Release Testing (IVRT)*

In vitro release testing is performed for semisolid drug products to assess the rate and extent to which a drug diffuses out from a pharmaceutical dosage form and so reflects the combined effect of several factors (Figure 1).

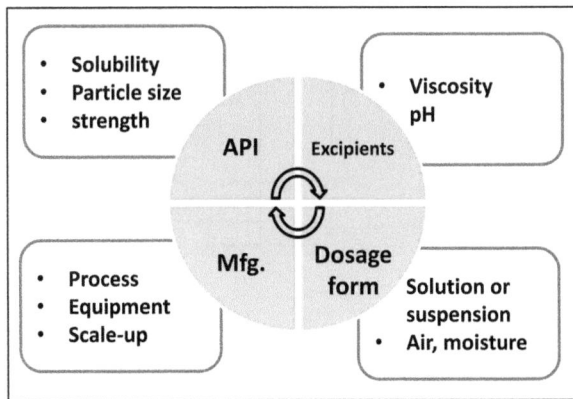

Figure 1 *Some factors which may have a combined effect of on in vitro release testing.*

In vitro release testing does not, however, directly predict the *in vivo* performance of drugs, as the primary factor that impacts bioavailability and clinical performance are the barrier properties of the epithelia to which the product is applied (epidermal or mucosal tissues). Although product performance tests do not directly measure bioavailability and relative bioavailability (bioequivalence), they can be used as a quality assurance tool to compare differences between formulations, minor differences in formulation composition, changes in particle size, and changes in manufacturing site. These changes may arise from changes in physicochemical characteristics of the drug substance and/or excipients or to the formulation itself, changes in the manufacturing process, shipping and storage effects, aging effects, and other formulation and/or process factors.

At present, IVRTs are used as product performance tests for creams, ointments, lotions, and gels. Currently, these tests are performed using several different types of equipment, including the Franz cell, the enhancer cell, and a special cell used with USP Apparatus 4 (Figure 2). In these methods, a thick layer of the semisolid product under evaluation is placed in contact with dissolution (receptor) medium in a reservoir via an inert, highly permeable support membrane. Membranes are chosen to offer the least possible diffusional resistance and not to be rate controlling. The formulation can either be occluded or unoccluded and the amount applied may be a finite dose or an infinite dose. The amount of drug released from the formulation into a receptor medium maintained at a constant temperature is monitored over a period of time [1]. The amount of drug released generally over six hours is plotted against the square root of time following the Higuchi equation as shown below and release rate specifications are prepared around the variation in slopes and intercepts:

$$M = \sqrt{2 \times Q \times D_m \times C_s \times t} \qquad (1)$$

The synthetic membrane, receptor media, and the equipment should be validated for a particular formulation. The method used for *in vitro* release and the analytical method used for assaying the drug should be simple, reliable, reproducible, sensitive and specific.

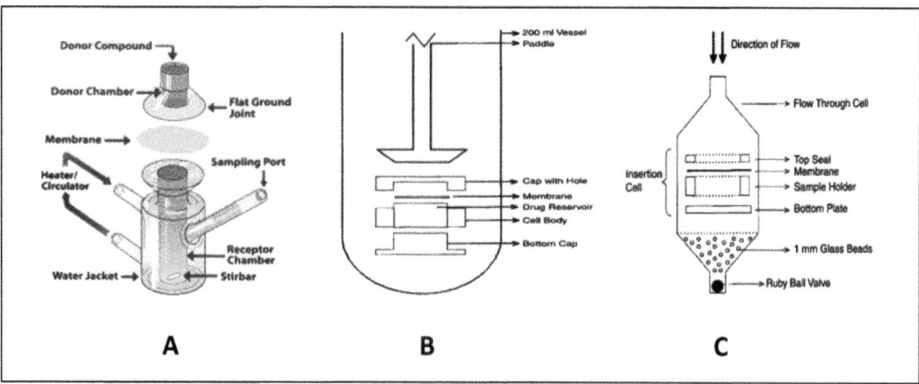

Figure 2 *Examples of in vitro release testing apparatus. A; Franz cell, B; Enhancer cell and C; Flow through cell.*

The major components of the development of an IVRT include the conditions under which the experiment should be performed including the selection of an appropriate membrane, receptor fluid and assay (Figure 3).

An IVRT needs to be developed keeping in mind that the test should assure batch-to-batch uniformity and be applicable to all products that are (and will be) marketed. It also should be discriminating enough to detect manufacturing changes that (may) influence product performance and where possible, should have proven relevance to *in vivo* product performance. Once an IVRT method is developed, it needs to be properly validated.

An IVRT Development Report should contain the rationale for the selection of the appropriate parameters for the proposed release method which should include (but are not limited to):

- Equipment/apparatus,
- Membrane
- Media (pH)
- Speed, temperature
- Assay
- Sink conditions, etc.

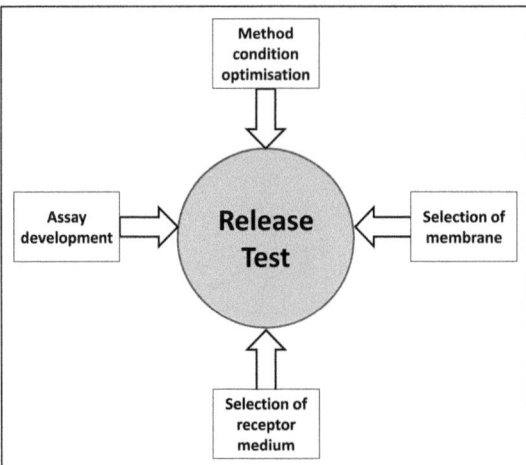

Figure 3 *Primary components of an in vitro release test system.*

Similarly, the developed IVRT method needs to be validated for parameters or factors such as (but not limited to):

- Precision
- Accuracy/Sameness
- Robustness
- Dose proportionality
- Sensitivity to changes in
    - Excipient type
    - Amount of Excipient
    - Size of Batch
    - Method of manufacture
- Mass Balance
- Analytical
- Back Diffusion of Alcohol in the Dosage Form
- Dependence of Release Rate on Temperature
- Differences in Instrumentation

The overall utility of IVRT can impact various areas including product development, quality control, post-approval changes and other areas (Figure 4). However, the current regulatory utility of the IVRT is limited mostly to determine the acceptability of minor process and/or formulation changes and stability testing in approved semisolid dosage forms in which any difference in delivery rate is undesirable. The evaluation or comparison of large formulation changes may provide meaningless results, unless extensive validation is performed to select test parameters that ensure that the sensitivity of the test is meaningfully correlated with *in vivo* performance.

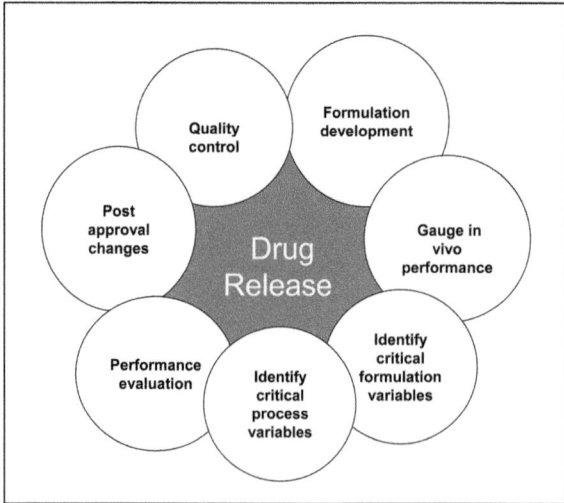

Figure 4 *Product development areas which may benefit from optimised in vitro release rate testing.*

Overall, there is a need to be cognisant that while IVRT is a useful test to assess product "Sameness" under scale up and post approval changes for semisolid products, it is neither as convincing as that for *in vitro* dissolution test as a surrogate for *in vivo* bioavailability of solid oral dosage forms, nor alone can it be a surrogate test for *in vivo* bioavailability or bioequivalence for topical products. Also, IVRT should not be used for comparing different formulations across manufacturers.

In summary, it appears that an IVRT method should be developed and validated to propose specification for routine batch to batch quality control of drug products. Once developed, validated and accepted by the Agency, the same method can be used for future post approval changes which in turn will be cost- and time-effective in the long run. Each manufacturer should establish/set specifications based on acceptable *in vivo* performance of the product from a batch approved by the Agency for marketing. It is also important to standardise reporting of acceptance criteria (i.e., *specification-sampling time points and specification values*) which should be uniform as the rate of release (slope value) with proper units ($mass/time^{1/2}$).

Release method development should start early in the IND phase for all topical submissions to propose release and stability specification in the NDA submission. There is a need to have the IVRT development and validation methodology standardised so that there will not be major differences among manufactures and/or products. While conducting and interpreting results of an IVRT study, we need to keep in mind that a failure does not necessarily signal that the formulation is clinically unsatisfactory (not efficacious or even different from the control product); it is, however, a call to look into the situation to identify the reason for the discrepant behavior as it may simply be an artifact of the test.

ii. In Vitro Skin Permeation

In Vitro Skin Permeation refers to the diffusion of active ingredients across the animal or human skin. For systemic topical products, *in vitro* skin penetration studies using a Franz cell can be performed during the formulation development stage to determine the amount of drug that penetrates into and permeates through the skin. Either animal or human cadaver skin can be used for this purpose. The formulation is applied to the skin surface and the amount of drug that permeates through the skin is measured by assaying the receptor solution. Removal of the skin layers by stripping and assay of the skin can provide valuable information on the mechanisms of skin penetration into the skin. However, no correlations can be made from this data regarding the efficacy of the product. It is not suitable for a Quality Control test due to the inherent high (skin) donor to donor variability. In essence, while in vitro skin permeation study results is a good tool for formulation development and screening, the regulatory utility of such results is currently limited.

5 BIOWAIVERS

A biowaiver is an exemption of clinical bioequivalence studies given to a drug product. Biowaiver requests for topical products are not that common as different strengths (e.g., 1% vs 5% API) of the product do not generally produce linear pharmacokinetics (PK), or pharmacodynamic (PD) relationship for topical products (for local as well as systemic formulations). However, a biowaiver may be requested for pre- and post-approval site and/or process and/or formulation changes based on IVRT data which will be subjected to the regulatory Agency's review.

6 CLINICAL PHARMACOLOGY

If the drug (also known as active pharmaceutical ingredient or API) is a New Molecular Entity (NME), full clinical pharmacology (CP) characterisation of the drug in addition to the topical product delivery characterisation is warranted. Examples of full CP characterisation include but not limited to evaluation of:

- Systemic Metabolism
- Exposure response analysis for efficacy and safety
- Drug drug interaction potential (Effect of test drug on concomitant drugs and vice versa)
- QT interval prolongation
- Special populations (Hepatic impairment, renal impairment etc)

However, if the drug is a previously approved entity and is otherwise well characterised from a CP perspective, assessment will involve (but not limited to):

- Relative bioavailability
- Single and Multiple dose pharmacokinetics(PK) and/or pharmacodynamics (PD)
- PK and/or PD of topical system under external factors that the system may be subjected during usage
- What extrinsic factors (drugs, herbal products, diet, smoking, and alcohol use influence exposure (PK usually) and/or response, and what is the impact of any differences in exposure on efficacy or safety responses?
- What dosage regimen adjustments, if any, are recommended for each of these groups?

CP studies specific to systemic gels include (but not limited to) the evaluation of impact of any differences in exposure on efficacy or safety responses for the following scenarios. The applicants are encouraged to design appropriate studies to address these issues.

- Differences in metabolism between skin versus oral route may result in differing
  - Parent/metabolite ratios
  - Drug-drug interaction profile

- Substantially different absorption and elimination characteristics
  - For lipophilic drugs (due to the depot effect);
    -Time to peak concentration can be relatively longer
    -Longer terminal half-life may be observed upon removal of patch

- Effect of external factors:
  - Potential for drug transfer where skin-to-skin contact with another person is possible
  - Heat pad, electric blankets, sunbathing, heat or tanning lamps, sauna, hot tubs or hot baths
  - Effect of sunscreen or moisturising lotion on absorption
  - Effect of Showering
  - Effect of exercise
  - Effect of site of application on the PK (outer arm, inner arm, chest, back, thigh, abdomen, shoulder etc.)

  - Safe time interval between reapplication to same site for consistent drug absorption

Based on the outcome of the above studies, recommendations for any dosage adjustments in general or to a specific population and/or specific usage instructions via labeling will be made. Information excerpted from the labels of two US approved systemic transdermal gels are provided in the following section for ready reference.

7 EXAMPLES OF DOSE ADJUSTMENT RECOMMENDATIONS

i.   Information from [AndroGel (Testosterone) 1.62%] Label:

Application during use of sunscreen or moisturising lotion: In a randomised, 3-way (3 treatment periods without washout period) crossover study in 18 hypogonadal males, the effect of applying a moisturising lotion or a sunscreen on the absorption of testosterone was evaluated with the upper arms/shoulders as application sites. For 7 days, moisturising lotion or sunscreen (SPF 50) was applied daily to the AndroGel 1.62% application site 1 hour after the application of AndroGel 1.62% 40.5 mg. Application of moisturising lotion increased mean testosterone Cavg and Cmax by 14% and 17%, respectively, compared to AndroGel 1.62% administered alone. Application of sunscreen increased mean testosterone Cavg and Cmax by 8% and 13%, respectively, compared to AndroGel 1.62% applied alone.

Application during showering: In this randomised, 3-way (3 treatment periods without washout period) crossover study in 24 hypogonadal men, on the 7th day of each treatment period, men took a shower with soap and water at either 2, 6, or 10 hours after drug application. The effect of showering at 2 or 6 hours post-dose on Day 7 resulted in 13% and

12% decreases in mean C_{avg}, respectively, compared to Day 6 when no shower was taken after drug application. Showering at 10 hours after drug application had no effect on bioavailability. The amount of testosterone remaining in the outer layers of the skin at the application site on the 7th day was assessed using a tape stripping procedure and was reduced by at least 80% after showering 2-10 hours post-dose compared to on the 6th day when no shower was taken after drug application.

Effect of Application Site: AndroGel 1.62% was evaluated in a multi-center, randomised, double-blind, parallel-group, placebo-controlled study (182-day double-blind period) in 274 hypogonadal men. Patients were randomised to receive active treatment or placebo using a rotation method utilising the abdomen and upper arms/shoulders for 182 days. Absorption of testosterone from AndroGel 1.62% from upper arms/shoulders were comparable to when AndroGel was applied using a rotation method utilising the abdomen and upper arms/shoulders (Figure 5).

In absence of any specific instruction in the label, it appears that application of moisturising lotion or sunscreen (SPF 50) or application site does not have a significant effect on systemic exposure of testosterone from AndroGel 1.62%. Also, no specific instruction based on the results of this showering study is included in the label signifying that decrease in systemic testosterone level after steady state (Day 7 or beyond) due to showering may not be clinically relevant. Comparison of the systemic level and amount remaining in the outer layers signifies that amount remaining on the outer layer does not correlate with the systemic absorption.

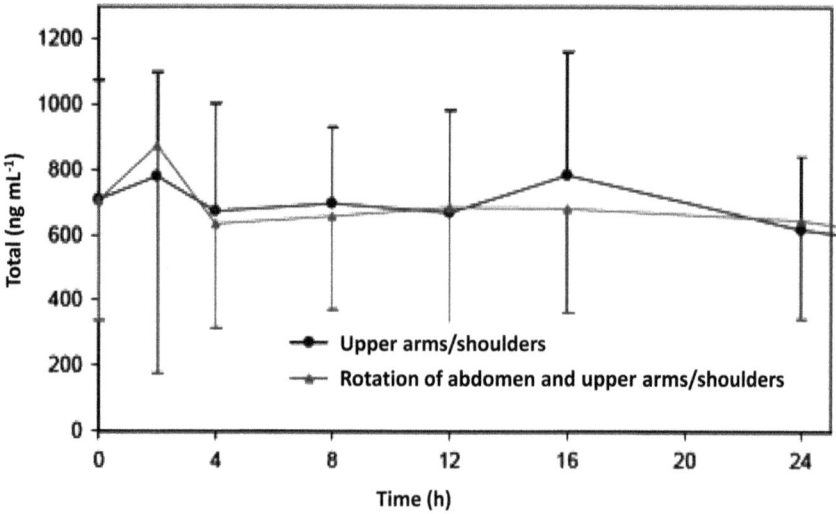

Figure 5 *Effect of application site and method of application on plasma levels of testosterone following topical administration of AndroGel™.*

In absence of any specific instruction in the label, it appears that application of moisturising lotion or sunscreen (SPF 50) or application site does not have a significant effect on systemic exposure of testosterone from AndroGel 1.62%. Also, no specific instruction based on the results of this showering study is included in the label signifying that decrease in systemic testosterone level after steady state (Day 7 or beyond) due to showering may not be clinically relevant. Comparison of the systemic level and amount remaining in the outer layers signifies that amount remaining on the outer layer does not correlate with the systemic absorption.

ii.   *Information from [Anturol (Oxybutynin) 3% Gel] Label:*

Application during use of sunscreen: 30 minutes before or 30 minutes after Anturol application was evaluated in a single-dose randomised crossover study (N=20). Concomitant application of sunscreen, either before or after Anturol application, had no effect on the systemic exposure of oxybutynin.

Effect of showering: The effect of showering on the absorption of oxybutynin was evaluated in a randomised, steady-state crossover study under conditions of no shower, or showering 1, 2 or 6 hours after Anturol application (N=22). The results of the study indicate that showering one hour after administration does not affect the overall systemic exposure to oxybutynin.

Effect of application site: The pharmacokinetic parameters and mean plasma concentrations during a randomised, crossover study of the three recommended application sites in 25 healthy men and women are shown in Table X and Figure Y, respectively. Absorption of oxybutynin is similar when Anutrol is applied to the abdomen, upper arm/shoulders or thighs. Pharmacokinetic Parameters and profiles (mean values) for Oxybutynin (84 mg/day) 3% Gel are described in Table 2

In the absence of any specific instruction in the label, it appears that application of sunscreen or showering or application site does not have significant effect on systemic exposure of oxybutynin from Anturol 3% gel.

Table 2 *Effect of application site on pharmacokinetic parameters for oxybutymim administered from Anturol Gel in human volunteers.*

Application Site	AuC_{0-t} (ng h^{-1} mL^{-1})	C_{max} (ng mL^{-1})	T_{max} (h)
Abdomen	284	6	24
Thigh	286	5	36
Upper arm/shoulder	329	9	24

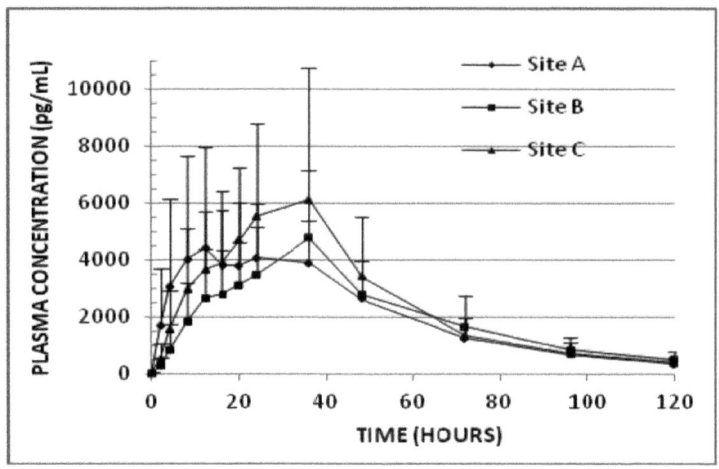

Figure 6 *Plasma concentration of oxybutynin following application of Anturol at three different sites on human volunteers.*

Sometimes, population PK/PD component may be added to the phase III safety and efficacy studies to gain further insight into effect of intrinsic factors such as age, gender, race, concomitant medications, hepatic and renal impairment. In addition to general safety studies, safety studies assessing dermal irritation and photosensitisation of the transdermal gel is required.

In summary, based on systemic exposure information from other dosage forms, some CP information for transdermal gels may be borrowed from previously approved products without repeating the studies. In general, assessment of PK/PD of the drug from the transdermal gel may facilitate the risk/benefit assessment and for informed labeling. Also, systemic topical gel specific studies as exemplified in the previous section (moisturising lotion, showering, application site etc.) are useful for informed labeling language.

8 UNIQUE CONCERN FOR GELS: PERSON-TO-PERSON TRANSFERENCE

Often, a key concern with a topical product, especially with a hydro-alcoholic gel, is the potential for the active ingredient to be exposed to subjects other than the intended user. The mechanism of transference is generally via skin-to-skin contact, for example partners making physical contacts, caregivers administering medicines or a parent hugging a child. It can also extend to contamination of surface areas with which subjects may come in contact, including clothing and bedding. At this time, there are no specific guidance documents on transference. However, the FDA and other Regulatory Agencies expect that data from properly designed transference studies be submitted for any dermal products for which there is a possibility of transference. That specifically includes systemic topical gels containing APIs like testosterone, estrogen and progestins which have high permeability through skin and as such transference could create an acute or chronic hazard to unintended human subjects. Examples of transference data can be found in the package inserts for all US approved systemic gel products such as AndroGel®, EstroGel®, Anturol® etc.

Unintended exposure via transference is usually evaluated using a clinical study with pharmacokinetic end points. The product is applied to the intended user and the potential for transference is determined by having an unintended user make contact by rubbing the area of application. Transference potential in terms of systemic exposure to the unintended person may be examined at short and long time points after application, and in the presence and absence of clothing. Appropriate labeling language is crafted based on the outcome of these studies.

The following section describes results from skin transference studies conducted for few systemic gel products as captured from the published labeling of the products.

Person-to-Person Transference (AndroGel 1.62%): The potential for testosterone transfer when it was applied only to upper arms/shoulders was evaluated in two clinical studies of males dosed with AndroGel 1.62% and their untreated female partners. In one study, 8 male subjects applied a single dose of 81 mg to their shoulders and upper arms. Two (2) hours after application, female subjects rubbed their hands, wrists, arms, and shoulders to the application site of the male subjects for 15 minutes. Serum concentrations of testosterone were monitored in female subjects for 24 hours after contact occurred. After direct skin-to-skin contact with the site of application, mean testosterone C_{avg} and C_{max} in female subjects increased by 280% and 267%, respectively, compared to mean baseline testosterone concentrations. In a second study, 12 male subjects applied a single dose of AndroGel 1.62% 81 mg to their shoulders and upper arms. Two (2) hours after application, female subjects rubbed their hands, wrists, arms, and shoulders to the application site of the male subjects for 15 minutes while the site of application was covered by a t-shirt. In this case, mean testosterone C_{avg} and C_{max} in female subjects increased by only 6% and 11%, respectively, compared to mean baseline testosterone concentrations.

As a result of these findings, the following Box Warning was included in the AndroGel 1.62% label.

WARNING: SECONDARY EXPOSURE TO TESTOSTERONE
- **Virilization has been reported in children who were secondarily exposed to testosterone gel (5.2, 6.2).**
- **Children should avoid contact with unwashed or unclothed application sites in men using testosterone gel (2.2, 5.2).**
- **Healthcare providers should advise patients to strictly adhere to recommended instructions for use (2.2, 5.2, 17).**

Also, it has been mentioned in the label that patients should wash hands immediately with soap and water after applying AndroGel 1.62% and cover the application site (s) with clothing after the gel has dried. Washing of the application site thoroughly with soap and water prior to any situation where skin-to-skin contact of the application site with another person is anticipated.

Person-to-Person Transference [Anturol (Oxybutynin) 3% Gel]: The potential for dermal transfer of oxybutynin from a treated person to an untreated person was evaluated in a single-dose study where subjects dosed with Anutrol engaged in vigorous contact with an untreated partner for 15 minutes, either with (N=14 couples) or without (N=14 couples) clothing covering the application area.

The untreated partners not protected by clothing demonstrated low detectable plasma concentrations of oxybutynin (mean C_{max} = 0.65 ng/mL). Only one of the 14 untreated subjects participating in the clothing-to-skin contact regimen had very low measurable oxybutynin plasma concentrations (C_{max} = 0.06 ng/mL) during the 24 hours following contact with treated subjects; oxybutynin was not detectable with the remaining 13 untreated subjects.

Regardless of the low exposure observed in this study, the label advises patients to cover the application site with clothing if skin-to-skin contact at the application site is anticipated and to wash hands immediately after product application.

9 SUMMARY

While topical dermal systems are mostly used to treat superficial and local skin diseases, with the right combination of high permeability of a drug substance and right choice of drug product formulation, a topical product can be formulated specifically for systemic delivery of the drug molecules. Few such STDFs are currently on the US market. They are topical products though their mode of action is similar to transdermal patches which deliver the drug molecules across the skin to the systemic circulation.

Whether it is topical or transdermal, effectiveness of the formulation depends on the release of the active drug from its dosage form to the skin surface. Once the drug is presented to the skin surface, either it stays mostly on the skin surface or penetrates through the stratum corneum to other cutaneous layers all the way up to systemic circulation to manifest its pharmacological action depending upon the location of the disease to be treated. Therefore, development of the right product with the drug molecule with appropriate physical-chemical and pharmacokinetic properties as well as optimisation of the product performance throughout the shelf-life is very important. This chapter outlined the general product development principles for these STDFs followed by biopharmaceutics and clinical pharmacology studies generally required to launch the product successfully in the US market with informed product labeling for the patient and the care-givers.

10 DISCLAIMER

The views expressed in this article are those of the author and do not reflect the official views of the FDA.

11 REFERENCE

1. M. Corbo, T. W. Schultz, G. K. Wong, and G. A. Van Buskirk, Development and Validation of In-Vitro Release Testing Methods for Semisolid Formulations, *Pharm. Technol.*, **17**(9), 112–128.

12 BIBLIOGRAPHY

Guidance for Industry: Nonsterile Semisolid Dosage Forms—Scale-Up and Postapproval Changes: Chemistry, Manufacturing, and Controls; In Vitro Release Testing and In Vivo Bioequivalence Documentation
(http://www.fda.gov/downloads/Drugs/GuidanceComplianceRegulatoryInformation/Guidance s/UCM070930.pdf)

Guidance for Industry: Guideline for Submitting Documentation for the Manufacture of and Controls for Drug Products
(http://www.fda.gov/downloads/Drugs/GuidanceComplianceRegulatoryInformation/Guidance s/UCM070630.pdf)

Guidance for Industry: Guideline for the Format and Content of the Human Pharmacokinetics and Bioavailability Section of an Application
(http://www.fda.gov/downloads/Drugs/GuidanceComplianceRegulatoryInformation/Guidance s/UCM072112.pdf)

USP Workshop on Topical and Transdermal Drug Products, Sept 14 – 15, 2009

Topically and Transdermal Drug Products—Product Quality Tests, Pharmacopeial Forum Vol. 37(4) [July–Aug. 2011]

Topical Drug Products - Performance Tests, Pharmacopeial Forum Vol. 37(4) [July–Aug. 2011]

Drugs@FDA: ANDROGEL 1.62% label http://www.accessdata.fda.gov/drugsatfda_ docs/label/2011/022309s000lbl.pdf)

Drugs@FDA: ANTUROL (oxybutynin) gel 3%, label (http://www.accessdata.fda.gov/ drugsatfda_docs/label/2011/202513s000lbl.pdf)

2011 FDA Drug Information Association (DIA) Meeting: Improved Development and Regulation of Transdermal Systems: *Presentations from the meeting can be purchased at* https://www.diahome.org/DIAHOME/Common/Templates/Login.aspx

25

CLASSIFICATION OF IRRITANT/ NON-IRRITANT AND CORROSIVE CHEMICALS USING EXCISED HUMAN SKIN *IN VITRO*

C A Stewart[1, 2], R B Turner[2], M B Brown[1, 2] and M J Traynor[1]

[1]School of Pharmacy, University of Hertfordshire, College Lane Campus, Hatfield, Herts, AL10 9AB, United Kingdom. [2]MedPharm Ltd, 50 Occam Road, Surrey Research Park, Guildford, Surrey, GU2 7AB, United Kingdom.

1 INTRODUCTION

Any chemical that is applied to the skin has the potential to induce a skin reaction. These reactions are complicated and are difficult to fully replicate *in vitro* and may exhibit species variations when reproduced *in vivo*. Skin irritation is the observed response (erythema, oedema, pain itching and heat) to chemicals that result in inflammation at the site of application. Skin irritation is defined as reversible damage of the skin following application of a chemical for up to 4 h [1]. By contrast, skin corrosion is defined as ''irreversible damage to the skin, namely visible necrosis through the epidermis and into the dermis, following the application of a test substance for a period of 3 min up to 4 h'' [2].

The skin is the largest human organ, with functions that far exceed its primary role as a protective barrier. The skin is the main target tissue for external toxins, provides protection from environmental hazards, UV-irradiation and water loss. The skin is composed of multiple layers serving multiple essential functions including metabolism of chemicals which come in contact with the skin. When the skin is exposed to xenobiotics, specific immunological and histological responses can occur. The skin is composed of three layers all of which are dominantly populated by keratinocytes therefore it is not surprising that keratinocytes are the key player in many skin functions including the immune responses resulting from chemical exposure. In response to physical or chemical stresses, keratinocytes produce and release cytokines including interleukins IL-1α, IL-8, IL-7, IL-15, tumour necrosis factor α (TNF-α), interferon induced protein 10 (IP-10), granulocyte/macrophage colony-stimulating factor (GM-CSF), transforming growth factor (TGF) and other signalling factors which initiate cutaneous inflammation [3].

The mechanism by which chemicals induce skin irritation can vary depending on the innate properties of the chemical. By far the most investigated mechanism by which chemicals initiate irritation is the mechanism that results from exposure to surfactants such as SDS, a commonly used model irritant [4, 5]. Surfactants initiate skin irritation by disruption of cellular membranes resulting in extracellular release of pro-inflammatory cytokines which normally reside within keratinocytes. This in turn results in an inflammatory cascade, from which skin irritation is the result [6]. A typical skin inflammatory response including the key cytokines and inflammatory mediators expressed during this reaction is depicted in Figure 1.

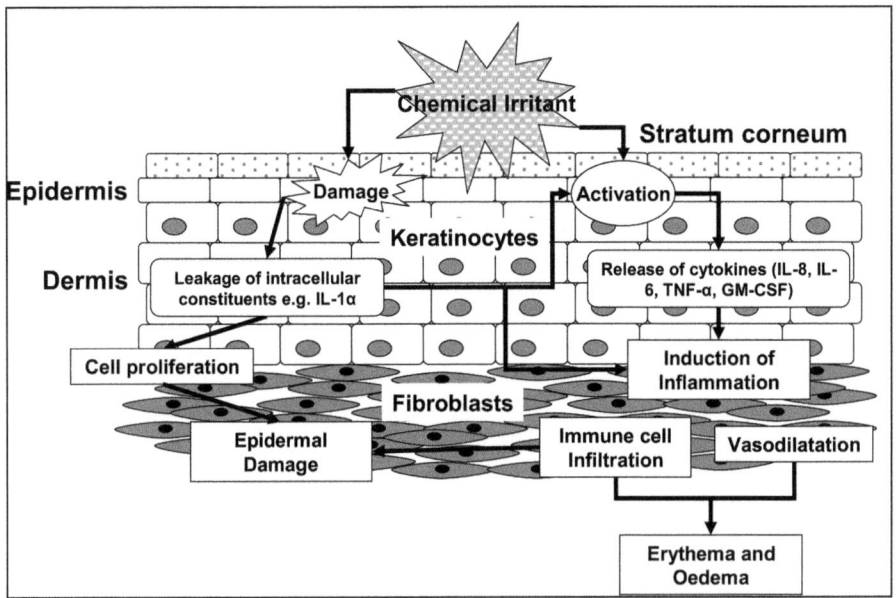

Figure 1 *Inflammatory responses in the skin.*

The potential for a chemical to cause skin irritation is an important safety consideration and therefore forms part of routine toxicological evaluation. Historically, the evaluation of the potential of a chemical to produce skin irritation has been carried out in rabbits using the Draize skin irritation test according to OECD TG 404 and Method B.4 of Annex V to Directive 67/548/EEC [7]. This involves the application of chemicals to the animal's skin and evaluation of visible changes such as erythema and oedema. The chemicals are then graded for erythema and oedema and a primary irritation index (PII) is calculated as shown in Equation 1.

$$\text{Primary Irritation Index (PII)} = \frac{\sum \text{erythema grades at 24/48/72/ h} + \sum \text{edema grades at 24/48/72/ h}}{3 \times \text{number of animals}}$$

(1)

Despite this method being the accepted regulatory technique for determining the irritancy of new chemicals it has come under a lot of scrutiny in recent years. From a scientific prospective the relevance of the rabbit test to estimate human skin irritation hazards has been seriously questioned due to species differences in both barrier properties and reactions to certain irritants [8]. Additionally, there is increasing pressure from a variety of sources to limit and replace this test particularly with regard to the testing of cosmetic products. This has lead to the development of a number of *in vitro* models to assess skin irritation *in vitro*, the predominate model type the Reconstructed Human Epidermis (RHE) models. Other possibilities for *in vitro* models which could be further investigated include keratinocyte cultures and *ex vivo* skin models.

In an attempt to develop new methods for assessment of the effect of chemicals on the skin, various reconstructed human equivalent models have been developed and used for skin irritancy testing including Skinethic RHE [9], EpiDerm [10], Episkin [11], Prediskin [10], RE-DED [12], Apligraf [13] Skin [14] and LSE / HSE [15]. There are two distinct types of models; the first are epidermal equivalents which are produced by culturing adult human keratinocytes on synthetic matrices. The second are full skin equivalents which consist of multilayed, differentiating human keratinocyte cultures grown on fibroblasts containing collagen matrices. Both model types use conditions which permit terminal differentiation and the reconstruction of an epidermis with a functional horny layer achieved by growing the tissue at the liquid air interface [6].

Three of these models (EpiDerm™ SIT (EPI-200), SkinEthic™ RHE and EpiSkin™) have validated protocols for the prediction of skin irritation [1] either as full replacements for animal testing or as a partial replacement test, within a tiered testing strategy depending on the specific country's regulations. The use of RHE models for the prediction of skin irritation testing involves the topical application of test materials to the surface of the epidermis, and the subsequent measurement of their cytotoxic effects, predominately by MTT assay [18, 19, 20].

Though the general structure, composition and biochemistry of these models bear a close resemblance to human skin, RHE models exhibit differences in terms of barrier function which leads to the potential for false positive results [19]. Several studies have investigated the reason for the deficit in barrier function, including the presence of unkeratinised microscopic foci [20] and the presence of desmosomal structures with preserved lamellar appearance demonstrating that the formation of corneosomes is compromised *in vitro* [21]. The weakness of the barrier function is improved in the full thickness models where a dermis is also present but despite this, deficiencies still persist in the barrier function of these models [22]. This means there is still both room for improvement in these models but also scope for development of new models.

Keratinocytes play a pivotal role in inflammatory responses in skin. A number of studies have investigated the possibility of using keratinocytes in culture as a model for the identification of dermal irritants. Eun *et. al.* (1994) compared the effects of irritants on human oral and skin keratinocytes by determining mitochondrial metabolism using the MTT assay and measuring membrane integrity by the LDH assay [23]. They showed that both assays were equally sensitive to the irritants tested. They concluded that skin keratinocytes would be a suitable model for evaluating if a chemical would by an irritant on either the skin or oral mucosa. Müller-Decker *et. al.* (1994) used pro-inflammatory mediators (IL-1α and proinflamatory eicosanoids) and cell viability in human keratinocytes to develop an assay for skin irritation [24]. They concluded that keratinocytes *in vitro* respond to chemicals of graded irritant potential with a graded release of proinflamatory mediators. They also suggested that for complete assessment of a compound, multiple mediator endpoints should be studied. Cohen *et. al.* (1991) studied the production and release of IL-1α and prostaglandin E2 (PGE2) in response to inflammatory stimuli following UVB irradiation and detergent injury by SDS [25]. They demonstrated that both inflammatory stimuli cause an increase in intracellular IL-1α and PGE2 levels followed by extracellular release. They suggested that the measurement of both markers both intra and extracellularly may provide an *in vitro* test to detect potentially irritant products. Gueniche and Ponec (1993) studied the effects of surfactants on keratinocytes and fibroblast cell cultures derived from human skin and SV40 transformed human keratinocytes (SVK14 cells) for *in vitro* screening of skin toxicity [26]. The results of this study showed that changes in the morphology and proliferation of cultured skin cells *in vitro* showed a similar ranking order to *in vivo* skin irritancy data and that nearly all of the surfactants tested increased IL-6 production. Together these studies show the enormous potential for using keratinocytes for the evaluation of potential skin irritants.

In addition to the cell-based systems already discussed, an alternative model for investigating skin irritation is using excised human skin cultures. An excised human skin culture has the advantages of the presence of differentiating keratinocytes and of the possibility of topical application of both water soluble and insoluble test compounds. The intact stratum corneum provides a physiological barrier between the chemical and living cells. In this organ culture model, *ex vivo* human skin discs are cultured on a micro porous membrane, which allows transport of culture medium through the dermis into the epidermis, whereas the epidermal side remains free of direct contact with culture medium. Therefore test substances can be applied, directly to the skin surface, in a manner similar to those encountered in *in vivo* exposure. Despite the potential of such a model there is little reference to this type of model using human tissue in the literature. However, there are references in the literature to animal derived skin irritancy cultures including rabbit skin [27] and skin from hairless mice [28]. Although these cultures may have their use, they have the same issues arising from *in vivo* animal testing, namely potential lack of correlation between animal and human skin and the ethical issue surrounding the use of animals/animal tissue for testing.

Pistoor *et. al.* (1996) investigated the potential use of human skin explant cultures for the identification of contact allergens [29]. In this study skin explants were dosed with two dermatological inactive compounds, five irritants and six contact allergens. Post treatment skin explants were immunohistochemically examined, and showed changes in immune cell distribution following exposure to contact allergens that were not seen following exposure to the other chemicals tested. Though this study did not deal directly with the potential of *ex vivo* skin for the identification of irritants, it did demonstrate the ability of *ex vivo* cultured skin to display pathophysiological responses to chemicals. Given the species specificity of such a model, this would in theory be an ideal *in vitro* method for irritancy testing.

Therefore, the aim of this study was to evaluate the use of both primary keratinocytes and *ex vivo* human skin to determine the irritant potential of chemicals.

2 METHODS AND MATERIALS

Keratinocytes Skin Irritation Model: HEKn cells were treated with chemicals corresponding to a wide range of primary irritation scores as determined by the Draize test (Table 1). The chemicals used in this study were primarily selected from the Chemicals Data Bank complied by the European Centre for Ecotoxicology and Toxicology of Chemicals [30]. Additional suitable compounds were selected from published literature [31]. HEKn cells were seeded into 96 well plates then allowed to adhere for 48h then the medium was replaced with medium containing the required treatments for 24h. The treatment concentrations used varied depending on the chemical be tested and were optimised to produce an IC_{50} value for each chemical. The viable cell number was assessed by MTT assay; average negative control values were subtracted and all results expressed as percentage of untreated control cell viability.

Excised Human Skin Model: Human testicular skin (ethics permission granted by South East London Research ethics committee 4; REC reference 10/H0807/51) was collected after elective surgery and transported on ice to the laboratory. The subcutaneous fat was mechanically removed taking care not to damage the skin, before punching full thickness skin discs. Skin explants were then placed dermal side down, mounted on inserts in a culture plate (see Figure 2), placed in media and cultured in an incubator at 37°C under 5% CO_2 humid atmosphere for the duration of the culture.

Table 1 *Summary of test compound properties, indicating risk phrases associated with the various compounds (R34: Causes burns, R35: Causes severe burns, R38: Irritating to skin). The primary irritation indices from Chemicals Data Bank complied by the European Centre for Ecotoxicology and Toxicology of Chemicals. (This table also summarises the data presented in Figures 5 and 6).*

R Phrase	Primary Irritation Index (PII)	Percentage Viability	IL-1α Release	IL-8 Release
NC	0.33	<50%	-	-
/	0.78	<50%	↑	↑
R38	2.92	<50%	↑	↑
R38	3.25	>50%	↓	↓
R38	3.63	>50%	↑	↑
R34	4	>50%	↓	↓
R34	4.11	<50%	↓	↓
/	4.33	<50%	↑	↑
R35	5.22	>50%	↓	↑
R38	6.78	>50%	↑	-

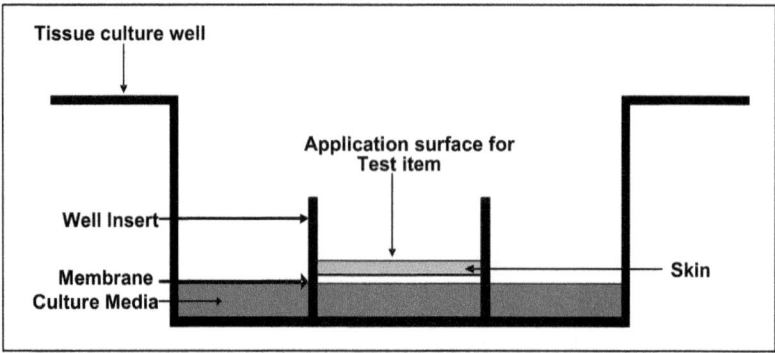

Figure 2 *Schematic representation of skin irritancy culture set-up.*

Ex Vivo Skin Irritation Assay. The test substance was applied to the epidermal side of the skin and incubated at room temperature. After a treatment period of skin discs were washed to remove all traces of test compounds. Subsequently the tissues were then placed in fresh culture medium and returned to the incubator for the post treatment incubation period at which point the conditioned media was removed for analysis of cytokines and an MTT assay was preformed. Production of IL-8 and IL-1α content was assessed by specific commercially available ELISA kits.

3 RESULTS AND DISCUSSION

When assessed with the HEKn keratinocyte model, the panel of test compounds displayed a wide range of IC_{50} values (0.026 to 67 mg ml^{-1}) with the most potent irritants displaying the lowest IC_{50} values (Figure 3A). The log IC_{50} values were then plotted against primary irritation scores in the available literature as determined by the Draize rabbit skin irritancy test (Figure 3B). The results showed a good correlation ($r^2 = 0.845$) between the primary irritation score and the IC_{50} values determined in this cell based model.

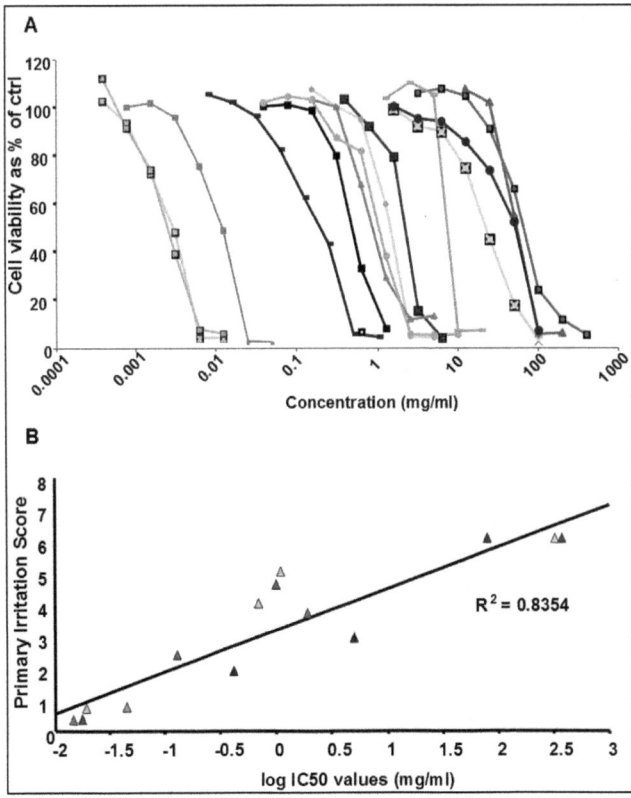

Figure 3 *Correlation between HEKn viability and primary irritation Score. HEKn cells were treated with test compounds over a range of concentrations to determine the IC_{50} values (n=3). A: concentration–viability curve for the chemicals tested. B: Plot of in vitro IC_{50} values against established in vivo primary irritation score.*

This correlation suggests that the *in vitro* assay holds potential as a model for skin irritation by predicting *in vivo* irritation from *in vitro* IC_{50} values in keratinocytes. However, the main limitation of this assay is that the test substance(s) must be soluble in the culture media. This is particularly important as often the chemicals used for topical applications are only marginally soluble in aqueous solution. In addition the model does not account for the barrier function of human skin therefore the potential for false positives is reasonably high in this model. Therefore an additional model is needed to evaluate chemicals that are not sufficiently soluble in the test media and with a barrier function close to human skin.

The model selected to investigate irritancy of chemicals which have low aqueous solubility in this study was *ex vivo* human skin. This model was chosen primarily due it is similarly to the *in vivo* situation particularly in terms of barrier function of the skin. Initially the viability of the skin was assessed to determine the suitability of the method i.e. could the viability of the tissue be maintained in the culture long enough to allow the intracellular increases in cytokines and their extra cellular release to occur.

Figure 4 show the results of the viability time course for excised skin and this showed that tissue viability could be maintained in the culture (above 60% of the 24h value at 96 hours) sufficiently long enough to allow for the dosing protocol, which required the tissue to be incubated for 48 hours post treatment. This time point was selected as starting point for this investigating to mimic the protocols used on RHE cultures which all use an endpoint close to 48 hours. Given the possibility that *ex vivo* skin might be less sensitive than a RHE model, it seemed unlikely that a lower incubation period would yield significant results.

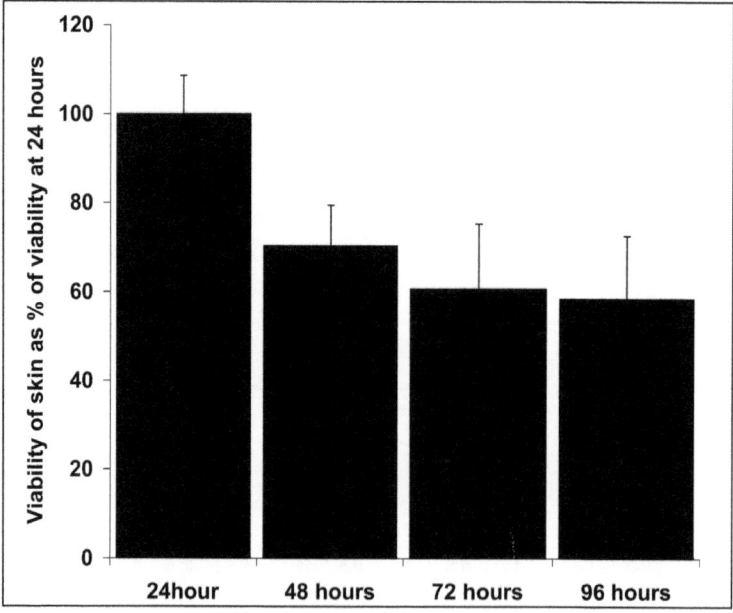

Figure 4 *Viability of ex vivo skin over 96 hour culture duration. Skins viability was assessed every 24 hours by MTT assay (n=5 ± SEM).*

Three end-points were investigated to assess the ability of *ex vivo* human skin in culture to determine the irritancy potential of chemical substances. The first was the viability of the tissue 48 hours after application of the test substance. The second was to measure the level of the inflammatory cytokine IL-1α in the conditioned media and the third was to measure the levels of IL-8 in the media. IL-1α was investigated as it is one of the most important cytokines in the irritation response. Whilst IL-1α is known to be constitutively produced in keratinocytes, its expression is increased in response to irritation and this is thought to be an essential primary event of the inflammation cascade. This release of IL-1α stimulates further release of secondary mediators including IL-8, a promoter of dendritic cell migration and recruitment of monocytes and neutrophils, all of which are vital components of the inflammation process [32].

The MTT assay results (Figure 5) showed that exposure to most of the irritants caused a significant decrease (> 50% compared to negative control) in cell viability. However, when chemicals with lower PII's were tested the MTT assay results only displayed slight changes in viability compared to the untreated control. There were two exceptions to this, both of which showed a reduction of only approximately 40% viability despite have PII greater than some of the other chemicals which caused viability reductions considerably higher. This suggests that tissue viability alone is not sufficient to determine if a chemical is an irritant alone in this assay.

Two cytokines (IL-8 and IL-1α) were also measured in this *ex vivo* cell model (Figure 6) and changes in the levels of both were detected following exposure to certain chemicals. However, not all chemicals which had elevated IL-8 had elevated IL-1α. The differences between the patterns of cytokines which were elevated in response to the different irritants could be suggestive of the mechanism of irritation induced by particular chemicals. As

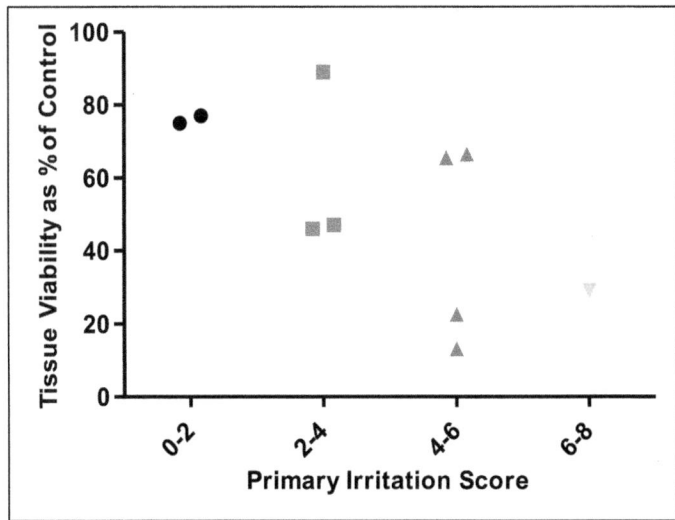

Figure 5 *Viability of Skin Post Treatment with Chemicals. Skin was treated with each chemical for the treatment period, skin was then cultured for an additional 48h before viability was determined by MTT assay (n=3).*

previously mentioned, surfactants initiate irritation by release of intracellular IL-1α as a result of disruption of cellular membranes. However, this is not the case for all irritants: some do not exhibit membrane-damaging characteristics and thus probably do not initiate the inflammatory response solely by the release of IL-1α [6].

Other possible mechanisms of initiation of irritation include the generation of reactive oxygen species (ROS) [33] or effects of compounds on transmembranous receptors resulting in altered signal transduction [34]. This suggests that monitoring a range of cytokines could be vital to the identification of all irritants regardless of the mechanism of action.

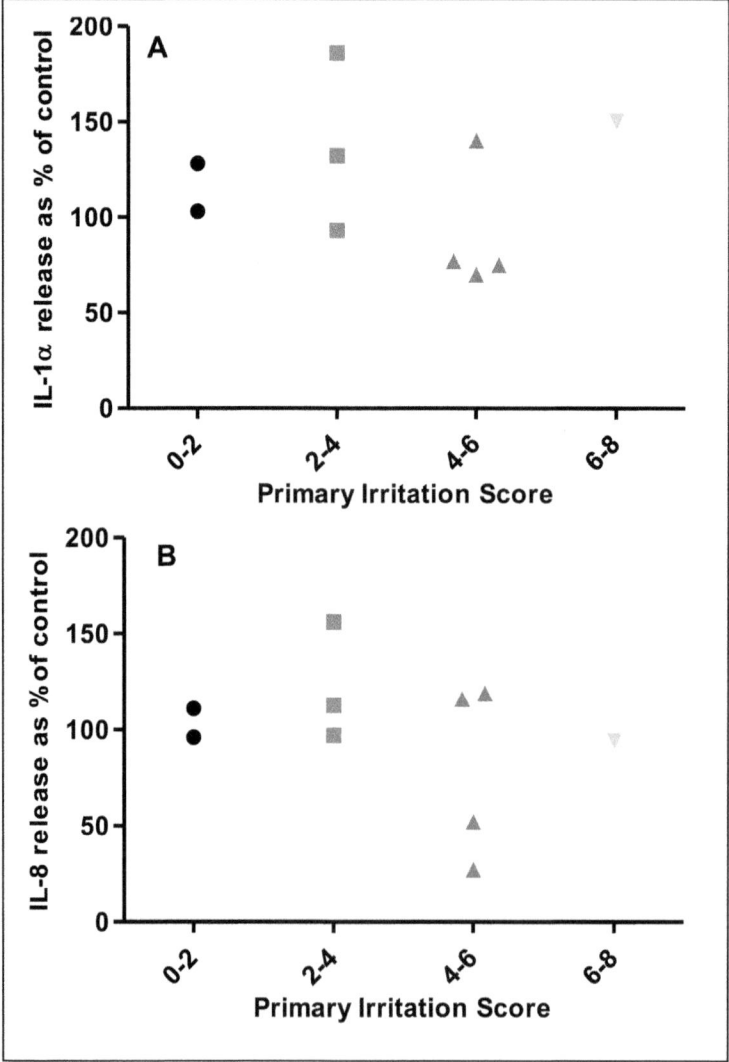

Figure 6 *Release of cytokines (IL1-α and IL-8) measured in conditioned media of ex vivo skin cultures 48 hours after exposure to dermal irritants.*

Additionally, the ratio of IL-1α and IL-8 can also be shown to be indicative of differences in terms of the potential of the compound to be a contact sensitiser. Coquette *et. al.* (2003) used reconstituted human epidermis to show that substances which act as contact sensitisers show higher levels of IL-8 than IL-1α, with the reverse being true of compounds that act only as irritants [33]. The results of this current study shows that one chemical, a mild skin irritant which is also known to be a mild skin sensitiser [34], demonstrates the highest response in terms of IL-8 release but only a modest increase in IL-1α release. In contrast, one of the strong irritants tested showed only an increase in IL-1α release with no effect on IL-8 release. Clearly, the low number of test compounds evaluated in this study precludes any firm conclusions regarding the full potential for this ex vivo skin culture model for identifying contact allergens.

The addition of the cytokine data can also be used to identify irritants which did not have significant effects on viability. For example, one of the chemicals which only induced a modest decrease in viability elicited an elevation in the production of both IL-1α and IL-8. This would suggest that analysing the cytokine release from this culture can identify irritants that would not otherwise be identified by analysing tissue viability alone.

4   CONCLUSIONS

The importance of developing more accurate and sensitive methods for classification of skin irritancy is particularly important due to the ban on animal testing for cosmetics and the cost and ethical considerations regarding the use of animal studies for pharmaceutical products.

The use of keratinocyte cultures is a rapid and convenient method due to the ease of procurement and use of commercially available test systems. The correlation shown in this study suggests that this method can be useful in determining the degree of risk in terms of irritation. The primary limitation of this method is that the test compounds must be soluble in the culture medium. Furthermore, this type of model does not take into consideration that the compound must first permeate through the stratum corneum before it reaches viable cells. Despite the limitations of the assay, the results of this investigation have demonstrated that there is value in this model due to the correlation achieved between *in vitro* and *in vivo* results. The lack of barrier in this test means that it tends to provide a conservative estimate of irritation potential and so could be used as a rapid and economic screen to indicate if further safety testing is appropriate.

The primary advantage of using human *ex vivo* skin is that it has an intact barrier layers and so more closely models the *in vivo* situation. Despite this advantage, validation of this type of method (for determining the dermal toxicity of single chemical entities) may be limited by the availability of three existing, fully validated protocols using reconstituted human skin models which already serve this purpose. However, there are currently no validated protocols for determining the irritancy of whole formulations (chemical mixtures) and this may be an area where an *ex vivo* skin model could have a significant advantage. The presence of the skin's natural barrier (stratum corneum) gives ex vivo skin models flexibility as it could easily be used for multiple dosing protocols to test topical formulations and its physical robustness would make it suitable for multiple washing-off procedures as it is not as fragile as alternatives like reconstituted human epidermis cultures. The main disadvantage to this model is the difficulty in procuring viable skin, but if a regular source can be obtained this assay could present a more appropriate means of identifying irritants than other methods. The results presented here represent only the initial stages of development but certainly support further evaluation and comparative analysis with the currently validated reconstituted human epidermis models to explore the full benefits of this type of *ex vivo* skin model.

5 REFERENCES

1. OECD, Test No. 439: In Vitro Skin Irritation: Reconstructed Human Epidermis Test Method, *OECD Guidelines for the Testing of Chemicals*, Section 4, OECD Publishing, 2010.

2. OECD, Test No. 404: Acute Dermal Irritation/Corrosion, *OECD Guidelines for the Testing of Chemicals*, Section 4, OECD Publishing, 2002.

3. I. R. Williams and T. S. Kupper, Immunity at the surface: homeostatic mechanisms of the skin immune system, *Life Sciences*, 1996, **58**, 1485–1507.

4. T. J. Hall-Manning, G. H. Holland, G. Rennie, P. Revell, J. Hines, M. D. Barratt, D. A. Basketter,. Skin irritation potential of mixed surfactant systems, *Food Chemical Toxicology*, 1998, **36**(3), 233–238.

5. S. Pappinen, S. Pasonen-Seppänen, M. Suhonen, R. Tammi and A. Urtti, Rat epidermal keratinocyte organotypic culture (ROC) as a model for chemically induced skin irritation testing, *Toxicol. Appl. Pharmacol.*, 2005, **208**(3), 233–241.

6. T. Welss, D. A. Basketter and K. R. Schröder, *In vitro* skin irritation: facts and future. State of the art review of mechanisms and models. *Toxicol. In Vitro*, 2004, **18**(3), 231–243

7. M. MacFarlane, P. Jones, C. Goebel, E. Dufour, J. Rowland, D. Araki, M. J. Costabel-Farkas, N. J. Hewitt, J. Hibatallah, A. Kirst, P. McNamee, F. Schellauf and J. Scheel, A tiered approach to the use of alternatives to animal testing for the safety assessment of cosmetics: Skin irritation, *Regul. Toxicol. Pharmacol.*, 2009, **54**, 188–196.

8. S. Golla, S. Madihally, R. L. Robinson Jr. and K. A. M. Gasem, Quantitative structure–property relationships modeling of skin irritation, *Toxicol. In Vitro*, 2009, **23**, 176–184.

9. A. De Brugerolle de Fraissinette, V. Picarles, S. Chibout, M. Kolopp, J. Medina, P. Burtin, M. E. Ebelin, S. Osborne, F. K. Mayer, A. Spake, M. Rosdy, B. De Wever, R. A. Ettlin, A. Cordier, Predictivity of an in vitromodel for acute and chronic skin irritation (SkinEthic) applied to testing of topical vehicles, *Cell Biol. Toxicol.*, 1999, **15**, 121–135.

10. J. H. Fentem, D. Briggs, C. Chesne, G. R. Elliott, J. W. Harbell, J. R. Heylings, P. Portes, R. Roguet, J. J. van de Sandt and P. A. Botham, A prevalidation study on *in vitro* tests for acute skin irritation results and evaluation by the management team, *Toxicol. In Vitro*, 2001, **15**(1), 57–93.

11. P. Portes, M. H. Grandidier, C. Cohen and R. Roguet, Refinement of the Episkin protocol for the assessment of acute skin irritation of chemicals follow-up to the ECVAM prevalidation study, *Toxicol. In Vitro*, 2002, **16**, 765–770.

12. M. Ponec and J. Kempenaar, Use of human skin recombinants as an in vitro model for testing the irritation potential of cutaneous irritants, *Skin Pharmacol.*, 1995, **8**(1–2), 49–59.

13. J. Medina, A. de Brugerolle de Fraissinette, S.-D. Chibout, M. Kolopp, R. Kammermann, P. Burtin, M.-E. Ebelin and A. Cordier, Use of Human Skin Equivalent Apligraf for *In Vitro* Assessment of Cumulative Skin Irritation Potential of Topical Products Toxicol, *Appl. Pharmacol.*, 2000, **164**(1), 38–45.

14. J. Demetrulias, T. Donnelly, V. Morhenn, B. Jessee, S. Hainsworth, P. Casterton, L. Bernhofer, K. Martin, D. Decker, Skin2– an *in vitro* human skin model the correlation between *in vivo* and *in vitro* testing of surfactants, *Exp. Dermatol.*, 1998, **7**(1), 18–22.

15. R. Gay, M. Swiderek, D. Nelson and A. Ernesti, The living skin equivalent as a modelin vitro for ranking the toxic potential of dermal irritants, *Toxicol. In Vitro*, 1992, **6**, 303–315

16. EpiSkin™ SOP, Version 1.8 (February 2009), ECVAM Skin Irritation Validation Study: Validation of the EpiSkin™ test method 15 min - 42 hours for the prediction of acute skin irritation of chemicals.

17. EpiDerm™ SOP, Version 7.0 (Revised March 2009), Protocol for: In vitro EpiDerm™ skin irritation test (EPI-200-SIT), For use with MatTek Corporation's reconstructed human epidermal model EpiDerm (EPI-200).

18. SkinEthic™ RHE SOP, Version 2.0 (February 2009), SkinEthic skin irritation test-42bis test method for the prediction of acute skin irritation of chemicals: 42 minutes application + 42 hours post-incubation.

19. M. Ponec, *In vitro* cultured human skin cells as alternatives to animals for skin irritancy screening, *Int J Cosmet Sci.*, 1992, **14**, 245–264.

20. V. H. W. Mak, M. B. Cumpstone, A. H. Kennedy, C. S. Harmon, R. H. Guy and R. O. Potts, Barrier Function of Human Keratinocyte Cultures Grown at the Air-Liquid Interface, *J. Invest. Dermatol.*, 1991, **96**, 323–327.

21. J. Vicanová, A. M. Mommaas, A. A. Mulder, H. K. Koerten, M. Ponec, Impaired desquamation in the *in vitro* reconstructed human epidermis, *Cell Tissue Res.*, 1996, **286**, 115–122.

22. F. Netzlaff, C.-M. Lehr, P. W. Wertz and U. F. Schaefer, The human epidermis models EpiSkin, SkinEthic and EpiDerm: an evaluation of morphology and their suitability for testing phototoxicity, irritancy, corrosivity, and substance transport, *Eur. J. Pharm. Biopharm.*, 2005, **60**, 167–178

23. H. C. Eun, J. H. Chung, S. Y. Jung, K. H. Cho and K. H. Kim, A comparative study of the cytotoxicity of skin irritants on cultured human oral and skin keratinocytes, *Br. J. Dermatol.*, 1994, **130**, 24–28.

24. K. Müller-Decker, G. Fürstenberger and F. Marks, Keratinocyte-derived proinflammatory key mediators and cell viability as in vitro parameters of irritancy: a possible alternative to the Draize skin irritation test, *Toxicol Appl Pharmacol.*, 1994, **127**, 99–108.

25. C. Cohen, G. Dossou, A. Rougier and R. Roguet, Measurement of inflammatory mediators produced by human keratinocytes in vitro: A predictive assessment of cutaneous irritation, *Toxicol. In Vitro.*, 1991, **5**, 407–410.

26. A. Gueniche and M. Ponec, Use of human skin cell cultures for the estimation of potential skin irritants, *Toxicol. In Vitro*, 1993, **7**(1),15–24.

27. J. J. M. Van de Sandt and A. A. J. J. L. Rutten, Release of Arachidonic and Linoleic Acid Metabolites in Skin Organ Cultures as Characteristics of *in vitro* Skin Irritancy, *Fundamental and Applied Toxicology*, 1995, **25**, 20–28

28. F. G. Bartnik, W. F. Pittermann, N. Mendorf, U. Tillman and K. Kunstler, Skin organ culture for the study of skin irritancy, *Toxicol. In Vitro*, 1990, **4**, 293–301.

29. F. H. M. Pistoor, A. Rambukkana, M. Kroezen, J.-P. Lepoittevin, J. D. Bos, M. L. Kapsenberg and P. K. Das, Novel predictive assay for contact allergens using human skin explant cultures, *American Journal of Pathology*, 1996, **149**, 337–343.

30. ECETOC, 1995. ECETOC, Skin Irritation and Corrosion: Reference Chemicals Data Bank. In: Technical Report No. 66, European Centre for Ecotoxicology and Toxicology of Chemicals, Brussels, 1995.

31. H. Kojima, Considerations Regarding the Importance of *In Vivo* Data for the Development of an Alternative to Conventional Skin Irritation Testing, *AATEX*, 2005, **11**(1), 31–38

32. J. V. Rogers, P. G. Gunasekar, C. M. Garrett, M. B. Kabbur and J. N. McDougal, Detection of oxidative species and low-molecular-weight DNA in skin following dermal exposure with JP-8 jet fuel, *J. Appl. Toxicol.*, 2001, **21**(6), 521–525

33. A. Coquette, N. Berna, A. Vandenbosch, M. Rosdy, B. De Wever and Y. Poumay, Analysis of interleukin-1alpha (IL-1alpha) and interleukin-8 (IL-8) expression and release in in vitro reconstructed human epidermis for the prediction of in vivo skin irritation and/or sensitization, *Toxicol. In Vitro*, **17** (3), 311–321.

34. M. Takeyoshi, S. Noda, S. Yamazaki, H. Kakishima, K. Yamasaki and I. Kimber, Assessment of the skin sensitization potency of eugenol and its dimers using a non-radioisotopic modification of the local lymph node assay, *J Appl Toxicol*, 2004, **24**(1), 77–81.

26

IS IVRT ALONE SUFFICIENT TO DEMONSTRATE 'SAMENESS' OF GENERIC PRODUCTS?

M J Traynor, S S Shetage, D G Wood, M B Brown

School of Pharmacy, University of Hertfordshire, College Lane, Hatfield, Herts, AL10 9AB, United Kingdom.

1 INTRODUCTION

Actinic keratosis (AK) also known as solar keratosis, is the cause of precancerous growths on the skin caused by UV radiation which correlates to cumulative exposure to UV, with a greater occurrence at higher exposure levels [1]. AK is one of the most prevalent dermatological conditions and can cause a number of symptoms from small (mm) sized rough lesions that are barely visible through to raised hyperkeratotic plaques several cm in length [2]. Over time these may progress into squamous [3] or basal cell carcinomas (originally not thought to be a result of AK progression). It is estimated that up to 16 % of people suffering with AK lesions progress on to squamous cell carcinoma with several severe cases progressing onto basal cell carcinoma and as such treatment at the first signs of AK is recommended [4].

There are numerous strategies available for the treatment of AK which can involve surgical or non-surgical interventions or a combination of both. Treatments include cryosurgery (often coupled with topical fluorouracil), curettage, chemical peels, laser therapy and photodynamic therapy in addition to numerous topical treatments alone such as; ingenol mebutate (Picato®), diclofenac sodium (Solaraze®), fluorouracil (e.g. Carac® cream and Efudex) and imiquimod (e.g. Aldara®, Zyclara® and Fougera®) [5,6]. Fluorouracil is the benchmark topical treatment for AK for which all other topical treatments are compared [7]. However, fluorouracil cream has been associated with skin irritation such as dryness, erythema and skin erosion. As such, it not appealing to many people suffering with AKs [6]. Other potential treatments such as diclofenac sodium have been suggested to be ineffective when following a treatment regime that is less than 3 months with twice daily application [8]. Furthermore, a clinical trial evaluating the use of ingenol mebutate, although an effective treatment, reported adverse effects of moderate intensity [9].

A successful treatment for AK with perceived improved patient compliance is the use of topically applied imiquimod (1-(2-methylpropyl)-1H-imidazo[4,5-c]quinolin-4-amine). This drug was approved by the Food and Drugs Administration (FDA) in 1997 under a NDA submitted by 3M (Application Number 020723) for the treatment of external genital warts [10] and subsequently received further approval from the FDA in 2004 for the treatment of AK and superficial basal cell carcinoma. Imiquimod is a topically active immune-modulatory agent that stimulates both the innate and acquired pathways of the immune response [11], exerting beneficial therapeutic effects as a result of immune activation through the TLR-7 signalling pathway [12]. The activation of this pathway results in the release of cytokines e.g. TNF, IL-6, IL-12 and interferon alpha in addition to natural killer cells which circulate in the lymphatic system and induce the adaptive immune response [13]. However, when applied topically systemic absorption has been demonstrated to be minimal with studies illustrating 97 % of the applied dose can be recovered from the application site [13].

Currently available commercial imiquimod creams include Aldara® 5 % w/w, Zyclara® 2.5 or 3.75 % w/w (originally Graceway Pharmaceutical LLC, now Medicis Corporation, USA) and Fougera® Imiquimod Cream 5.0 % w/w (Fougera Pharmaceuticals Inc., part of Nycomed Inc., USA). The later was recently approved by the FDA under an abbreviated new drug approval (ANDA) as a generic for Aldara®. The original imiquimod cream, Aldara® and the generic Fougera® Imiquimod Cream both have a dosing regimen of 2 times per week for a treatment period of up to 16 weeks [14]. In comparison, Zyclara® was approved by the FDA in 2010 as a lower strength product that could deliver an efficacious dose with a dosing regimen involving daily application for two weeks followed by 2 weeks of no treatment, followed by 2 weeks of daily use. This shorter treatment regimen, potentially offers a significant advantage over the 16 week treatment time associated with the original product Aldara® which would only improve patient compliance.

Zyclara® was approved by the FDA as it offered a more intuitive treatment regimen which is shorter than Aldara® with no safety concerns due to the lower levels of imiquimod in comparison to the 5 % product [15]. In contrast, Fougera® imiquimod cream was approved as a generic by the FDA as the product demonstrated bioequivalence to the existing product, Aldara® [16]. The FDA guidance on *in vitro* release testing (IVRT) and *in vivo* bioequivalence documentation [17] provides current scientific opinion intended to lower regulatory barriers, allowing products to be approved for the market assuming they can demonstrate comparable effectiveness. Historically, physical and chemical tests such as solubility, particle size, crystalline form, homogeneity and viscosity have been used to suggest comparable performance. However, more recent *in vitro* release testing has been employed to comprehensively demonstrate the delivery of the active pharmaceutical ingredient. Despite guidance on *in vitro* tests to demonstrate 'sameness' the exact methods are not defined which can result in the criticism that the methodology used can be manipulated to achieve the desired result of 'sameness'.

Although generic formulations approved by the FDA are required to have identical routes of administration, safety, intended use and equivalent drug concentrations, the level of excipients within the formulation may be changed by up to 10 %. Considering these requirements for the approval of a generic, the published information available on patient information leaflets highlights a difference between the composition of Aldara® and Fougera® imiquimod creams (excipient composition listed in Table 1) [14]. Where Aldara® is observed to contain isostearic acid and Fougera®, the generic product appears to contain oleic acid as a replacement. Despite declaration of a difference, the exact % w/w composition of each formulation is unknown.

There is limited information on the approval of Fougera® available in the public domain with regards to the demonstration of bioequivalence and subsequent approval of Fougera®. As such, this paper aims to evaluate the use of *in vitro* release testing (IVRT) and skin permeation as a method for demonstration of 'sameness' for generic products using the topical imiquimod formulations Aldara®, Fougera® and Zyclara®. During this study

Table 1 *Excipients within commercially available imiquimod creams [14].*

Aldara®	Zyclara®	Fougera®
Isostearic acid, benzyl alcohol, cetyl alcohol, stearyl alcohol, white petrolatum, polysorbate 60, sorbitan monostearate, glycerin, methyl paraben, propyl paraben, purified Water, xanthan gum	Isostearic acid, cetyl alcohol, stearyl alcohol, white petrolatum, polysorbate 60, sorbitan monostearate, glycerin, xanthan gum, purified water, benzyl alcohol, methyl paraben, and propyl paraben	Oleic acid, benzyl alcohol, polysorbate 60, sorbitan monostearate, cetyl alcohol, stearyl alcohol, petrolatum, propylparaben, purified water, glycerin, methylparaben, xanthan gum

silicone membrane was used to model drug release, as defined in the SUPAC guidance, as this membrane presents a confluent homogenous barrier and has been repeatedly used to model drug permeation through skin. *Ex vivo* skin was employed as a comparator to the silicone membrane to evaluate the 'sameness' results and confirm if IVRT would be representative of human skin.

2   METHODS

Imiquimod was obtained from HBCChem, Inc. (California, USA). Isostearic acid was purchased from Uniqema (Wilmington, USA). Acetonitrile (high performance liquid chromatography (HPLC) grade), water (HPLC grade), triethylamine and hydrochloric acid were all obtained from Fisher Scientific (Loughborough, UK). Octyl sodium sulfate was obtained from VWR (UK). Tritiated water (specific activity 37.0 MBq g^{-1}) was obtained from Perkin Elmer (Massachusetts, USA). Silicone membrane (polydimethyl siloxane) was supplied by BioPlexus (California, USA). Human abdominal skin (post abdominoplasty) was obtained with informed consent and ethical approval from the University of Hertfordshire, Pharmacy and Postgraduate Medicine Ethics committee with Delegated Authority (approval number: PHAEC/09-23).

Imiquimod was quantified using the HPLC method adapted from US patent application 2011/0263635 A1. The system consisted of a Waters 265 Alliance Separations module (Waters, USA), Water 996 photodiode array detector (Waters, USA). Millenium[32] Version 4 software was used for data acquisition. A flow rate of 2.0 ml min^{-1} was used. The mobile phase was 28 % acetonitrile and 72 % aqueous solution (1 % v/v triethylamine, 0.2 % v/v octyl sodium sulfate, remainder deionised water, adjusted to pH 2.0 with phosphoric acid). The stationary phase was a Supelcosil LC-8-DB (5 μm, 150 x 4.6 mm) used with a Supelguard LC-8-DB (Sigma-Aldrich, UK) maintained at 25 ± 2° C. The injection volume was 20 μL. The retention time of imiquimod was ca. 6 min detected at 258 nm. Calibration curves were constructed from a series of standard concentrations prepared by serial dilution between 0.1 – 250 μg mL^{-1}. The assay was shown to be 'fit for purpose' in accuracy and precision (CV < 2.0 %) which meets the standards described by the International Conference on Harmonisation guidelines [18].

Sheets of human epidermis were prepared in accordance to the method devised by Kligman and Christophers [19]. Subcutaneous fat was removed by blunt dissection. The skin was rinsed and placed into a glass beaker containing hot (60 ± 3 °C) deionised water for 45 seconds. The skin was removed from the water and laid dermal side down on aluminium foil. The epidermis was carefully rolled off the dermis using the thumb and transferred to a tray of deionised water, SC facing upwards. Filter paper was floated underneath and the epidermal sheet was removed and frozen until required. Prior to use the epidermal sheet was completely defrosted and mounted onto the Franz cells as required.

A saturated solution of imiquimod was prepared in isostearic acid and used as a control for the drug transport studies. The remaining three test formulations are all commercially available and were purchased as required; Aldara® (Graceway LLC, USA), Zyclara® (Graceway LLC, USA) and Fougera® (Nycomed Inc., USA).

Two barriers were employed for the transport experiments: silicone membrane (polydimethyl siloxane) and epidermal sheet. Each barrier was mounted in small Franz diffusion cells which were individually calibrated to allow use of their precise dimensions in subsequent calculations and maintain an identical internal fluid volume during the experiments. The cells displayed an approximate diffusional area of 0.6 cm^2 and receiver volume of 2.0 mL. Details of the experimental parameters for each barrier are defined in the subsequent Sections ('silicone membrane permeation' and epidermal membrane permeation'). Following the release experiment (silicone membrane) the cumulative amounts of drug per sample area (μg cm^{-2}) was plotted against time, in addition to the square root of time (as per SUPAC-SS guidelines). The permeation of imiquimod through epidermal membrane per sample area (μg cm^{-2}) was plotted against time (h) and steady state flux was taken from the linear portion of the curve ($\geq$ 5 points) with linearity of greater than R$^2 \geq 0.96$).

Silicone membrane was selected for the drug release testing after preliminary experiments with other membranes showed significant back diffusion of the receiver fluid (data not shown). After assembly, cells were filled with receiver fluid and cell integrity was checked visually by inversion. Cells were placed in a water bath and allowed to equilibrate for 30 minutes prior to the application of the test formulation. An infinite dose of 250 μL of each formulation was applied to each cell following equilibration. Imiquimod has poor solubility in both aqueous and hydro-alcoholic medium (as suggested in SUPAC guidelines) although it is known to have high solubility within isosteric and oleic acid however use of this within a receiver fluid would potentially favour the permeation of imiquimod from Aldara® or Fougera®, respectively. As such, in accordance to SUPAC guidelines on receptor medium, dilute hydrochloric acid solution (0.1 N) was used as the receiver fluid in an attempt to ensure sink conditions were maintained during the transport experiments. Samples (200 μL) were taken for HPLC analysis at 0, 15, 30, 60, 120, 240, 300 and 360 minutes. Following sampling each receiver compartment was replenished with 200 μL of pre-warmed, de-gassed receiver fluid.

Following assembly and upon completion of each experiment, skin (epidermal) membrane integrity was assessed using tritiated water over a 2 h period. Where the permeation of tritium was observed to be significantly higher than the mean permeation prior to the experiment, cells were excluded. Upon selection, cells were dried, filled with receiver fluid (0.1 N HCl solution) and allowed to equilibrate for 30 minutes prior to the application of the test formulation. A finite dose (10 mg cm^{-2}) of formulation was applied to each cell following equilibration. Samples (200 μL) were taken for HPLC analysis at 0, 1, 4, 8, 24, 24, 32, 48 and 56 h. Pre-warmed receiver fluid was replenished in each cell following sampling.

The mean cumulative amount (μg cm^{-2}) of imiquimod permeated (through either silicone membrane or epidermal membrane) ± SEM was plotted against time (h). Steady state flux was calculated with respect to most linear part of the permeation profile between 0.25 - 6 h for silicone membrane and 8 – 56 h for epidermal membrane. The lag phase was calculated by extrapolation of the steady state portion of the permeation profile to determine the intercept with the x-axis (which approximates to the lag time). Statistical analysis of the permeation (cumulative amount (μg cm^{-2}) was performed for all formulations using ANOVA, with post hoc Tukey's HSD test at a 95% confidence level.

3   RESULTS

Imiquimod transport through the silicone membrane vs. time gave a linear profile ($r^2 >$ 0.96) for all three commercial formulations and the saturated solution over the 6 h experimental period (Figure 1).

The steady state flux of imiquimod through silicone membranes was observed to be highest from the saturated isostearic acid (1.64 ± 0.11 μg cm^{-2} h^{-1}). There was no significant difference (p > 0.05) in steady state flux between the saturated isostearic acid and Fougera$^{®}$ which was 1.36 ± 0.10 μg cm^{-2} h^{-1}. Both the saturated isostearic acid and Fougera$^{®}$ had significantly higher (p < 0.05) steady state flux in comparison to Aldara$^{®}$ (0.52 ± 0.19 μg cm^{-2} h^{-1}) and Zyclara$^{®}$ (0.53 ± 0.09 μg cm^{-2} h^{-1}) (Table 2) between which no statistical difference (p > 0.05) was observed (Table 2).

No statistical difference (p > 0.05) was observed in lag times between the three commercial formulations and saturated isosteric acid (Table 2).

Following comparison of transport through silicone membrane using standard statistical methods, imiquimod transport was compared following SUPAC-SS guidance (SUPAC-SS, 1997) and the same trends in data were observed. In accordance with the first step of SUPAC-SS, when individual release values of Fougera$^{®}$ were calculated in comparison to Aldara$^{®}$, values were observed to fall outside the 90 % confidence intervals (i.e. not within 75 – 133.33 %).

Table 2 *Steady state release of imiquimod from saturated isostearic acid, Zyclara$^{®}$, Aldara$^{®}$ and Fougera$^{®}$ through silicone membranes. Each value of steady state transport is represented by the mean ± SEM (n ≥ 5). Steady state transport is calculated with respect to most linear part of the permeation profile (between 0.25-6 h). Lag time is calculated from the x-intercept of the slope at steady-state flux. Each value is represented by the mean ± SEM (n ≥ 5).*

Formulation	Steady state transport (μg cm^{-2} h^{-1})	Lag phase (h)
Isostearic acid (saturated)	1.64 ± 0.11	0.51 ± 0.13
Zyclara$^{®}$ (3.75 % w/w)	0.53 ± 0.09	0.88 ± 0.14
Aldara$^{®}$ (5.0 % w/w)	0.52 ± 0.19	1.21 ± 0.24
Fougera$^{®}$ (5.0 % w/w)	1.36 ± 0.10	0.51 ± 0.03

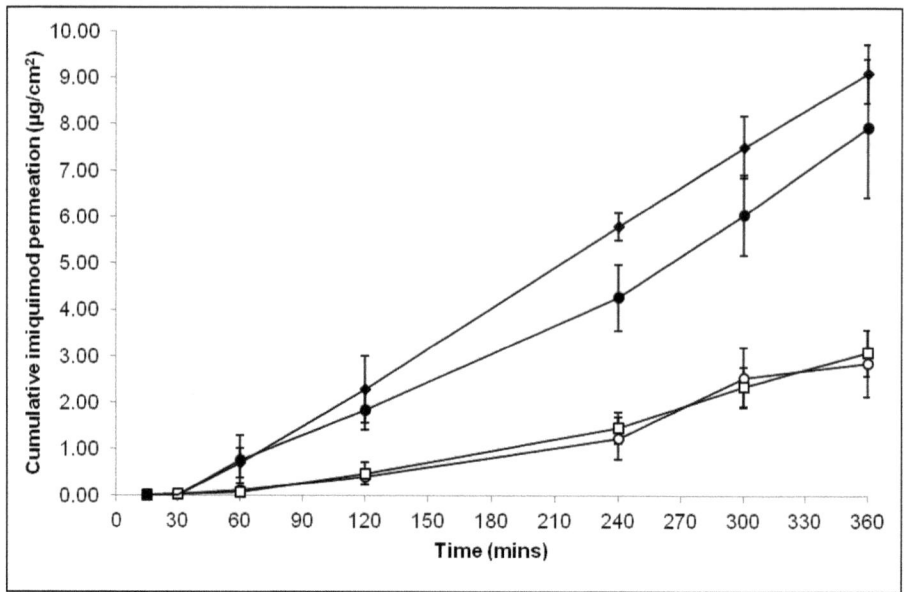

Figure 1 *Imiquimod release from a saturated isostearic acid (♦), Zyclara® (□), Aldara® (○) and Fougera® (●) into receiver fluid (0.1 N HCl solution) through silicone membrane with a surface temperature of 32 °C versus time (min). Each point represents the mean cumulative mass per area ± SEM (n ≥ 5).*

Imiquimod transport vs. time through *ex vivo* human epidermal sheet was relatively linear ($r^2 > 0.98$) for the three commercial test formulations. Different trends were exhibited in terms of imiquimod permeation (Figure 2) in comparison to the transport through silicone membrane.

No statistical difference (p>0.05) was observed in lag times between the three formulations (mean ± SEM, Fougera®, 9.26 ± 1.96 h; Zyclara®, 8.58 ± 2.60 h; Aldara®, 13.58 ± 2.31 h. No statistical difference (p>0.05) in steady state flux was observed between Zyclara® (0.34 ± 0.04 µg cm^{-2} h^{-1}) and Aldara® (0.58 ± 0.17 µg cm^{-2} h^{-1}, Table 3). Furthermore, no statistical differences (p > 0.05) in steady state flux was observed between Aldara® and Fougera® (0.83 ± 0.13 µg cm^{-2} h^{-1}).

4   DISCUSSION

Comparison of imiquimod transport through silicone membrane using Franz diffusion cells and standard statistical analysis methods illustrated that Aldara® and Fougera® delivered non-equivalent amounts of imiquimod. Fougera® and saturated isostearic acid were observed to be statistically the same (p>0.05) and had higher transport rates in comparison to Zycala® and Aldara®, which were also observed to be equivalent. Unsurprisingly, the same findings were observed when the SUPAC guidelines were followed, where again Fougera® was not found to be equivalent to Aldara®. Despite SUPAC guidelines attempting to define appropriate testing methods, the guidance provided is vague and could potentially

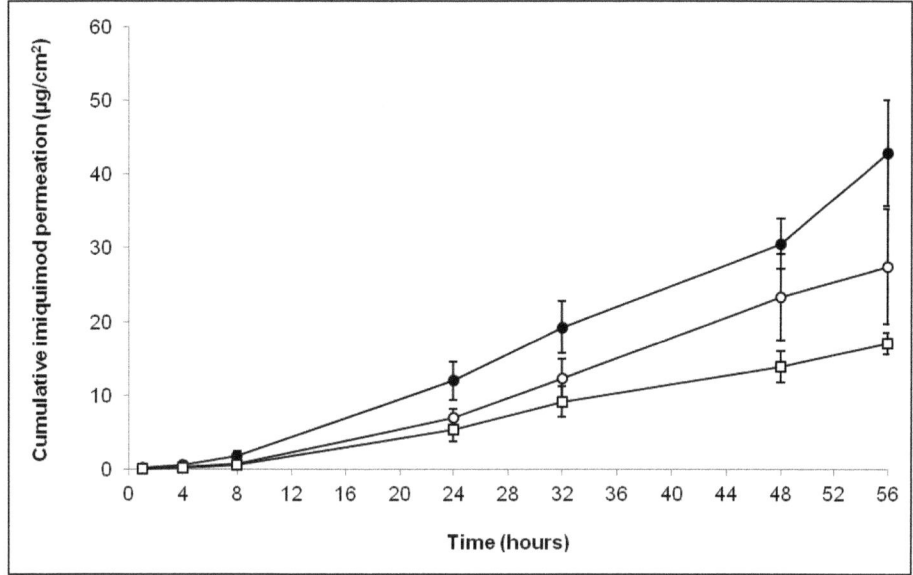

Figure 2 *Imiquimod permeation from Zyclara® (□), Aldara® (○) and Fougera® (●) into receiver fluid (0.1 N HCl solution) through human epidermal sheet with a surface temperature of 32 °C. Each point represents the mean cumulative mass per area ± SEM (n ≥ 4).*

be exploited to manipulate results. Appropriate testing conditions are defined as; 'a diffusion cell with a standard open cap ground glass surface with a 15 mm diameter orifice and a total diameter of 25 mm', a receptor medium that is 'appropriate' and selection can be justified, in addition the number of replicates (n=6), the amount of sample to be applied to the membrane, sampling times and synthetic membrane are also suggested. The guidance on synthetic membranes is superficial, suggesting 'appropriate inert and commercially available synthetic membranes of appropriate size to fit the diffusion cell diameter'.

Silicone membrane can be employed as a model for human skin due to the similar hydrophobicities of the barrier and the SC [20, 21]. Silicone used in membranes is structurally homogenous and does not form into an ordered bilayered lipid structure as found in skin. This has advantages if drug transport needs to be investigated in the absence of major structural changes, however for this reason should be employed to support the data from human skin transport and not replace it. Currently silicone membrane provides the most representative barrier to skin and in addition it is relatively inert, has good chemical, pH and temperature resistance and is widely commercially available [22 – 25]. Alternative membranes may provide a better representation to human skin such as Strat-M™ membrane however as it is a relatively new product there is little data to validate its drug transport profiles in relation to skin.

The differences observed between Fougera® and Aldara®, in terms of transport through silicone membrane could be explained by numerous reasons. A possible explanation could be because of a differing influence of the two formulations on the membrane itself. Both Aldara® and Fougera® are typical oil-in-water creams with believed to be similar compositions with the exception of isosteric or oleic acid with both being demonstrated to act as effective penetration enhancers through skin [26 – 28]. However, the membrane thickness was measured prior to and after the silicone experiment and was found to remain constant during this study which suggests that both the formulations had no swelling effect on the membrane which would otherwise result in alteration of the diffusional path length and variation in transport profiles. Possibly the most plausible reason for the differences in permeation observed between Fougera and Aldara® is a different thermodynamic activity of imiquimod between the two formulations. According to the Higuchi papers published in the 1960's [29 – 31] and assuming that there is no interaction between the active pharmaceutical ingredient and the membrane, the rate of a drug traversing the barrier is directly proportional to the thermodynamic activity of the drug within the formulation (Equation 1).

$$\frac{dm}{dt} = A \frac{D\alpha}{h\gamma_{bar}} \tag{1}$$

Where the amount of drug permeation per unit time (dm/dt) at steady-state, is related to the diffusional surface area (A), the diffusion coefficient of the drug (D), its thermodynamic activity in the vehicle (α), its activity coefficient in the diffusional path (γ_{bar}) and the thickness of the diffusional path (h).

The solubility of imiquimod in isostrearic and oleic acid are similar (17 mg g^{-1} and 20 mg g^{-1}, respectively [32]) and although both formulations contain the same drug concentration the relative amounts of the two excipients is unknown. However, the saturated solution of imiquimod in isosteric acid (thermodymanic unity) and Fougera® resulted in a similar imiqiumod transport which was much higher than that for Aldara® indicating an equivalent thermodynamic activity of the first two systems. The equivalence of imiquimod transport between Zyclara® and Aldara® suggests that the level of isostearic acid in Zyclara® had been reduced in comparison to Aldara® to ensure equal thermodynamic activities with lower drug content whilst achieving the same steady state transport rate. In this study, the second stage of SUPAC testing was not performed, however due to the scale of differences observed it was postulated that the non-equivalence would continue be observed had this work proceeded onto the second stage of testing (further *in vitro* testing yielding an additional 12 slopes for each product (18 slopes in total including the first stage of testing) after which, the 90% confidence interval should fall within the limits of 75 % to 133.33 % [17].

Assessment of the imiquimod permeation from three commercial formulations through epidermal membranes highlighted different trends in comparison to the IVRT and as would be expected the permeation of imiquimod was notably lower through skin in comparison to silicone membrane. Although not as marked as in the membrane study, the permeation of imiquimod was notably higher from Fougera® in comparison to Aldara® despite the difference not being significant. Such findings suggest that *in vitro* skin permeation studies are perhaps not as sensitive to differing drug thermodynamic activities and/or formulation changes as synthetic membrane studies. Both fatty acids within the formulations are known penetration enhancers [26 – 28]. Oleic acid is believed to act by perturbing both the polar and non-polar lipids within the stratum corneum [33], disordering the highly packed SC intercellular domain lipids or by disruption of the lamellar solid fluid phase separation [34]. Although there is little information available on the action of isostearic acid due to its limited use in commercially available products it would be logical to assume that it would act as an enhancer in a similar manner to oleic acid. However, these findings would suggest that isostearic acid may be a more potent enhancer as it could be argued that its presence in Aldara® is to some degree neutralising the enhanced drug thermodynamic activity, demonstrated in the membrane study for the oleic acid containing Fougera® formulations. As with the membrane studies, Zyclara® was observed to have equivalent steady state permeation through epidermal membrane as Aldara® but at a lower imiquimod concentration suggesting that the former formulation has been further optimised to achieve equivalence.

5 CONCLUSION

The discrepancy observed between the *in vitro* release test of the imiquimod formulations and the permeation of imiquimod through epidermal skin in this study questions the relevance of IVRT as a standalone test to demonstrate 'sameness'. The differences observed in this study would suggest drug permeation experiments through human skin must be conducted in conjunction with IVRT when attempting to use the data for approval of generic products.

6 REFERENCES

1. S. J. Salasche, Epidemiology of actinic keratoses and squamous cell carcinoma, *J. Am. Acad. Dermatol.*, 2000, 42, 4–7.

2. J. Anwar, D.A. Wrone, A. Kimyai-Asadi and M. Alam, The development of actinic keratosis into invasive squamous cell carcinoma: Evidence and evolving classification schemes, *Clin. Dermatol.*, 2004, **22**(3), 189–196.

3. A. A. Memon, J. A. Tomenson, J. Bothwell and P. S. Friedmann, Prevalence of solar damage and actinic keratosis in a Merseyside population, *Br. J. Dermatol.*, 2000, **142**, 1154–9.

4. A. N. Persaud, E. Shamuelova, D. Sherer, W. Lou, G. Singer and C. Cervera, Clinical effect of imiquimod 5% cream in the treatment of actinic keratosis, *J. Am. Acad. Dermatol.*, 2002, **47**, 553-6.

5. C. Perras, Imiquimod 5% cream for actinic keratosis, *Issues Emerg. Health Technol.*, 2004, **61**, 1–4.

6. W. J. Mcintyre, M. R. Downs and S. A. Bedwel, Treatment Options for Actinic Keratoses, *Am. Fam. Physician*, 2007, **76**, 667–72.

7. J. M. Weinberg, Topica therapy for actinic keratoses: Current and evolving therapies, *Reviews on Recent Clinical Trials*, 2006, **1**, 53–60.

8. J. M. Spencer, M. Henry, Medscapre Reference, Actinic Keratosis Treatment & Management 2012, http://emedicine.medscape.com/article/1099775. Accessed 12 May 12.

9. M. Lebwohl, N. Swanson, L. L. Anderson, A. Melgaard, Z. Xu and B. Berman, Ingenol mebutate gel for actinic keratosis, *N. Engl. J. Med.*, 2012, **366**(11), 1010–1019.

10. R. A. Moore, J. E. Edwards, J. Hopwood and D. Hicks, Imiquimod for the treatment of genital warts: a quantitative systematic review, *BMC Infect. Dis.*, 2001, **1**, 3.

11. D. N. Saunder, Imiquimod: modes of action, *Br. J. Dermatol.*, 2003, **149**(Supplement 66), 5–8.

12. A. A. Gaspari, S. K. Tyring and T. Rosen, Beyond a decade of 5% imiquimod topical therapy, *J. Drugs Dermatol.*, 2009, **8**(5), 467–74.

13. C. Caperton and B. Berman. Safety, efficacy, and patient acceptability of imiquimod for topical treatment of actinic keratosis, *Clinical, Cosmetic and Investigational Dermatology*, 2011, **4**, 35–4.

14. Daily Med - Current Medical Information, 2012, http://dailymed.nlm.nih.gov/, accessed 24th February 2012.

15. FDA Drug Approval Package. Zyclara (imiquimod) Cream, 3.75%., http://www.accessdata.fda.gov/. Accessed 12 May 12.

16. FDA approval correspondence, 2010. ANDA 078548 – FDA Approval letter to Nycomed U.S., Inc. http://www.accessdata.fda.gov/drugsatfda_docs/appletter/2010/078548s000ltr.pdf. Accessed 12 May 12.

17. SUPAC-SS, 1997. FDA Guidance for industry. Nonsterile semisold dosage forms. Scale-up and post approval changes: Chemistry, Manufacturing, and Controls: In vitro release testing and bioequivalence documentation, May 1997.

18. ICH, 1996. International conference of harmonisation of technical requirements for registration of pharmaceuticals for human use. Q2(R1): Validation of analytical procedures: Text and methodology.

19. A. M. Kligman and E. Christophers, Preparation of isolated sheets of human stratum corneum, *Arch. Dermatol.*, 1963, **88**, 702–705.

20. S. Geinoz, S. Rey, G. Boss, A. L. Bunge, R. H. Guy, P.-A. Carrupt, M. Reist and B. Testa, Quantitative Structure–Permeation Relationships for Solute Transport across Silicone Membranes, *Pharm Res*, 2002, **19**, 1622–1629.

21. S. C. Wasdo, J. Juntunen, H. Devarajan and K. B. Sloan, A correlation of flux through a silicone membrane with flux through hairless mouse skin and human skin in vitro, *Int. J. Pharm.*, 2009, **373**, 62–67.

22. J. M. Haigh and E. W. Smith, The selection and use of natural and synthetic membranes for *in vitro* diffusion experiments, *Eur. J. Pharm. Sci.*, 1994, **2**(5–6), 311–330.

23. T. S. Radhakrishnan, New Method for Evaluation of Kinetic Parameters and Mechanism of Degradation from Pyrolysis–GC Studies: Thermal Degradation of Polydimethylsiloxanes, *J. Appl. Polym. Sci.*, 1998, **73**, 441–450.

24. G. Camino, S. M. Lomakin and M. Lazzari, Polydimethylsiloxane thermal degradation Part 1, Kinetic aspects, *Polym.*, 2001, **42**, 2395–2402.

25. K. Moser, K. Kriwet, C. Froehlich, A. Naik, Y. N. Kalia and R. H. Guy, Permeation enhancement of a highly lipophilic drug using supersaturated systems, *J. Pharm. Sci.*, 2001, **90**(5), 607–616.

26. T. Loftsson, N. Gildersleeve and N. Bodor, The effect of vehicle additives on the transdermal delivery of nitorglycerin, *Pharmaceutical Research*, 1987, **4**(5), 436–437.

27. B. J. Aungst, Structure/Effect Studies of Fatty Acid Isomers as Skin Penetration Enhancers and Skin Irritants, *Pharmaceutical Research*, 1988, **6**(3), 244–247.

28. B. M. Elyan, M. B. Sidhom and F. M. Plakogiannis, Evaluation of the effect of different fatty acids on the percutaneous absorption metaproterenol sulphate, *J. Pharm. Sci.*, 1996, **85**(1), 101–105.

29. T. Higuchi, Physical chemical analysis of percutaneous absorption process from creams and ointments, *Journal of the Society of Cosmetic Chemists*, 1960, **11**, 85–97.

30. T. Higuchi, Rate of release of medicaments from ointment bases containing drugs in suspension, *Journal of Pharmaceutical Sciences*, 1961, **50**, 874–875.

31. T. Higuchi, Analysis of data on the medicaments release from ointments, *Journal of Pharmaceutical Sciences*, 1962, **51**, 802–804.

32. J. L. Chollet, M. J. Jozwiakowski, K. R. Phares, M. J. Reiter, P. J. Roddy, H. J. Schuitz, Q. V. Ta and M. A. Tomal, *Pharmaceutical Development and Technology*, 1999, **4**(1), 35–43.

33. S. J. Jiang and X. J. Zhou, Examination of the mechanism of oleic acid-induced percutaneous penetration enhancement: an ultrastructural study, *Biol. Pharm. Bull.*, 2003, **26**, 66–68.

34. E. Touitou, B. Godin, Y. Karl, S. Bujanover and Y. Becker, Oleic acid, a skin penetration enhancer, affects Langerhans cells and corneocytes, *J. Contrl. Rel.*, 2002, **80**(1–3), 1–7.

EXPERIMENTAL AND MATHEMATICAL MODELS

27

INTRODUCTION SECTION V – EXPERIMENTAL AND MATHEMATICAL
MODELS

G P Moss

The School of Pharmacy, Keele University, Keele, Staffordshire, ST5 5BG, United
Kingdom.

While it is apparent that the wide acceptability of research work in a particular field is
based on standardised methods, or accepted variations therein, it is also the case that
research into the development of novel methods for evaluating percutaneous absorption
improves and refines our understanding of skin permeability. This applies both
experimentally and theoretically, with the latter focusing on the use of new mathematical
and statistical methods which increase our ability to understand the mechanism of skin
permeation, while also providing improved predictions of permeation for new chemicals.
This chapter focuses on recent experimental and theoretical work which improves our
understanding of these processes.

Skin visualisation techniques have been employed for a number of years to qualitatively or
quantitatively map the ingress of chemicals to the skin. In some cases, such methods have
attempted to replace conventional Franz-type diffusion cells [1, 2]. Microscopic imaging
techniques have also been used to map dermal permeation; these methods include infra-red
and Raman spectroscopy and confocal microscopy. These methods have significant
advantages but also substantial disadvantages, including low spatial resolution and the
need for specific chemical activity within a penetrant. For example in some cases,
fluorescent tagging has been employed, but this may affect the distribution and permeation
of a topically applied chemical. Mass spectrometric methods remove many of these
problems but the most common method, Matrix-Assisted Laser Desorption/Ionization
Imaging Mass Spectrometry (MALDI-IMS), while a significant tool for quantitatively
mapping the dermal absorption of specific penetrants without the need for tagging or other
modifications, is limited by a poor resolution (approximately 100 μm) which is an artefact
of the laser diameter [3]. Therefore, Judd *et al.* (Chapter 28) have described the use of
Time-of-Flight Secondary Ion Mass Spectrometry (ToF-SIMS) in mapping the dermal
absorption of a topically applied drug, chlorhexidine. This technique is capable of
providing spatial imaging of a range of penetrants, identified by mass fragments, within the
skin. By comparison with other imaging methods used to map dermal absorption, ToF-
SIMS provides high sensitivity and specificity.

Further, Xiao *et al.* (Chapter 29) describe the use of capacitive-based contact imaging to
examine solvent penetration through a silicone membrane and porcine skin. They
demonstrate that contact imaging using capacitance fingerprint sensors can clearly and
quantitatively discriminate between a range of model permeants, and that this method
provides a powerful tool for studying solvent penetration through their chosen membranes.

Clearly, imaging techniques provide a substantial benefit in understanding the process of skin permeability. However, the vast majority of *in vitro* skin permeation studies still employ Franz-type vertical diffusion cells and these have been described elsewhere in considerable detail [4, 5]. Indeed, Edwards *et al.* (Chapter 31) demonstrate the significance of such methods in determining bioequivalence of dosage forms and in the quality control of topically applied drug delivery systems. Dalton *et al.* (Chapter 30) describe the effect of altering the receptor compartment of *in vitro* static diffusion cell experiments and how it can influence the permeation of topically applied chemicals. They conclude that the experimental design is a function of the required outcome for the experiment, and that the method chosen should reflect this, suggesting perhaps that there is no single experimental protocol that should be adopted by skin researchers. Such findings have significance not only for experimental studies but also for mathematical modelling studies, such as those described in Chapters 32 and 33, as the data used in these studies comes from a range of experiments reported in the literature.

The development of mathematical models has progressed from the use of mathematics to quantify the phenomena of percutaneous absorption (elegantly reviewed by Roberts *et al.* [6]) to the development of mathematical relationships between the physicochemical properties of a penetrant and its measured permeability which are based on the development of quantitative structure-activity (or permeability) relationships (QSARs, or QSPRs).

In silico models are increasingly employed, not just experimentally but also in a regulatory context, to predict skin permeability and to assess the risks and hazards that may be associated with exposure of human skin to topically applied chemicals in a range of forms including pharmaceutical, cosmetic and industrial products (e.g. pesticides). As well as providing predictions of permeation, most models developed from QSPRs result in the production of an equation with discrete terms representing specific and statistically significant physicochemical properties such as molecular weight and lipophilicity as per the "Potts and Guy" equation [7]. This gives an insight into the mechanism governing absorption of the data modelled and may also provide opportunities for the design of novel chemicals whose properties are optimised with regard to maximising or minimising percutaneous absorption as a particular application requires.

Modelling of percutaneous absorption provides efficient, viable and ethical alternatives to laboratory experimentation. Classically, while many researchers have elucidated aspects of the mechanism of skin absorption based on experimental findings (for example, the delipidisation of the stratum corneum barrier to demonstrate the importance of the lipid domain in governing percutaneous absorption), the quantification of this process has attracted substantial interest, most notably since the early 1990s when Flynn collated a dataset of experimental permeability values that was used to develop semi-quantitative relationships between structure and permeability [8]. This work was subsequently developed by Potts and Guy (1992), who developed their much-quoted and seminal QSPR which related skin permeability to the lipophilicity and molecular weight of a penetrant [7].

The following years saw a plethora of models published on this subject, which, in itself, demonstrated a fundamental issue with the field – the models were either specific to particular groups of chemicals or were "global" in scope, focusing on simple representations of percutaneous absorption for a wide range of chemicals – those which, in effect, fitted within the "chemical space" of the dataset used to construct them. Such generalisations, while providing important mechanistic information for percutaneous absorption, as well as the ability to predict the permeability of new chemicals, did not address specific groups of chemicals (as, for example, Potts and Guy did when specifically modelling nonelectrolytes [9]). Such models have been extensively modified with particular emphasis on non-linearity [10].

The modelling work reported in this chapter deals with some of the recent advances in this field. Gaussian Process Regression is defined as a collection of random variables that, jointly, have a Gaussian distribution and which is characterised completely by its mean and covariance. The mean is usually defined as the 'zero everywhere' function, and the covariance function, $k(x_i, x_j)$, allows for specifying *a priori* knowledge from a training dataset [11, 12]. To make a prediction, y^*, at a new input, x^*, the conditional distribution, $p(y^*|y_1 . . . ,y_N)$, is computed on the observed vector $[y_1 . . . ,y_N]$. Since the model being applied is a Gaussian Process, this distribution is also Gaussian in nature and is therefore completely defined by its mean and covariance. A recent study on Gaussian Process modelling is described in Chapters 32. This chapter, which models permeability across a polydimethylsiloxane membrane, suggests a significant difference in the mechanism of permeability across this membrane compared to mammalian skin. However, they do also comment that the nature and quality of the data may significantly impact on model development.

One shortcoming of Gaussian models is that they currently do not yield a specific functional output – that is, an equation [13]. However, this was addressed by Lam *et al.*, who used Gaussian Process methods coupled with feature selection methods to statistically determine which combination of parameters provided the best model. This is further developed in Chapter 33 by Moss *et al.*, who use the Non-linear Auto-Regressive Moving Average with eXogenous inputs (NARMAX) model to address this issue. These recent studies on the development of quantitative models for skin permeability follow the recommendations of Cronin and Schultz on the multi-disciplinary nature of developing such models; comments that have informed the OECD Principles for the Validation of (Q)SAR Models [14, 15].

Therefore, this chapter provides an insight to novel experimental and mathematical methods that aim to enhance our understanding of the process of skin absorption, and in doing so demonstrates the links between experimental and theoretical aspects of this field.

REFERENCES

1. Y. Yokomizo, Effect of phosphatidylcholine on the percutaneous penetration of drugs through the dorsal skin of guinea pigs *in vitro*; And analysis of the molecular mechanism, using attenuated total reflectance Fourier transform infrared (ATR-FTIR) spectroscopy, *Journal of Controlled Release*, 1996, **42**, 249–262.

2. H. A. Ayala-Bravo, D. Quintanar-Guerrero, A. Naik, Y. N. Kalia, J. M. Cornejo-Bravo, A. Ganem-Quintanar, Effects of sucrose oleate and sucrose laureate on *in vivo* human *stratum corneum* permeability, *Pharmaceutical Research*, 2003, **20**, 1267–1273.

3. S. A. Schwartz, M. L. Reyzer, R. M. Caprioli, Direct tissue analysis using matrix-assisted laser desorption/ionization mass spectrometry: practical aspects of sample preparation, *Journal of Mass Spectrometry*, 2003, **38**, 699–708.

4. R. L. Bronaugh, H. L. Hood, M. E. K. Kraeling and J. J. Yourick, Determination of percutaneous absorption by in vitro techniques, in *Percutaneous Absorption – Drugs, Cosmetics, Mechanisms, Methodology*, ed. R. L. Bronaugh and H. I. Maibach, 3rd edn, Marcel Dekker, New York, 1999, pp. 229–234

5. R. L. Bronaugh and H. L. Hood, Will cutaneous blood levels of absorbed material be systemically absorbed?, in *Percutaneous Absorption – Drugs, Cosmetics, Mechanisms, Methodology*, ed. R. L. Bronaugh and H. I. Maibach, Marcel Dekker, New York, 3rd edn, 1999, pp. 235–240

6. M. S. Roberts, Y. G. Anissimov and R. A. Gonsalvez, Mathematical models in percutaneous absorption, in *Percutaneous Absorption – Drugs, Cosmetics, Mechanisms, Methodology*, ed. R. L. Bronaugh and H. I. Maibach, Marcel Dekker, New York, 3rd edn, 1999, pp. 3–56.

7. R. O. Potts and R.H. Guy, Predicting skin permeability, *Pharmaceutical Research*, 1992, **9**, 663–669.

8. G. L. Flynn, Physicochemical determinants of skin absorption, in *Principles of Route-to-Route Extrapolation for Risk Assessment*, ed. T. R. Gerrity and C. J. Henry, Elsevier, New York, 1990, pp. 93 – 127.

9. R. O. Potts and R. H. Guy, A predictive algorithm or skin permeability – the effects of molecular size and hydrogen-bond activity, *Pharmaceutical Research*, 1995, **12**, 1628–1633.

10. A. Wilschut, W. F. ten Berge, P. J. Robinson and T. E. McKone, Estimating skin permeation — the validation of 5 mathematical skin permeation models, *Chemosphere*, 1995, **30**, 1275–1296.

11. G. P. Moss, Y. Sun, N. Davey, R. Adams, W. J. Pugh and M. B. Brown, The application of Gaussian Processes to the prediction of percutaneous absorption, *Journal of Pharmacy and Pharmacology*, 2009, **61**, 1147–1153.

12. G. P. Moss, Y. Sun, S. C. Wilkinson, N. Davey, R. Adams, G. P. Martin, M. Prapopoulou and M. B. Brown, 2011, *Journal of Pharmacy and Pharmacology*, **63**, 1411–1427.

13. L. T. Lam, Y. Sun, N. Davey, R. Adams, M. Prapopoulou, M. B. Brown and G. P. Moss, The application of feature selection to the development of Gaussian Process

models for percutaneous absorption, *Journal of Pharmacy and Pharmacology*, 2010, **62**, 738–749.

14. M. T. D. Cronin and T. W. Schultz, Pitfalls in QSAR, *Journal of Molecular Structure: THEOCHEM*, 2003, **622,** 39–51.

15. http://www.oecd.org/env/ehs/risk-assessment/validationofqsarmodels.htm   [Accessed 30th August 2013].

28

VISUALISATION OF THE PERMEATION OF CHLORHEXIDINE WITHIN SKIN USING TIME-OF-FLIGHT SECONDARY ION MASS SPECTROMETRY (TOF-SIMS).

A Judd[1], D Scurr[2], J Heylings[3], D Griffiths[4], K-W Wan[4] and G P Moss[1]

[1]School of Pharmacy, Keele University, Staffordshire, United Kingdom. [2]The School of Pharmacy, University of Nottingham, Nottingham, United Kingdom. [3]Dermal Technology Laboratory Ltd, Keele, Staffordshire, United Kingdom, [4]School of Pharmacy, University of Central Lancashire, Preston, United Kingdom.

1   INTRODUCTION

The stratum corneum, the outermost layer of the skin, is the rate-limiting barrier to the ingress of exogenous chemicals [1]. To improve the bioavailablility of future topically applied formulations, the permeation pathway and distribution throughout the skin strata must be understood and permeation into and across the skin characterised. Thus, this study outlines the development of a Time-of-Flight Secondary Ion Mass Spectrometry (ToF-SIMS) method in order to visualise drug permeation within the skin and its potential applications in skin permeation studies.

Conventionally, extraction methods are used to determine a drug dose within the skin due to their ease and low cost. The drug concentration can be determined using a sensitive analytical method, most commonly HPLC for cold compounds *i.e.* clobetasol propionate [2], or if radio-labelled [3], liquid scintillation counting or scinitigraphic techniques such as autoradiography. Tape stripping has several limitations, including method standardisation, incomplete removal of the stratum corneum, inconsistent removal of skin layers [4], inconsistent pressure applied and nature of the adhesive tape used [5]. With each sequential tape strip taken, the amount of the stratum corneum material removed decreases. To prevent this problem from affecting the distribution profile within the stratum corneum the quantity of removed corneocytes must be determined by weighing (gravimetric analysis) or other means such as spectroscopic analysis [5] or transepidermal water loss (TEWL) measurements [6 – 8]. Horizontal sectioning of skin does not provide an accurate means of identifying within which histological skin compartment the drug is localised as there may be overlap between the epidermis and the dermis within the skin sections. The horizontal sectioning method is also time consuming and laborious.

There are numerous methods available for visualising the distribution of compounds within the skin. Autoradiography, for example, allows the distribution of a radio-labelled compound to be visualised by means of exposing a radio-sensitive film to the tissue and allowing a long exposure time for the development process. The technique itself is expensive and can require a long exposure period for the radiogram to develop [9]. The handling, transport and high cost of radio-labelled drugs are all additional and significant limitations.

Various spectroscopic techniques have been used to map the passage of exogenous chemicals into the skin. Attenuated Total Reflectance Fourier Transform Infrared Spectroscopy (ATR-FTIR) has been utilised as an alternative to conventional Franz-type diffusion cell studies [10, 11]. The stratum corneum is placed onto a reflectance cell and is dosed with an infrared (IR) active compound. The presence of, and any changes to, the characteristic peaks of interest is monitored over the designated contact time [12]. Further, IR microscopic imaging techniques have been proposed as a method that may provide insight into the permeation pathway of an IR active compound and have been used to track the permeation of phospholipids into the skin [13]. IR microscopic imaging allows for larger skin areas to be analysed but suffers from a low spatial resolution of around 10 -12 μm.

Raman microscopy has been utilised as a complementary vibrational spectroscopic technique to FTIR methods to observe the permeation of pharmaceutical drugs across the stratum corneum [13, 14]. Confocal Raman microscopy allows for the analysis of depths of around 100 μm within skin tissue with a pixel size of around 2-3 μm [15]. Raman microscopy has been employed to analyse biomolecules within the skin and to attempt to elucidate the mechanism of penetration enhancers [16, 17]. Raman microscopy is dependent on the compound being Raman-active and has a spatial resolution of around 5 μm [18]. Both Raman and FTIR spectroscopic techniques have *in vivo* applications and are highly complementary.

Confocal Laser Scanning Microscopy (CLSM) is a widely used technique that provides high resolution images [19]. It allows for the visualisation of the upper skin strata *in situ* without the need for fixation or mechanical sectioning. Liposomes with a fluorescent dye encapsulated within their interior have been applied to human skin and their localisation within optical sections observed by CLSM, allowing determination of the distribution of the released dye [20]. However, the limitations of fluorescence microscopy are that the compound must fluoresce when excited by the laser or, alternatively, it must be tagged with a fluorescent probe. If fluorescent labelling is required, the CLSM images are illustrating the localisation of the fluorophore tag and not necessarily the native compound as the addition of a label may alter the permeation characteristics of the compound. Autofluorescence of biological samples may be problematic as it can substantially increase background noise thus reducing the contrast of the recorded images [21]. The main limitation of CLSM is that the signal and thus resolution decreases with increasing depth into the skin due to scattering and absorption of both the laser excitation light and the emitted fluorescence [22].

Matrix-Assisted Laser Desorption/Ionisation Imaging Mass Spectrometry (MALDI-IMS) has emerged as a powerful tool for imaging molecular compounds within sections of complex biological tissue. A key advantage of mass spectral techniques, compared to the imaging techniques discussed above, is that they can map the distribution of many biologically- or pharmaceutically-relevant compounds simultaneously [23]. For example, Bunch *et al.* mapped the distribution of ketoconazole, a constituent of medicated shampoos, after application to porcine epidermis using MALDI-IMS [24]. The resulting mass spectral image of ketoconazole within the skin tissue section was then superimposed onto a histological image. This spatial molecular information allowed for the permeation of the drug to be mapped and quantified within a porcine skin tissue section. A significant disadvantage of MALDI-IMS is its resolution – approximately 100 μm – which is due to the laser diameter [25].

There is an increasing interest in the development of imaging techniques that do not require radiolabelling or complicated sample preparation. One such imaging technique with high mass and spatial resolution is Time-of-Flight Secondary Ion Mass Spectrometry (ToF-SIMS). ToF-SIMS is capable of imaging the spatial distribution of a compound within the surface of complex tissue, which may yield information on the distribution of the permeant with the skin and, potentially, its mechanism of permeability [26].

ToF-SIMS chemical imaging analysis involves the rastering of a pulsed primary ion beam, in this case Bi_3^+ ions, over a surface of the tissue samples leading to a collision cascade [27]. Secondary ions are sputtered from the sample surface and are accelerated into the flight path in which they are separated by their 'Time-of-Flight' which is proportional to the square root of the mass of the ion. Chemical maps of selected secondary ions can then be retrospectively reconstructed for the analysed areas. In ToF-SIMS analysis, all secondary ions liberated from the sample for a set mass/charge range are detected in parallel. More comprehensive details and principles of the ToF-SIMS technique can be found elsewhere [28]. ToF-SIMS analysis provides high sensitivity with the detection of trace elements within the ppm range, including the parallel detection of their isotopes, spatial localisation for many biomolecules simultaneously and the ability to retrospectively analyse data, where each pixel within the ToF-SIMS distribution map corresponds to a full mass spectrum [29]. The technique is not dependent on the penetration of a reagent or probe reactivity and is not adversely affected by background staining [30].

ToF-SIMS has been used to explore the effects of photo-ageing on human skin. Lee *et al.* used ToF-SIMS analysis of *ex vivo* skin to demonstrate spatial alterations of major skin constituents, such as collagen and lipids within the epidermis and dermis, following exposure to UVB irradiation [31]. Furthermore, ToF-SIMS has been utilised to show the distribution and localisation of synthetic pseudoceramides within a cosmetic formulation topically applied to porcine epidermis [32, 33].

The aims of this study were to characterise the distribution of a topically applied drug, chlorhexidine, using ToF-SIMS and to compare this technique to conventional tape stripping methods.

2   METHODOLOGY

Skin was prepared from the flank of six week old pigs collected from an abattoir prior to any steam cleaning treatment. The skin was cleaned and dermatomed to a thickness of 400 μm using an electric dermatome and mounted on to a static Franz cell with physiological saline (0.9% w/w NaCl) as the receptor fluid. Skin integrity was checked using transepithelial electrical resistance. Any skin sample that was shown to have abnormal skin permeability properties was discarded according to established criteria [34]. The surface of intact skin membranes was dosed with an aqueous 2% solution (w/v) of chlorhexidine digluconate (CHG) for 24 h, or in the case of the control, ultrapure water. After the 24h contact time the CHG was removed by agitating 10 mL of ultrapure water across the skin surface followed by blot drying. The skin samples were removed from the diffusion chambers then immediately plunged into liquid nitrogen and placed on dry ice for transport. The snap-frozen porcine skin samples were mounted on a cryostat (Leica, CM1850, temperature of cryostat chamber -28°C) using Optical Cutting Temperature (OCT) embedding material (Fisher scientific, UK). Vertical cross-sections of skin, each of 8 μm thickness, were placed onto a clean glass cover slip (1 cm × 1 cm). The cover slips were first rinsed in ultra-pure water followed by methanol and lastly chloroform, before the skin sample was loaded. For the tape strip investigation, 21 tape strips were taken to remove the stratum corneum. The strips were then freeze-dried and placed on a solvent-cleansed microscope slide ready for analysis.

ToF-SIMS was performed using a ToF-SIMS IV instrument (IONTOF, GmbH, Münster, Germany) using a Bi_3^+ cluster source and a single-stage reflectron analyser. The cover slip containing the samples was mounted onto a cryostage with a cold finger mechanism and frozen to -80°C using a liquid nitrogen cooling system. The mounted skin samples were then exposed to an ultra-high vacuum. A primary ion energy of 25 kV, along with a pulsed target current of approximately 1 pA and post-acceleration energy of 10 kV were employed throughout the analysis. The primary ion dose density was maintained at less than 1×10^{12} ions per cm^2 throughout to ensure static conditions. Spectra were acquired in both positive and negative mode and imaging data was recorded over a sample area ranging from 100 - 500 μm^2 by raster scanning a primary beam over the sample. Data processing was performed using SurfaceLab 6 (IONTOF GmbH) software for spectroscopy and image analysis. Although all data was collected in the positive and negative ion mode, only data within the negative mode will be presented here.

HPLC analysis of tape strips was conducted on a Shimadzu UCFC system with an SPD M20 diode array detector using an isocratic mobile phase consisting of a methanol:water mixture (75:25) with 0.005 M sodium octane sulphonate and 0.1% (v/v) triethylamine. The pH was adjusted with glacial acetic acid to pH 4. The flow rate was 1.5 mL per minute on a MetLab ODS-H reverse phase column (150 × 4.6 mm, 5μm). Mobile phase was used for all drug extractions from tape strips.

3   RESULTS AND DISCUSSION

The ToF-SIMS technique generates mass spectral data in which fragment ions are detected and identified. Identified fragment ions are added to a peak list that is used to retrospectively build a spectral image for each fragment ion from the rastered area of sample analysed, providing a highly detailed mass spectral map of the biological sample [35]. Biological material can yield hundreds of different secondary ions (Figure 1) making the interpretation difficult. This intricacy occurs as a result of fragmentation of larger secondary ions liberated from the sample surface and these complex spectra hold detailed chemical information about the composition, distribution and surface order of the sample [36]. In order to identify and assign secondary ion peaks, it is imperative to run reference samples *e.g.* drug sample, embedding and substrate materials, in order to understand characteristic spectral differences. Control biological tissue should also be analysed in parallel [37]. To study the deposition of CHG within the skin it is important to first elucidate characteristic fragment ions unique and indicative to the presence of CHG. The overview of the negative polarity spectra obtained (Figure 1a) illustrates the secondary ion peaks detected within the mass range 1 - 160 (amu) from the surface of dosed 2% (w/v) solution of CHG and the negative control porcine skin.

Chlorhexidine (Figure 2b) consists of two chlorinated aromatics, therefore chlorine might be an anticipated ideal marker of the drug within the tissue. This was found not to be the case and chlorine was only used as supplementary supporting data as the element is abundant within tissue and is not unique to the presence of chlorhexidine. Chlorine is observed in the negative control skin surface as highlighted in the survey spectra (Figure 1a) and associated chemical images (Figure 2). A fragment ion thought to be characteristic of chlorhexidine was observed at 151 amu, with the proposed secondary ion assignment $C_7H_4N_2Cl^-$ (Figure 1). The magnification of the overview spectra for this fragment ion (Figure 1c) highlights a potential complication of analysing biological tissue in that a low intensity endogenous peak is detected within the skin tissue itself at 150.8 amu. Great care must be taken integrating the 151-$C_7H_4N_2Cl^-$ fragment ion to prevent overlap, resulting in noise within the spectral image (Figure 2). The overview spectrum (Figure 2a) highlights the $C_{22}N_{10}H_{30}Cl_2^-$ [M-H]$^-$ ion peak for chlorhexidine (505 amu). This molecular ion peak is clear evidence for the identification and localisation of chlorhexidine, but, due to the high molecular weight the intensity is relatively low. Such low signal intensity amongst the noise and interference of complex biological substrates results in low chemical contrast and thus impedes the ability to obtain detailed chemical images.

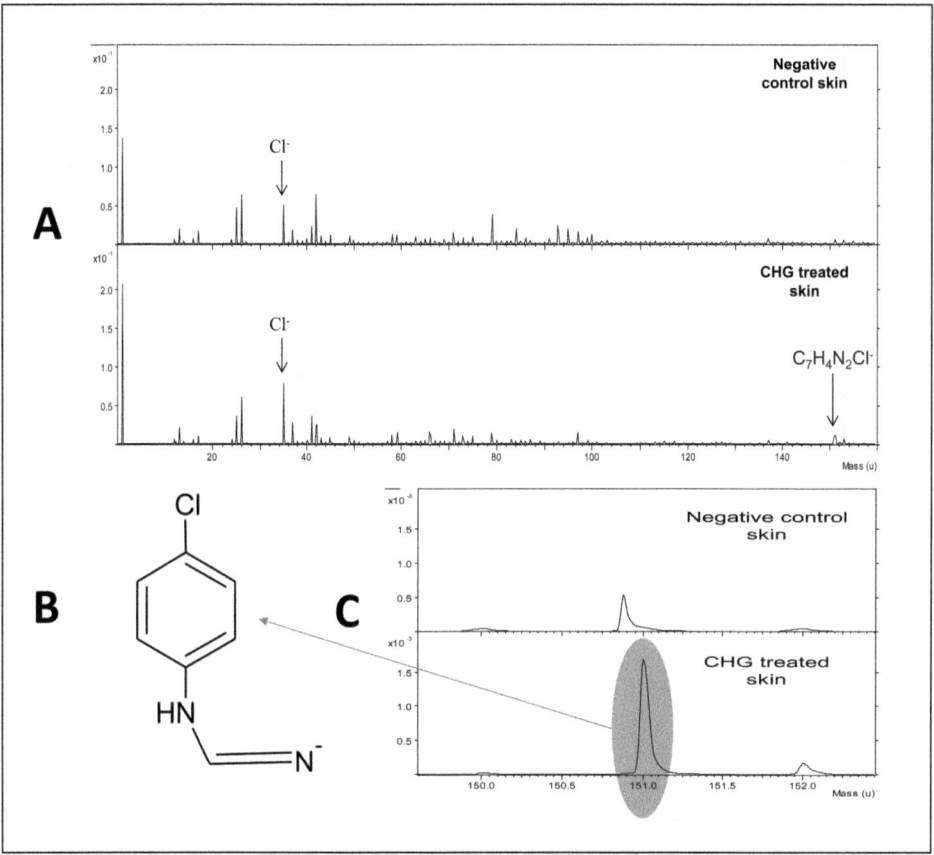

Figure 1 *(A) ToF-SIMS negative polarity survey spectra of the negative control porcine skin and skin dosed with a 2% chlorhexidine digluconate (w/v) solution after a 24 h contact time and wash. The CHG markers chlorine (Cl⁻) and $C_7H_4N_2Cl^-$ fragment ion are indicated. (B) Structure of fragment ion ($C_7H_4N_2Cl^-$) unique to chlorhexidine diglcuonate identified from negative polarity overview spectrum of CHG treated porcine skin. (C) Magnification of chlorhexidine digluconate unique fragment ion within ToF-SIMS spectrum for negative control porcine skin and skin dosed with a 2% CHG (w/v) solution after a 24h contact time and wash.*

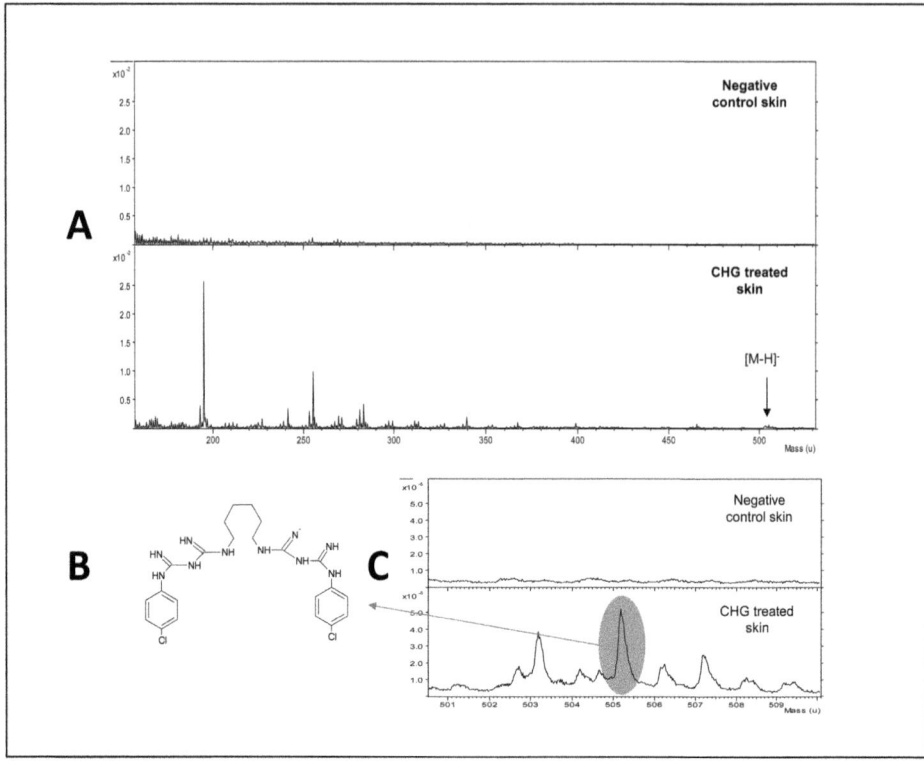

Figure 2 *(A) ToF-SIMS negative polarity survey spectra of negative control porcine skin and skin dosed with a 2% chlorhexidine digluconate (w/v) solution after a 24 h contact time and wash (CHG molecular ion is indicated by [M-H]⁻). (B) Molecular ion of chlorhexidine [M-H]- identified within the overview spectra of the porcine skin dosed with a 2% chlorhexidine digluconate (w/v) solution. (C) Magnification of chlorhexidine [M-H]⁻ within the ToF-SIMS spectrum for the negative control porcine skin and skin dosed with a 2% chlorhexidine digluconate (w/v) solution after a 24h contact time and wash.*

Relevant chemical images were generated after the secondary ion peaks were identified and assigned as CHG markers. These spectral images visualise the distribution of chlorhexidine upon the surface of dosed and negative control skin. The skin was analysed post-wash, and the presence of such a high intensity signal for ions characteristic of CHG was observed (Figure 3). This suggests that the cationic CHG strongly binds to the superficial layers of the skin. The suggestion of a strong binding affinity of CHG the skin protein has been described elsewhere [38]. It is clear from Figure 3 that Cl⁻ is not a strong candidate as a characteristic CHG marker due to its abundance upon the skin surface, in contrast to the other selected markers, $C_7H_4N_2Cl^-$ and [M-H]⁻.

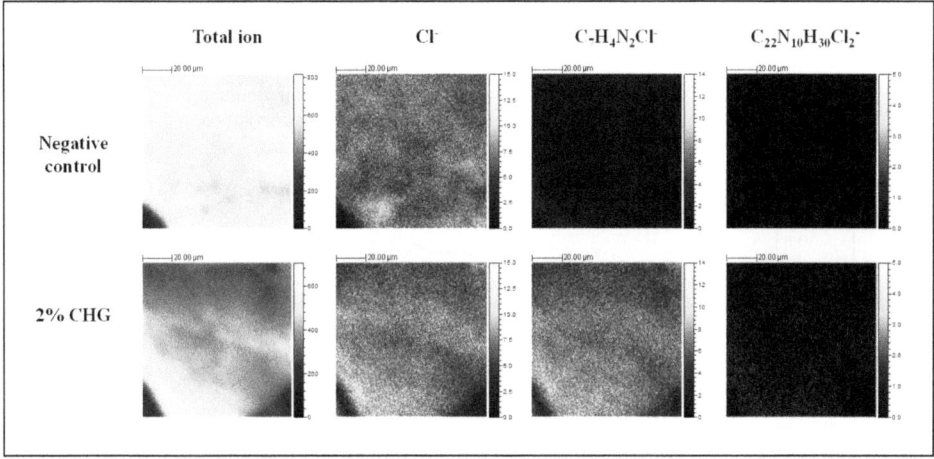

Figure 3 *ToF-SIMS secondary ion images of various secondary ions (Cl⁻, C₇H₄N₂Cl⁻ and [M-H]⁻) of the negative control and 2% CHG (w/v) dosed porcine dermatomed skin (400 μm thick). Data taken from skin surface after washing negative control and dosed skin surface with 10 mL of ultrapure water. Scale bar equates to 20 μm.*

ToF-SIMS analysis can be undertaken in two operational modes: the bunched mode, with high mass resolution and low spatial resolution, or the burst alignment mode, with high spatial resolution and low mass resolution. The differences in mass and spatial resolution achievable in the two different operational modes are illustrated in Figures 4 and 5, respectively. The decrease in mass resolution associated with burst alignment mode can result in the inability to discriminate one peak from another, as exemplified in Figure 4 where the ^{37}Cl⁻ and C₃H⁻ ion peaks cannot be resolved in the burst alignment spectrum. Figure 4 illustrates the mass spectra obtained in both operational modes for chlorine on the surface of the skin. The two modes can be complementary, offering both high mass resolution and high spatial resolution. Sodhi (2004) discussed the principles behind both operating modes at length [39].

Cryosections of skin were analysed to determine the penetration depth and the distribution of CHX within the tissue. A series of anomalous results (as a consequence of incorrect sample preparation) highlighted that this stage is of paramount importance to the integrity and quality of the ToF-SIMS data generated. Figure 5 illustrates one such artefact identified from the cryosectioning process which was that the stratum corneum had 'curled' when adsorbed onto the solvent cleaned glass cover slip resulting in the stratum corneum and upper viable epidermis folding over onto the dermis. This issue is easily identifiable using the ToF-SIMS technique as the change in topography is readily observable. The ToF-SIMS system can also act as a scanning electron microscope (SEM) that can be utilised to check the tissue integrity. Furthermore, complementary histology such as haematoxylin and eosin staining can be performed post ToF-SIMS analysis (Figure 6) to confirm tissue integrity or skin structures of interest.

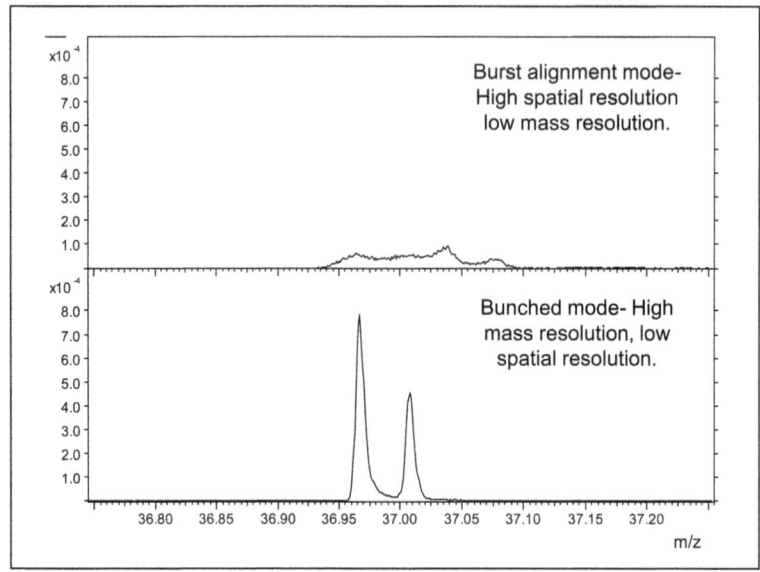

Figure 4 *ToF-SIMS spectra of* $^{37}Cl^-$ *present in cryosections (8 μm thick) of untreated control dermatomed porcine skin (400 μm) and skin dosed with 2% CHG (w/v). The spectra were obtained in negative polarity using burst alignment and then bunched mode to demonstrate the difference in observed mass resolution.*

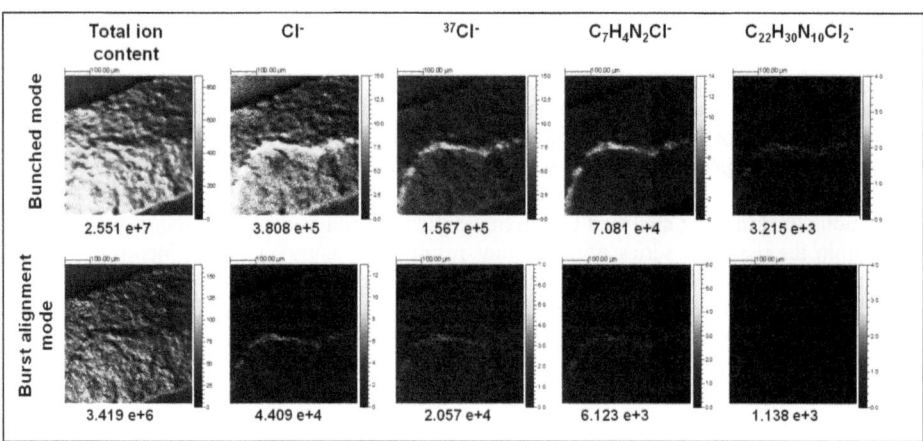

Figure 5 *Secondary ion images of 2% CHG (w/v) dosed porcine dermatomed skin (400 μm thick) vertically cryo-sectioned into 8 μm thick sections. Curling of the sample occurred when adsorbing the section onto solvent cleaned coverslip (10 mm × 10 mm). Scale bar equates to 100 μm, number corresponds to the total ion count for the specific ions of interest (TC).*

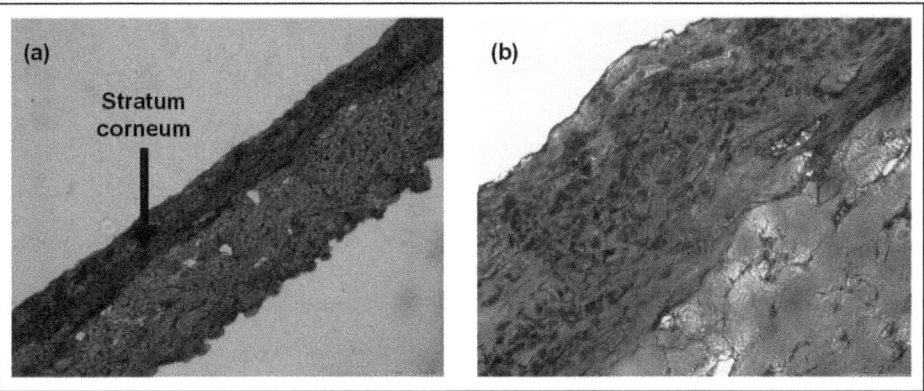

Figure 6 *Haematoxylin and eosin stained cryosection of dermatomed porcine skin (400 µm) post dosing with 2% CHG, illustrating the folding of the stratum corneum and viable epidermis onto the dermis. (a) ×100 magnification (b) × 400 magnification.*

Although the sectioning of the skin sample in Figure 5 was found to be substandard due to tissue curling, the drug markers assigned from the mass spectra analysed in the bunched mode on the skin surface were found to be of high enough intensity to map the deposition and distribution of chlorhexidine within the skin. All data presented here was recorded in negative polarity but the data obtained in the positive polarity (including the [M]+ ion) also supported these observations.

To complement the ToF-SIMS cryosectioned tissue study, a tape stripping technique was investigated to compare the data and validate the obtained distribution profile. An advantage of using ToF-SIMS to analyse the drug content within the stratum corneum material on the tape strip samples is the high sensitivity of the technique (within the femtomolar region).

Conventional tape strip techniques would require tape strips to be pooled together to ensure that the signal is above the limit of quantification for detection systems such as HPLC. ToF-SIMS analysis also reveals the spatial distribution of this drug within the removed tissue material. Additionally, owing to the retrospective data analysis capabilities of ToF-SIMS, the ion intensity can be exported exclusively from tissue material where samples exhibit non-homogeneous coverage of the adhesive tape; that is, they include stratum corneum material only in the analysis.

A semi-quantitative depth profile was obtained from the ToF-SIMS analysis and can be compared to the depth profile obtained from extracting the CHG from the tape strips and subsequent HPLC analysis (Figure 7).

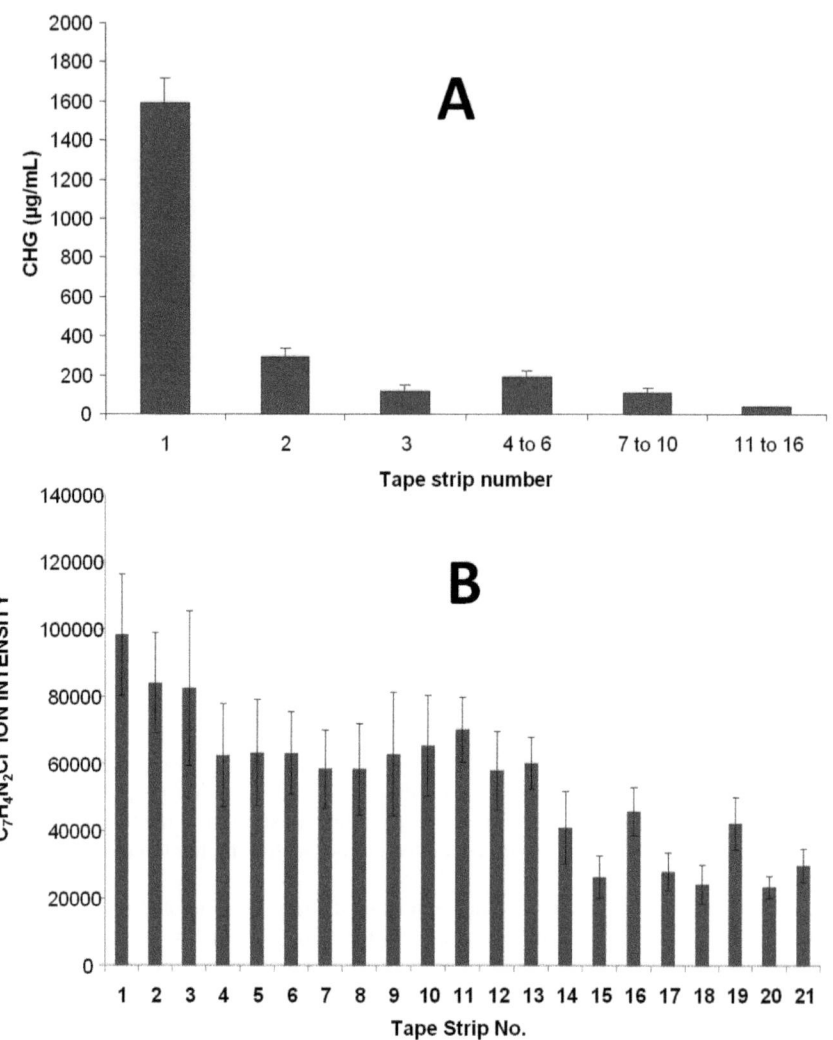

Figure 7 *(A) Concentration of chlorhexidine diglcuonate extracted from tape strips and analysed by HPLC (n=14). (B) Fragment ion (C₇H₄N₂Cl) assigned to chlorhexidine diglucConate intensity for 21 consecutive tape strips taken from CHG dosed (2% w/v) porcine skin (n=4). Ion intensity normalised to total ion count of stratum corneum material on each tape strip.*

The ToF-SIMS analysis yields semi-quantitative data only but, due to the inherent sensitivity of the method, the 'pooling' of the tape strips is not required. This advantage can aid the elucidation of the depth profile of the drug within the stratum corneum as observed in Figure 7. The depth profile revealed by solvent extraction followed by HPLC analysis of tapes and ToF-SIMS analysis of tape is in good agreement with the concentration of CHG which decreases with increasing depth into the stratum corneum.

4   CONCLUSION

This study has detailed the method development process for the utilisation of ToF-SIMS to analyse the permeation and absorption of CHG, a model topical compound used extensively as an antibacterial agent, within porcine skin. It was possible to identify and detect secondary ions unique to CHG and obtain spatial information to map the distribution within the skin following a 24-hour contact time. Analysis of tape strips by ToF-SIMS illustrated the same general trend as those analysed by conventional solvent extraction-HPLC methods but, due to the low limit of detection, avoided the requirement for the 'pooling' of samples often associated with techniques such as tape stripping.

5   REFERENCES

1. A. C. Williams, *Transdermal and topical drug delivery from theory to clinical practice*. London, Pharmaceutical Press, 2003.

2. H. J. Weigmann, J. Lademann, R. V. Pelchrzim, W. Sterry, T. Hagemeister, R. Molzahn, M. Schaefer, M. Lindscheid, H. Schaefer, V. P. Shah, Bioavailability of clobetasol propionate - Quantification of drug concentrations in the *stratum corneum* by dermatopharmacokinetics using tape stripping, *Skin Pharmacology and Physiology*, 1999, **12**, 46–53.

3. K. R. Brain, K. A. Walters, D. M. Green, S. Brain, L. J. Loretz, R. K. Sharma, W. E. Dressler, Percutaneous penetration of diethanolamine through human skin *in vitro*: Application from cosmetic vehicles, *Food and Chemical Toxicology*, 2005, **43**, 681–690.

4. R. G. van der Molen, F. Spies, J. M. van't Noordende, E. Boelsma, A, M. Mommaas, H. K. Koerten, Tape stripping of human *stratum corneum* yields cell layers that originate from various depths because of furrows in the skin, *Archives of Dermatological Research*, 1997, **289**, 514–518.

5. E. Marttin, M. T. A. NeelissenSubnel, F. H. N. DeHaan and H. E. Bodde, A critical comparison of methods to quantify *stratum corneum* removed by tape stripping. *Skin Pharmacology*, 1996, **9**, 69–77.

6. C. S. King, Barton, S. Nicholls and R. Marks, Change in Properties of the Stratum-Corneum as a Function of Depth, *British Journal of Dermatology*, **100**, 1979, 165–172.

7. Y. N. Kalia, F. Pirot and R. H. Guy, Homogeneous transport in a heterogeneous membrane: Water diffusion across human *stratum corneum in vivo*, *Biophysical Journal*, 1996, **71**, 2692–2700.

8. K. L. Trebilcock, J. R. Heylings and M. F. Wilks, In-Vitro Tape Stripping as a Model for In-Vivo Skin Stripping, *Toxicology in Vitro*, 1994, **8**, 665–667.

9. E. Touitou, V. M. Meidan and E. Horwitz, Methods for quantitative determination of drug localized in the skin, *Journal of Controlled Release*, 1998, **56**, 7–21.

10. Y. Yokomizo, Effect of phosphatidylcholine on the percutaneous penetration of drugs through the dorsal skin of guinea pigs *in vitro*; And analysis of the molecular mechanism, using attenuated total reflectance Fourier transform infrared (ATR-FTIR) spectroscopy, *Journal of Controlled Release*, 1996, **42**, 249–262.

11. H. A. Ayala-Bravo, D. Quintanar-Guerrero, A. Naik, Y. N. Kalia, J. M. Cornejo-Bravo, and A. Ganem-Quintanar, Effects of sucrose oleate and sucrose laureate on *in vivo* human *stratum corneum* permeability, *Pharmaceutical Research*, 2003, **20**, 1267–1273.

12. K. Moser, K. Kriwet, A. Naik, Y. N. Kalia, R. H. Guy, Passive skin penetration enhancement and its quantification *in vitro*, *European Journal of Pharmaceutics and Biopharmaceutics*, 2001, **52**, 103–112.

13. C. H. Xiao, D. J. Moore, M. E. Rerek, C. R. Flach and R. Mendelsohn, Feasibility of tracking phospholipid permeation into skin using infrared and Raman microscopic imaging, *Journal of Investigative Dermatology*, 2005, **124**, 622–632.

14. B. Gotter, W. Faubel and R. H. H. Neubert, FTIR microscopy and confocal Raman microscopy for studying lateral drug diffusion from a semisolid formulation, *European Journal of Pharmaceutics and Biopharmaceutics*, 2010, **74**, 14–20.

15. J. L. Bruneel, J. C. Lassegues and C. Sourisseau, In-depth analyses by confocal Raman microspectrometry: experimental features and modeling of the refraction effects, *Journal of Raman Spectroscopy*, 2002, **33**, 815–828.

16. P. J. Caspers, G. W. Lucassen, E. A. Carter, H. A. Bruining, and J. G. Puppels, *In vivo* confocal Raman microspectroscopy of the skin: Noninvasive determination of molecular concentration profiles, *Journal of Investigative Dermatology*, 2001, **116**, 434–442.

17. P. J. Caspers, A. C. Williams, E. A. Carter, H. G. M. Edwards, B. W. Barry, H. A. Bruining and G. J. Puppels, Monitoring the penetration enhancer dimethyl sulfoxide in human *stratum corneum in vivo* by confocal Raman spectroscopy, *Pharmaceutical Research,* 2002, **19**, 1577–1580.

18. P. J. Caspers, G. W. Lucassen, H. A. Bruining, G. J. Puppels, Automated depth-scanning confocal Raman microspectrometer for rapid *in vivo* determination of water concentration profiles in human skin, *Journal of Raman Spectroscopy*, 2000, **31**, 813–818.

19. R. Alvarez-Roman, A. Naik, Y. Kalia, R. H. Guy and H. Fessi, Skin penetration and distribution of polymeric nanoparticles, *Journal of Controlled Release*, 2004, **99**, 53–62.

20. D. D. Verma, S. Verma, G. Blume and A. Fahr, Liposomes increase skin penetration of entrapped and non-entrapped hydrophilic substances into human skin: a skin penetration

and confocal laser scanning microscopy study, *European Journal of Pharmaceutics and Biopharmaceutics*, 2003, **55**, 271–277.

21. S. W. Paddock, Principles and practices of laser scanning confocal microscopy, *Molecular Biotechnology*, 2000, **16**, 127–149.

22. A. J. Hoogstraate, C. Cullander, J. F. Nagelkerke, S. Sengel, J. C. Verhoef, H. E. Junginger and H. E. Bodde, Diffusion Rates and Transport Pathways of Fluorescein Isothiocyanate (FITC)-Labeled Model Compounds through Buccal Epithelium, *Pharmaceutical Research*, 1994, **11**, 83–89.

23. Y. Hsieh, J. Chen and W. A. Korfmacher, Mapping pharmaceuticals in tissues using MALDI imaging mass spectrometry, *Journal of Pharmacological and Toxicological Methods*, 2007, **55**, 193–200.

24. J. Bunch, M. R. Clench and D. S. Richards, Determination of pharmaceutical compounds in skin by imaging matrix-assisted laser desorption/ionisation mass spectrometry, *Rapid Communications in Mass Spectrometry*, 2004, **18**, 3051–3060.

25. S. A. Schwartz, M. L. Reyzer and R. M. Caprioli, Direct tissue analysis using matrix-assisted laser desorption/ionization mass spectrometry: practical aspects of sample preparation, *Journal of Mass Spectrometry*, 2003, **38**, 699–708.

26. A. Benninghoven, Chemical-Analysis of Inorganic and Organic-Surfaces and Thin-Films by Static Time-Of-Flight Secondary-Ion Mass-Spectrometry (Tof-SIMS), *Angewandte Chemie-International Edition in English*, 1994, **33**, 1023–1043.

27. J. Vickerman, ToF-SIMS – Surface Analysis by Mass Spectrometry, in *Surface Analysis by Mass Spectrometry*, ed. J. Vickerman, D. Briggs, 2nd edn, 2001, IMP Publications, Chichester.

28. A. Benninghoven, Chemical-analysis of inorganic and organic-surfaces and thin-films by static Time-of-Flight Secondary-ion mass-spectrometry (Tof-SIMS), *Angewandte Chemie-International Edition in English*, 1994, **33**, 1023–1043.

29. J. M. Chabala, K. K. Soni, J. Li, K. L. Gavrilov and R. Levisetti, High-resolution chemical imaging with scanning ion probe SIMS, *International Journal of Mass Spectrometry and Ion Processes*, 1995, **143**, 191–212.

30. P. Malmberg, E. Jennische, D. Nilsson and H. Nygren, High-resolution, imaging ToF-SIMS: novel applications in medical research, *Analytical and Bioanalytical Chemistry*, 2011, **399**, 2711–2718.

31. T. G. Lee, J.-W. Park, H. K. Shon, D. W. Moon, K. Li and J. H. Chung, Biochemical imaging of tissues by SIMS for biomedical applications, *Applied Surface Science*, 2008, **255**, 1241–1248.

32. M. Okamoto, N. Tanji, Y. Katayama and J. Okada, ToF-SIMS investigation of the distribution of a cosmetic ingredient in the epidermis of the skin, *Applied Surface Science*, 2006, **252**, 6805–6808.

33. N. Tanji, M. Okamoto, Y. Katayama, M. Hosokawa, N. Takahata and Y. Sano, Investigation of the cosmetic ingredient distribution in the *stratum corneum* using NanoSIMS imaging, *Applied Surface Science*, 2008, **255**, 1116–1118.

34. D. Davies, R. Ward and J. Heylings, Multi-species assessment of electrical resistance as a skin integrity marker for *in vitro* percutaneous absorption studies, *Toxicology in Vitro*, 2004, **18**, 351–358.

35. L. Wu, X. Lu, K. S. Kulp, M. G. Knize, E. S. F. Berman, E. J. Nelson, J. S. Felton and K. J. J. Wu, Imaging and differentiation of mouse embryo tissues by ToF-SIMS, *International Journal of Mass Spectrometry*, 2007, **260**, 137–145.

36. D. J. Graham, M. S. Wagner and D. G. Castner, Information from complexity: Challenges of TOF-SIMS data interpretation, *Applied Surface Science*, 2006, **252**, 6860–6868.

37. A. Piwowar, J. Fletcher, N. Lockyer and J. Vickerman, Top-down approach to studying biological components using ToF-SIMS, *Surface Interface Analysis*, 2011, **43**, 265–268.

38. K. Lim and P. C. A. Kam, Chlorhexidine - pharmacology and clinical applications, *Anaesthesia and Intensive Care*, 2008, **36**, 502–512.

39. R. N. S. Sodhi, Time-of-flight secondary ion mass spectrometry (ToF-SIMS): versatility in chemical and imaging surface analysis, *Analyst*, 2004, **129**, 483–487.

29

MEMBRANE SOLVENT PENETRATION MEASUREMENTS USING CONTACT IMAGING

P Xiao[1], H Abdalghafor[2] and M E Lane[2]

[1] Faculty of ESBE, London South Bank University, London SE1 0AA, United Kingdom.
[2] School of Pharmacy, University College London, London WC1N 1AX, United Kingdom.

1 INTRODUCTION

Contact imaging using silicon fingerprint sensors, originally designed for biometric applications, has shown potential for skin hydration imaging and surface analysis [1-4]. Previous studies have shown that such sensors are also sensitive to solvents with high dielectric constant values, ε, which may make them useful for solvent penetration measurements [5]. This study presents recent investigations on solvent penetration through a silicone membrane and porcine tissue samples using contact imaging.

2 MATERIALS AND METHODS

Silicone membrane (thickness ~ 80μm) was a gift from Dow Corning (Seneffe, Belgium). Porcine skin tissue was isolated from the ear (donated from a local abattoir) and prepared with a thickness of approximately 900μm.

Figure 1 shows the Fujitsu Fingerprint sensor used in the study, and the measurement setup. The sensor has 256 x 300 capacitance-sensing pixels, with 50μm x 50μm spatial resolution and 8-bit grayscale level resolution.

A small piece of membrane or tissue was cut and placed on the sensor surface. A droplet of solvent was placed on the top of the sample (Figure 1). The sensor then recorded images continuously over a defined period of time. The time-dependent grayscale values of the images were then used for subsequent permeability analysis.

For solvent penetration through a membrane, a time-lag method was used to analyse the time-dependent image grayscale curves and to determine the permeability and diffusion coefficient of the membrane. If the solvent concentration is assumed to be a constant (C_0) at one side of the membrane, at the surface of the other side of the membrane, the solvent concentration $C(t)$ at time t can be expressed as [6]:

$$\frac{C(t)}{C_0} = 1 - \frac{4}{\pi}\sum_{n=0}^{\infty}\frac{(-1)^n}{2n+1}\exp\left(\frac{-D(2n+1)^2\pi^2 t}{4L^2}\right) \tag{1}$$

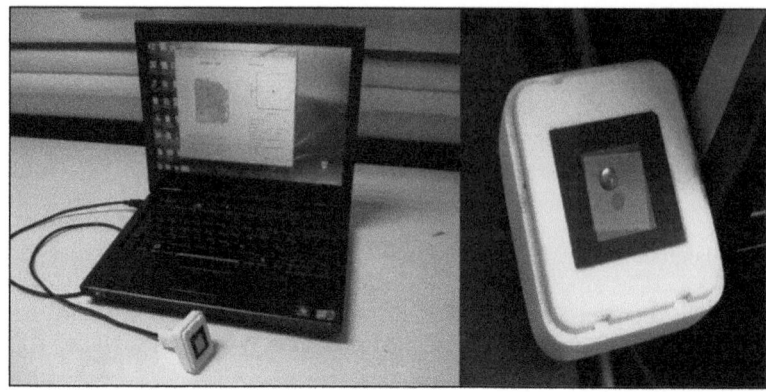

Figure 1 *Fujitsu Fingerprint sensor (left) and solvent penetration through silicone membrane (right).*

Where **D** is the diffusion coefficient of the membrane, and **L** is the thickness of membrane. Using least squares fitting, the experimental data may be fitted to Equation (1) to arrive at values for **D/L**^{2} or **D** if the thickness of the membrane (**L**) is known.

4   RESULTS AND DISCUSSIONS

In this study, three solvents were examined: water, undiluted alcohol (EtOH) and undiluted dimethyl sulfoxide (DMSO). Water, EtOH and DMSO have dielectric constants of 80, 24.3 and 46.7, respectively.

Figure 2 shows contact images of water, EtOH and DMSO penetrating through a silicone membrane at different time points. The results show that water penetrates through the membrane much faster than DMSO, while DMSO penetrates faster than EtOH.

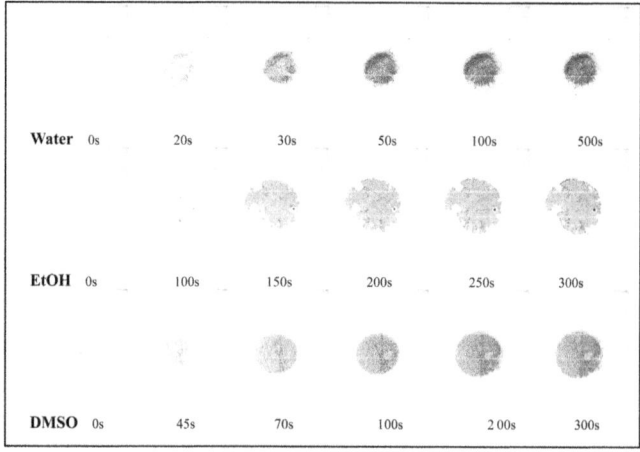

Figure 2 *Time-dependent capacitive contact images of water, undiluted alcohol and DMSO penetration through a silicone membrane.*

Figure 3 shows the corresponding time-dependent grayscale curves of water, EtOH and DMSO images in Figure 2. Again, the results confirm that water penetrates through the membrane faster than EtOH and DMSO. EtOH is also expected to evaporate from the surface of the membrane and this may explain the lower profile for this solvent compared with the other solvents examined [7].

By using Equation (1) to fit the data shown in Figure 3 it is possible to determine the permeability of the membrane. Calculation of the diffusion coefficient of the individual solvents in the membrane is also possible using the thickness of the silicone membrane (**L**), which is 80 μm. Figure 4 shows the least squares fitting results for the water penetration data shown in Figure 3. The results show that, for water, $D/L^2 = 0.024$ s^{-1} and $D = 1.54 \times 10^{-10}$ m^2 s^{-1}; corresponding values for EtOH are: $D/L^2 = 0.0038$ s^{-1} and $D = 2.4 \times 10^{-11}$ m^2 s^{-1}; for DMSO, $D/L^2 = 0.012$ s^{-1} and $D = 7.6 \times 10^{-11}$ m^2 s^{-1}.

The water penetration through different layers of the same silicone membrane was measured in order to study the membrane thickness effect during the water penetration. Figure 5 shows the time dependent grayscale value curves of water penetration through 1 layer, 2 layers, and 3 layers of the same silicone membrane. Figure 6 and Table 1 show the corresponding Least Squares Fitting results. The results for D/L^2 suggest that, as the

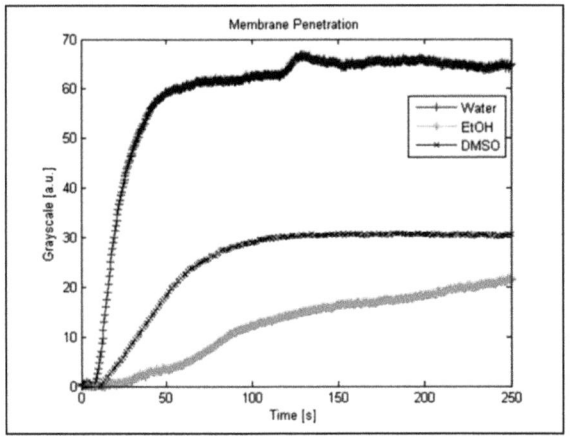

Figure 3 *Corresponding time dependent grayscale curves of water, EtOH and DMSO images in Figure 2.*

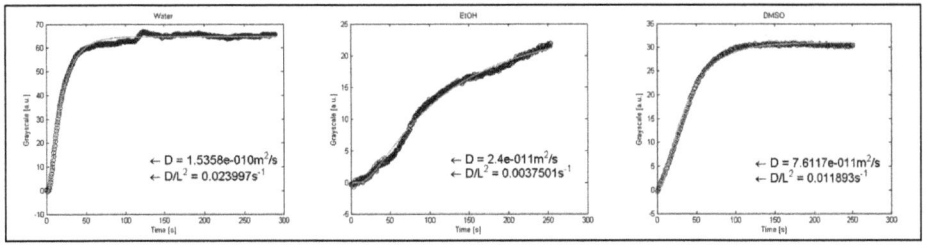

Figure 4 *Least Squares Fitting results for water, EtOH and DMSO penetration curves.*

number of layers increases, it takes longer for water to penetrate through the membrane. The diffusion coefficient of the membrane, however, stays more or less the same, as expected. The discrepancy in the diffusion coefficient values for different layers of the membrane, especially for the 2 layers, is likely due to the imperfect contact between the layers.

Figure 7 shows the outside and inside surface images of porcine tissue, as well as the images at different times during water penetration. The outside porcine tissue image clearly shows the porcine skin surface texture and hairs. Figure 8 shows the corresponding time-dependent grayscale value curves for the images shown in Figure 7. The results

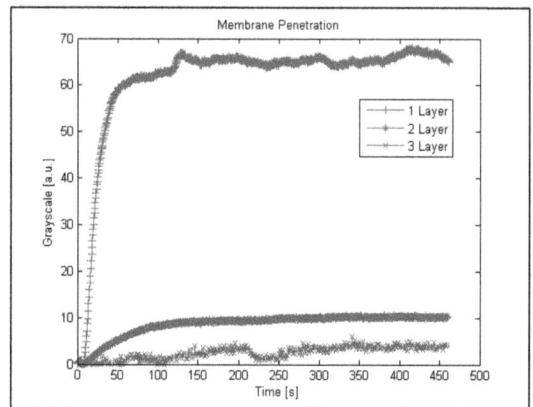

Figure 5 *Time dependent grayscale value curves for water penetration through 1 layer of silicone membrane (top), 2 layers of silicone membrane (middle), and 3 layers of silicone membrane (bottom).*

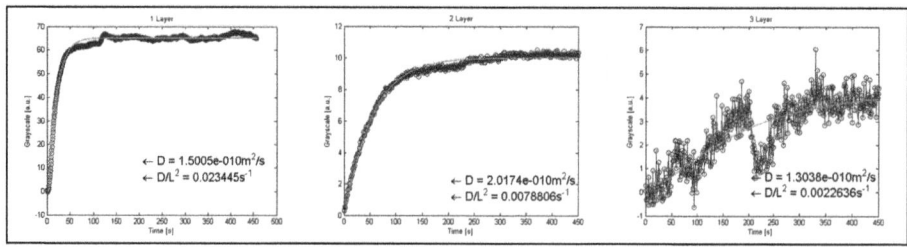

Figure 6 *The Least Squares Fitting results for the time dependent water penetration through 1 layer (left), 2 layers (centre), and 3 layers of silicon membrane (right).*

Table 1 *Permeability of water through different silicone membrane layers.*

	D/L^2 [s^{-1}]	D [m^2 s^{-1} x 10^{-10}]
1 layer	0.023	1.5
2 layers	0.0079	2.0
3 layers	0.0023	1.3

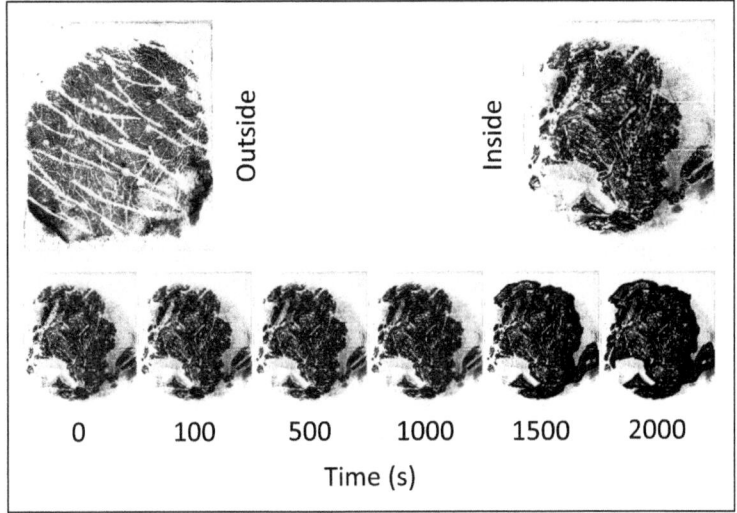

Figure 7 *Capacitive contact images of the outside surface (top left), and the inside surface (top right) of in-vitro porcine tissue, and images of water penetration through the in-vitro porcine tissue (bottom) at different times (0 – 2,000 s).*

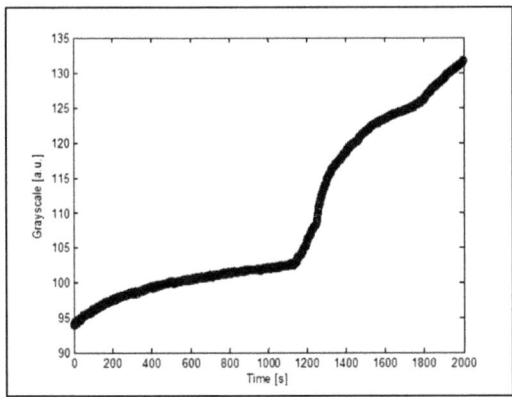

Figure 8 *The corresponding time dependent grayscale value curves for the water penetration images in Figure 7.*

suggest that water permeates through the porcine tissue in a different manner compared the silicone membrane. It appears that water penetrates through porcine tissue in two stages. Initially, water diffuses through the tissue at a slow rate, but, after approximately 20 minutes (1200 seconds, Figure 8) it starts to diffuse at a faster rate. The reason for this sudden change in the rate of diffusion is unclear, but may potentially be due to the status of the applied water within the porcine tissue, e.g. whether it is present as bound water or free water. One possible hypothesis is that, as water begins to diffuse through porcine tissue, it also hydrates this tissue; initially all the water is bound, diffusing slowly and, after a certain threshold, it becomes free water and diffuses more rapidly.

5   CONCLUSIONS

Contact imaging using capacitance fingerprint sensors is a powerful tool for studying solvent penetration through membranes, as well as *in vitro* skin samples. The capacitive fingerprint sensors are not only sensitive to water, but also to other solvents which have relatively large dielectric constants. By calculating the grayscale values of the images, it is possible to estimate the quantity of solvent that has penetrated through the membrane and, from this, diffusion coefficients may be calculated. The water penetration through porcine skin shows different characteristics from water penetration through the silicone membrane. The results have implications for advancing our understanding of excipients in model membranes and ultimately, in skin.

6   ACKNOWLEDGEMENTS

We wish to thank the EPSRC and London South Bank University for financial support for this project.

7 REFERENCES

1. J L Leveque and B Querleux, SkinChip - a new tool for investigating the skin surface *in vivo, Skin Research and Technology*, 2003, **9**, 343-347.

2. D. Batisse, F. Giron and J. L. Leveque, Capacitance imaging of the skin surface, *Skin Research and Technology*, 2006, **12**, 99-104.

3. P. Xiao, H. Singh, X. Zheng, E. P. Berg and R. E. Imhof, *In-vivo* Skin Imaging For Hydration and Micro Relief Measurements, Stratum Corneum V conference, July 11-13, 2007, Cardiff, UK.

4. H. Singh, P. Xiao, E. P. Berg and R. E. Imhof, Skin Capacitance Imaging for Surface Profiles and Dynamic Water Concentration Measurements, ISBS Conference, Seoul, Korea, May 7-10, 2008.

5. P. Xiao, H. Singh, X. Ou, A. R. Caparnagiu, G. Kramer and R. E. Imhof, *In-vivo* Solvent Penetration Measurement using Contact Imaging and Skin Stripping, SCC Annual Scientific Meeting & Technology Showcase, New York, 2011.

6. J. A. Cowen, *In vivo* opto-thermal transdermal diffusion measurement, PhD thesis, London South Bank University, 1999.

7. P. Santos, M. Machado, A. C. Watkinson, J. Hadgraft, M. E. Lane, The effect of drug concentration on solvent activity in silicone membranes, *Int J Pharm.*, 2009, **377**(1-2), 70-5.

30

INFLUENCE OF RECEPTOR MEDIA ON GD SKIN PENETRATION *IN VITRO*[†]

C H Dalton[1,2], S J Graham [1], O J Payne[1], J K Chipman[2], R P Chilcott[3] and J Jenner[1]

[1]Biomedical Sciences, Dstl Porton Down, Salisbury, United Kingdom. [2]School of Biosciences, University of Birmingham, Birmingham, United Kingdom. [3]Department of Pharmacy, University of Hertfordshire, Hatfield, United Kingdom.

1 INTRODUCTION

When designing *in vitro* static diffusion studies the choice of receptor media is important as it may influence measured penetration rates. OECD guidelines [1] regarding suitable receptor fluids state that "The use of a physiologically conducive receptor fluid is preferred although others may also be used provided that they are justified". Further to this is the statement, "Adequate solubility of the test chemical in the receptor fluid should be demonstrated so that it does not act as a barrier to absorption." Guidance [2] for acceptable receptor fluids for non-viable skin preparations is split between the evaluation of water soluble compounds and lipohilic test substances. For the former saline receptor solutions are preferred, whilst for lipophilic test substances it is suggested that "the receptor fluid can contain organic solvents such as 1:1 ethanol: water or 6% polyethylene glycol 20 oleyl ether in water."

More recently OECD guidance included the use of 5% bovine serum albumin for use with lipophilic penetrants [3] and acknowledged that use of 50% aqueous ethanol may enhance the absorption. The guidance notes [3] clearly state that "Compounds such as anionic surfactants or other solvents must be added to the receptor fluid in order to increase the uptake of lipophilic compounds."

The current study investigated whether choice of receptor fluid media altered the steady state percutaneous penetration rate (J_{SS}) and lag time of the chemical warfare agent soman (GD; Table 1) through split thickness abdominal flank pig skin using Franz type static diffusion cells. Three typical receptor fluids were used, phosphate buffered saline (PBS), phosphate buffered saline with 5% bovine serum albumin (PBS-BSA) and 50% aqueous ethanol. Of these, PBS is a standard choice of receptor fluid for a penetrant that has suitable solubility characteristics within an aqueous matrix. For penetrants of limited aqueous solubility, such as GD, which has a solubility of 21 mg ml[-1] in aqueous solution [4], the addition of solubility enhancers within the receptor fluid may be required to prevent it becoming a rate limiting barrier to absorption.

[†] © Crown Copyright 2013.

Table 1 *Identity and physicochemical properties of soman adapted from Munro [4]. Where log K_{OW} is the log of the octanol water partition coefficient, log K_{OC} is the log of the organic carbon partition coefficient. Log K_{OW} values obtained from [14][a] and [15][b].*

Soman (GD)				
Formula	**CAS No.**	**Molecular weight**	**State**	**Colour**
$C_7H_{16}FO_2P$	96-64-0	182.18	Liquid	Colourless
Melting Point	**Boiling point**	**Density**	**Vapour Pressure**	**Volatility**
- 42 °C	198 °C	1.022 @ 25 °C	0.40 @ 25 °C	3900 mg m^{-3} @ 25 °C
Vapour Density	**Water Solubility**	**Hydrolysis Rate**	**Log K_{OW}**	**Log K_{OC}**
6.3 (air =1)	21 g L^{-1} @ 20 °C	45 h $T_{1/2}$ @ pH 6.6	1.78[a], 1.824[b]	1.17

2 MATERIALS AND METHOD

The synthesis, use and destruction of GD in this study was conducted in accordance with the Chemical Weapons Convention (1996) to which the UK is a signatory state. Radiolabelled (^{14}C) pinacoyl methylfluorophosphonate (GD) was synthesised by TNO (Rijswijk, Netherlands) and had a radiochemical purity >97% (as determined by radiometric HPLC analysis). The chemical purity of unlabelled GD was reported to be >97% (by NMR). Both radiolabelled and cold agent were mixed in appropriate proportions to give a nominal activity of approximately 0.5 µCi µl^{-1}.

Liquid scintillation counting (LSC) materials (Soluene-350^{TM}, Ultima Gold and opaque plastic vials) were purchased from Perkin-Elmer (Chandler's Ford, Hampshire). All other chemicals were analytical grade and were purchased from the Sigma Chemical Company (Poole, Dorset) or VWR International (Lutterworth, Leicestershire).

The use of animals in this study was conducted in accordance with the Animals (Scientific Procedures) Act 1986. Eight weanling pigs (large white strain, weight range 20 - 30 kg) were purchased from a local supplier. Animals were group housed and given 24 h access to food and water. After one week acclimatisation, each animal was sedated with Hypnovel® (Midazolam, 6ml i.m., 5 mg ml^{-1}) and culled with an overdose of Euthatal™ (sodium pentobarbitol, 6ml i.v., 200 mg ml^{-1}). The whole abdominal skin flank (approximately 40 x 30 cm) was excised from each animal. The skin was stored flat between sheets of aluminium foil at -20°C for up to three months prior to use. Prior to study commencement, skin samples were removed from cold storage and thawed in a refrigerator at 5°C for approximately 24 hours. The skin was close-clipped and subcutaneous fat removed before being dermatomed (Humeca Model D42, Eurosurgical Ltd, Guildford, UK) to a nominal thickness of 500 µm prior to insertion into diffusion cells.

Percutaneous absorption experiments were performed with Franz-type glass static diffusion cells [5] with an area available for diffusion of 0.19 cm^2. Sections (3 x 3 cm) of dermatomed pig skin was placed, epidermal side up, between the donor (upper) and receptor (lower) chamber. Each receptor chamber was filled with 5 ± 1 ml receptor fluid (phosphate buffered saline (PBS), PBS with added 5% bovine serum albumin (BSA) or 50% aqueous ethanol) to a level that ensured the meniscus of the receptor fluid in the sampling arm was level with the skin surface. The diffusion cells, once assembled, were placed into a metal holder incorporating a circulating water supply upon a magnetic stirrer block. Stirring of the receptor fluid was achieved via a Teflon®-coated iron bar situated within the receptor chamber. The skin surface within each diffusion cell was maintained at a temperature of approximately 32°C (as confirmed by infrared thermography; FLIR Model P640 camera, Cambridge, UK) using water pumped at approximately 37°C through the metal holder via a circulating water heater (Model GD120, Grant Instruments, Cambridge, UK). All diffusion cells were set up in a fume cupboard and once assembled, were allowed to equilibrate for approximately 24 hours. Baseline samples (100µl) were taken, with replacement by fresh receptor media, from each receptor chamber prior to commencement of the experimental protocol.

Topical dosing was performed by the direct application of undiluted, ^{14}C-radiolabelled GD (100 µl) onto the centre of the skin surface within the donor chamber. Skin samples were mounted over (i) 50% aqueous ethanol, (ii) PBS or (iii) PBS containing 5% BSA. Each treatment group comprised n=8 diffusion cells, with each diffusion cell within each treatment group containing skin from different animals. Immediately after dosing with 100 µl neat ^{14}C-GD, the cells were occluded with a piece of tin foil which was affixed to the glass rim of the donor chamber using a layer of inert perflourinated barrier cream (AG-7, Dstl, Wiltshire, UK). Following application of GD, samples (100µl) of receptor chamber fluid were removed into 5 ml of scintillation fluid (Ultima Gold, Perkin Elmer LAS (UK) Ltd, Buckinghamshire, UK) at regular intervals up to 24 h post exposure. Each sample was replaced with an equivalent volume of fresh receptor fluid. Twenty four hours post-exposure a dose distribution was performed as follows; the receptor chamber fluid and tinfoil were removed from each diffusion cell prior to dry cotton wool swabbing of the skin surface. After swabbing, the skin from each diffusion cell was removed. At each stage, the component removed was placed into a glass vial to which was added either 20 ml isopropanol (tinfoil and cotton wool swabs) or 10 ml Soluene-350 (skin). Dose distribution vials were then stored at room temperature (with occasional shaking) until the skin had dissolved, after which aliquots (250 µl) were removed into 5 ml of scintillation fluid. The amount of radioactivity in each sample was measured using a Perkin Elmer Tri-Carb liquid scintillation counter (Model 2810 TR), using the manufacturer's ^{14}C-quench curve library set to exclude single-photon (non-radioactive) events. The amount of radioactivity in each sample was converted to amount of GD by comparison to standards (containing known quantities of ^{14}C-GD) prepared and measured simultaneously. Cumulative ^{14}C-GD penetration profiles for the 24 h study period were generated for each diffusion cell. Steady state (J_{SS}) values were calculated by plotting the amount of GD penetrated against time and calculating the gradient of the slope. Skin that had a GD absorption rate with a linear regression correlation coefficient of < 0.98 or skin that exhibited a large penetration of GD during the diffusion lag-time were removed from the analysis as it was likely that the integrity of the skin had been compromised over the course of the study. Lag times were estimated using the x-axis intercept of the J_{SS}. Statistical analysis of J_{SS} values was performed using either a one way ANOVA with Tukey's multiple comparisons post-test (different receptor fluids over the same time range) or a paired t-test (same receptor fluid over the two different time ranges). A predetermined alpha level of 0.05 was used for each statistical calculation.

3 RESULTS

Cumulative penetration profiles of ^{14}C-GD through pig skin are shown in Figure 1. All values were mean ± standard deviation of n=8 diffusion cells with skin acquired from 8 individual pigs with samples being taken at the time points shown. Penetration of ^{14}C-GD into 50% aqueous ethanol receptor media increased steadily over the 24 hour study duration, whereas for the other two receptor media types (PBS and PBS-BSA), there was an apparent reduction in penetration from 9 hours.

Due to the apparent biphasic nature of the cumulative penetration curves for the PBS receptor fluids, J_{SS} values and lag times were calculated between 3 and 9 hours and 12 and 24 hours (Table 2). The value of the J_{SS} for ^{14}C-GD skin penetration into the 50% aqueous ethanol receptor media was similar (439 ± 158 and 415 ± 250 $\mu g.cm^{-2}.h^{-1}$) for the two evaluated time ranges. Conversely, for the PBS receptor fluid, a statistical difference in J_{SS} (219 ± 195 and 119 ± 116 $\mu g.cm^{-2}.h^{-1}$) was shown between the evaluated time ranges. Although a similar trend in penetration rate reduction for the measured J_{SS} values for ^{14}C-GD penetration into the PBS-BSA receptor media was apparent (166 ± 176 and 84 ± 78 $\mu g.cm^{-2}.h^{-1}$), it was not statistically significant.

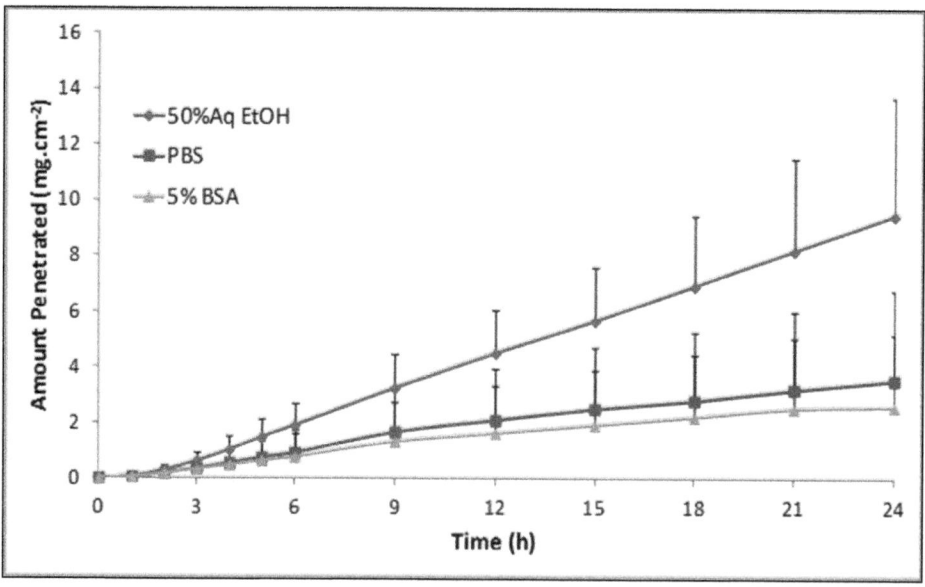

Figure 1 *Cumulative penetration profiles for ^{14}C-GD through pig skin using 3 different receptor media (50% aqueous ethanol, phosphate buffered saline and 5% bovine serum albumin in phosphate buffered saline). All values are mean ± standard deviation of n=8 diffusion cells.*

Table 2 *Penetration rate (J_{SS}) and lag times for the penetration of GD through 500 μm undamaged pig skin into 3 different types of receptor media. For J_{SS} values, statistical significance ($p<0.05$) is indicated between same labelled parameters. Data obtained from measurements performed between 3 and 9 or 12 and 24 hours ("calculation period"). Negative lag times are a consequence of the reduced calculated J_{SS} values later in the experimental time course.*

Receptor fluid	50% Aqueous Ethanol		PBS		PBS with 5% BSA	
Calculation period:	3–9 h	12–24 h	3–9 h	12–24 h	3–9 h	12–24 h
J_{SS} (μg cm^{-2} h^{-1})	439 ± 158^A	$415 \pm 250^{B,C}$	219 ± 195^X	$119 \pm 116^{B,X}$	166 ± 176^A	84 ± 78^C
Lag Time (h)	1.7 ± 0.5	-0.4 ± 4.5	1.6 ± 0.3	-6.5 ± 4.2	1.7 ± 0.9	-4.6 ± 6.5

Comparison of ^{14}C-GD penetration into the three different receptor fluids showed a significantly higher J_{SS} with 50% aqueous ethanol than both the PBS-containing receptor fluids when measured between 12 and 24 hours. However, when the J_{SS} was evaluated between 3 and 9 hours, significance was only shown between 50% aqueous ethanol and PBS-BSA.

The lag time measured for ^{14}C-GD penetration into each of the receptor fluids was similar when evaluated between 3 and 9 hours. Evaluation of lag times between 12 and 24 hours generated substantial negative lag times for the PBS-containing receptor fluids which, whilst not realistic, are a consequence of the reduced penetration rates measured for these receptor fluids later in the experimental time course.

A dose distribution was carried out for all skin samples (Figure 2). Total recoveries of GD were 91.8 ± 7.6 % (50% aqueous ethanol receptor fluid), 89.2 ± 10.5 % (PBS receptor fluid) and 95.7 ± 7.6 % (PBS with 5% BSA receptor fluid). In all cases, the majority (greater than 80 %) of ^{14}C-GD was recovered from the surface of the skin indicating that infinite dosing conditions had been maintained throughout the study duration. Smaller amounts of ^{14}C-GD were recovered from within the skin and from the tin foil.

4 DISCUSSION

One of the major considerations in the experimental design of an *in vitro* percutaneous penetration study is the choice of receptor fluid. The primary aim of the receptor fluid is to collect material that has penetrated the skin in a manner analogous to absorption through skin and into the peripheral vasculature *in vivo*. For studies where the maintenance of viable skin is necessary, such as metabolism protocols, the overriding factor is for a receptor fluid to be able to maintain skin viability within the diffusion cell system. Of equal importance is a requirement that the penetrant must be adequately soluble within the chosen receptor media. Should this requirement not be met it is likely that the receptor fluid could limit diffusion from the underside of the skin into the receptor fluid and lead to a gross underestimation in penetration rates. The current study used previously frozen, non-viable tissue and, therefore, did not need to utilise receptor fluids that would maintain

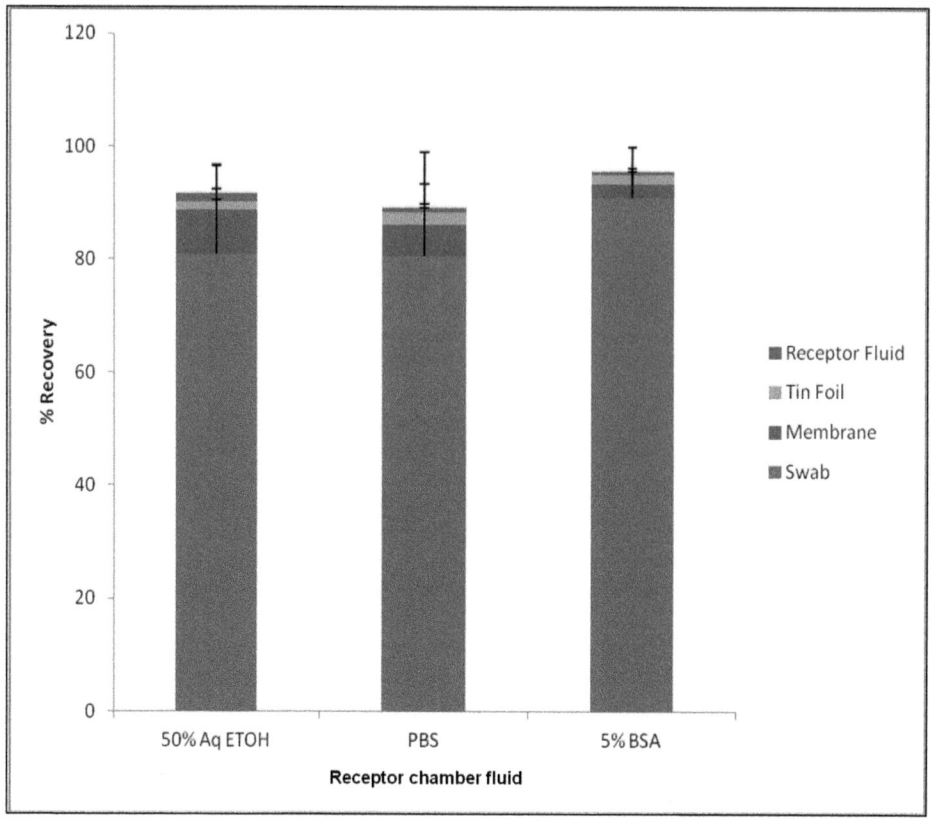

Figure 2 *Dose distribution expressed as % recovery for* ^{14}C*-GD recovered from each of the named compartments at 24 hours for the 3 different receptor media. All values are mean ± standard deviation of n=8 diffusion cells.*

tissue viability. Of concern was whether GD, being a relatively hydrophobic compound, would exhibit suitable solubility characteristics within a traditional aqueous receptor fluid. The current study used both ethanol and bovine serum albumin (BSA) within the receptor fluid to increase penetrant solubility by offering a more lipohilic environment.

Previous work carried out into the percutaneous absorption of chemical warfare agents has used 50% aqueous ethanol as a receptor fluid to ensure that the receptor fluid did not become a rate-limiting step in the diffusion process [6 – 9]. However, the skin penetration of GD has not previously been determined using our *in vitro* diffusion cell methodology. The solubility of GD in water (21g L^{-1}) and log K_{OW} value of 1.8 suggested that a standard phosphate buffered saline (PBS) receptor fluid may have offered sufficient sink conditions to be an appropriate receptor fluid for GD.

The penetration profiles indicated that of the three evaluated receptor fluids, only 50% aqueous ethanol offered suitable sink conditions for GD. Steady state (J_{SS}) flux was observed for GD penetrating into 50% aqueous ethanol after the initial lag phase, but, not for either of the PBS containing receptor fluids. For GD penetration into the PBS containing receptor fluids, the effect was similar, an initial flux of higher magnitude between 3 and 9 hours which then reduced over the twenty four hour time course, indicating that neither of the PBS-based receptor fluids were able to solvate GD sufficiently.

It is of interest why PBS with 5% BSA did not offer more suitable sink conditions for GD than PBS on its own. Studies by other researchers have shown the addition of BSA to the receptor media increases the solubility of lipophilic penetrants such as methyl paraben [10]. Wilkinson and co-workers also evaluated an ethanolic receptor fluid alongside PBS and a BSA containing receptor fluid, with maximum absorption rates falling in the order of: aqueous ethanol > BSA > PBS. The trend for the current study was: aqueous ethanol > BSA = PBS. Ethanol is able to interact with and solvate lipophilic molecules due to its two-carbon chain and is also able to hydrogen bond with hydrophilic molecules due to the presence of the hydroxyl group. This broad solvation ability allows ethanol to act as a general organic solvent for lipohphilic compounds including GD. Conversely, the current study has shown that BSA is not a suitable solubility enhancing excipient for all organic molecules. BSA is typically chosen as a solubility enhancer for receptor fluids where it is necessary to maintain skin viability [11]. For studies where maintenance of skin viability is not of concern, this work has shown that 50% aqueous ethanol offers superior sink conditions, although mention must also be made of other receptor fluids containing excipients, such as polyethylene glycol (PEG) 20 oleyl ether which have recently been recommended due to its ability to facilitate sink conditions for hydrophobic chemicals [12].

One of the questions over the use of an ethanolic receptor fluid is as to whether ethanol disrupts the skins barrier layer and whether it is this disruption that leads to the observed increase in penetration rates compared to PBS-based receptor fluids. The current study shows that use of an ethanol water receptor fluid did lead to increased measured penetration rates. Whilst other researchers have ascribed increases in penetration rate by ethanol to lipid extraction [13], this is not supported by the current study, as it would be expected that a disruption of this type would cause a reduction in the measured lag time when comparing aqueous ethanol to the PBS containing receptors fluids. As all the lag times measured between 3 and 9 hours were similar, this offers evidence that use of an ethanolic receptor fluid causes minimal barrier disruption.

Ultimately, as long as a receptor fluid is able to maintain skin viability (if appropriate) and suitable sink conditions, experimental choice as to the most appropriate receptor fluid comes down to achieving experimental aims. For the quantification of penetration rates of chemical warfare agents through skin, any underestimation in skin penetration rate could lead to the development of inappropriate medical countermeasures with potentially life-threatening consequences. Using a receptor fluid such as 50% aqueous ethanol gives assurance that any experimental measure of dermal absorption models the "worst case" scenario, and that medical countermeasures developed against such worst case scenarios should be adequate for deployment in the field.

In conclusion, the data from the current study has shown that the most suitable receptor fluid (from those evaluated) for maintaining skin conditions during the *in vitro* percutaneous absorption of GD was 50% aqueous ethanol.

5 ACKNOWLEDGEMENTS

This work was funded by the Health Protection Agency (HPA) and conducted in support of the "Haemostatic Decontaminants for Penetrating Injuries Contaminated with CW Agents" project (DTRA Project ID Number 2.F0026_08_RC_C) carried out at facilities operated by the Defence Science and Technology Laboratory (Dstl).

6 REFERENCES

1. *OECD Test Guideline 428: Skin Absorption:* in vitro *Method*, Paris, Organisation for Economic Co-operation and Development, 2004.

2. *OECD Guidance Document for the Conduct of Skin Absorption Studies*, OECD Series on Testing and Assessment Number 28, Paris, Organisation for Economic Co-operation and Development, 2004.

3. OECD OECD Guidance Notes on Dermal Absorption, *Paris, Organisation for Economic Co-operation and Development*, 2010.

4. N. B. Munro, S. S. Talmage, G. D. Griffin, L. C. Waters, A. P. Watson, J. F. King and V. Hauschild, The sources, fate, and toxicity of chemical warfare agent degradation products, *Environ. Health Perspect.*, 1999, **107**(12), 933-974.

5. T. J. Franz, Percutaneous Absorption - Relevance of *In vitro* Data, *Journal of Investigative Dermatology*, 1975, **64**(3), 190-195.

6. R. P. Chilcott, C. H. Dalton, Z. Ashley, C. E. Allen, S. T. Bradley, M. P. Maidment, J. Jenner, R. F. Brown, R. J. Gwyther and P. Rice, Evaluation of barrier creams against sulphur mustard: (II) *in vivo* and *in vitro* studies using the domestic white pig, *Cutaneous and Ocular Toxicology*, 2007, **26**(3), 235-247.

7. C. H. Dalton, M. P. Maidment, J. Jenner and R. P. Chilcott, Closed cup vapor systems in percutaneous exposure studies: What is the dose?, *Journal of Analytical Toxicology*, 2006, **30**(3), 165-170.

8. C. H. Dalton, I. J. Hattersley, S. J. Rutter and R. P. Chilcott, Absorption of the nerve agent VX (O-ethyl-S-[2(di-isopropylamino)ethyl] methyl phosphonothioate) through pig, human and guinea pig skin *in vitro*, *Toxicology in Vitro*, 2006, **20**(8), 1532-1536.

9. I. J. Hattersley, J. Jenner, C. Dalton, R. P. Chilcott and J. S. Graham, The skin reservoir of sulphur mustard, *Toxicology in Vitro*, 2008, **22**(6), 1539-1546.

10. R. Wilkinson, C. Jewell, S. C. Wilkinson, F. M. Williams and P. G. Blain. The effect of *in vitro* receptor media choice on the percutaneous absorption of methyl paraben. *Toxicology*, 2007, 240, (3), 178-179.

11. S. W. Collier, N. M. Sheikh, A. Sakr, J. L. Lichtin, R. F. Stewart and R. L. Bronaugh, Maintenance of Skin Viability During *In vitro* Percutaneous Absorption-Metabolism Studies, *Toxicology and Applied Pharmacology*, 1989, **99**(3), 522-533.

12. W. J. Fasano and J. N. Mcdougal, In vitro dermal absorption rate testing of certain chemicals of interest to the Occupational Safety and Health Administration: Summary and evaluation of USEPA's mandated testing, *Regulatory Toxicology and Pharmacology*, 2008, **51**(2), 181-194.

13. D. Van Der Merwe and J. E. Riviere, Comparative studies on the effects of water, ethanol and water/ethanol mixtures on chemical partitioning into porcine stratum corneum and silastic membrane, *Toxicology in Vitro*, 2005, **19**(1), 69-77.

14. H. P. Benschop and H. C. Wesselman, Pharmacokinetics of the soman simulant 1,2,2-trimethylpropyl dimethylphosphinate (PDP) in rats, *Arch.Toxicol.*, 1989, **63**(3), 238-243.

15. S. E. Czerwinski, D. M. Maxwell and D. E. Lenz, A method for measuring octanol : water partition coefficients of highly toxic organophosphorus compounds, *Toxicology Methods*, 1998, **8**(2), 139-149.

31

EVALUATION OF METHYLPHENIDATE PERMEATION FROM DAYTRANA™ PATCHES ACROSS SILICONE AND HUMAN EPIDERMAL MEMBRANES

A EDWARDS[1], F LIU[1], M B BROWN[1, 2], W J MCAULEY[1]

[1]School of Pharmacy, University of Hertfordshire, Hatfield, Hertfordshire, United Kingdom. [2]MedPharm Ltd, 50 Occam Road, Surrey Research Park, Guildford, Surrey, United Kingdom.

1   INTRODUCTION

Methylphenidate (MPH) is a mild central nervous system stimulant and has been a common drug used for treating attention deficit hyperactivity disorder (ADHD) for numerous years [1]. MPH inhibits re-uptake of dopamine into the pre-synaptic nerve by blocking the dopamine transporter, leading to an increase in dopamine within the synapse and further stimulation of the dopamine receptors on the post-synaptic nerve [2]. In patients with ADHD this has a beneficial effect on behaviour, cognition, short-term memory, reaction time, vigilance and learning [3]. MPH is currently available in a range of dosage forms from oral controlled or immediate release formulations through to Daytrana™ a controlled release transdermal matrix patch [4].

To enable successful delivery through the skin, a drug must first be released from the formulation. This release from the formulation can be used to control the rate at which the drug permeates through the skin [5]. The advantage of controlled released medications can be seen with drugs such as MPH where, as a result of its short half-life, the use of oral immediate-release formulations is problematic with the drug having to be administered to children two or three times a day, requiring supervision from school staff members due to the legally controlled status of the drug. Daytrana™ permits sustained absorption of the free base form of the drug through the skin into the bloodstream from a controlled release matrix patch, as well as allowing easier administration of the drug for adolescents in comparison to controlled release tablets and capsules for oral administration. As such Daytrana™ offers a convenient method to avoid repeated administration, tablet swallowing and/or accidental chewing of control release capsule formulations with some parents preferring it to oral dosage forms [6].

A previous study has shown that there is a potential for the MPH containing adhesive of Daytrana™ transdermal patches to adhere to the release liner which can result in the loss of drug available for transdermal delivery [7]. A further potential issue with drug-in-adhesive patches is the potential for crystallisation of drug within the matrix. This crystallisation can occur during storage, potentially affecting drug release and therapeutic effect [8].

To determine whether there is significant variation in drug transport between different Daytrana[TM] patches, or across individual patches, MPH permeation through a silicone membrane was assessed. The silicone membrane was used in this study as it can provide an understanding of the drug transport from the formulation, which can be difficult to determine from studies using skin because of the inherent biological variation in its properties [9]. In addition, permeation studies across skin using human epidermis from two different donors were performed to provide an indication of inter-subject variability in the absorption of MPH across skin.

2   MATERIALS AND METHODS

Daytrana[TM] patches from a single batch were supplied from Masters Pharmaceuticals. Silicone membranes were obtained from Bioplexus. All other chemicals were supplied by Fisher Scientific.

A Kinetex™ 2.6 μm XB-C18 100 Å LC Column 100 x 4.6 mm from Phenomenex was selected. The HPLC method used a mobile phase of 74.9:12.5:12.5:0.1 Phosphate Buffer:Acetonitrile:Methanol:Triethylamine, adjusted to pH 3 with a flow rate of 0.6ml/min, an injection volume of 20 μl and a wavelength of 206 nm

Drug transport studies across silicone and human epidermal membranes were performed using static Franz-type diffusion Cells (Figure 1) with a receptor volume and diffusional area of approximately 3 ml and 1 cm^2, respectively. Franz-type cells were placed within a water bath that was heated so that the membrane temperature was maintained at 32°C. Six circular regions from approximately the same positions were cut from all four Daytrana[TM] patches. The receiver fluid was a pH 3 0.1M phosphate buffer, and samples were taken over a 24 hour period. Silicone membrane of 0.005 inch thickness was used.

Full thickness abdominal skin was obtained from male Caucasian donors with ethical approval granted by the Pharmacy and Postgraduate Medicine Ethics Committee (PHAEC/09-23) and samples were stored in a freezer at -20°C. The human epidermal membranes for the permeation studies were prepared by the heat separation method reported previously [10].

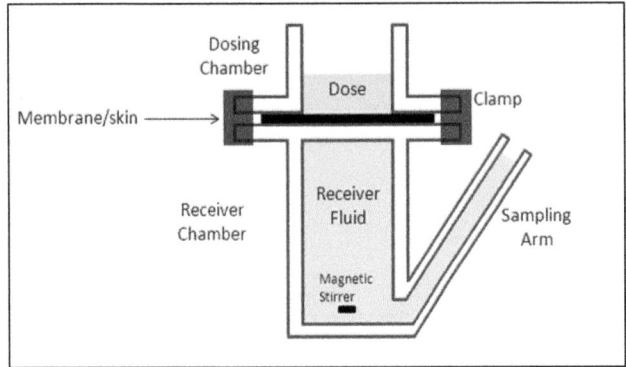

Figure 1 *Schematic of a Franz-type diffusion cell.*

The data were assessed for normal distribution using a Kolmogorov-Smirnov test. Statistical analyses were then performed using analysis of variance (ANOVA) for the drug transport across silicone membrane and an independent t-test for drug permeation through human epidermis. The tests were performed to determine if MPH transport across silicone or human epidermal membrane was the same for different samples or skin donors over an 8 or 24 hour period.

The drug transport data across silicone membrane was modelled using the Higuchi equation by plotting the cumulative amount of MPH permeated against the square-root of time [11-13]. Linear regression analysis was performed on the results.

3   RESULTS AND DISCUSSION

The cumulative amount of MPH permeating through the silicone membrane from Daytrana[TM] is shown in Figure 2. Within one circular region of the patch there is approximately 2.9 mg of MPH, and over 24 hours, approximately 100% of the drug was released. Statistical analyses comparing drug transport across the membrane at 8 and 24 hours (ANOVA) indicated no significant difference between patches or within the six regions of the four patches from the single batch tested.

Plots of the cumulative amount permeated against the square-root of time (Figure 3) were linear over 6 hours for all patches and the obtained r^2 values were 0.9913, 0.9964, 0.9980 and 0.9986 for the four tested patches, A to D respectively. This was consistent with the Higuichi model of drug release from matrix systems. If the 8 hour time point is included, the regression declines with r^2 values of 0.9835, 0.9912, 0.9954 and 0.9974, respectively.

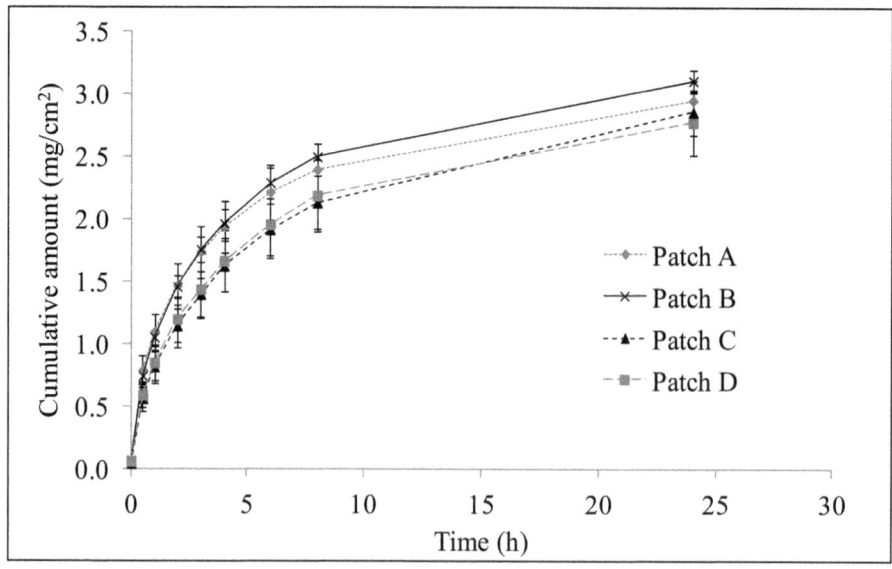

Figure 2 *Cumulative amount of MPH base permeated through silicone membrane from four Daytrana[TM] patches over a 24 hour period (n=6 ± SEM)*

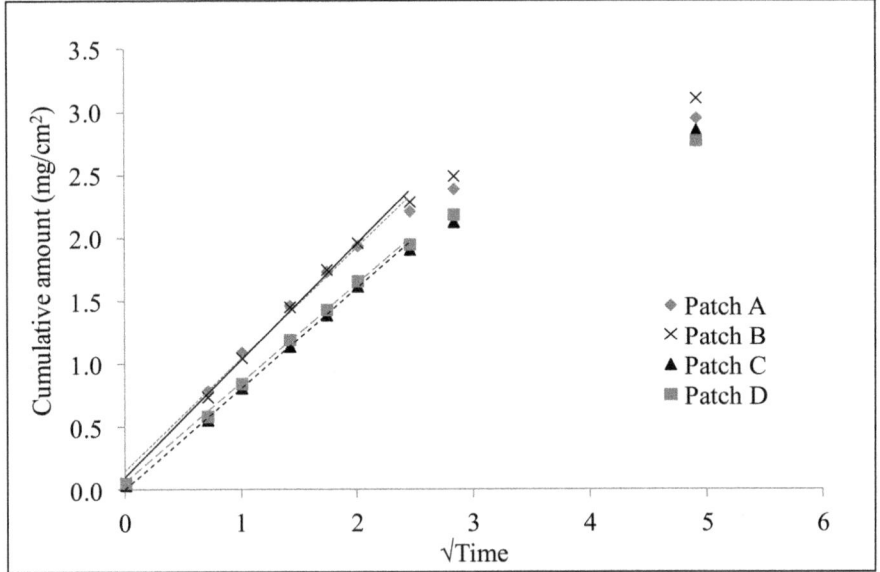

Figure 3 *Cumulative amount of MPH base permeated through silicone membrane from four Daytrana^{TM} Patches plotted against the square-root of time.*

The reduction in linearity at later time points is likely to be due to dose depletion from the patch and is also consistent with the Higuchi model, which is only applicable for the first 60% of drug release, demonstrating that the results follow the Fickian diffusion process [11-13]. Given that these data are acquired from different regions of each patch and from samples of separate patches, there appears to be no issue with the Daytrana^{TM} adhesive sticking to the release liner [7] or variability in drug release from the different patch regions that would have suggested the occurrence of drug crystallisation during storage [8].

As drug transport across the silicone membrane from the four different Daytrana^{TM} patches was similar, permeation through human epidermal membranes was tested using two different skin donors in order to provide some indication of inter-subject variability. These data are presented in Figure 4. Statistical analyses (independent t-tests) indicated that the amount of drug permeating at 8 hours across the two skin samples was not significantly different. However, at 24 hours, a significant difference was observed. Approximately 25 and 36% of the available drug had permeated across the skin after 8 hours from donors 1 and 2, respectively, which increased to 36 and 66 % of drug at 24 hours.

After 8 and 24 hours, there was still a large quantity of the applied dose of drug that had not permeated the epidermal membrane, suggesting that there may be a large quantity of drug remaining within the patch or the skin after use. This is undesirable as a large quantity the of drug remaining in the dosage form will increase cost of the formulation and potentially have implications for the safety and disposal of discarded patches whilst a large quantity of drug remaining within the skin after removal of the patch may result in unwanted systemic delivery of drug. Similar behaviour has been observed with other

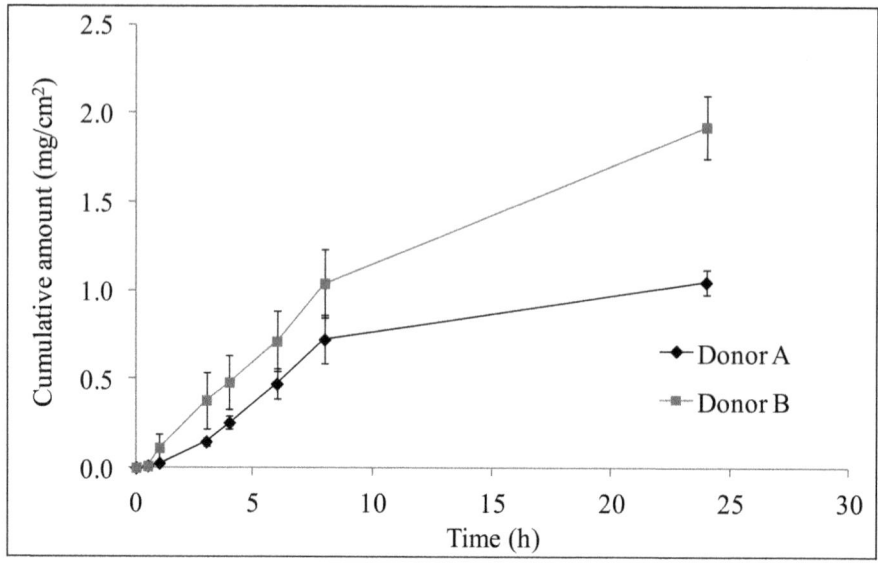

Figure 4 *Cumulative amount of MPH base permeated from Daytrana^{TM} through human epidermis from two separate skin donors over a 24 hour period (n=6 ± SEM).*

transdermal patches; for example a fentanyl transdermal system was shown to have approximately 65% of the applied drug remaining in the dosage form and approximately 2-3% remaining within the skin after use [14].

4   CONCLUSIONS

The cumulative amount of MPH that permeates silicone membranes from the Daytrana^{TM} patch is not statistically different over an 8 or 24 hour period, indicating that there was no significant variability in drug release from the different patches, or across the individual patches tested. Thus, for the sample tested there appeared to be no issue with the Daytrana^{TM} adhesive sticking to the release liner, or variability in drug release from different regions within the patch which may have suggested that drug crystallisation occurred during storage. Drug permeation through human epidermal membrane from two different donors was similar over 8 hours but was significantly different over 24 hours.

5 REFERENCES

1. J. Setlik, G. R. Bond and M. Ho, Adolescent prescription ADHD medication abuse is rising along with prescriptions for these medications, *Pediatrics*, 2009, **124**(3), 875.

2. C. L. Brandon *et al.*, Enhanced reactivity and vulnerability to cocaine following methylphenidate treatment in adolescent rats, *Neuropsychopharmacology*, 2001, **25**(5), 651–661.

3. V. R. Anderson and L. J. Scott, Methylphenidate transdermal system: in attention-deficit hyperactivity disorder in children, *Drugs*, 2006, **66**(8), 1117–1126.

4. D. A. Parasrampuria *et al.*, Do Formulation Differences Alter Abuse Liability of Methylphenidate?: A Placebo-Controlled, Randomized, Double-Blind, Crossover Study in Recreational Drug Users, *J Clin Psychopharm*, 2007, **27**(5), 459–467.

5. Y. N. Kalia and R. H. Guy, Modeling transdermal drug release, *Advanced Drug Delivery Reviews*, 2001, **48**(2–3), 159–172.

6. A. Lloyd *et al.*, Methylphenidate delivery mechanisms for the treatment of children with attention deficit hyperactivity disorder: Heterogeneity in parent preferences, *International Journal of Technology Assessment in Health Care*, 2011, **27**(3), 215–223.

7. A. M. Wokovich *et al.*, Evaluating elevated release liner adhesion of a transdermal drug delivery system (TDDS): a study of Daytrana methylphenidate transdermal system, *Drug Development and Industrial Pharmacy*, 2011, **37**(10), 1217–24.

8. R. Lipp, Selection and use of crystallization inhibitors for matrix-type transdermal drug-delivery systems containing sex steroids, *Journal of Pharmacy and Pharmacology*, 1998, **50**(12), 1343–1349.

9. P. Santos *et al.*, The effect of drug concentration on solvent activity in silicone membranes, *International Journal of Pharmaceutics*, 2009, **377**(1–2), 70–75.

10. A. M. Kligman and E. Christophers, Preparation of isolated sheets of human stratum corneum, *Archives of Dermatology*, 1963, **88**(6), 702.

11. P. Macheras and A. Iliadis, *Modeling in biopharmaceutics, pharmacokinetics, and pharmacodynamics: homogeneous and heterogeneous approaches*, 2006, Springer Verlag, New York, vol. 30.

12. T. Higuchi, Rate of release of medicaments from ointment bases containing drugs in suspension, *Journal of Pharmaceutical Sciences*, 1961, **50**(10), 874–875.

13. J. Siepmann and N. Peppas, Modeling of drug release from delivery systems based on hydroxypropyl methylcellulose (HPMC), *Advanced Drug Delivery Reviews*, 2001, **48**(2–3), 139–157.

14. G. Oliveira, J. Hadgraft and M. E. Lane, Toxicological implications of the delivery of fentanyl from gel extracted from a commercial transdermal reservoir patch, *Toxicology in Vitro*, 2012.

32

THE APPLICATION OF GAUSSIAN PROCESSES IN THE PREDICTION OF PERMEABILITY ACROSS A POLYDIMETHLYSILOXANE MEMBRANE.

G P Moss[1], Y Sun[2], N Davey[2], R G Adams[2], S C Wilkinson[3], D R Gullick[4].

[1]School of Pharmacy, Keele University, Keele, United Kingdom. [2]School of Engineering & Information Science, University of Hertfordshire, Hatfield, United Kingdom. [3]Medical Toxicology Centre, Institute for Cellular Medicine, Newcastle University, Newcastle-upon-Tyne, United Kingdom. [4]The University of Georgia College of Pharmacy, Athens, GA, United States of America.

1 INTRODUCTION

Polydimethylsiloxane (PDMS) silicone membranes, such as Silastic®, have been used widely in place of mammalian tissue in the determination of percutaneous absorption. While many experiments have shown correlations between the permeability across both membranes, Moss *et al.* demonstrated in a systematic study that PDMS membranes tend to exhibit greater permeability than mammalian skin, and that the relationship between permeability across both membranes was not found when the lipophilicity of the penetrant was greater than 3 [1]. Further, it was shown previously that when five commonly used physicochemical descriptors were applied to human, pig and rodent membranes they cannot represent the main characteristics of the PDMS dataset when using Gaussian Process (GP) regression to predict skin permeability [2]. However, the previous study in which this process was modelled employed a small dataset (n=19, as part of the wider aims of that work to investigate the effect of dataset construction on model quality).

The aims of the current study were to apply and validate GP regression methods to a new PDMS dataset. Permeability data for absorption across a polydimethylsiloxane (Silastic®) membrane includes more chemical compounds than used in previous studies (n = 31) [2, 3]. Simple linear regression was applied to provide a comparison to the Machine Learning models.

2 MATERIALS AND METHODs

The synthetic membrane dataset consists of 31 polydimethylsiloxane (PDMS, Silastic®) compounds collated from a small range of literature sources [5, 6, 8 – 10]. Previously, it was shown that using five physicochemical features (molecular weight (MW), solubility parameter (SP), log P, counts of the number of hydrogen bonding acceptor (HA) and donor groups (HD))

Table 1 *Physicochemical descriptors employed in this study.*

Physicochemical Descriptors	Molecular weight (MW)	log P	log PK_{owwin}	Solubility parameter (SP)	HA	HD	Melting Point (MPt)
Five descriptors (including log P) **5f log P**	X	X	-	X	X	X	-
Five descriptors (including log PK_{owwin}) **5f log P_{kowwin}**	X	-	X	X	X	X	-
Six descriptors (including log P) **6f log P**	X	X	-	X	X	X	X
Six descriptors (including log PKowwin) **6f log PKowwin**	X	-	X	X	X	X	X

can produce better predictions when compared to using only lipophilicity and molecular weight alone for human, pig and rodent membranes, but cannot represent the main characteristics of the Silastic® dataset when using Gaussian process regression to predict skin permeability [2]. In this new PDMS dataset, melting point, denoted as MPt, is also provided and included in the analysis, in order to determine if its inclusion improves the quality of analysis. Moreover, there are two values for the lipophilicity (log P) descriptor: one is an experimental value, denoted as log P; the other one is the software-calculated log P values, denoted as log PK_{owwin}. Thus, there are four datasets used in this study with either five or six features (see Table 1) and are denoted as 5f log P, 5f log PK_{owwin}, 6f log P and 6f log PK_{owwin}, respectively.

Each dataset was normalised so that all features had a zero mean and unit variance. Next, principle component analysis (PCA) was applied individually to the four normalised datasets. The compounds were plotted using the corresponding log K_p values against the first two principal components to represent the variation in the five or six features of all chemical compounds (Figures 1 and 2).

Analysis of data was carried out as described previously, using a combination of simple regression methods, single layer networks and Gaussian Process Machine Learning methods [4 – 6].

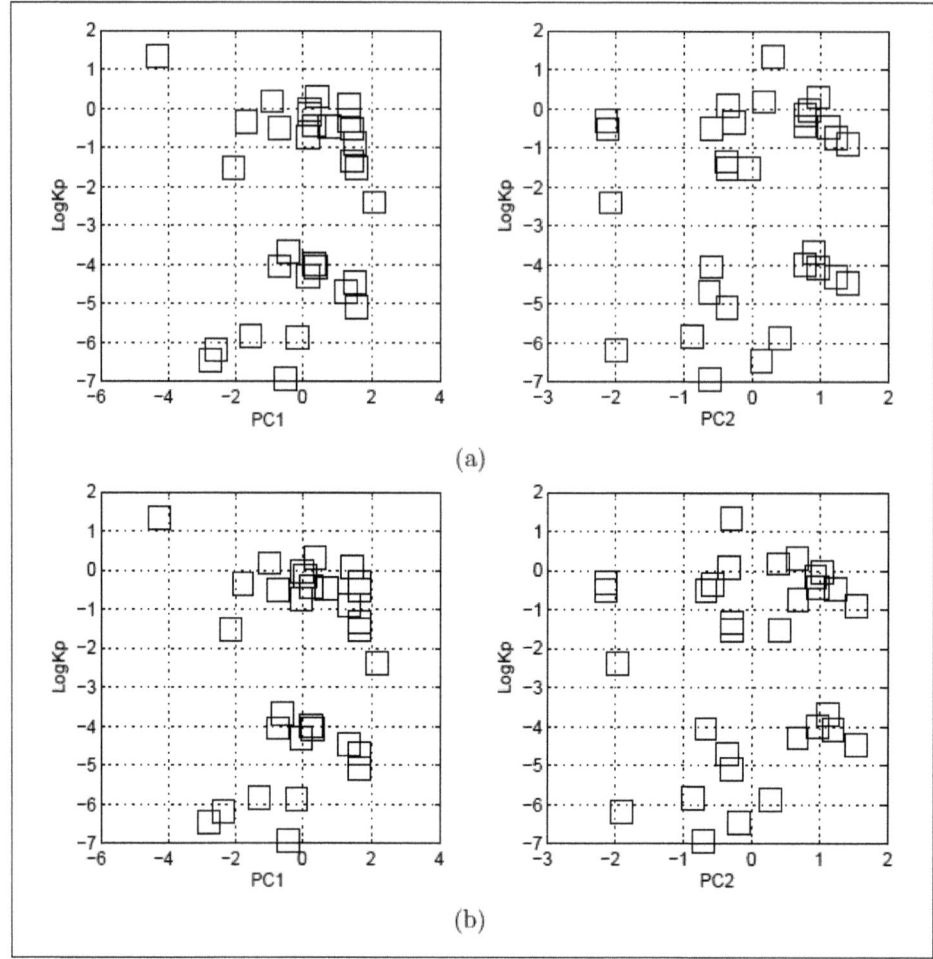

Figure 1 *The relationship between Log Kp and the PCA space of chemical compounds: (a) Five features with log P; (b) Five features with log PK$_{owwwin}$.*

A *Gaussian process* (GP) is defined as a collection of random variables which, jointly, have a Gaussian distribution. It is characterised completely by its mean and its covariance function. Usually, the mean function is considered to be the "zero everywhere" function. The covariance function, $k(x_i, x_j)$, allows for the specification of *a-priori* knowledge from a training dataset. It defines "nearness", or similarity, between the values of $f(x)$ at the two points, x_i and x_j.

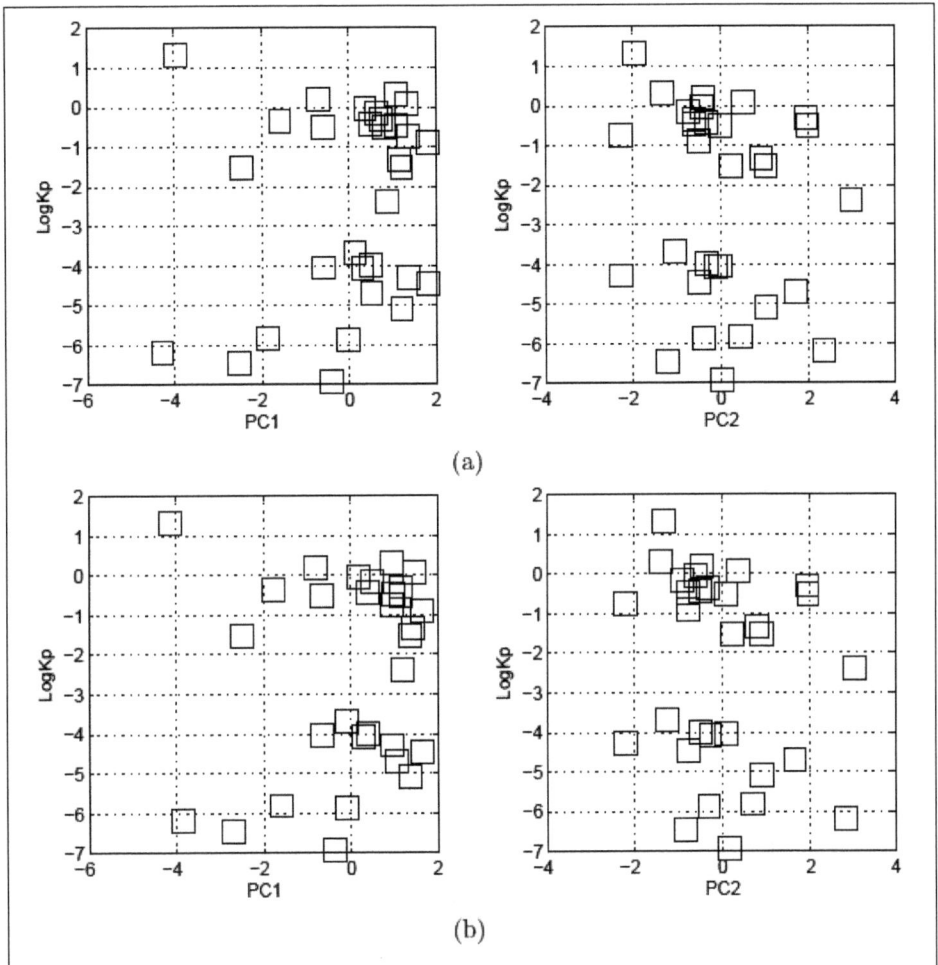

Figure 2 *The relationship between Log Kp and the PCA space of chemical compounds: (a) Six features with log P; (b) Six features with log PK$_{owwwin}$.*

To make a prediction y$_*$ at a new input x$_*$ the conditional distribution $p(y_*|y_1,......y_N)$ on the observed vector [y$_1$,......y$_N$] needs to be computed. Since the model is a Gaussian process, this distribution is also Gaussian and is, again, defined completely by its mean and variance. By applying standard linear algebra, the mean and variance at x$_*$ are given by:

$$E[y_*] = k_*^T (K + \sigma_n^2 I)^{-1} y$$

(1)

and

$$var[y_*] = k(x_i, x_j) - k_*^T(K+\sigma_n^2 I)^{-1}k_*$$

(2)

where k$_*$ denotes the vector of covariances between the test point and the N training data; K denotes the covariance matrix of the training data; σ_n^2 denotes the variance of an independent identically distributed Gaussian noise (which means observations are noisy); y denotes the vector of training targets; and k(x$_i$, x$_j$) denotes the variance of y$_*$. As is normally the case, mean values were used as predictions, and the variance was used as error bars on the predictions.

The squared exponential covariance function, the neural network covariance function, the rational quadratic covariance function, and two members from the *Matern* class of covariance function are applied to the analysis of the data sets [7]. In each case, an independent noise contribution is incorporated into the covariance function. Statistical performance measures employed to compare model quality are mean squared error (MSE), improvement over naïve (ION), negative log estimated predictive density (or negative log loss, NLL) and the Pearson correlation coefficient (CORR). The MSE measures the average squared difference between model predictions and the corresponding targets. The ION measures the degree of improvement of the model over the naïve predictor, whose value is always the same value, namely the mean of log K$_p$ in the training set. CORR measures the correlation between predictions and targets. For comparison, a model should have low values of both MSE and NLL, as well as high values of both ION and the correlation coefficient (CORR) on a given test dataset. Two experiments were carried out. Firstly, the Gaussian Process methods were validated for the PDMS membrane using the log P descriptor. The most suitable covariance functions were investigated for percutaneous absorption, and models with five and six descriptors were compared. Secondly, the use of log PK$_{owwin}$ and log P was compared.

3 RESULTS AND DISCUSSION

On average, the first principal component accounts for 44.33% of the total variance, and the second for 25.28%. Figures 1 and 2 show that there is no linear relation between log K$_p$ and the descriptors examined, suggesting the presence of more complex non-linear structures in the data. It can also be seen that there are two clusters in all PCA plots, one where log Kp is greater than -3 and the other where it is less than -3. Further, there is little difference (comparing Figures 1 and 2) when log PK$_{owwin}$ is used instead of log P. The overall structure of the plots does not change when melting point is added as a sixth descriptor. The nature of the dataset is important in the development of any model. A clear issue with this study is the size of the dataset (n=31), although this is offset slightly by the small number of studies from which this data was collated. Further, this dataset contains data for the same chemical from different experiments. In such cases each experiment has reported different permeability coefficients, reflecting the range of experimental conditions used by researchers – the inclusion of such multiples of data is not uncommon, although the size of the dataset may affect the impact such variance has on any analysis [11].

In the first experiment, the Gaussian Process regression model was validated for the synthetic PDMS membrane; the most suitable covariance functions were determined and models with five and six descriptors were compared. It was found that the different covariance functions examined yielded similar results, and as such only the squared exponential covariance function is reported. The results demonstrated that Gaussian Process models with different covariance functions do not, in general, provide statistically sound models using the same descriptors as in previous studies for mammalian skin (molecular weight, lipophilicity (as log P), the count of hydrogen-bond acceptor and donor groups and the solubility parameter). Addition of a sixth descriptor, melting point, does not improve the quality of models significantly (Table 2). In some cases the results are worse than the naïve model.

The second experiment investigated whether Gaussian Process models trained on the descriptor sets, including log PK_{owwin} are able to provide comparable predictions to those trained using the experimentally determined log P values. The results of this study are shown in Table 3. There are few significant differences observed between analyses using log P or log PK_{owwin} (Table 3). Again, as in the first experiment, some of the results are worse than the naïve model. Such results suggest that either the physicochemical parameters employed in this study might not be appropriate for developing models that represent this membrane, or the synthetic membrane may not respond in a similar way to similar compounds, a subject that has been addressed previously when PDMS membranes were shown to provide reasonably similar permeability to mammalian skin, but only up to a point – specifically, the relationship between permeability across both membranes as increasingly dissimilar as the lipophilicity of the penetrant increased [1].

Table 2 *Leave-one-out results on the PDMS dataset for models employing five or six physical descriptors. MSE; mean squared error. NNL; Negative log estimated predictive density. ION; improvement over naïve model. CORR; correlation coefficient.*

Model		MSE	ION	CORR	NLL
Naïve model		6.23	0	-1	-
Gaussian Process model	5 descriptors	6.43	-3.02	-0.51	2.37
	6 descriptors	6.34	-1.60	-0.64	2.36
Simple linear regression model	5 descriptors	11.61	-86.14	-0.25	-
	6 descriptors	12.78	-104.92	0.20	-

Table 3: *Leave-one-out results on the synthetic datasets comparing log P and log PK_{owwin} with five (log P or log PK_{owwin}, MW, HA, HD, SP) or six descriptors (log P or log PK_{owwin}, MW, HA, HD, SP) used.*

Model		Mean Squared Error (MSE)	Improvement over naïve model (ION)	Correlation coefficient (CORR)	Negative log estimated predictive density (NLL)
Naïve model		6.23	0	-1	-
Model with 5 descriptors	log P	6.43	-3.02	-0.51	2.37
	log PK_{owwin}	6.44	-3.34	-0.46	2.37
Model with 6 descriptors	log P	6.34	-1.60	-0.64	2.36
	log PK_{owwin}	6.37	-2.19	-0.56	2.36

In conclusion, the results of this study are consistent with those shown in the previous studies, including those using smaller datasets and suggest that either the measured molecular features are not suitable for predicting the permeability of synthetic membranes such as PDMS, or that synthetic membranes such as PDMS are not mechanistically viable alternatives to mammalian skin [2, 3]. The size of the dataset, and the diversity of the data therein, may also impact on the quality of the model produced and this, as with all models, should be taken into account when assessing the quality and significance of any model produced.

4 REFERENCES

1. L. T. Lam, Y. Sun, N. Davey, R. Adams, M. Prapopoulou, M. B. Brown and G. P. Moss, The application of feature selection to the development of Gaussian Process models for percutaneous absorption, *Journal of Pharmacy and Pharmacology*, 2010, **62**, 738–749.

2. G. P. Moss, Y. Sun, N. Davey, R. Adams, W. J. Pugh and M. B. Brown, The application of Gaussian Processes to the prediction of percutaneous absorption, *Journal of Pharmacy and Pharmacology*, 2009, **61**, 1147–1153.

3. G. P. Moss, Y. Sun, S. C. Wilkinson, N. Davey, R. Adams, G. P. Martin, M. Prapopoulou and M. B. Brown, *Journal of Pharmacy and Pharmacology*, **63**, 2011, 1411–1427.

4. Y. Sun, G. P. Moss, N. Davey, R. Adams, M. B. Brown, The application of stochastic machine learning methods in the prediction of skin penetration, *Applied Soft Computing*, 2011, **11**, 2367–2375.

5. G. P. Moss, D. R. Gullick, P. A. Cox, C. Alexander, M. J. Ingram, J. D. Smart and W. J. Pugh, Design, synthesis and characterisation of captopril prodrugs for enhanced percutaneous absorption, *Journal of Pharmacy and Pharmacology*, 2006, **58**, 167.

6. S. Geinoz, S. Rey, G. Boss, A. L. Bunge, R. H. Guy, P.-A. Carrupt, M. Resit and B. Testa, Quantitative structure-permeability relationships for solute transport across silicone membranes, *Pharmaceutical Research*, 2002, **19**, 1622–1629.

7. C. E. Rasmussen and C. K. I. Williams, *Gaussian Processes for Machine Learning*, The MIT Press, Cambridge, MA, 2006.

8. G. Ottaviani, S. Martel and P.-A. Carrupt, Parallel artificial membrane permeability assay: A new membrane for the fast prediction of passive human skin permeability, *Journal of Medicinal Chemistry*, 2006, **49**, 3948–3954.

9. A. D. Woolfson, D. F. McCafferty, K. H. McClelland and V. Boston. *Percutaneous local anaesthesia: comparison of in vitro predictions with clinical response*, in *Prediction of Percutaneous Penetration*, ed. R.C. Scott, R. H. Guy and J. Hadgraft, IBC Technical Services, London, 1989, pp. 192–198.

10. R. E. Baynes, J. D. Brooks, M. Mumtaz, J. E. Riviere, Effect of chemical interactions in pentachlorophenol mixtures on skin and membrane transport, *Toxicological Sciences*, 2002, **69**, 295–305.

11. B. M. Magnusson, Y. G. Anissimov, S. E. Cross, M. S. Roberts, Molecular size as the main determinant of solute maximum flux across the skin, *Journal of Investigative Dermatology*, 2004, **122**, 993–999.

33

THE APPLICATION OF NON-LINEAR AUTO-REGRESSIVE MOVING AVERAGE WITH EXOGENOUS INPUT (NARMAX) TO MODELLING THE ABSORPTION ACROSS HUMAN SKIN

G P Moss[1], T Kyriacou[2], S C Wilkinson[3]

[1]School of Pharmacy and [2]School of Computing & Mathematics, Keele University, Keele, United Kingdom. [3]Medical Toxicology Centre, Institute for Cellular Medicine, Newcastle University, Newcastle-upon-Tyne, United Kingdom.

1 INTRODUCTION

Cronin and Schultz commented that biological processes were seldom linear in nature and that modelling of such processes was unlikely to be successful without the consideration of non-linearity when constructing models [1]. Non-linear models are sometimes prone to overfitting, resulting in modelling the error present in the data. As skin permeability datasets are constructed from numerous literature sources, often using diverse experimental protocols, past considerations of non-linear modelling have been understandably cautious but, ultimately, reasonable, and have included substantial revisions and modifications to linear models in attempts to explain permeability at the extremes of datasets [2]. Recently, non-linear methods, including Machine Learning techniques, have been successfully employed to model both percutaneous absorption and to classify the enhancement effect of chemical enhancers of percutaneous absorption have been developed [3 – 6]. While such models have clearly demonstrated that they are more accurate both statistically and in terms of the predictions they output (for example, as demonstrated by the comparisons in Lam *et al.* [3] with the Potts and Guy model [6]) than previous models of percutaneous absorption, they are often criticised due to their lack of transparency as they do not produce an explicit functional output. However, this was addressed by Lam *et al.* who used the technique of feature selection, coupled with statistical analysis, to produce a specific functional output [3]. Their work resulted in an output that demonstrated the inter-changeability of certain molecular descriptors employed in their study; this meant that models of similar statistical and predictive qualities could be obtained using different descriptors, underpinning that the *perception* of such parameters as discrete functions is not correct. Further, Moss *et al.* demonstrated that model quality was also dependent upon the size of the dataset, and the nature of the data used to construct the model from [5].

The development of a relationship between input and output for a system should be straightforward, cost efficient and, where possible, relevant to the field of interest. For linear systems, such as discrete-time systems, it is known that a linear difference equation model exists that involves only a fixed and finite number of calculations at each stage if, for example, the Hankel matrix of the system has finite rank, resulting in a concise description of the output than for an impulse response function. Analogous to this are non-linear systems, to which the non-linear difference equation model, known as the Non-Linear Auto-Regressive Moving Average with eXogenous inputs (NARMAX) model, may be applied [7]. The NARMAX model provides a unified representation for a wide range of non-linear systems, clearly offering significant practical advantages over representations such as, for example, the Volterra series [8]. Unlike a number of Machine Learning methods that have recently been successfully applied to the problem domain of percutaneous absorption, the NARMAX method expresses its output as an explicit functional relationships of the statistically significant parameters in the form of discrete functional representations – that is, an equation [3 – 5]. Further, Chen and Billings demonstrated that the NARMAX method may provide a natural representation for a wide class of non-linear systems, and they included a number of well know input-output models as examples [8].

Clearly, the composition of the dataset and the method of analysis would appear to influence the model output. The aim of this study was to develop further modelling approaches to percutaneous absorption by applying Non-Linear Auto-Regressive Moving Average with Exogenous Input (NARMAX) methods to a well-established skin permeability database of maximum flux (J_{max}) values [9], and to compare this analysis to previous methods [10].

2 MATERIALS AND METHODS

The dataset used was that published by Magnusson *et al.* [9]. It was modified only to exclude members with aqueous solubility values given as an arbitrary benchmark, i.e. "greater than" or "less than". The Non-Linear Auto-Regressive Moving Average with Exogenous Input (NARMAX) [11, 12] method was applied using MatLab (R2010a). The influence of physicochemical parameters of the dataset members as well as experimental parameters, notably the temperature at which each experiment was conducted, on J_{max} was modelled. The output of each study was presented as an explicit functional relationship of statistically significant parameters.

3 RESULTS AND DISCUSSION

Initial results of the study (Equation 1) demonstrate good agreement with Magnusson's original study (Eq. 2), yielding a linear model as its explicit output:

$$log \, J_{max} = -0.014MW - 4.53$$

(1)

[Sum squared error = 357.20; $r^2 = 0.72$; $p<0.001$].

This model is very similar to that reported by Magnusson *et al.* [9] ($r^2 = 0.69$):

$$log\ J_{max} = -0.014MW - 4.52$$

$$(2)$$

Slight differences in the models and their descriptive statistics may be due to alterations to the dataset described above; that is, the removal of chemicals whose aqueous solubilities were listed in the original dataset (and excluded from this study) as "greater than" or "less than" and not as a specific value. However, the new model is still in excellent agreement with Magnusson's original equation. Further analysis of the data, by the addition of experimental parameters to the analysis, yielded a new model (Equation 3) in which one experimental parameter, the experimental temperature (T_{expt}) was returned as a highly significant parameter:

$$log\ J_{max} = 0.071T_{exp} - 0.01MW - 0.086MPt - 29$$

$$(3)$$

[Sum squared error = 243.85; $r^2 = 0.90$; p<0.001]

In addition, Equation 3 sees the inclusion of melting point (MPt) in the final equation, resulting in a model with improved statistical performance (i.e. an increase in the correlation coefficient, r^2, 0.90). This result also mirrors Magnusson's original findings [9].

It should also be noted that the use of the NARMAX method addresses the perceived limitations of previous Machine Learning methods [3 – 5, 11] as, applied to the modelling of percutaneous absorption, it yields an explicit functional output (Equations 1 and 3), resulting in a model whose mechanistic significance may be readily seen and interpreted – although Lam *et al.* specifically addressed this issue in the context of Machine Learning methods when they applied feature selection methods to their human skin dataset [5]. In doing this they were able to specify the significance of each physicochemical descriptor used – alone or in combination with any other descriptor used in their study – but did not produce a specific equation to represent their system. Differences with previous non-linear models (specifically, the output from the NARMAX analyses of linear models) may be attributed to the different nature of the datasets used, particularly with regard to the distribution of data points within each set [4, 5, 10]. When compared to Magnusson's study the NARMAX methods have produced very similar results, and have shown the significance of experimental parameters (specifically, experimental temperature). This is interesting in the context of non-linear modelling, in that NARMAX analysis of this dataset yields a linear output, which is perhaps an artefact of the nature of the dataset [13].

Moss *et al.* recently examined two datasets used for the construction of mathematical models of skin absorption [13]. One dataset has been used to develop models using Gaussian Processes and other Machine Learning methods [2 – 5], while the other was that used by Magnusson *et al.* [9]. Comparison of the two datasets used in the Machine Learning and Magnusson studies suggest significant differences in the composition of both datasets. For example, they showed that the medians of the molecular weight (MW) in each dataset are significantly different (p<0.05), with the median MW for the Magnusson dataset being approximately 100 lower than the median MW for the Machine Learning dataset. They suggested that the nature of the datasets plays a substantial role in the nature of the models output from each study. Thus, in applying the NARMAX method to the Magnusson dataset a linear model was obtained which, while validating this method against a known dataset, may suggest that the construction of the dataset plays a significant role in the nature of the output, particularly as the Machine Learning methods reported previously demonstrated the fundamental non-linearity of the dataset used in those studies [2 – 5].

The results of this study suggest that the NARMAX method may be applied to the modelling of skin permeability data as it yields highly correlated, transparent models and addresses the criticism of previous Machine Learning methods in producing a readily comprehensible output that is similar in its nature to previous models. Further, this technique has demonstrated the significance of experimental conditions – represented by the presence of the experimental temperature – as a significant descriptor in the final model (Equation 3) and suggests that such parameters should be more extensively considered when constructing models, particularly as the data used may originate from a range of experimental sources – such an approach may also suggest how subsets may be constructed, or collated, in further analyses.

Therefore, this study has shown that the NARMAX method can be successfully applied to a human skin dataset. The NARMAX model was able to produce models that were very similar to those produced by Magnusson [9], although the addition of experimental temperature to the output was a significant development, and suggested the significance of terminology not before analysed in this way in skin absorption studies. This study shows that the NARMAX method provides a useful tool for the analysis of a skin permeability dataset, and that it could be readily applied to the analysis of similar datasets.

4 REFERENCES

1. M. T. D. Cronin and T. W. Schultz, Pitfalls in QSAR, *Journal of Molecular Structure: THEOCHEM*, 2003, **622,** 39–51.

2. S. Mitragotri, Y. G. Anissimov, A. L. Bunge, H. F. Frasch, R. H. Guy, J. Hadgraft, G. B. Kasting, M. E. Lane and M. S. Roberts, Mathematical models of skin absorption: an overview, *International Journal of Pharmaceutics*, 2011, **418,** 115–129.

3. L. T. Lam, Y. Sun, N. Davey, R. Adams, M. Prapopoulou, M. B. Brown, G. P. Moss, The application of feature selection to the development of Gaussian Process models for percutaneous absorption, *Journal of Pharmacy and Pharmacology*, 2010, **62,** 738–749.

4. G. P. Moss, Y. Sun, S. C. Wilkinson, N. Davey, R. Adams, G. P. Martin, M. Prapopoulou and M. B. Brown, *Journal of Pharmacy and Pharmacology*, 2011, **63,** 1411–1427.

5. Y. Sun, G. P. Moss, N. Davey, R. Adams and M. B. Brown, The application of stochastic machine learning methods in the prediction of skin penetration, *Applied Soft Computing*, 2011, **11,** 2367–2375.

6. R. O. Potts and R. H. Guy, Predicting skin permeability, *Pharmaceutical Research*, 1992, **9,** 663–669.

7. I. J. Leontaritis and S. A. Billings, Input-output parametric models for non-linear systems. Part 1: Deterministic non-linear systems; Part 2: Stochastic non-linear systems. *International Journal of Control*, 1985, **41,** 303–344.

8. S. Chen and S. A. Billings, Representations of non-linear systems: the NARMAX model. *International Journal of Control*, 1989, **49,** 1013–1032.

9. B. M. Magnusson, Y. G. Anissimov, S. E. Cross and M. S. Roberts, Molecular size as the main determinant of solute maximum flux across the skin, *Journal of Investigative Dermatology*, 2004, **122,** 993–999.

10. G. P. Moss, A. J. Shah, R. G. Adams, N. Davey, S. C. Wilkinson, W. J. Pugh and Y. Sun, The application of discriminant analysis and Machine Learning methods as tools to identify and classify compounds with potential as transdermal enhancers, *European Journal of Pharmaceutical Sciences*, **45,** 2012, 116–127.

11. S. Chen, S. A. Billings, C. F. N. Cowan and P. M. Grant, Practical identification of NARMAX models using radial basis functions, *International Journal of Control*, 1990, **52,** 1327–1350.

12. R. Iglesias, T. Kyriacou, U. Nehmzow and S. A. Billings, Task identification and characterisation in mobile robotics through non-linear modelling, *Robotics and Autonomous Systems*, 2007, **55,** 267–275.

13. G. P. Moss, A. J. Shah, R. G. Adams, N. Davey, M. B. Brown, G. P. Martin, M. Prapopoulou and Y. Sun. Identification and classification of potential transdermal enhancers by discriminant analysis and Machine Learning methods, in *Perspectives in Percutaneous Penetration 13th International Conference*, ed. K. R. Brain and K. A. Walters, La Grande Motte, France, 2012, p. 50.

SECTION VI: SKIN PROTECTION

F Hafeez and H I Maibach

Dermatology Department, University of California, San Francisco, San Francisco, United States of America.

Human skin provides a partially effective barrier against potentially toxic chemicals [1, 2]. Much of this barrier protection can be attributed to the stratum corneum (SC), the uppermost and nonviable layer of the epidermis because the rate-limiting step for the penetration of most substances is diffusion through the SC [1,2].

Often the most readily available decontamination system is washing with soap and water or water only [3 – 5]. The mechanism of action of decontamination with water include the following: (1) dilution of the chemical agent, (2) rinsing off the chemical substance, (3) decreasing the rate of chemical reaction, (4) decreasing the rate of tissue metabolism (thus minimising the inflammatory reaction), (5) decreasing the hygroscopic effects of the chemicals and (6) restoring the skin's normal pH in acid and alkali burns [6]. However, a traditional soap and water wash are relatively ineffective at removing methylene bisphenyl isocyanate, a potent contact sensitiser, from the skin, and thus, this traditional method may not be the most effective means of skin decontamination for all contaminants, particularly lipophilic ones [7]. The method of decontamination needs to be carefully selected and optimised because chemicals left on the skin after traditional washing procedures can have toxic consequences, and moreover, percutaneous absorption of contaminants can be enhanced inadvertently through the ''wash-in'' effect [8].

Skin decontamination is the primary intervention needed in chemical warfare agent exposures [9]. Skin decontamination involves the immediate removal of the contaminant from the skin performed in the most efficient way possible, and it can be achieved by various means: physical removal of the contaminant, solvating or emulsifying the contaminant in a liquid vehicle, transferring the contaminant to another media through absorption/adsorption of the chemical, through chemical alteration of the contaminant, and rubbing materials to dislodge the contaminant (friction) [10 - 12]. The five subsequent chapters (36-40) investigate the role of decontamination in the removal of chemical warfare agents from the skin and hair using *in vitro* models, and one of the chapters (38) interrogates the validity of using frozen tissue samples to predict percutaneous absorption of chemical warfare agents *in vitro*.

The effectiveness of decontaminants is influenced by the chemical properties of the contaminant, the amount of contaminant on the skin, the timing of decontamination, duration of decontamination, and the anatomical site [2, 12]. Rolland *et al.* in Chapter 36 investigated the decontamination of one anatomical site (the human scalp) after a chemical warfare agent exposure. They chose to study the human scalp because the scalp is often unprotected compared to other body regions covered with clothes, and so they reasoned that it might serve as a preferential route of entry in case of a chemical warfare agent exposure. It had already been established that the thickness of the stratum corneum varies in different parts of the body; for example, the palm is seven times thicker than that of the eyelid [2]. This variation in thickness partially correlates with a differential rate in penetration and absorption of chemicals. For example, areas such as the face, head, scalp, and neck absorb two to six times more than the forearm [13]. The stated purpose of Rolland *et al.* in Chapter 36 was threefold: (1) investigate the validity of an *in vitro* scalp model in studying the decontamination of VX nerve agent, (2) determine if skin decontamination could still be effective even when performed after a delay between time of exposure and time of decontamination (since decontamination often occurs an hour or more after a chemical exposure in the civilian world) and (3) elucidate the role of hair and hair follicles in contamination/decontamination. Towards this goal, the authors determined that unclipped pig skull roof skin could be a suitable model of hairy human scalp for hair contamination and decontamination studies. The authors showed that about 75% of the applied quantity of VX nerve agent was recoverable on the skin surface even 2 hours after skin exposure, thus demonstrating that usefulness of decontamination despite a time delay between exposure and decontamination. Moreover, they demonstrated that hairs modified the percutaneous penetration of VX by binding to the chemical agent. Finally, the authors determined that Reactive Skin Decontamination Lotion (RSDL), an oxime-based lotion currently in use for military and civilian security organisations, and Fuller's Earth (FE), an adsorbing powder used by many different militaries, were active in skin decontamination 45 min post-exposure, but RSDL was more effective in decreasing the amount of VX on both the skin and the hairs in the *in vitro* model.

In Chapter 37, Josse *et al.* compare the effectiveness of standard and widely available absorbents on *in vitro* skin decontamination of VX nerve agent. Though RSDL and Fuller's Earth (FE), essentials of military decontamination kits, have been shown to be effective in immediately decontaminating VX and sulfur mustard, these agents are not widely available in the civilian world, and in the unfortunate circumstance in which a chemical warfare agent exposure has occurred, time is of the essence [14]. As a result, the authors sought to compare the decontamination effectiveness of widely-available standard absorbents (disposable paper tissue, cotton towel, a standard T-shirt [80% polyamide, 20% elastane tissue], standard non-woven absorbent tissue, and polyvalent [water and oil]) absorbent) in decontaminating the nerve agent VX from skin. They chose FE associated with a standard towel as a reference decontamination kit. The authors used an *in vitro* model of pig ear skin for their experiments, and they optimised dosing and the timings of exposure and decontamination to reflect conditions that would likely present in the field. FE + cotton tissue application was found to be the most effective skin decontamination strategy in removing VX; however, the authors also demonstrated that any of the other widely available absorbent materials often available to first rescuers and the victims themselves that were tested (paper tissue, cotton towels, etc.) are helpful and useful emergency decontaminants before more thorough but delayed decontamination of VX is performed with FE.

In Chapter 38, Payne *et al.* sought to understand the effect freezing had upon the integrity of the skin barrier function of *in vitro* sulfur mustard (SM)-exposed pig skin because freezing skin for storage is a common practice for laboratories for convenience and practical purposes. Neither the percutaneous penetration of therapeutic substances through SM-lesioned skin nor the effect of freezing SM lesions on percutaneous penetration had been previously investigated. The authors aimed to assess barrier layer integrity of SM-exposed fresh pig skin and SM-exposed frozen pig skin by measuring water flux through the skin in terms of permeability coefficient (Kp), steady state penetration (Jss), and lag time (T_L). Previously, Wester *et al.* [15] determined that human cadaver skin stored at 4° C can sustain viability (as determined by anaerobic metabolism in which glucose is converted to lactose) for 8 days following donor death. Payne *et al.* demonstrated that skin stored frozen did result in a small but statistically significant (p < 0.05) reduction in barrier function of the skin to water permeation as evidenced by an increase in Kp and Jss and a reduction in lag time of tritiated water. However, they concluded that storing skin under frozen conditions does not cause substantial damage to the barrier function of skin because gross leakage of water across the tissue was not observed.

It has been previously established that up to 80% of an applied liquid droplet of sulfur mustard (SM) can evaporate from the skin surface before absorption and devastating skin lesions develop [16]. Knowing that barrier creams that limit partitioning of SM into the skin can potentially increase evaporation of SM from the skin and minimise absorption, Chilcott *et al.* (Chapter 39) sought to determine if topical administration of fatty acids (palmitic and stearic acids) or cholesterol decreased skin absorption of SM under non-occluded conditions. Previously, in both many *in vivo* and *in vitro* studies, barrier creams have been found to reduce irritant or allergic contact dermatitis. However, their benefit is *sub judice* in clinical trials and inappropriate barrier cream use may be more detrimental than beneficial [17]. Though barrier creams have been recommended not be used as the primary protection against high-risk substances and corrosive agents, Chilcott *et al.* determined that pre-treatment of non-occluded skin with cholesterol can cause a four-fold decrease in skin absorption of sulfur mustard.

In many industrial and laboratory environments, gloves and other forms of protective clothing are used to protect their wearers from toxic exposures, but in certain contexts, like chemical warfare agent releases as occurred in the 1995 Tokyo sarin incident, removal of clothing is considered a crucial step in the decontamination process [18]. Matar *et al.* in Chapter 40 aimed to investigate the effect of common clothing fabrics (overcoat [100% nylon], T-shirt [100% cotton], jumper [100% acrylic], denim jeans [100% cotton]) on the percutaneous absorption of three chemical warfare agents (sulfur mustard, soman, and VX nerve agents) and one chemical warfare stimulant (methyl salicylate). They found that the clothing offered the most protection against VX nerve agent and the least protection against sulfur mustard (which penetrated clothing with relative ease). They also determined that the protective effect of the clothing decreased exponentially with time to disrobing and so concluded that it was important to remove contaminated clothing as early as possible following exposure.

Chemical warfare agent exposures pose a significant and ever-present threat to nations worldwide. These agents lead to serious dermatological consequences, and by permating through the skin's barrier layer, they can rapidly reach toxic levels in the body and induce death. This threat has spurred research by militaries and nations worldwide to better develop effective decontamination strategies in the event of an exposure. Each of the chapters presented in this section add an important piece to the much-needed literature documenting the efficacies of various decontamination strategies for skin and hair against toxic materials and, taken together, these chapters contribute to our increasing knowledge of the complexity [19, 20] of protecting and decontaminating human skin.

REFERENCES

1. F. Dreher, B. S. Modjtahedi, S. P. Modjtahedi and H. I. Maibach, Quantification of stratum corneum removal by adhesive tape stripping by total protein assay in 96-well microplates, *Skin Research and Technology*, 2005, **11**, 97–101.

2. D. Lee and T. Korzun, Skin decontamination, in *Textbook of Pediatric Emergency Procedures*, ed. C. King and F. M. Henretig, Wolters Kluwer Health/Lippincott Williams and Wilkins, Philadelphia, PA, 2nd edn, 2008, pp. 1179–1184.

3. P. H. Moore Jr and F. A. Mettler Jr, Skin decontamination of commonly used medical radionuclides, *Journal of Nuclear Medicine*, 1980, **21**, 475–476.

4. R. C. Wester, (1995) Twenty absorbing years, in *Exogenous Dermatology (Advances in Skin-Related Allergology, Bioengineering, Pharmacology, and Toxicology)*, C. Surber, P. Elsner and A. J. Bircher, Karger, Basel, 1995, pp. 112–113.

5. H. Zhai, S. Barbadillo, X. Hui and H. I. Maibach, In vitro model for decontamination of human skin: formaldehyde, *Food and Chemical Toxicology*, 2007, **45**, 618–621.

6. A. Hall and H. I. Maibach, Water decontamination of chemical skin/eye splashes: a critical review, *Cutaneous and Ocular Toxicology*, 2006, **25**, 67–83.

7. R. C. Wester, X. Hui, T. Landry and H. I. Maibach, In vivo skin decontamination of methylene bisphenyl isocyanate (MDI): soap and water ineffective compared to polypropylene glycol, polyglycol-based cleanser, and corn oil, *Toxicol. Sci.*, 1999, **48**, 1–4.

8. R. P. Moody and H. I. Maibach, Skin decontamination: Importance of the wash-in effect, *Food Chem. Toxicol.*, 2006, **44**, 1783–1788.

9. M. Houston and R. G. Hendrickson, Decontamination, *Critical Care Clinics*, 2005, **21**, 653–667.

10. C. R. Hurst, Decontamination, in *Medical Aspects of Chemical and Biological Warfare*, ed. F. R. Sidell, E. T. Takafuji and D. R. Franz, Office of the Surgeon General, United States Army, Washington, DC, 1997, pp. 351–360

11. W. K. Loke, S. H. U, J. S. Lim, G. S. Tay and C. H. Koh, Wet decontamination-induced stratum corneum hydration–effects on the skin barrier function to diethylmalonate, *Journal of Applied Toxicology*, 1999, **19**, 285–290; M. Boeniger, *Skin Decontamination of*

Chemical Exposures, National Institute for Occupational Health and Safety, Cincinnati, OH, 2005.

12. M. Boeniger, *Skin Decontamination of Chemical Exposures*, National Institute for Occupational Health and Safety, Cincinnati, OH, 2005.

13. R. J. Feldmann and H. I. Maibach, Percutaneous penetration of some pesticides and herbicides in man, *Toxicology and Applied Pharmacology*, 1974, **2**, 88–99.

14. L. Taysse, S. Daulon, S. Delamanche, B. Bellier and P. Breton, *Hum. Exp. Toxicol.*, 2007, **26**, 135.

15. R. C. Wester, J. Christoffel, T. Hartway, N. Poblete, H. I. Maibach and J. Forsell, Human Cadaver Skin Viability for in Vitro Percutaneous Absorption: Storage and Detrimental Effects of Heat-separation and Freezing, *Pharmaceutical Research*, 1998, **15**(1), 82–84.

16. R. P. Chilcott, J. Jenner, W. Carrick, S. a. M. Hotchkiss and P. Rice, Human Skin Absorption of Bis-2-(chloroethyl)sulphide (sulphur Mustard) in Vitro, *Journal of Applied Toxicology*, 2000, **20**(5), 349–355.

17. H. Zhai and H. I. Maibach, Barrier Creams – Skin Protectants: Can You Protect Skin?, *Journal of Cosmetic Dermatology*, 2002, **1**(1), 20–23.

18. R. P. Chilcott, CBRN Contamination, in *Textbook of Environmental Medicine*, ed. J. Ayres, R. Harrison, G. Nichols and R. L. Maynard, Hodder Arnold, London, 2010.

19. H. P. Chan, H. Zhai, X. Hui and H. I. Maibach, Skin Decontamination: Principles and Perspectives, *Toxicology and Industrial Health*, 2012.

20. M. A. Ngo and H. I Maibach, Dermatotoxicology: Historical Perspective and Advances, *Toxicology and Applied Pharmacology*, 2010, **243**(2), 225–238.

35

HUMAN SCALP DECONTAMINATION AFTER CHEMICAL WARFARE AGENT EXPOSURE

P Rolland[1,2], M-A Bolzinger[1,2], D Josse[3,4] and S Briançon[1,2]

[1] Université de Lyon, F-69622, Lyon, France, Université Lyon 1, Villeurbanne, CNRS UMR5007, Laboratoire d'Automatique et de Génie des Procédés. [2] Université de Lyon, F-69373, Lyon, France, Université Lyon 1, Lyon, Laboratoire de Dermopharmacie et Cosmétologie, Institut des Sciences Pharmaceutiques et Biologiques. [3] Institut de Recherche Biomédicale des Armées, Département de Toxicologie et Risques Chimiques, La Tronche, France. [4] Service de Santé et de Secours Médical, SDIS06, Villeneuve-Loubet, France.

1 INTRODUCTION

The occurrence of exposure to chemical warfare agents is not limited to the military. The Tokyo subway attack in 1995 showed that civilians could also be exposed to highly toxic nerve agents [1, 2]. The primary routes of exposure to these compounds are inhalation and skin absorption. The persistent nerve agent VX is one of the most highly toxic compounds following skin exposure (estimated human $LD_{50} \sim 10$ mg) [3]. It has a low volatility, i.e. 8-10 mg m^{-3} at 25°C [4], and thus mainly presents a skin contact hazard. VX was shown to rapidly penetrate the skin both using *in vivo* [5] and *in vitro* skin models [6, 7]. The clinical signs of nerve agent poisoning (resulting from inhibition of acetylcholinesterase) were demonstrated to rapidly appear in domestic swine challenged with VX, with death occurring within one to two hours [8, 9].

Due to high the density of terminal hair follicles compared to other anatomical sites, the human scalp could be an important pathway for these toxic agents. Actually, hair follicles offer a large surface area for topically applied substances which increase the dermal absorption of a molecule [10 – 12]. The human scalp has consistently been shown to be highly permeable to chemicals both *in vivo* [13 – 15] and *in vitro* [16, 17]. Moreover, the hair itself might represent a potential reservoir for VX, since hair is an effective trap for chemical agents [18 – 21]. As the toxicity of chemical warfare agents precludes the use of human subjects for studying skin permeation and decontamination, a human scalp model has to be developed. Pig, and more precisely pig-ear, has previously been shown to be the most relevant animal model for human when performing skin permeation studies of nerve agents both *in vitro* and *in vivo* [5 – 7] and the structure of pig-ear skin is similar to that of human skin [22 – 25]. However, to our knowledge, there is no validated animal model for human scalp skin. Whilst related studies have focussed on identifying the relative importance of different transport pathways (appendages and epidermis) [26, 27], no work appears to have addressed the issue of decontaminating hairy skin. Thus, our first goal was to determine whether pig skin from the same anatomical site, i.e. skull roof, or from the ear could be relevant models for human scalp. Our second goal was to validate these scalp models by establishing the kinetics and skin distribution of VX after several exposure times during the lag time period. The knowledge of these parameters is of crucial importance when designing decontamination procedures since the amount of VX still accessible to the decontaminant should be present in the hair, on the skin surface and in the upper layers of the stratum corneum.

Fuller's Earth (FE) and the Canadian reactive skin decontaminant lotion (RSDL) can be viewed as reference decontamination systems for a use against chemical warfare agents. FE is an adsorbing powder fielded for the military in many countries. RSDL is an oxime-based lotion currently in service for military and civilian security organisations. Skin decontamination with both systems can be effective in preventing intoxication when performed within the first few minutes after contamination with VX [28 – 30]. Delayed decontamination with these systems can also be effective 30 min [8, 28] to 45 min [9] post exposure. Finally, our third challenge was to establish the effectiveness of FE and RSDL for skin and hair decontamination in standard conditions, i.e. 45 min post-exposure to VX, in order to complete the validation of the scalp skin model.

2 MATERIALS AND METHODS

O-ethyl-s-[2(di-isopropylamino)ethyl]methyl phosphonothioate (VX, 97.7% pure, CAS Registry number 50782-69-9) was synthesised by the Centre d'Etudes du Bouchet (CEB, Vert-le-petit, France). The receptor fluid was composed of Hanks's Balanced Salt Solution (HBSS, pH 7.4) containing 1% of penicillin-streptomycin. Horse butyrylcholinesterase and butyrylthiocholine iodide were provided by Sigma (Saint Quentin Fallavier, France).

Human scalp and abdominal skin samples were purchased from Biopredic International (Rennes, France). They were obtained after surgical intervention with the full consent of each patient, both male and female (age range 23-63). Pig skin samples from different anatomical sites (ear and skull-roof area), were collected from a slaughterhouse (Saint-Egrève, France) or from our animal facility immediately after animals were culled. Skin samples were kept at 4°C during transportation to our laboratory. The hypodermis was removed from the skin using a scalpel. Skin samples were cut into 9.42 cm² circular pieces, and thickness measured with a micrometer gauge (Palmer®). They were stored at -20°C for a maximum period of one year.

On the day of the experiment, skin samples were thawed at room temperature and mounted on Franz-type glass diffusion cells (Laboratoire Verre Labomodula, Corbas, France) maintained at 36°C in a water bath in order to keep a skin surface temperature of 32 ± 1°C. The skin samples were divided into groups of 5 or 6 biological replicates. The membrane area available for diffusion was 1.13 cm². After one hour equilibration with the receptor fluid, the skin integrity was assessed by measuring transepidermal water loss (Tewameter® TM210, Courage and Khazaka). After which 5.7 µl of VX was loaded on the centre of the skin surface. This resulted in an applied dose (Q_0) of 5 mg cm⁻². The donor compartment remained open for the duration of the experiment and were conducted in a fume hood (air extraction velocity 0.5 m s⁻¹) at ambient temperature $(22 \pm 2$°C) and with a relative humidity of $48 \pm 6\%$.

Three protocols were used depending on the aim of the experiment:

Protocol 1: Human scalp skin permeability to VX was compared to other skin membranes. The experiment was performed up to 24h in order to underline the differences of permeability between several species (human vs. pig) and body regions. Samples of receptor fluid (400 µl) were collected manually from the receptor compartment at various times (2h, 4h, 6h, 8h, 10h, 12h, 22h and 24h) after exposure in order to establish the absorption kinetics. Each sample removed from the receptor chamber was replaced with an equivalent volume of fresh receptor solution.

Protocol 2: Pig ear skin distribution of VX was determined after several exposure durations (5 min - 2h). The experiment was conducted up to 2h as the apparent lag time of VX percutaneous penetration through pig ear skin is about 1.7h [31]. Three groups of unclipped skin samples were used. Each group had different exposure duration: 5 min, 45 min or 2h, before collection and treatment of samples in order to establish the VX distribution in the skin layers. Data for 24h exposure, taken from a previous experiment, were added in order to extend our findings after 2h [31].

Protocol 3: Comparison of pig skull-roof clipped vs. unclipped skin distribution of VX. The experiment was conducted up to 6h as the apparent lag time of VX percutaneous penetration through pig skull roof skin is about 5h [31]. Three groups of skin were used per exposure duration: 1 group was clipped and 2 groups were unclipped. Before collection of samples, one of the unclipped groups was gently wiped with absorbent paper. Hence, the hair fraction recovery indicates only the VX fraction bound to the hairs, the fraction just present at the hair surface being removed by wiping.

The schematic procedure for the skin decontamination is given in Figure 1. Unclipped skin samples (pig ear skin and pig skull roof models) were exposed to VX and then decontaminated 45 min later either with fuller's earth (Sté Paul Boyé, France) or reactive skin decontamination lotion (RSDL® (E-Z-EM, Inc., Lake Success, NY, USA)). The decontamination process consisted pouring either 70 mg of fuller's earth (FE) or 0.5 ml of RSDL on the skin surface, then rubbing the site in a circular motion for 10 sec with a 1.5 x 1.5 cm pad (from the FE or RSDL package). Excess decontaminant was removed with a second pad. The VX permeation study was then continued until a total 6h experiment time, in order to determine the impact of the decontamination on each model.

At the end of the different exposure durations, samples of receptor fluid (400 µl) were collected manually and stored at -20°C for later quantification ("receptor fluid fraction") after which the skin samples were removed from the diffusion cells. The surface of skin samples was gently wiped 3 times with a piece of absorbent paper (Wypall, Kimberly Clark) in order to recover the fraction of VX remaining on the skin surface. The absorbent paper was then placed with the donor compartment in 25 ml aqueous ethanol (70%). After overnight incubation, aliquots of 1 ml were collected and stored at -20°C for later quantification ("skin surface fraction").

The skin samples were then stripped in order to remove the stratum corneum using cyanoacrylate glue (Loctite SuperGlue-3, Henkel) spread on a glass plate, according to the method of cyanoacrylate surface biopsies [32 – 34]. After separation, the viable skin and stratum corneum (attached to the glass plate) were placed separately in 15 ml ethanol for

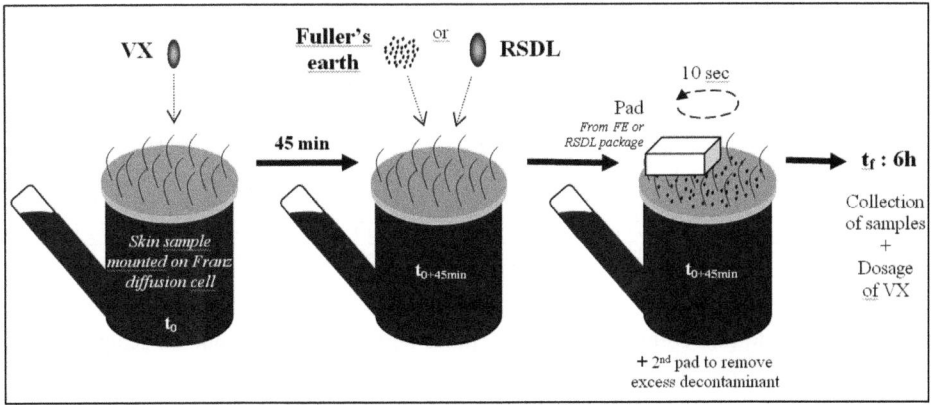

Figure 1 *Overview of procedure for decontamination with fuller's earth (FE) and RSDL following skin exposure to VX.*

overnight incubation at room temperature. The extraction of VX from each skin layer was further promoted by sonication for 20 min at 60 Hz. Aliquots of 1 ml were collected and stored at -20°C for later quantification ("stratum corneum fraction" and "viable skin fraction"). Hairs from the unclipped skull roof skins were cut and placed in 5 ml ethanol. The extraction was conducted as previously described for "hair fraction".

VX is a cholinesterase inhibitor. Therefore, the amount of VX present in each skin sample was determined by an enzymatic method based on the titration of a fixed amount of horse serum butyrylcholinesterase (BChE) active sites, as described by Loke [35]. Since the chemical stability of VX is one of the highest among the chemical warfare agents [36], we can reasonably assume that these organophosphorous (OP) cholinesterase inhibitors were mainly represented by intact VX.

The amount of VX was expressed as a percentage of the applied dose (% Q_0) recovered in the receptor fluid, the viable skin, the stratum corneum, the skin surface and the hairs. Penetration rate values were calculated from the slope of the curve % Q_0= f(time) determined after equilibrium was reached and the intercept of this slope with the x-axis corresponded to the apparent lag time (λ). The means of all data are indicated with their standard error of the mean (SEM). Statistical analysis was performed by using non-parametric Mann and Whitney's test. The level of significance was p < 0.05.

3 RESULTS

We compared the skin and stratum corneum thicknesses, hair follicle density and diameter of the skin samples since they could contribute to the skin permeability and reservoir capacity. As indicated in Table 1, there were significant differences across body regions for both human and pig skin (Mann–Whitney test, p < 0.05). For human scalp skin, the follicular density and diameter was about 12 fold higher and 2 fold larger, respectively, than that of abdomen. The hair density of human scalp was similar to that of pig skull roof but its hair diameter was about 1.6 times lower. The hair follicle diameter of human scalp was similar to that of pig-ear but its density was about 6 times higher. Pig skull roof skin

Table 1 *Characterisation of skin samples used in the study. Results are mean ± SEM.*

Skin Type		Skin structure				Distribution & Kinetics of VX			
		Skin thickness (mm)	Stratum corneum thickness (μm)	Follicle diameter (μm)	Follicle density (cm^{-2})	Permeated dose at 24h (% Q$_0$)	Absorbed dose at 24h (% Q$_0$)	Apparent lag time (h)	Penetration rate (%Q$_0$h^{-1})
Human	Scalp	2.1 ± 0.2	4.5 ± 0.2	95 ± 3	92 ± 3	29.1 ± 6.9	56.9 ± 10.3	2.0 ± 0.3	1.4 ± 0.3
	Abdomen	1.7 ± 0.1	5.1 ± 0.2	54 ± 2	7 ± 1	15.8 ± 3.3	21.3 ± 4.2	7.0 ± 0.7	0.9 ± 0.2
Pig	Ear	1.7 ± 0.1	4.9 ± 0.2	110 ± 3	16 ± 1	40.8 ± 1.9	68.6 ± 5.6	1.7 ± 0.1	2.0 ± 0.2
	Skull roof	2.3 ± 0.1	9.2 ± 0.6	153 ± 4	82 ± 2	2.5 ± 0.4	81.6 ± 2.8	4.9 ± 0.4	0.2 ± 0.0

had the thickest stratum corneum. It was ~ 2 times larger than that of human scalp, human abdomen and pig-ear skin, which were of similar thickness.

The percutaneous penetration kinetic profiles of VX through human and pig skin are presented in Figure 2. The corresponding kinetic parameters are detailed in Table 1. Data for human abdominal skin were taken from studies performed in the same laboratory according to the same protocols [7]. The results showed that VX penetrated through human scalp about 1.5 times quicker than through abdomen. The apparent lag time was 3.5 times lower for the scalp than for the abdomen. Pig skull roof skin permeability to VX was the lowest when compared to that of human scalp, abdomen and pig-ear skin. Human scalp and pig-ear had similar permeability to VX. As indicated in Table 1, the total absorbed fraction of VX, which corresponded to the sum of the skin fraction and the receptor fluid fraction, was similar for human scalp and pig-ear skin (Mann-Whitney test, p > 0.05).

The skin distribution of VX after exposure durations of 5, 45 min, 2 and 24 h is given in Figure 3. After a 2-hour skin exposure, VX remained mostly on the skin surface (75%), and less than 0.1% of VX has reached the receptor fluid. The amount of VX in the stratum corneum (~10%) did not significantly change during the first 2 hours of exposure (Mann–Whitney test, p > 0.05). Thus, the unabsorbed fraction of VX (i.e. skin surface and stratum corneum) is about 85% up to 2 h of exposure. The amount of VX recovered in the viable skin, which is assumed to be potentially absorbed, increased from 5% to 28% between 5 min and 2 h post-exposure. It was not significantly different between 2 h and 24 h post-exposure (Mann–Whitney test, p > 0.05), suggesting that the viable skin was saturated with VX 2 h following dosing.

The skin distribution of VX after exposure durations of 5, 45 min, 2 and 24h is given in Figure. 3. After a 2-hour skin exposure, VX remained mostly on the skin surface (75%), and less than 0.1% of VX has reached the receptor fluid. The amount of VX in the stratum corneum (~10%) did not significantly change during the first 2 hours of exposure (Mann–Whitney test, p > 0.05). Thus, the unabsorbed fraction of VX (i.e. skin surface and stratum corneum) is about 85% up to 2h of exposure. The amount of VX recovered in the viable skin, which is assumed to be potentially absorbed, increased from 5% to 28% between 5 min and 2h post-exposure. It was not significantly different between 2h and 24h post-exposure (Mann–Whitney test, p > 0.05), suggesting that the viable skin was saturated with VX 2h following the exposure.

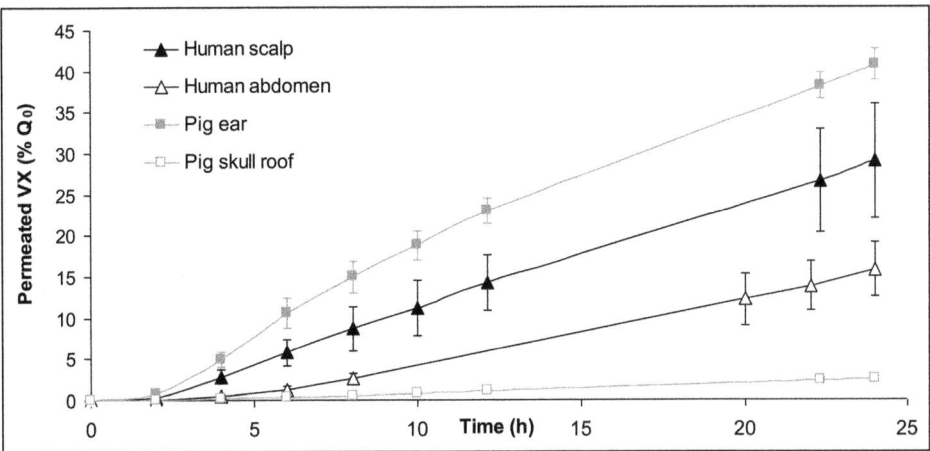

Figure 2 *In vitro percutaneous penetration of VX (applied at a dose of 5 mg cm^{-2}) across unclipped full-thickness skin samples. Results expressed as percentage of applied dose (mean ± SEM).*

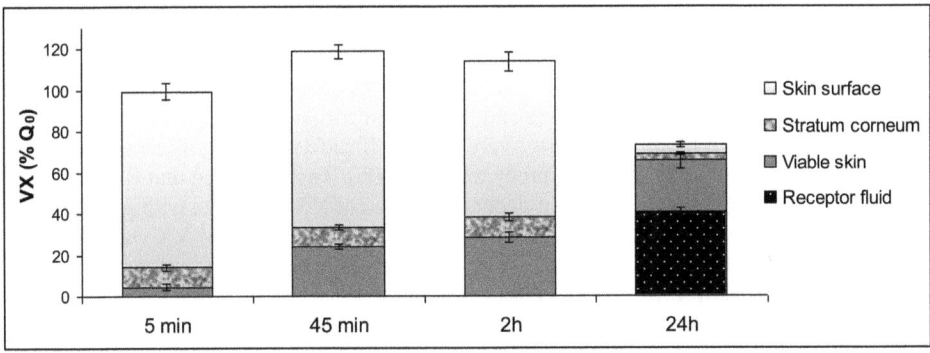

Figure 3 *In vitro distribution of VX (applied at a dose of 5 mg cm^{-2}) through pig-ear full-thickness unclipped skin at 5 min, 45 min, 2 h and 24 h. Results expressed as percentage of applied dose (mean ± SEM).*

The distribution of VX between skin surface and hairs after 2 h and 6 h exposure durations is shown in Figure. 4. The goal of this experiment was to determine the affinity of VX for hair. Following the conditions, i.e. skin surface wiped out or not, the fraction of VX recovered on the hairs 2h and 6h post-exposure was not significantly different (Mann–Whitney test, $p > 0.05$). When the skin surface was wiped out, the fraction of VX recovered on the hairs (~10%) was half of the fraction recovered without any wiping (~20%).

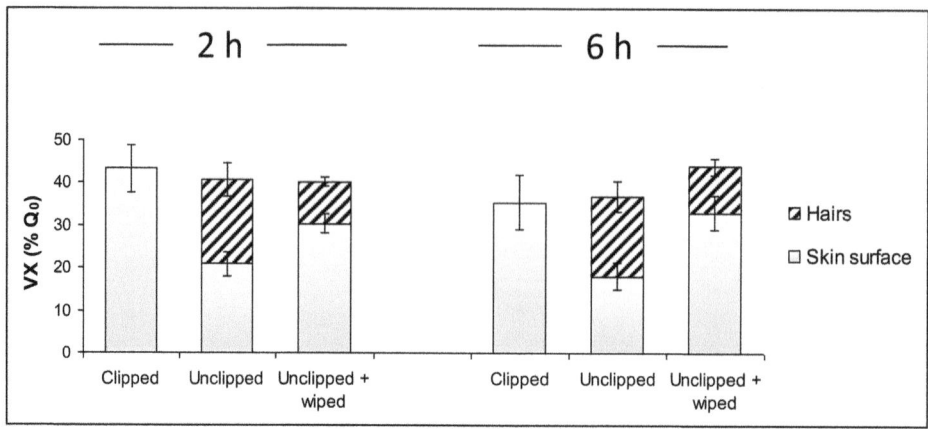

Figure 4 *In vitro distribution of VX (applied dose 5 mg cm^{-2}) recovered from pig-skull roof skin surface and hairs at 2 h and 6 h, either clipped or unclipped, with or without any wiping of the skin surface. Results expressed as percentage of applied dose (mean ± SEM).*

As shown in Figure 5, skin decontamination with fuller's earth (FE) or RSDL 45 min post-exposure to VX was very efficient: decontamination efficacy was greater than 80%. More specifically, decontamination efficacy of the absorbed fraction (i.e. viable skin and receptor fluid) from pig-ear skin was 83 and 94% whereas decontamination efficiency of the unabsorbed fraction (i.e. skin surface and the stratum corneum) was 89 and 98% following FE or RSDL treatment, respectively. These efficacies were significantly higher following RSDL treatment (Mann–Whitney test, $p < 0.05$). This indicated that RSDL was more effective than FE for delayed skin decontamination of VX. Our data with pig skull roof skin showed that the decontamination efficacy of the hair fraction of VX was 92 and 99%, following FE or RSDL treatment, respectively. RSDL was more effective than FE for hair decontamination of VX (Mann–Whitney test, $p < 0.05$).

4 DISCUSSION

Ex vivo skin samples of human scalp being uncommon and expensive, a suitable model had to be designed for subsequent VX percutaneous penetration and decontamination studies. The choice of a model, especially *in vitro* model, is a sensitive issue; a model cannot exactly match the original and different models could be necessary to consider different characteristics such as permeability, the skin reservoir capacity and presence of hair.

Abdominal skin is usually used as an *in vitro* model for human skin, mainly for availability reasons. Our results clearly showed that the reservoir capacity and permeability of human scalp for VX were much higher than that of human abdominal skin. Since the stratum corneum thicknesses of human scalp and abdominal skin were similar, this suggested that the hair follicles density and diameter, both of which being higher for the scalp, significantly contributed to the skin reservoir effect. This important difference was

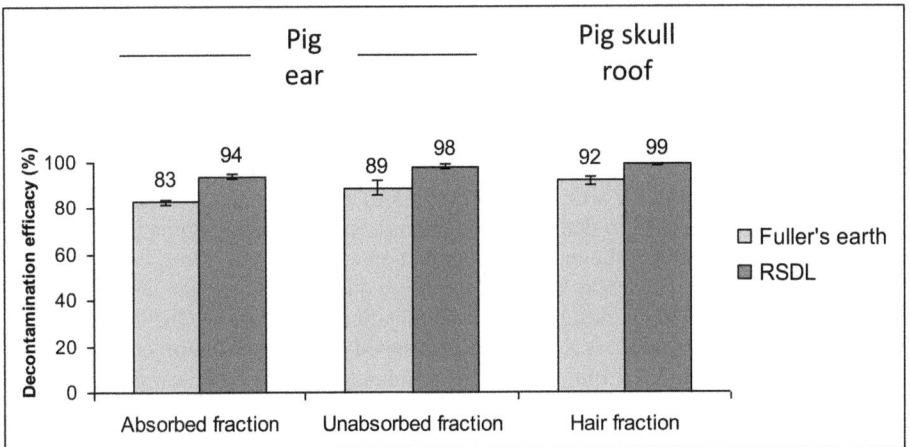

Figure 5 *Decontamination efficacy of Fuller's earth and RSDL used 45 min following VX exposure through pig-ear (absorbed + unabsorbed fractions) and pig skull roof skin (hair fraction). Results are mean ± SEM.*

previously reported by several authors who demonstrated the contribution of appendages to percutaneous penetration [26, 27]. These authors showed that the permeation process of hydrophilic permeants or nanoparticles through appendages was considerable. For example, follicles can account for 50% of the total absorption of caffeine [37]. The permeation process was also shown to depend on follicular density and diameter. Consistently, using caffeine as a drug model, a shorter lag time was found in human skin with a higher follicle density comparatively to human epidermis with lower follicular density or reconstituted human epidermis without any follicle (30 min vs. 2 h). These important differences have to be taken into account in the development of effective skin models for decontamination purposes.

Among animal models, porcine skin is recommended and often used in studies of percutaneous penetration as a substitute for human skin [38]. We performed a quantitative comparison between primary hairs in pig skin and terminal hair follicles in human scalp. In accordance with a previous study, we found similar hair shaft diameters between human terminal hair (mean 92 μm) and pig ear hair (82 μm for Jacobi et al [24] and 110 μm in the present study), but different densities. Our results showed that human scalp skin and pig-ear skin had similar skin reservoir capacity and permeability to VX, even if the follicular density was different. Therefore, the follicular diameter may be a more crucial parameter in the prediction of percutaneous penetration. The reservoir effect may also be partly attributed to the follicle infundibulum, generally similar or larger in pig that in human skin depending on the region [24, 33]. Pig-ear skin could thus be a relevant *in vitro* model when evaluating the percutaneous penetration of VX through human scalp skin.

However, the hair itself might represent a potential reservoir for VX. As a result, hair shaft density should be taken into account when evaluating skin permeation of VX through human scalp. The human and pig hairs were shown to have a similar macroscopic structure [24, 39]. The interactions between chemicals and the hair fibre could thus be assumed to be similar for human and pig hair. Our results showed that compared to human scalp, pig skull roof skin had similar follicular density and slightly higher absorption capacity of VX. In this respect, pig skull roof skin, unclipped, could be a suitable model of hairy human scalp when evaluating the importance of hair contamination and decontamination.

We have established that pig-ear skin could be used as a model for human scalp when evaluating the permeation of VX or the effectiveness of decontaminants. Alternatively, pig skull roof skin could be used to determine the importance of hair contamination following VX exposure and to evaluate the hair decontamination effectiveness. First, the knowledge of the skin distribution of VX after different exposure durations during the lag time period is of crucial importance when designing decontamination procedures. Our data indicated that, with the pig-ear model, VX quickly permeated into the stratum corneum until saturation (which occurred ~ 5 min. As VX has a mid-range octanol-water partition coefficient (log P = 0.675) [40], this toxic agent can rapidly dissolve in lipids and permeate into the stratum corneum [41 – 43]. This was in agreement with other studies which demonstrated that the stratum corneum reservoir was quickly saturated when the lipophilic drugs curcumin (log P = 2.5) or flufenamic acid (log P = 2.7) were topically applied [44, 45].

With *in-vivo* models, we can reasonably assume that (i) skin decontamination may be effective as long as the contaminant is present on the skin surface and in the stratum corneum reservoir, and (ii) that VX within the viable skin fraction is no longer accessible to decontamination [38]. In a civilian context, decontamination might not start in the first hour following exposure. With the pig model *in vivo*, it was demonstrated that RSDL could be effective up to 45 min following skin exposure to lethal doses (5 x LD_{50}) of VX [9]. This suggested that a significant amount of VX was still unabsorbed, i.e. present on the skin surface and in the stratum corneum, and accessible to RSDL up to at least 45 min post-exposure. Our results showed that, after 2h, 85% of the VX applied dose was still unabsorbed and that the absorbed dose was much lower than the human lethal dose. This crucial point indicated that it is worth decontaminating the skin even if contamination occurred 2h before.

In our study, we focused on the scenario which involved hair contamination arising from the external environment. However, the incorporation of drugs into hair may also occur *via* diffusion from blood to the growing follicle during formation of the hair shaft or *via* secretions of the apocrine and sebaceous glands [46]. Previous studies have demonstrated that drugs could bind to the hair surface via electrostatic interactions and then diffuse into the cortex [19 – 21]. Additionally, the role of melanin was evidenced by the importance of hair colour in cocaine binding [19]. Cationic chemicals could be attracted to anionic sites on melanin and protein, i.e. ionised carboxylic acid groups [47] as previously reported with positively charged dye rhodamine B (log P = 1.95) which can enter the hair fibre at the scale edges between the cuticle cells and then diffuse radially and intercellularly along the cell membrane complex [20]. The nerve agent VX, which has a pKa of 9 [48], is positively charged at skin pH, and so may conceivably bind to anionic sites on hair fibres.

The distribution of VX between hair and the skin surface could be an indicator for the importance of hair contamination [49]. With the pig skull roof model, our results showed that a constant fraction of about 20% of VX was recovered in the hair fraction 2h post-exposure and for at least 6h after exposure. Half of this fraction could be removed by a simple wiping; the other half remained bound to the hair for at least 6h. This means that VX could have a high affinity for the hair surface and that a significant fraction of VX could remain bound on hair if decontamination is not performed. As a consequence of this affinity for VX, hair may have an impact on dermal absorption. Indeed, a previous study demonstrated that more VX was resorbed and recovered in viable skin when hairs were clipped [49]. Moreover, an *in vivo* comparison between hairless and hairy rat showed that resorption of topically applied drugs (hydro- or lipophilic) were greater through hairless rat skin [50]. In our case, it seems that the influence of hair is essentially related to their ability to bind the nerve agent VX which resulted in a decreasing penetration. Hair may thus represent a potential binding site for VX, meaning that it has to be taken into account in the decontamination process.

Two decontaminant systems are currently in use for immediate skin decontamination: Fuller's Earth (FE) and the Canadian Reactive Skin Decontaminant Lotion (RSDL). Their form and mechanism of action are different. FE is an adsorbing powder with a large specific surface area (> 180 m²/g) and selectivity for lipophilic molecules. When spread on contaminated skin, it adsorbs toxic agents. Then, the contaminated powder is removed from the skin surface with a specific glove. Several studies have shown that absorbing materials may be more effective than a water washing process for decontamination of the skin, since the latter involves skin hydration and massage which may enhance dermal absorption of the contaminant [51]. In contrast, RSDL is a basic (pH~10) lotion which was specifically designed for solubilising lipophilic molecules and neutralising toxic agents. It contains an oxime which can react and neutralise certain toxic agents, including organophosphates [52]. In our study, we applied exactly the same decontamination procedure for both products; it included a rubbing step after having applied the powder or the lotion on the skin surface, and then a final wiping to remove the excess contaminant. Our results showed that RSDL was more effective than FE for decontaminating human scalp skin and hair models. This is consistent with previous data established *in vivo* with domestic swine, where FE and RSDL, used 45 min post-VX exposure were both effective in preventing animals death but those treated with FE exhibited much more severe signs of poisoning [9]. In contrast, RSDL and FE used 5 min after domestic swine exposure to VX had similar effectiveness [29]. As shown in our study, most of the contaminant was present 5 minutes post exposure on the skin surface and in the stratum corneum, whereas after 45 min, a significant fraction of VX had already penetrated in the viable skin. This suggested that FE and RSDL have similar capability to interact with VX present on the skin surface and in the upper layers of the stratum corneum. One of the main advantages of RSDL over FE would come from its ability to penetrate deeper in the skin and neutralise VX present in the skin reservoir, potentially in the viable skin. Our analysis method did not make it possible to distinguish what VX fraction was removed and what VX fraction was destroyed. Thus, we cannot conclude clearly on the real mechanism of action of RSDL.

Our results indicated that it is worth decontaminating the skin and hair with FE or even better with RSDL since every VX fraction was extensively reduced. This could be true not only 45 min following VX exposure, but also as suggested from our skin distribution studies following an exposure of several hours, given that hairs can potentially trap a significant fraction of VX. The liquid form of RSDL containing detoxifying agents was assumed to be better adapted for this purpose as the higher RSDL pH level (~ 10) (compared to VX pKa value ~ 9), the nerve agent present on the skin surface might not be cationic anymore if RSDL was topically applied. Consequently, binding with hair could become weaker, leading to easier removal of VX. This alkaline medium could also lead to an alteration of the hair fibre structure, resulting in a decrease or a loss of formerly incorporated VX [53]. This hypothesis is consistent with a previous study where a commercially available bleaching solution was demonstrated to remove the total amount of drugs from hair fibres following incubation with opiates solutions [54].

5 CONCLUSIONS

In the present study, we established that it is worth decontaminating the skin 2h following skin contamination with a chemical warfare agent, since a significant amount of VX was still present on the skin surface. Nevertheless, we chose to evaluate skin decontamination 45 min following VX exposure for the validation of our skin models under standard conditions. *In vitro*, using the two skin models for human scalp (i.e. pig ear and pig skull roof), our results indicated that stratum corneum and hairs were both reservoirs for VX and were quickly saturated and persistent. RSDL and fuller's earth were both active in the delayed decontamination of the two skin models, but RSDL was generally more efficient. Other liquid decontaminants may have potential utility in extracting contaminants from the superficial skin layers or to neutralise contaminants *in situ*. For example, a calixarene-loaded nanoemulsion was demonstrated to be effective for decontamination of uranium 30 min post-exposure, which would correspond to the lag time of the penetration of uranium through pig ear full thickness [55]. In this context, we could assume that oxime-based microemulsions, well known system promoting a fast percutaneous absorption of drugs [56, 57], may be an efficient decontaminant system.

REFERENCES

1. T. Okumura, K. Taki, K. Suzuki and T. Satoh, in *Handbook of toxicology of chemical warfare agent*, ed. R. Gupta, 2009, p. 25.

2. N. Yanagisawa, H. Morita and T. Nakajima, *J. Neurol. Sci.*, 2006, **249**, 76.

3. N. Munro, *Environ. Health Perspect.*, 1994, **102**, 18.

4. C. Bertrand, C. Ammirati and C. Renaudu, *Les principaux agents chimiques*, Elsevier, 2006.

5. R. P. Chilcott, C. H. Dalton, I. Hill, C. M. Davison, K. L. Blohm, E. D. Clarkson and M. G. Hamilton, *Hum Exp Toxicol*, 2005, **24**, 347.

6. C. Dalton, I. Hattersley, S. Rutter and R. Chilcott, *Toxicol In Vitro*, 2006, **20**, 1532.

7. V. Vallet, C. Cruz, J. Licausi, A. Bazire, G. Lallement and I. Boudry, *Toxicology*, 2008, **246**, 73.

8. M. G. Hamilton, I. Hill, J. Conley, T. W. Sawyer, D. C. Caneva and P. M. Lundy, *Mil Med*, 2004, **169**, 856.

9. S. Bjarnason, J. Mikler, I. Hill, C. Tenn, M. Garrett, N. Caddy and T. W. Sawyer, *Hum Exp Toxicol*, 2008, **27**, 253.

10. F. Knorr, J. Lademann, A. Patzelt, W. Sterry, U. Blume-Peytavi and A. Vogt, *Eur J Pharm Biopharm*, 2009, **71**, 173.

11. J. Lademann, F. Knorr, H. Richter, U. Blume-Peytavi, A. Vogt, C. Antoniou, W. Sterry and A. Patzelt, *Skin Pharmacol Physiol*, 2008, **21**, 150.

12. A. Vogt, N. Mandt, J. Lademann, H. Schaefer and U. Blume-Peytavi, *J. Investig. Dermatol. Symp. Proc.*, 2005, **10**, 252.

13. U. Blume-Peytavi, L. Massoudy, A. Patzelt, J. Lademann, E. Dietz, U. Rasulev and N. Garcia Bartels, *Eur J Pharm Biopharm*, 2010, **76**, 450.

14. F. Hueber-Becker, G. J. Nohynek, W. J. A. Meuling, F. Benech-Kieffer and H. Toutain, *Food Chem. Toxicol.*, 2004, **42**, 1227.

15. J. Lademann, H. Richter, U. Jacobi, A. Patzelt, F. Hueber-Becker, C. Ribaud, F. Benech-Kieffer, E. Dufour, W. Sterry, H. Schaefer, J. Leclaire, H. Toutain and G. Nohynek, *Food Chem. Toxicol.*, 2008, **46**, 2214.

16. T. Ogiso, T. Shiraki, K. Okajima, T. Tanino, M. Iwaki and T. Wada, *J Drug Target*, 2002, **10**, 369.

17. W. A. Ritschel, A. Sabouni and A. S. Hussain, *Methods Find Exp Clin Pharmacol*, 1989, **11**, 643.

18. S. F. DeLauder and D. A. Kidwell, *Forensic Sci. Int.*, 2000, **107**, 93.

19. R. Joseph, W. Tsai, L. Tsao, T. Su and E. Cone, *J. Pharmacol. Exp. Ther.*, 1997, **282**, 1128.

20. L. Pötsch and M. R. Moeller, *J. Forensic Sci.*, 1996, **41**, 121.

21. G. Ran, Y. Zhang, Q. Song, Y. Wang and D. Cao, *Colloids Surf B Biointerfaces*, 2009, **68**, 106.

22. G. A. Simon and H. I. Maibach, *Skin Pharmacol. Appl. Skin Physiol.*, 2000, **13**, 229.

23. W. Meyer, N. Zschemisch and S. Godynicki, *Pol J Vet Sci*, 2003, **6**, 17.

24. U. Jacobi, M. Kraiser, R. Toll, S. Mangelsdorf, H. Audring, N. Otberg, W. Sterry and J. Lademann, *Skin Research and Technology*, 2007, **13**, 19.

25. I. P. Dick and R. C. Scott, *J. Pharm. Pharmacol.*, 1992, **44**, 640.

26. A. Teichmann, N. Otberg, U. Jacobi, W. Sterry and J. Lademann, *Skin Pharmacol Physiol*, 2006, **19**, 216.

27. B. W. Barry, *Adv. Drug Deliv. Rev.*, 2002, **54 Suppl 1**, S31.

28. E. H. Braue, K. H. Smith, B. F. Doxzon, H. L. Lumpkin and E. D. Clarkson, *Cutan Ocul Toxicol*, 2011, **30**, 15.

29. L. Taysse, S. Daulon, S. Delamanche, B. Bellier and P. Breton, *Hum Exp Toxicol*, 2007, **26**, 135.

30. L. Taysse, F. Dorandeu, S. Daulon, A. Foquin, N. Perrier, G. Lallement and P. Breton, *Hum Exp Toxicol*, 2010, 1.

31. P. Rolland, M. Bolzinger, C. Cruz, S. Briançon and D. Josse, *Toxicol In Vitro*, 2011, **25**, 1974.

32. N. Otberg, H. Richter, A. Knuttel, H. Schaefer, W. Sterry and J. Lademann, *Laser Phys Lett*, 2004, **1**, 46.

33. N. Otberg, H. Richter, H. Schaefer, U. Blume-Peytavi, W. Sterry and J. Lademann, *J. Invest. Dermatol.*, 2004, **122**, 14.

34. M. Förster, M. A. Bolzinger, M. R. Rovere, O. Damour, G. Montagnac and S. Briançon, *Skin Pharmacol Physiol*, 2011, **24**, 103.

35. W. K. Loke, B. Karlsson, L. Waara, A. G. Nyberg and G. E. Cassel, *Anal. Biochem.*, 1998, **257**, 12.

36. M. D. Crenshaw, T. L. Hayes, T. L. Miller and C. M. Shannon, *J Appl Toxicol*, 2001, **21 Suppl 1**, S3.

37. S. Trauer, J. Lademann, F. Knorr, H. Richter, M. Liebsch, C. Rozycki, G. Balizs, R. Büttemeyer, M. Linscheid and A. Patzelt, *Skin Pharmacol Physiol*, 2010, **23**, 320.

38. OECD, OECD, Paris, 2004.

39. M. Mowafi and R. Cassens, *Journal of Animal Science*, 1975, **41**, 1281.

40. S. E. Czerwinski, J. P. Skvorak, D. M. Maxwell, D. E. Lenz and S. I. Baskin, *J. Biochem. Mol. Toxicol.*, 2006, **20**, 241.

41. H. Benson, *Curr. Drug Deliv.*, 2005, **2**, 23.

42. T. Klinger, *J Chem Health Saf*, 2007, 11.

43. B. Magnusson, W. Pugh and M. Roberts, *Pharm. Res.*, 2004, **21**, 1047.

44. A. Teichmann, S. Heuschkel, U. Jacobi, G. Presse, R. Neubert, W. Sterry and J. Lademann, *Eur J Pharm Biopharm*, 2007, **67**, 699.

45. H. Wagner, K. H. Kostka, C. M. Lehr and U. F. Schaefer, *J Control Release*, 2001, **75**, 283.

46. G. L. Henderson, *Forensic Sci. Int.*, 1993, **63**, 19.

47. E. J. Cone, *Ther Drug Monit*, 1996, **18**, 438.

48. E. Le Guevel, *Substitution nucléophile de composés organophosphorés par les ions oximate*, Application en décontamination douce, Université de Versailles-Saint-Quentin-En-Yvelines, 1997.

49. P. Rolland, M. Bolzinger, C. Cruz, D. Josse and S. Briançon, *Toxicol In Vitro*, 2012, under review.

50. A. C. Lauer, J. T. Elder and N. D. Weiner, *J Pharm Sci*, 1997, **86**, 13.

51. J. Lademann, A. Patzelt, S. Schanzer, H. Richter, I. Gross, K. H. Menting, L. Frazier, W. Sterry and C. Antoniou, *Skin Pharmacol Physiol*, 2011, **24**, 87.

52. T. W. Sawyer, D. Parker, N. Thomas, M. T. Weiss and R. W. Bide, *Toxicology*, 1991, **67**, 267.

53. G. Skopp, L. Pötsch and M. R. Moeller, *Forensic Sci. Int.*, 1997, **84**, 43.

54. L. Pötsch and G. Skopp, *Forensic Sci. Int.*, 1996, **81**, 95.

55. A. Spagnul, C. Bouvier-Capely, G. Phan, G. Landon, C. Tessier, D. Suhard, F. Rebière, M. Agarande and E. Fattal, *Eur J Pharm Biopharm*, 2011, **79**, 258.

56. M. Kreilgaard, E. J. Pedersen and J. W. Jaroszewski, *J Control Release*, 2000, **69**, 421.

57. M. Bolzinger, S. Briançon, J. Pelletier, H. Fessi and Y. Chevalier, *Eur J Pharm Biopharm*, 2008, **68**, 446.

36

IN VITRO SKIN DECONTAMINATION EFFECTIVENESS OF VX BY USING ABSORBENTS

D Josse[1,2], G Barrier[1], R Bifarella[2] and C Cruz[2]

[1]Service Départemental d'Incendie et de Secours (SDIS) 06, 140 av Maréchal de Lattre de Tassigny, BP99, 06271 Villeneuve Loubet – Cedex, France. [2]Institut de Recherche Biomédicale des Armées, Antenne de La Tronche, Département de Toxicologie, Protection et décontamination de la peau, 24 avenue des Maquis du Grésivaudan, 38700 La Tronche, France.

1 INTRODUCTION

In a scenario of military or civilian exposure to chemical warfare agents (CWA), emergency procedures include immediate body surface decontamination [1, 2]. The objectives of immediate decontamination are to remove as much contamination as possible from the body surface, including clothes. The resulting level of contamination should be low enough to allow the victims survival and to limit irreversible damage of victims' skin, eyes, and mucosal membranes.

The decontaminants RSDL® or Fuller's Earth (FE) have been shown to be effective for immediate skin decontamination of VX and sulphur mustard [3]. They are part of emergency decontamination kits for the military. However, in a civilian context, these might not be quickly available to first responders. Consequently, since the critical element is time and early physical removal of the contaminant is the most important, it is commonly agreed to perform immediate decontamination with any adsorbent systems available on site [2]. However, there is a lack of scientific evidence for the decontamination effectiveness of these systems. The design and validation of a self decontamination procedure susceptible to be applied in this context was studied by the French Civilian Security in collaboration with the French Army.

In this work, our goal was to compare the skin decontamination effectiveness of the nerve agent VX by using standard absorbents that might be quickly available at the site of the incident. Among those available, we chose and selected absorptive products which might belong to the victims themselves such as disposable paper tissue, cotton towel, standard T-shirt (80% polyamide, 20% elastane tissue) or those which might be brought on-scene by first responders such as standard non-woven absorbent tissue and a polyvalent absorbent normally used to limit the expansion of pollution in the environment. FE associated with a standard towel was chosen as reference decontamination kit. *In vitro* skin models and decontamination procedures were specifically designed to match as close as possible those that would be performed on the field.

2   MATERIALS AND METHODS

O-ethyl-s-[2(di-isopropylamino)ethyl]methyl phosphonothioate (VX, 99% pure, CAS Registry number 50782-69-9) was synthesized and provided by DGA-MNRBC (Vert-le-petit, France). Hanks' Balanced Salt Solution (HBSS) from Invitrogen was used as receptor fluid (RF) in the diffusion cells. Horse butyrylcholinesterase and butyrylthiocholine iodide were provided by Sigma (Saint Quentin Fallavier, France).

Decontaminant products were: paper tissue (Kleenex®, Kimberly-Clark) (**A**), non woven absorbent tissue (PGI NONWOVENS, Cuijk, Netherlands) (**B**), cotton towel (**C**), Polyvalent absorbent TL 100 (DEP SYSKO, Bourgoin Jallieu, France) (**D**), standard T-shirt (80% polyamide, 20% elastane tissue) (**E**) and Fuller's earth + cotton tissue (**F**).

In vitro studies were conducted within a fume hood (face flow rate of 0.7 m s^{-1}) at room temperature (20 ± 2°C) and relative humidity of 50 ± 10%. Pig ears were collected from a slaughterhouse (Saint-Egrève, France) immediately after the animals were killed. Skin samples were kept at 4° C during the transportation to our laboratory. After cleaning with tap water, the pig ears were close-clipped and excised as full-thickness skin samples then stored for a maximum duration of 6 months at -20°C. On the day of the experiment, the skin samples were thawed at room temperature, dermatomed to a thickness of 500 ± 50 μm, then visually inspected for integrity. Split-thickness skin samples were mounted in Franz-type static glass diffusion cells maintained at 40°C in a water bath in order to get a skin surface temperature of 32 ± 1°C. The skin samples were divided into groups of 4 to 6 biological replicates. The skin area available for diffusion was 1.13 cm². After one hour equilibration with the receptor fluid, the skin integrity was assessed by measuring transepidermal water loss (Tewameter TM210, Courage and Khazaka). In agreement with the OECD guidelines (2004), only the skin samples that had a TEWL value between 3 and 10 g h-^{1} m-^{2} were used in this work.

VX (5.7 μl) was placed on the centre of each skin surface. This resulted in an applied dose (Q_0) of 5 mg cm^{-2}. The donor compartment remained open. Non-decontaminated skin samples were used as controls.

Skin decontamination was performed 5 or 15 min following VX exposure. The following skin decontamination principles were followed: the products were blotted on the skin to avoid the spreading of contaminant on the skin surface and any enhancement of skin permeation. The decontamination procedure preserved the skin integrity as much as possible and was repeated several times (using clean products). Fuller's earth (70 mg) was spread on the skin surface then removed 15 s later using a cotton wipe. Absorbent tissues (9.6 cm²) were blotted on the skin surface for 3 sec with a force of 100 g cm^{-2}. This was repeated 3 times.

At the end of 6 hours exposure, the surface of skin samples was gently wiped 3 times with absorbent paper (Wypall, Kimberly Clark) in order to recover the fraction of VX remaining on the skin surface. The absorbent paper was then placed with the donor compartment in 25 ml aqueous ethanol (70%). After overnight incubation, aliquots of 1 ml were collected and stored at -20°C for later quantification ("unabsorbed fraction").

The skin samples were removed from the diffusion cells and incubated overnight in 10 ml ethanol at room temperature after which extraction of VX was further promoted by sonication for 20 min at 60 Hz. According to this protocol, the VX extraction yield from the skin was less than 100%, *i.e.* 68% ± 12%. However, the extraction of VX from skin samples was performed exactly according to the same procedure and so loss of VX during the extraction process can be assumed to be similar for each skin sample. Aliquots of 1 ml were collected and stored at -20° C for later quantification ("skin fraction"). Samples of receptor fluid (400 µl) were collected manually from the receptor compartment at various times (0.5, 1, 1.5, 3, 4.5 and 6 h) after exposure. It was followed by the replenishment of an equivalent volume of fresh solution.

The amount of VX present on the skin surface (unabsorbed fraction), in the skin (skin fraction (SF)) and that penetrated through the skin (receptor fluid fraction (RF)) was quantified. Although the chemical stability of VX is one of the highest among the chemical warfare agents, we could not exclude that some metabolites of VX could contribute to the inhibition of butyrylcholinesterase. Therefore, the results were expressed as VX equivalent (VX_{eq}) corresponding to the sum of VX and its active metabolites, *i.e.* cholinesterase inhibitors. The amount of VX_{eq} was determined by an enzymatic method based on the titration of a fixed amount of horse serum butyrylcholinesterase (BChE) active sites as described by Loke et al., 1998 [4]. Samples were diluted with HBSS by an automatic liquid handling system (Freedom Evo, Tecan, Männedorf, Switzerland) and cholinesterase activity was measured in a microplate reader at 420 nm (Safire, Tecan, Männedorf, Switzerland). VX calibration standards (1.5 to 25 nM) were prepared in the HBSS buffer and used to titrate BChE. Our calibration curve (ln (BChE activity) = f (VX)) was linear from 0 to 25 nM. Standards and sample dilutions were incubated for 2h at 20°C with BChE then the residual enzyme activity was measured. The calibration curve was used to determine the amount of VX_{eq} in the samples dilutions.

The cumulative amount of VX_{eq} was expressed as a percentage of the applied dose (%Q0) recovered in the receptor fluid. The decontamination effectiveness was determined from the ratio E (%) = ((SF+RF)$_{controls}$ − (SF+RF)$_{decontaminated}$)/ (SF+RF)$_{controls}$.

Statistical analysis was performed by using non-parametric Kruskal-Wallis one way analysis of variance on ranks. Multiple Comparisons were performed by the Dunn's Method. The level of significance adopted was p < 0.05.

3   RESULTS

As indicated in Figure 1, VX could be detected in the receptor fluid (RF) from 30 min following exposure. Six hours post-exposure, about 11% of the initial dose (Q0) had penetrated through the skin.

VX was mainly recovered in the skin (46% of Q_0; Figure 2) and unabsorbed fractions (34% of Q0).

As shown in Figure 3, when skin decontamination was performed 5 min post-exposure to VX, the fractions of VX recovered in the RF and in the skin were reduced in comparison with those of non-decontaminated skin samples. This reduction was greatest for the skin fractions, ranging from a ~10-fold (following skin decontamination with products A or B)

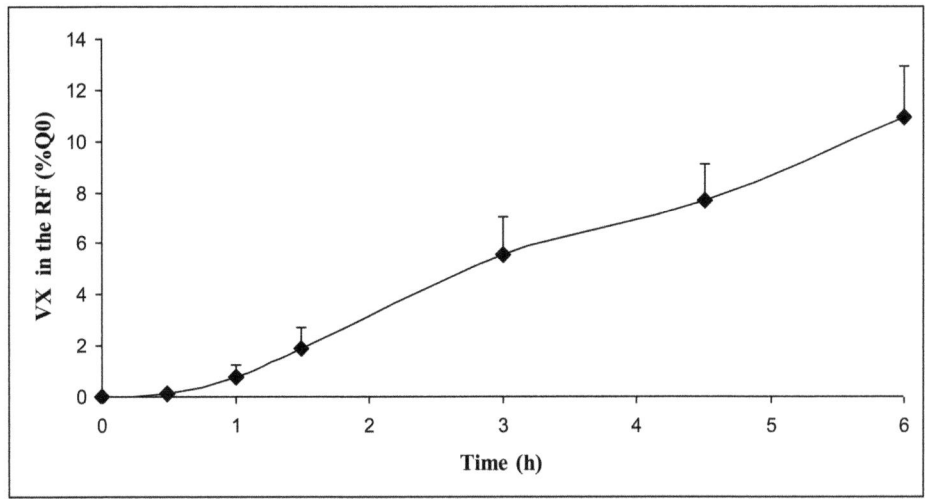

Figure 1: *In vitro percutaneous penetration of VX through split thickness pig-ear skin; n = 6, Q_0 = 5 mg cm^{-2}, RF = Receptor Fluid.*

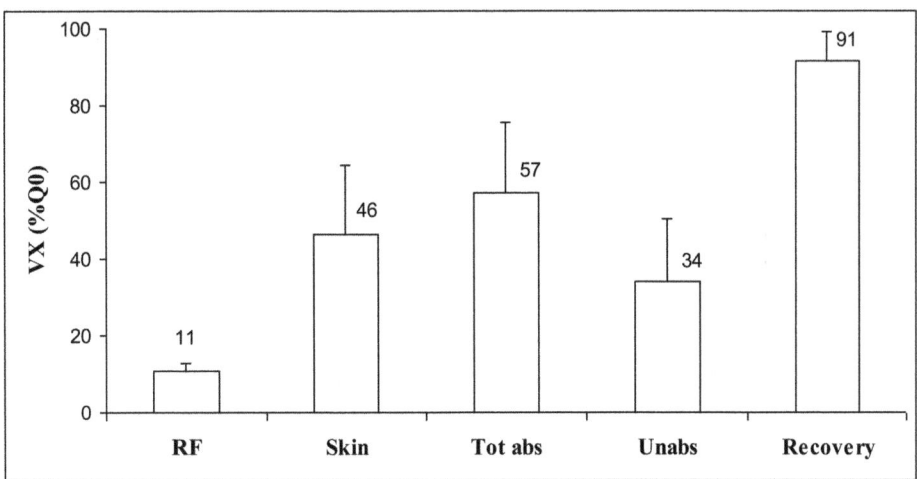

Figure 2 *Distribution of VX 6 hours following skin exposure; Tot abs = total absorbed (sum of the skin and RF fractions of VX); Unabs = unabsorbed.*

up to a 50-fold (following skin decontamination with F). The difference with the non-decontaminated skin samples was significant ($p < 0.05$) following skin decontamination with C, D or F. The fractions of VX recovered in the RF were about 2 times lower after skin decontamination with either A, B, C, D or E ($p > 0.05$) while it was about 10 times lower following skin decontamination with F ($p < 0.05$). There was a ~ 30 fold decrease of total absorbed fraction of VX when decontamination was performed with F ($p < 0.05$).

Relative to the non-decontaminated samples, skin decontamination with F led to a significant decrease of the RF, skin and totally absorbed fractions of VX. When compared to the A or B skin decontaminant, F led to a significantly reduced total absorbed fraction of VX. The RF fraction of VX was also significantly reduced after skin decontamination with F when compared to A, B or E.

As shown in Figure 4, skin decontamination with C, D or F 15 min post-exposure led to a significant reduction of the amount of VX in the RF, skin and totally absorbed fractions. The amount of VX found in the RF and totally absorbed fractions was significantly lower following skin decontamination with F than with B or E.

As shown in Figure 5, the decontamination effectiveness varied from 79% to 97%. C, D and F were the most effective decontaminants when used 5 or 15 min following skin exposure to VX. F was more effective than A and B when used 5 min post-exposure to VX, and than B and E when used 15 min post-exposure to VX.

When the exposure duration to VX increased from 5 to 15 min, there was no significant change of effectiveness for a given decontaminant

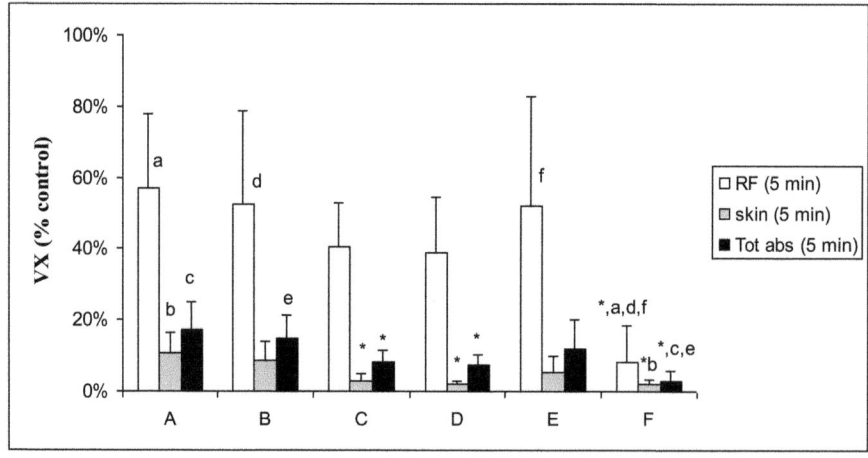

Figure 3 *Distribution of VX 6 hours following skin exposure for decontaminated samples. Decontamination was performed 5 min after skin exposure; * and letters indicate significant differences of decontaminants with the non-decontaminated samples (controls) and between same labelled parameters, respectively. Capital letters stand for: paper tissue (A), non woven absorbent tissue (B), cotton towel (C), Polyvalent absorbent (D), standard T-shirt (80% polyamide, 20% elastane) (E) and Fuller's earth + cotton tissue (F).*

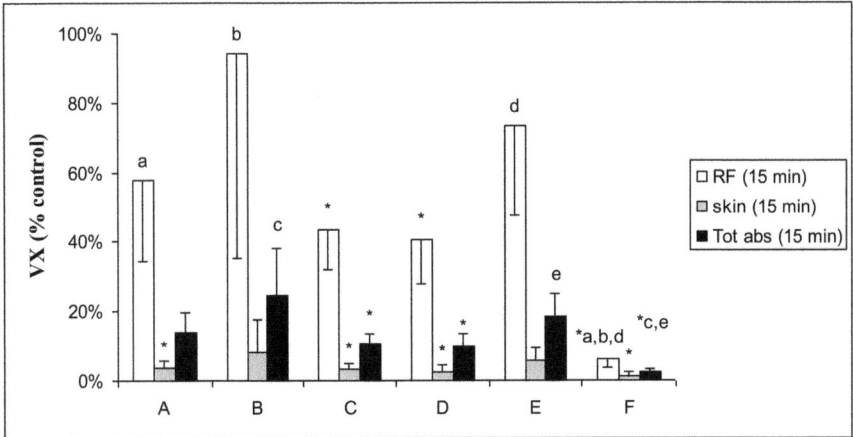

Figure 4 *Distribution of VX 6 hours following skin exposure for decontaminated samples; decontamination was performed 15 min after skin exposure; * and letters indicate significant differences of decontaminants with the non-decontaminated samples (controls) and between same labelled parameters, respectively.*

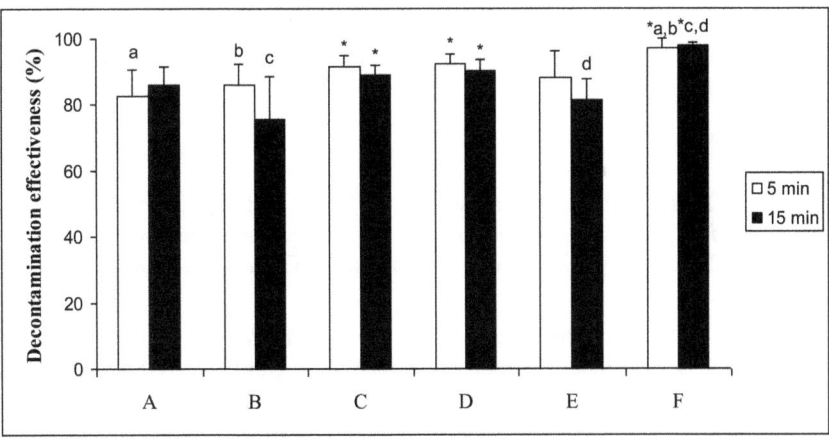

Figure 5 *Skin decontamination effectiveness; skin samples were decontaminated 5 or 15 min post-exposure to VX; * and letters indicate significant differences of decontaminants with the non-decontaminated samples (controls) and between same labelled parameters, respectively.*

4   DISCUSSION AND CONCLUSIONS

In a civilian context of exposure to highly toxic chemicals, emergency decontamination should ideally be performed with the shortest possible delay following contamination [5]. It should firstly consist in the removal of contamination [1]. It should also be easy to perform in order to facilitate personal self-decontamination, even for untrained people without supervision. In this work, we compared the skin decontamination effectiveness of standards absorbents. *In vitro* skin models were found to be quite suited and relevant to quickly compare the effectiveness of skin decontaminants against VX.

Our results indicated that the use of any potentially absorbent tissue available from the victims themselves or from the first rescuers could be valuable and strongly recommended as an emergency decontamination procedure before thorough and delayed decontamination is performed. This was consistent with previous animal studies performed by van Hooidonk [6] who found that when performed within 4 min following skin contamination, wiping the skin with a dry adsorbent object such as paper, gauze or a towel was effective for removing both nerve agents and mustard.

Consistently with previous *in vitro* and *in vivo* decontamination studies performed on VX or sulphur mustard, FE application on the skin surface followed by removal of the contaminated powder with a cotton tissue was found to be the most effective skin decontamination procedure when performed 5 or 15 min post-exposure to VX. This could be related to the high surface area and, as a result, adsorption capacity of VX by FE when compared with the absorbing capacity of standard tissues. Further experiments are required to determine the rank of absorbent efficiencies for the skin decontamination of other highly toxic chemicals such as soman or sulphur mustard.

5 ACKNOWLEDGEMENTS

This work was supported by Service de Santé des Armées, Direction Générale de la Sécurité Civile et de la Gestion des Crises and Service Départemental d'Incendie et de Secours des Alpes-Maritimes.

6 REFERENCES

1. B. J. Lukey, H. F. Slife, E.D. Clarkson, C. G Hurst, and E. H. Braue, in *Chemical warfare agents: chemistry, pharmacology, toxicology, and therapeutics*, ed. J. A. Romano and B. J. Lukey, CRC Press, Taylor/Francis Group, Boca Raton, USA, 2008, ch. 21, p. 611.

2. E. H. Braue, C. H. Boardman, and C. G. Hurst, in *Medical aspects of chemical warfare*, ed. S. D. Tuorinsky, Borden Institute, Washington, D.C., USA, 2009, ch. 16, p. 527.

3. L. Taysse, S. Daulon, S. Delamanche, B. Bellier, and P. Breton, *Hum. Exp. Toxicol.*, 2007, **26**, 135.

4. W. K. Loke, B. Karlsson, L. Waara, A. G. Nyberg and G. E. Cassel, *Anal. Biochem.*, 1998, **257**(1), 12.

5. C. G. Hurst, in *Medical Aspects of Chemical and Biological Warfare*, ed. R. Zajtchuk and R.F. Bellamy, Borden Institute, Washington, D.C., USA, 1997, p. 351.

6. C. Van Hooidonk, B. I. Ceulen, J. Boc, and J. van Genderen, in *Agents and the skin*, Proceedings of an International Symposium on Protection against chemical warfare agents – Stockholm (Sweden), June 6-9, 1983, p. 153.

37

THE EFFECT OF SULFUR MUSTARD EXPOSURE AND FREEZING ON TRANSDERMAL PENETRATION OF TRITIATED WATER THROUGH *EX VIVO* PIG SKIN

O J Payne[1], S J Graham[2], C H Dalton[2], P M Spencer[1], R Mansson[1], J Jenner[2], J Azeke[3], E Braue[3]

[1]Detection Department, DSTL, Porton Down, Salisbury, United Kingom. [2]Biomedical Sciences Department, DSTL, Porton Down, Salisbury, United Kingdom. [3]Medical Toxicology Branch, US Army Medical Research Institute of Chemical Defense, Aberdeen Proving Ground, Maryland, United States of America.

1 INTRODUCTION

Sulfur mustard (SM) is a well-known and potent vesicating agent [1–4] that has been used in many conflicts during the twentieth century [5]. Cutaneous sulfur mustard injuries are often painfully debilitating, requiring surgical intervention (*e.g.,* tissue debridement) and specialist burn care support. Wounds can take months to heal and place a substantial burden on the medical chain [6–8]. Topical pharmaceutical formulations as medical countermeasures to cutaneous sulfur mustard injuries carry important advantages, including ease and non-invasiveness of dermal application, and effectiveness in reducing SM associated skin damage [6, 8, 9].

Potential topical therapies can be screened by measuring skin penetration *in vitro*. Such studies are valuable for understanding transdermal drug penetration because experimental variables can be controlled and pharmacokinetic data may be determined prior to conducting more costly and invasive *in vivo* studies. Both animal and human *in vitro* skin models are used to predict percutaneous penetration in humans *in vivo* [10 – 12], though many *in vitro* studies can only be practically conducted using tissue that has been stored frozen [11, 13–16].

The effectiveness of modern candidate therapies to SM injury can only be assessed *in vivo* using animal models because human volunteer studies would be unethical. Two animal models have been widely used: the mouse ear vesicant model (MEVM) and the pig. The MEVM has been used as a simple rapid screen of efficacy using reduction of oedema and erythema as a measure of SM-associated injury [1, 6, 9]. However the structural and morphological similarities between porcine and human skin [8, 17 – 18] make the pig a better indicator of penetration of drugs through the skin [19, 20]. When it is not possible to conduct *in vivo* animal studies and skin tissue cannot be used fresh for *in vitro* studies, then freezing skin for storage is a common practice for laboratories conducting diffusion measurements [3, 15, 16, 21].

Understanding the effect that freezing has on the skin barrier function is vital to appropriate interpretation of percutaneous drug penetration. Testing skin integrity when performing penetration studies is also in accordance with Organisation for Economic Co-operation and Development (OECD) test guideline 428 [22]. Knowing the effect that freezing skin has on drug penetration through SM-injured skin is also essential for appropriate interpretation of potential topical therapies. The percutaneous penetration of therapeutic substances through SM-lesioned skin has not previously been published. The effects of freezing SM lesions upon the barrier layer to percutaneous penetration has also not been assessed, though an established method for generating lesions in the large-white pig has previously been reported [23]. The purpose of the study was to generate dermal absorption data on sulfur mustard exposed pig skin using fresh skin tissue and tissue that had been stored frozen. Assessment of the barrier layer integrity was determined by measuring water flux through the skin and expressed in terms of permeability coefficient (Kp), steady state penetration (Jss) and lag time (T_L).

2 MATERIALS AND METHODS

Animal experiments were carried out in accordance with the UK Animal (Scientific Procedures) Act (1986). Six healthy, female large white pigs; weight range 15 – 20 Kg) were singly housed in a controlled environment ($22 \pm 2°C$, $53 \pm 2\%$ relative humidity) and subject to a 12-hour light/dark cycle with food and water provided *ad libitum*. The animals were habituated to human contact by twice daily contact sessions for 7 days prior to SM exposures.

Twelve sites on each abdomen of six large-white pigs (*Sus scrofa domesticus*) were exposed to pure liquid SM; 12 other abdominal sites on each animal were designated as naïve control sites. Site numbers within each group (SM-exposed and naïve control) were then randomised. Half of the SM-dosed and negative control sites were allocated for freezing; the other half were designated for use as freshly excised skin. Six animals were used to produce tissue for each skin test group (*e.g.,* frozen SM-exposed skin) with six intra-animal replicates used for each group. This means that each test group comprised a maximum of 36 skin samples.

Prior to anaesthesia, each animal was given ~4 ml Hypnovel® (midazolam hydrochloride, 2 mg ml^{-1}, Roche Products Ltd., Hertfordshire, UK) intra-muscularly. General anaesthesia was induced by 5% isoflurane (Isoflurane-VET®, Merial Animal Health Ltd, Essex, UK), 4 – 6 L min^{-1} O$_2$, 0.5 L min^{-1} NO$_2$, and maintained with isoflurane (up to 1-5%), 4 – 6 L min^{-1} O$_2$, 0.5 L min^{-1} NO$_2$. Whilst anaesthetised, each animal underwent pulse oximetry monitoring using a Propaq physiological monitoring system (Propaq 104*encore*, Propaq Systems, Inc., Beaverton, Oregon, USA). Twenty-four hours prior to SM exposure, abdominal hair was carefully close clipped using Wahl clippers (Wahl Clipper Corporation, Sterling, Illinois, USA). Following exposure to SM, each animal was administered approximately 1 ml Temgesic® (buprenorphine hydrochloride (0.3 mg ml^{-1}, Schering-Plough, Hertfordshire, UK), allowed to recover and returned to its home pen prior to culling.

Distilled sulfur mustard was synthesised by the Defence Science Technology Laboratory at Porton Down (>99% pure by NMR analysis) and used in accordance with the Chemical Weapons Convention [24]. Dosing of SM was conducted using a modification of an established method [23]. Twelve circular dosing templates (4.9 cm^2 each) fixed to the abdomen of each animal were individually dosed with 292 µl of pure SM. The modified method used smaller dosing templates than previously described, thus enabling more dose templates to be applied to each animal, increasing the information gained from each animal and working in line with the National Centre for Replacement, Refinement and Reduction of Animals in research (NC3Rs).

Dose templates were adhered to the pig using self adhesive sticky tape incorporated within the template. Pure liquid SM was applied to the whole dose area (comprised of Whatman EPM2000 filter paper) using a 1 ml Gilson MICROMAN® (Gilson Scientific Ltd, Luton, Bedfordshire, UK) positive displacement pipette and left in contact with the skin for eight minutes. A Teflon disc, rubber stopper and 300 g brass weight were placed on top of the filter paper area within the template to ensure proper contact with the skin. On completion of SM dosing, the rubber stopper, Teflon discs and dosing templates were removed using plastic disposal forceps and placed into a decontamination solution (sodium hypochlorite 14% available chlorine) for 24 h. The area exposed to liquid SM was blotted with a sterile gauze pad by running a wooden tongue depressor lightly along the pad (for 15 seconds), thereby swabbing any surface residues of SM. The depressor and gauze pad were then placed into decontamination solution for 24 h. Exposure sites were then covered with activated charcoal cloth to trap any desorbing SM vapour until the animals were euthanised.

Animals were euthanised 3 hours after SM exposure by administration of Euthatal (10 ml) whilst under sedation and respiratory anaesthesia (isofluorane ~5%, 4–6 $L.min^{-1}$ O_2, 0.5 $L.min^{-1}$ NO_2). Un-exposed skin (control) and SM-exposed skin lesions were excised post mortem using a scalpel. All skin was then cut to a nominal 500 µm thickness using a Zimmer™ air dermatome (Zimmer LTD, Dover Ohio 44622, USA). Split thickness skin sections for use as fresh samples were then mounted directly within Franz-type glass static diffusion cells within three hours of excision. Split thickness skin samples to be used after freezing were placed flat onto aluminium foil sheets, supported by plastic boards and wrapped with aluminium foil before being placed within a freezer. Skin samples were maintained at -20°C for up to two weeks before use.

For each penetration experiment using freshly excised skin (< 3 h after excision), twenty four Franz-type glass static diffusion cells were assembled using skin from two pigs. Six pieces of control skin and six pieces of SM-exposed skin were used from each animal. These experiments were repeated three times until fresh tissue from a total of six animals (n=6) had been used. All skin taken from the six animals that was stored frozen (12 skin pieces per animal) was defrosted under ambient room temperature conditions (approximately 21°C) for approximately one hour on the morning of the study, mounted onto 72 diffusion cells and processed within a single penetration experiment. Receptor chambers of all diffusion cells for fresh/frozen comparisons were filled with 5 ml of ethanol: water 1:1 (v: v) receptor fluid and stirred by externally driven, Teflon-coated magnetic bars. Receptor fluid levels were adjusted to ensure that the meniscus in the sampling arm was level with the skin surface. Once assembled diffusion cells were left for approximately 16 hours to equilibrate on heated stirrer plates to maintain skin surface temperatures of 32 ± 1°C.

Baseline receptor fluid samples (20 µl) were withdrawn from each diffusion cell prior to the dosing of skin surface with radiolabelled water, and a replacement volume of fresh receptor fluid (20 µl) was inserted. Tritium (^{3}H) labelled water, obtained from New England Nuclear Corporation USA, was diluted with non-radiolabelled water to achieve an activity of 0.185MBq ml^{-1}, and a 1 ml aliquot was applied to the surface of each skin sample held within a Franz-type diffusion cell. Contact between the ^{3}H$_2$O and the skin surface was taken as the start of the experiment (time zero). The top rim of each donor chamber was sealed with aluminium foil held in place with a perfluorinated cream to produce occluded infinite dose conditions. Receptor fluid samples (20 µl) were taken half an hour after dosing and then every hour until six hours post-dosing. Thereafter samples were taken every three hours up to 24 hours post-dosing. All samples removed from receptor chambers were replaced with an equivalent volume of fresh receptor fluid. Penetration was assessed by liquid scintillation counting (LSC) using a Wallac 1214 Rackbeta scintillation counter (Wallac Oy, Turku, Finland), using manufacturer's ^{3}H-quench curve library set to exclude single-photon non-radioactive events. Samples removed for analysis were added to 5 ml of scintillation fluid (Ultima Gold scintillation fluid, Perkin Elmer, Massachusetts, USA) and analysed by LSC. Water flux (^{3}H$_2$O permeability) was expressed as the permeability coefficient (Kp [cm h^{-1}]) and steady state penetration rate (Jss [g cm^{-2} h^{-1} x 10^{-6}])

It was assumed that under infinite sink conditions the penetrant concentrations in the receptor compartment would be negligible compared to that in the donor chamber. The skin permeability coefficient (Kp) indicates the rate at which a specific chemical permeates through the skin. The initial concentration of water used as a penetrant is equivalent to its density, *i.e.*, 1 g cm^{-3}; therefore in this study Kp is numerically identical to the Jss. The rate of penetration at steady state (Jss) was taken as the gradient of the linear portion of the plot of amount penetrated against time. The plot was considered linear if the correlation coefficient between amount penetrated and time was equal to or greater than 0.95. The lag time (t_L) was taken as the x-intercept as a regression line through the linear portion of this plot.

The relationship between the penetration parameters (Jss, Kp and t_L) and the skin challenge and storage conditions was investigated using a general linear model (GLM). The distribution of the penetration parameters was found to be normal using an Anderson-Darling test (Figure 1) on the model residuals derived from the GLM. The GLM compared the average parameters for the skin challenge and storage conditions (a total of four groups) while taking account of the variation within and between the six animals used in the experiment. The model output is used to determine whether there is a statistically significant difference between these four groups.

Data presented are average values given as mean (± standard deviation, SD) together with the number of observations for each group (n). Six inter-animal replicates were used for each skin group (treatment/condition) tested, with each group having up to six intra-animal replicates. Probabilities of less than 5 % (P < 0.05) were considered significant.

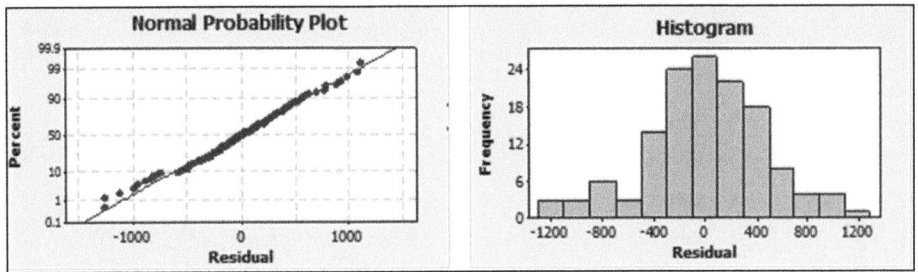

Figure 1 *Anderson-Darling test output for distribution of study data.*

3 RESULTS

The percutaneous absorption of tritiated water (^3H_2O) through sulfur mustard (SM) exposed abdominal pig skin was measured using *in vitro* Franz-type static diffusion cells. The barrier function to water permeation following exposure to liquid SM for 8 minutes and excision 3 hours later did not change significantly (Figure 2). A statistically significant difference (p <0.05) in steady state penetration (Jss), permeability coefficient (Kp) and lag time (t_L) of ^3H_2O was observed between fresh skin and skin stored frozen (-20°C) for up to two weeks. Steady-state penetration and Kp values were significantly higher (p <0.05) in skin stored frozen compared with fresh skin. Fresh naïve skin had an average Kp of 1.65 x 10^{-3} cm h^{-1}, whereas frozen naïve skin was 2.04 x 10^{-3} cm h^{-1}. Fresh SM-exposed skin had a mean Kp of 1.72 x 10^{-3} cm h^{-1}, whereas frozen SM-exposed skin had a mean Kp of 2.31 x 10^{-3} cm h^{-1} (Table 1). Lag times were also significantly shorter in skin that had been stored frozen (Figure 3, Table 2). Exposing pig abdominal skin to pure liquid SM for 8 minutes and excising 3 hours later did not significantly affect the barrier layer to tritiated water permeation compared with unexposed skin excised within 3 hours of termination (Table 1). The permeation profiles of tritiated water through fresh and frozen pig skin, with and without SM challenge, were similar in shape and magnitude (Figure 2); however, statistically significant differences in penetration parameters were observed. Eight cells did not pass the integrity test and were not used in the study.

Fresh skin demonstrated significantly lower mean Kp than skin stored frozen (Table 1). There was no significant difference in average Kp between the SM-exposed and non-exposed skin. The Kp was higher in SM-challenged skin than in non-challenged skin.

Table 1 *Summary of Kp values of ^3H_2O through skin of six animals (n = 6 per animal) unless otherwise stated. * denotes significant difference (p < 0.05) from fresh skin group. [a]Used within three hours of excision. [b]Stored at -20°C for 7-14 days.*

Skin treatment or condition	Kp (cm^{-1} h^{-1} x 10^{-3})			
	Mean	Median	SD	N
Control (unexposed) fresh[a]	1.65	1.38	0.72	33
Control (unexposed) frozen[b]	2.04*	2.00*	0.70	36
SM-exposed fresh[a]	1.72	1.63	0.76	34
SM-exposed frozen[b]	2.31*	2.43*	0.83	33

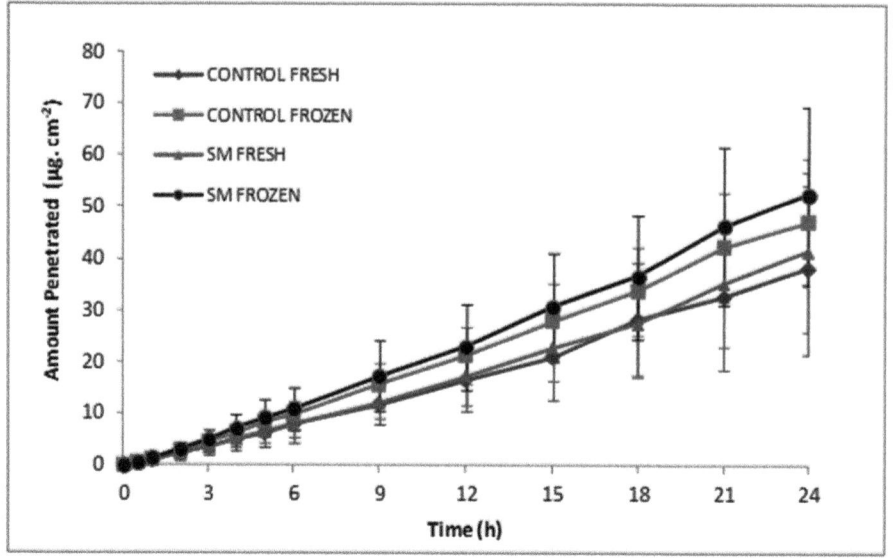

Figure 2 *Cumulative penetration profiles against time for ³H₂O through pig skin. All values are mean ± standard deviation of six biological replicates; n= 6 with up to six technical replicates per animal, maximum, 36 diffusion cells. Naïve control fresh skin, 33 skin samples; naïve control frozen, 36 skin samples; SM-exposed fresh, 34 skin samples; SM-exposed frozen, 33 skin samples.*

However, there was statistical evidence of an interaction between the pigs and the tissue storage condition applied; *e.g.,* fresh skin of a pig demonstrating low Kp will not necessarily show low Kp following storage frozen.

Because the density of water was taken to be 1 g.cm⁻³, the Jss metric was identical to the Kp metric, and as such, results were exactly the same as those for Kp analyses. This result highlighted the need for careful selection of animal numbers when conducting *in vitro* drug penetration studies and extrapolating results to effects that may be observed within man.

Lag time (T$_L$) values were derived from the linear portion of penetration data *i.e.,* the steady state penetration (data from 2 h-24 h), with the intercept of the x-axis used as the lag time to reach steady-state penetration. To determine the Jss values, the linear portion of each penetration profile was found using the correlation coefficient function in Microsoft Excel, with correlation coefficients of 0.95 or above deemed to be sufficiently linear for inclusion in the analysis. Output from the GLM showed statistical evidence (p < 0.05) of fresh skin demonstrating a longer average lag time before Jss was reached – compared to skin that had been stored frozen (Figure 3, Table 2). There was no evidence of any difference in average lag time between skin treatment groups, *e.g.,* SM- exposed and naïve control (P-value > 0.05).

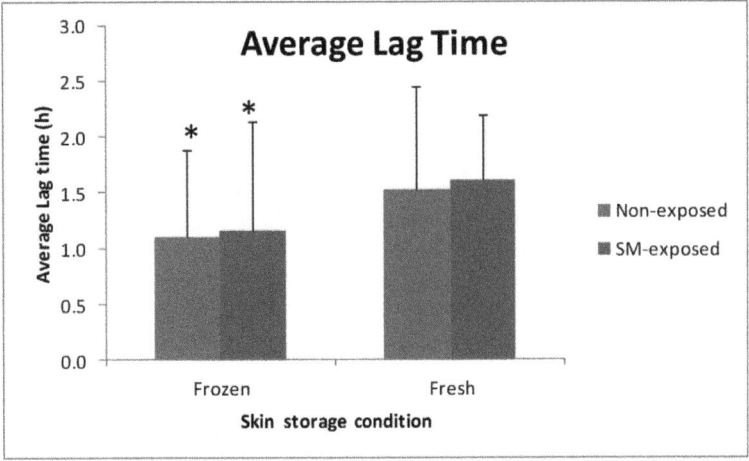

Figure 3 *Average lag time of ^3H_2O through pig skin to reach steady state penetration. All values are mean ± standard deviation of n= 6 biological replicates, with up to six technical replicates per animal, maximum 36 diffusion cells. Naïve control fresh skin, 33 skin samples; naïve control frozen, 36 skin samples; SM- exposed fresh, 34 skin samples; SM-exposed frozen, 33 skin samples. * denotes significant difference from fresh skin group.*

Table 2 *Summary of lag time to reach steady state penetration (mean ± S.D.) through skin of six animals under different treatment and storage conditions. ^{a} denotes use within three hours of excision. ^{b} denotes storage at -20°C for 7-14 days. * denotes significant difference from fresh skin group (P < 0.05).*

Skin treatment or condition	Lag Time (h)			
	Mean	Median	SD	N
Control (unexposed) fresh[a]	1.54	1.35	0.93	33
Control (unexposed) frozen[b]	1.11*	1.00	0.78	36
SM-exposed fresh[a]	1.61	1.68	0.58	34
SM-exposed frozen[b]	1.16*	1.30	0.97	33

4 DISCUSSION

One of the major considerations in the experimental design of an *in vitro* penetration study is the choice of membrane. If the results are to be related to likely effects observed in man, human skin should be used whenever possible. The human tissue used in such studies is generally stored frozen before use. When human skin is not available, an appropriate surrogate must be selected. Likewise, use of fresh tissue is preferred whenever possible, but previously frozen tissue is a viable alternative, accepting certain caveats. There have been conflicting findings published as to whether freezing skin affects percutaneous penetration. Some studies have reported no increase in permeation of water through split-thickness skin that has been stored frozen [3, 5, 14]. Other studies have demonstrated an increased permeation of water [25] and other penetrants following storage of skin tissue under frozen conditions [26 – 28]. It is difficult to directly compare these studies as they were not run under the same conditions. Many experimental factors may affect percutaneous penetration, which include but are not limited to physical characteristics of the penetrant, the drug vehicle, receptor fluid type, skin species, skin age and anatomical site, skin pre-treatment and storage conditions [29]. It is essential therefore to select an appropriate skin model, carefully control experimental conditions and ensure that the experimental design adequately controls these factors enabling the analysis of as many variables as possible to determine the critical determinants of penetration data. Where appropriate, experiments should be conducted in line with OECD guidelines and consistently with other published research methods. What is deemed to be an appropriate experimental design needs careful consideration and may be dependent upon numerous experimental factors and issues of practicality.

The study presented here showed that storing skin frozen (-20°C) did result in a small but statistically significant (P < 0.05) increase in Kp and Jss and a reduction in lag time of ^3H_2O, equating to a small reduction in barrier function of the skin to water permeation. Penetration of ^3H_2O under occluded infinite dose conditions was conducted because it has been proposed as a more rigorous test of the skin barrier function than non-occluded tests [29-30]. Substantial damage to the skin barrier layer would lead to gross leakage of water across the tissue, something not observed in the present study. It can be concluded that storing skin under frozen conditions as reported here caused some damage to the barrier to percutaneous penetration, but this should not preclude the use of *in vitro* techniques if the experiments take storage degradation into account when interpreting data. This study used six animals, and some animals demonstrated a smaller change than others in water permeability due to storage under frozen conditions. To be more confident of the effect of freezing in general, a range of penetrants should be assessed.

There are several possible mechanisms that may explain the changes in the barrier properties of the skin that allow for increased water and drug penetration following the freezing process. These mechanisms include perturbation of the lamellar lipid bilayers; formation of macroscopic fractures and cracks exposing polar groups within lipid bilayers; and, loosening of corneocyte cohesiveness because of degradation/modification of structural proteins in the tissue interstices [28].

Freezing porcine abdominal skin that had been exposed to pure liquid sulfur mustard *in vivo* for eight minutes did not overtly affect the barrier layer to tritiated water permeation using *in vitro* Franz-type static diffusion cells. It is noteworthy that there were increased variations in average Jss values and lag times in skin that had been exposed to SM. The time chosen to excise skin following exposure to SM (3 hours post-exposure) was critical to the findings upon the barrier layer assessment presented here. Had the SM lesions been allowed to progress beyond three hours then gross pathological damage would have developed and led to perturbation of the stratum corneum [31]. This would most likely have resulted in a large increase in ^3H_2O flux. Frozen SM-exposed porcine abdominal skin may be used for *in vitro* penetration studies, as long as effects of treatment and storage on the barrier layer are taken into account. It would be useful to know how long after exposure to SM the skin barrier layer starts to degrade.

Many factors are known to affect the progression of chemically induced skin lesions (exposure period, skin temperature, moisture, anatomical location, *etc*) [32 – 33]. For the purposes of this study it was important to identify an exposure period that would produce lesions with a consistent level of damage. The excision time was based on previous work supporting the effectiveness of therapies applied within 4 hours of exposure to SM [6]. The exposure period and excision time chosen should be readily reproducible within other laboratories, enabling other harmonised data sets to be constructed.

Freezing skin for storage is a common practice in laboratories carrying out diffusion measurements [3, 15 – 16, 21]; it is recommended that experimenters consider measuring the effects that their storage method has on the percutaneous barrier function of skin.

5 DISCLAIMER

The views, opinions and/or findings contained in this report are those of the authors and should not be construed as an official US Department of the Army position, policy or decision unless so designated by other documentation.
The views, opinions and/or findings contained in this report are those of the authors and should not be construed as an official Department of the Army position, policy or decision unless so designated by other documentation.

6 ACKNOWLEDGEMENTS

This work was supported by the US Army Medical Research and Materiel Command under contract No. W81XWH-08-C-0070. The authors would like to thank the technical assistance kindly provided by Rosi Perrott and Stephen Rutter.

7 REFERENCES

1. M. C. Babin, K. Ricketts, J. P. Skvorak, M. Gazaway, L. W. Mitcheltree and R. P. Casillas, Systemic Administration of Candidate Antivesicants to Protect Against Topically Applied Sulfur Mustard in the Mouse Ear Vesicant Model (MEVM), *Journal of Applied Toxicology,* 2000, **20**, 141–144.

2. M. P. Shakarjian, P. Bhat, M. K. Gordon, Y.-K. Chang, S. L. Casbohm, T. L. Rudge, R. C. Kiser, C. L. Sabourin, R. P. Casillas, P. Ohman-Stricklan, D. J. Riley, D. R. Gerecke, Preferential expression of matrix metalloproteinase-9 in mouse skin after sulfur mustard exposure, *Journal of Applied Toxicology,* 2006, **26**, 239–246.

3. M. P. Shakarjian, D. E. Heck, J. P. Gray, P. J. Sinko, M. K. Gordon, R. P. Casillas, N. D. Heindel, D. R. Gerecke, D. L. Laskin and J. D. Laskin, Mechanisms Mediating the Vesicant Actions of Sulfur Mustard after Cutaneous Exposure, *Toxicological Sciences*, 2010, **114**, 5–9.

4. W. Smith, Therapeutic options to treat sulfur mustard poisoning – The road ahead. *Toxicology*, 2009, **263**, 70–73.

5. D. Evison, D. Hinsley and P. Rice, Clinical review: Chemical weapons, *British Medical Journal*, 2002, **324**, 332–335.

6. S. Dachir, E. Fishbeine, Y. Meshulam, R. Sahar, S. Chapman, A. Amir and T. Kadar, Amelioration of sulfur mustard skin injury following a topical treatment with a mixture of a steroid and a NSAID, *Journal of Applied Toxicology*, 2004, **24** 107–113.

7. S. G. Mellor, P. Rice and G. J. Cooper, Vesicant burns, *Br J Plas Surg.*, 1991, **44**, 434–437.

8. J. S. Graham, R. P. Chilcott, P. Rice, S. M. Milner, C. G. Hurst and B. I. Maliner, Wound Healing of Cutaneous Sulfur Mustard injuries: Strategies for the Development of Improved Therapies, *Journal of Burns and Wounds*, 2005, **4**.

9. R. P. Casillas, R. C. Kiser, J. A. Truxall, A. W. Singer, S. M. Shumaker, N. A. Niemuth, K. M. Ricketts, L. W. Mitcheltree, L. R. Castrejon and J. A. Blank, Therapeutic Approaches to Dermatotoxicity by Sulfur Mustard I. Modulation of Sulfur Mustard induced Cutaneous Injury in the Mouse Ear Vesicant Model, *Journal of Applied Toxicology*, 2000, **20**, 145–151.

10. W. Diembeck, H. Beck, F. Benech-Kieffer, P. Courtellemont, J. Dupuis, W. Lovell, M. Paye, J. Spengler and W. Steiling, Test Guidelines for *In Vitro* Assessment of Dermal Absorption and Percutaneous Penetration of Cosmetic Ingredients, *Food and Chemical Toxicology*, 1999, **37**, 191–205.

11. F. P. Schmook, J. G. Meingassner, A. Billich, Comparison of human skin or epidermis models with human and animal skin in in-vitro percutaneous absorption, *International Journal of Pharmaceutics*, 2001, **215**, 51–56.

12. B. Godin and E. Touitou, Transdermal skin delivery: Predictions for humans from *in vivo, ex vivo* and animal models. *Advanced Drug Delivery Reviews*, 2007, **59**, 1152–1161.

13. S. M. Harrison, B. W. Barry and P. H. Dugard, Effects of freezing on human skin permeability, *J. Pharm. Pharmacol.*, 1984, **36**, 261–262.

14. G. S. Hawkins and W. G. Reifenrath, Influence of Skin Source, Penetration Cell Fluid, and Partition Coefficient on In Vitro Skin Penetration, *Journal of Pharmaceutical Sciences*, 1986, 75, **4**, 378–81.

15. R. L. Bronaugh, R. F. Steward and M. Simon, Methods for In Vitro Percutaneous Absorption Studies VII: Use of Excised Human Skin, *Journal of Pharmaceutical Sciences*, 1986, 75, **11**, 1094–1097.

16. D. J. Davies, R. J. Ward and J. R. Heylings, Multi-species assessment of electrical resistance as a skin integrity marker for in vitro percutaneous absorption studies, *Toxicology in Vitro.*, 2004, **18**, 351–358.

17. U. Jacobi, M. Kaiser, R. Toll, S. Mangelsdorf, H. Audring, N. Otberg, W. Sterry and J. Lademann, Porcine ear skin: an in vitro model for human skin, *Skin research and Technology*, 2007, **13**, 19–24.

18. R. Kong and R. Bhargava, Characterization of porcine skin as a model for human skin studies using infrared spectroscopic imaging, *Analyst*, 2011, **136**, 2359–2366.

19. W. G. Reifenrath, S. G. Hawkins and M. S. Kurtz, Percutaneous Penetration and Skin Retention of Topically Applied Compounds: An In Vitro-In Vivo Study, *Journal of Pharmaceutical Sciences*, 1991, 80, **6**, 526–532.

20. N. Sekkat, Y. N. Kalia and R. H. Guy, Biophysical Study of Porcine Ear Skin In Vitro and Its Comparison to Human Skin In Vivo, *Journal of Pharmaceutical Sciences*, 2002, **91**, 2376–2381.

21. A. M. Barbero and F. H. Frasch, Pig and guinea pig as surrogates for human *in vitro* penetration studies: A quantitative review, *Toxicology in Vitro*, 2009, **23**, 1–13.

22. OECD Guideline for the testing of chemicals, No 428: Skin Absorption: *In vitro* Method, Organisation for Economic Co operation and Development, Paris, 2004.

23. J. S. Graham, K. T. Schomaker, R. D. Galtter, C. M. Briscoe, E. H. Braue and K. S. Squibb, Bioengineering methods employed in the study of wound healing of sulphur mustard injury, *Skin Res. Technol.*, 2002, **8**, 57–69.

24. Organisation for the Prohibition of Chemical Weapons (OPCW), Convention on the prohibition of the development, production, stockpiling and use of chemical weapons and on their destruction, 1997.

25. L. A. Ahlstrom, S. E. Cross and P. C. Mills, The effects of freezing skin on transdermal drug penetration kinetics, *Journal of Veterinary Therap.*, 2007, **30**, 456–463.

26. J. Swarbrick, G. Lee and J. Brom, Drug Permeation through Human Skin: I. Effect of Storage Conditions on Skin, *The Journal of Investigative Dermatology*, 1982, **78**, 63–66.

27. K. R. Brain, K. A. Walters, D. M. Gree, S. Brain, L. J. Lortez, R. K. Sharma and W. E. Dressler, Percutaneous penetration of diethanolamine through human skin in vitro: Application from cosmetic vehicles, *Food and Chemical Toxicology*, 2005, **43**, 681–690.

28. A. C. Sintov and S. Botner, Transdermal delivery using microemulsion and aqueous systems: Influence of skin storage conditions on the in vitro permeability of diclofenac from aqueous vehicle systems, *International Journal of Pharmaceutics*, 2006, **311**, 55–62.

29. *Principles and Practice of Skin Toxicology*, ed. R. Chilcott and S. Price, John Wiley & Sons, Ltd, The Atrium, Southern Gate, Chichester, West Sussex, 2008, p. 98.

30. S. E. Cross, M. Russell, I. Southwell and M. Roberts, Human skin penetration of the major components of Australian tea tree oil applied in its pure form and as a 20% solution in vitro, *European Journal of Pharmaceutics and Biopharmaceutics*, 2008, **69**, 214–222.

31. J. S. Graham, R. S. Stevenson, L. W. Mitcheltree, M. Simon, T. Hamilton, R. R. Deckert, R. B. Lee, Improved Wound healing of Cutaneous Sulfur mustard Injuries in a Weanling Pig Model, *Journal of Burns and Wounds*, 2006, **5**, 46–65.

32. B. Renshaw, Observations on the role of water in the susceptibility of human skin to injury by vesicant vapors, *The Journal of Investigative Dermatology*, 1974, 75–85.

33. K. Kehe, F. Balszuwiteit, J. Emmler, H. Kreppel, M. Jochum and H. Thiermann, Sulfur Mustard Research – Strategies for the Development of Improved Medical Therapy, *Journal of Plastic Surgery*, 2008, **8**, 312–332.

38

MODULATION OF SULPHUR MUSTARD SKIN ABSORPTION BY TOPICAL DELIVERY OF FATTY ACIDS AND CHOLESTEROL[†]

R P Chilcott, O J Hutton and J Jenner

Biomedical Sciences Department, Defence Science and Technology Laboratory, Porton Down, Salisbury, United Kingdom.

1 INTRODUCTION

Sulphur mustard (SM) is a chemical warfare agent that causes severe skin lesions [1]. Previous work has shown that 80% of an applied liquid droplet of SM will vaporise from the skin surface before absorption can occur [2]. There are a number of potential methods to increase the vapour loss of SM from the skin surface (thereby decreasing the dose received by the skin). These include use of an appropriate barrier cream [3] or topical application of compounds that limit partitioning of SM into the skin.

The purpose of this study was to determine if topical administration of fatty acids (palmitic and stearic acids) or cholesterol decreased skin absorption of SM into human skin under unoccluded conditions. These candidate compounds were chosen because they are known to influence skin permeability [4] and, being normally resident within the stratum corneum, may be less likely to provoke an irritant reaction.

This study was conducted in three stages. (1) A kinetic profile of the skin absorption / desorption characteristics of each candidate compound was measured to determine the optimum pre-treatment time and longevity of each compound within the skin. (2) In consideration of the safety issues of handling skin tissue contaminated with SM, the effect of the candidate compounds on human skin partition coefficients (Km) of a series of corticosteriods (Log P 0.54 – 3.75) was measured. (3) The most effective candidate pre-treatment was then evaluated against SM using human skin in vitro.

2 MATERIALS AND METHODS

Materials : Radiolabelled (^{3}H) corticosteroids (testosterone, β-oestradiol, hydrocortisone and progesterone) were obtained from the Sigma Chemical Co. and were reported to be >98% pure by TLC analysis. Radiolabelled (^{14}C) cholesterol, palmitic and stearic acid were purchased from Amersham International and were reported to be >99% pure. Unlabelled Cholesterol and palmitic acid were obtained from Sigma Chemical Co. (95% and 99% pure, respectively) and stearic acid and unlabelled steroids were purchased from ICN Biochemicals (99% purity).   Radiolabelled (^{35}S) sulphur mustard (SM) was

---

[†] © Crown Copyright 2013.

synthesised at Dstl Porton Down and was 98% pure by HPLC analysis. All other reagents were at least AR grade.

Skin Absorption / Desorption Profiles: Discs (2.5 cm^2) of human (abdominal) epidermal membranes were placed into 5ml 50% aqueous ethanol solutions of unfiltered, saturated ^{14}C-radiolabelled cholesterol, palmitic or stearic acid. Skin samples were removed at regular intervals and the concentration of each pre-treatment compound in the skin was measured radiometrically by dissolving the skin in Soluene-350™ prior to counting in a RackBeta 1215 liquid scintillation counter. After 36h, the remaining skin was removed, rinsed in distilled water, blotted and transferred to 5ml 50% aqueous ethanol solution. Concentrations of each pre-treatment desorbed from the skin were measured by liquid scintillation counting of 20 µl samples of the solution bathing the skin. Concentrations of palmitic and stearic acid were also measured by HPLC (to account for effects such as metabolism or exchange labeling within the skin). Samples of fatty acids (50 µl) were subject to derivatisation with a mixture of bromomethyl-7-methoxy-coumarin (1 ml saturated solution in ethanol), acetone (1 ml), 18-crown-6-ether (saturated solution in 1 ml acetone) and potassium carbonate (200 mg) for 4h at 60° C, from which samples (20 µl) were injected onto a Supelcosil 15cm x 2.1mm C18 HPLC column (Supelco, USA) heated to 50° C with a 75% MeOH: 25% H$_2$O mobile phase, at a flow rate of 2.5 ml.min^{-1} with UV detection (254nm), using a Perkin-Elmer ISS2000 HPLC system and LKB Bromma 2151 variable wavelength UV monitor.

Membrane Partition Coefficients: Corticosteroid Km values were measured as previously described [5] using heat separated [6] human (abdominal) epidermal membranes that had been soaked in 50% aqueous ethanol (control) or saturated solutions of pre-treatment compounds in 50% aqueous ethanol for 36 h at 30° C.

Skin Absorption Experiments: Skin absorption of liquid SM was conducted as previously described [2] using static diffusion cells [7] containing heat separated human abdominal skin. The upper surface of the membrane was pre-treated with 200 µl ethanol (vehicle control), 200 µl saturated cholesterol in ethanol or untreated (control) for 36 h before addition of 25 mg liquid SM.

3 RESULTS

Absorption / Desorption: Human epidermal membrane absorption of pre-treatments was highest for palmitic acid and lowest for stearic acid (Table 1). After a further 36 h desorption, at least 75% of the delivered pre-treatment remained within the skin.

Table 1 *Amounts of pre-treatment absorbed / desorbed from human epidermal membranes. All values are mean ± SD of n=3 determinations.*

Pre-treatment	Absorbed Dose (µg.mg skin^{-1})	Desorption (% absorbed dose)
Palmitic acid	3.2 ± 0.3	4 ± 1
Stearic acid	6.6 ± 0.7	22 ± 7
Cholesterol	14.3 ± 0.8	24 ± 3

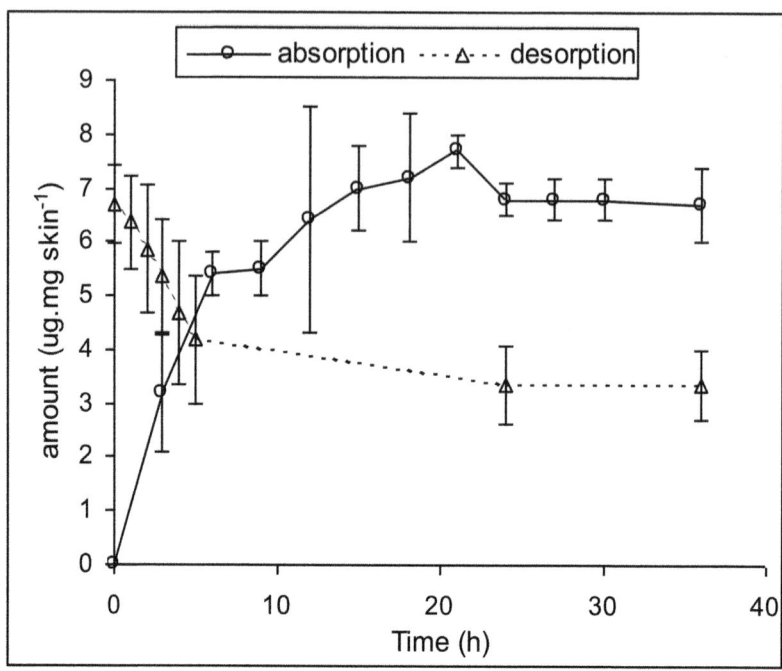

Figure 1 *Absorption / desorption profile of stearic acid. All points are mean ± SD of n=3 pieces of skin obtained from a total of 12 individuals.*

The absorption or desorption of each compound mainly occurred over the first 10-20h, with no further detectable changes in skin concentrations after 30 - 36h (for example, Figure 1).

Membrane Partition Coefficients: All pre-treatments caused significant ($p<0.05$) changes in corticosteroid Km values (Figure 2). Pre-treatment with stearic acid increased all Km values. Palmitic acid increased the majority of corticosteroid Km's whereas only cholesterol caused a consistent decrease in Km.

Skin Absorption: Pre-treatment of epidermal membranes with cholesterol caused a significant ($p<0.05$) decrease in the total amount of SM penetrated under unoccluded conditions, but did not significantly affect the skin absorption rate in comparison with untreated (control) membranes (Figure 3). Pre-treatment with vehicle only (ethanol) caused a significant increase in skin absorption rate, but no significant difference in total amount penetrated (Table 2).

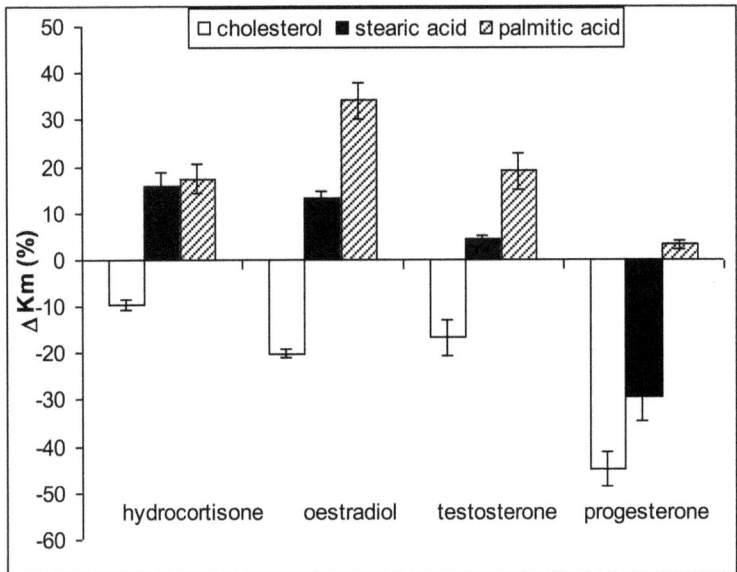

Figure 2 *Effect of pre-treatments (cholesterol, stearic and palmitic acid) on membrane partition coefficient (Km) of hydrocortisone (Log P=0.54), oestradiol (Log P=2.69), testosterone (Log P=3.22) and progesterone (Log P=3.75), expressed as percentage change in comparison to untreated (control) membranes. All values are mean ± standard deviation of n=6 skin discs obtained from separate individuals*

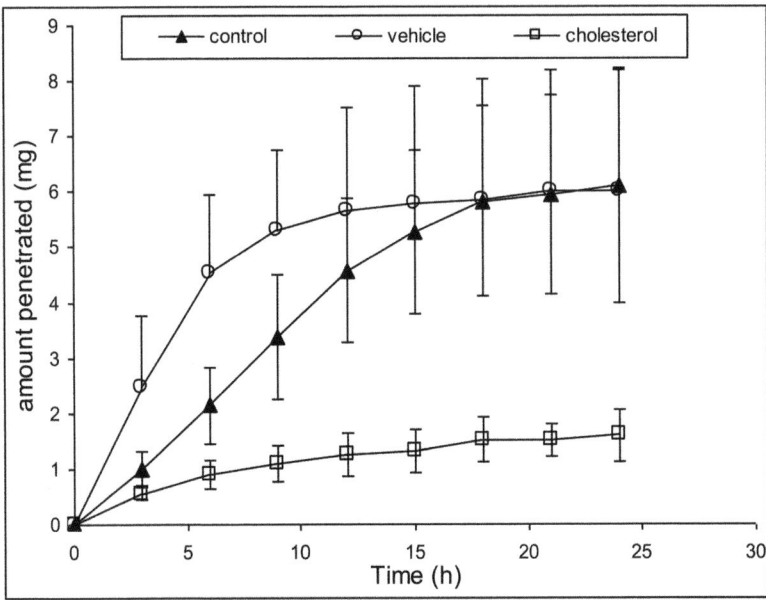

Figure 3 *Skin absorption of SM through human epidermal membranes pre-treated with ethanol ("vehicle"), cholesterol in ethanol ("cholesterol") and untreated ("control"). All values are mean ± standard deviation of n=6 diffusion cells containing skin from one individual.*

Table 2 *Skin absorption rates (Jmax) and percentage applied dose of SM penetrated (%D) through human (abdominal) epidermal membranes. All values are mean ± standard deviation of n=6 diffusion cells (containing skin from one individual). Asterisks indicate value is significantly different from control (p<0.01).*

Pre-treatment	Jmax	%D
Control	56 ± 32	24 ± 9
Vehicle	107 ± 37*	24 ± 8
Cholesterol	40 ± 16	6 ± 2*

4 DISCUSSION AND CONCLUSIONS

This study has demonstrated that (1) cholesterol, palmitic and stearic acid can be delivered to, and predominantly retained within, human skin. (2) Pre-treatment of skin with cholesterol leads to a significant reduction in the ability of a range of corticosteroids to partition into skin. (3) Pre-treatment of human skin with cholesterol can cause a 4-fold decrease in skin absorption of SM under unoccluded conditions.

Membrane partition coefficients of SM were not measured, as SM is known to rapidly hydrolyse in aqueous solutions [8]. However, the Log P value of progesterone (3.75) is similar to that calculated for SM (3.8) and thus it was assumed that the changes in Km observed for progesterone following pre-treatment with cholesterol would model those of SM. The fact that there was a reduction in the amount of SM penetrating cholesterol pre-treated skin supports this assumption.

It is unlikely that pre-treatment with cholesterol altered the diffusional resistance of the skin, since fluxes of SM through cholesterol pre-treated and control skin were the same, indicating that after partitioning, diffusion through the skin was relatively unaffected. This may imply that cholesterol pre-treatment will be ineffective against SM under occluded conditions, as there is no competition between vapour loss and partitioning. Thus, the entire applied dose will eventually permeate the skin at a rate equivalent to untreated skin. This hypothesis was supported in an additional series of (unreported) experiments where no significant change in the skin absorption rate of progesterone was measured through cholesterol pre-treated skin under occluded conditions in vitro.

Interestingly, the effect of pre-treatments appeared to be inversely proportional to the solubility of each corticosteroid when expressed as P (octanol-water partition coefficient, Figure 4). If the change in partitioning was purely due to addition of the lipophilic pre-treatments, the opposite effect would be expected in that Km of hydrocortisone (Log P=0.54) would decrease more than that of progesterone (Log P=3.75). These data imply that the effect of pre-treatments on skin partitioning of corticosteroids are not due to changes in membrane solubility, but are consistent with the hypothesis that compounds such as cholesterol contribute to the barrier function of human skin by decreasing the fluidity of lipid layers within the stratum corneum.

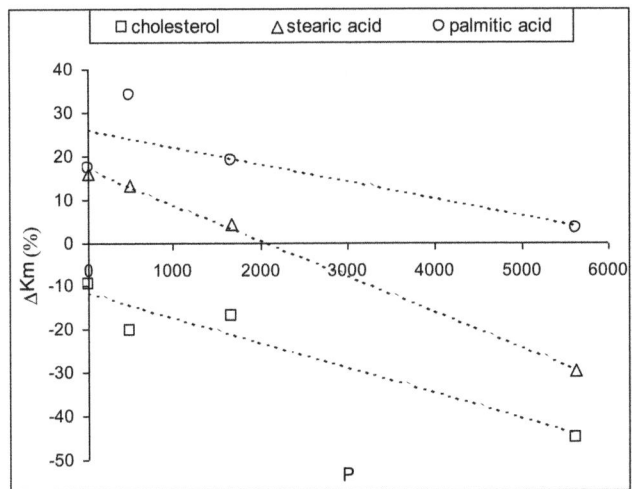

Figure 4 *Effect of pre-treatments (cholesterol, stearic and palmitic acid) on corticosteroid membrane partitioning (Km, expressed as percentage change) against solubility (expressed as P). Dotted lines indicate linear trend for cholesterol ($r^2=0.92$), stearic acid ($r^2=0.99$) and palmitic acid ($r^2=0.64$).*

Previous studies have shown that (oil-in-water (o/w) based) barrier cream formulations are protective against SM under occluded conditions but may enhance skin absorption of SM under unoccluded conditions [9]. The results of this study have shown that some protection may be afforded against SM under unoccluded conditions by modulating the partitioning characteristics of the skin. However, the two protective measures are not compatible, since topical application of an o/w barrier cream onto skin pre-treated with cholesterol would negate the partitioning effects of the cholesterol.

5 ACKNOWLEDGEMENTS

The authors would like to thank Mr E.T. Stephenson for his technical assistance. This work was funded by the Ministry of Defence (MoD) at facilities operated by the Defence Science and Technology Laboratory (Dstl).

6 REFERENCES

1. T. C. Marrs, R. L. Maynard and F. R. Sidell, Mustard gas, in *Chemical Warfare Agents: Toxicology and Treatment*, Wiley, UK, 1996, pp. 139–173.

2. R. P. Chilcott, J. Jenner, W. Carrick, S. A. M. Hotchkiss and P. Rice, In vitro human skin absorption of bis-2-(chloroethyl)sulphide (sulphur mustard), *Journal of Applied Toxicology*, 2000, **20**, 349–355.

3. R. P. Chilcott, J. Jenner, S. A. M. Hotchkiss and P. Rice, "Evaluation of barrier creams against sulphur mustard: (I) In vitro studies using human skin", *Skin Pharmacology and Applied Physiology*, 2002 **15**(4) 225–235.

4. H. Schaeffer and T. E. Redelmeier, *Skin barrier*, Karger, Switzerland, 1996.

5. R. P. Chilcott, S. A. M. Hotchkiss and J. Jenner J, In vitro percutaneous penetration of steroids under comparable thermodynamic conditions, in *Perspectives in Percutaneous Penetration*, ed. K. R. Brain, V. J. James and K. A. Walters, STS Publishing, Cardiff, 1997, vol. 5b, pp. 242–245.

6. A. M. Kligman and E. Christophers, Preparation of isolated sheets of human stratum corneum, *Arch Dermatol*, 1963, **88**, 70–73.

7. T. J. Franz, In vitro vs in vivo percutaneous absorption: on the Relevance of in Vitro Data, *J Invest Dermatol*, **64**, 190–195.

8. R. I. Tilley, The hydrolysis of bis(2-chloroethyl)sulfide (sulphur mustard) in aqueous mixtures of ethanol, acetone and dimethylsulfoxide, *Australian Journal of Chemistry*, 1993, **46**, 293–300.

9. R. P. Chilcott, J. Jenner, C. H. Dalton, Z. Ashley, C. Allen and S. Bradley, Evaluation of barrier creams against sulphur mustard: In Vitro and In Vivo studies using pig skin, *Journal of Cutaneous and Ocular Toxicology*, 2007, **26**, 235–247.

39

PROTECTIVE AND TEMPORAL EFFECTS OF CLOTHING ON THE *IN VITRO* ABSORPTION OF CHEMICAL WARFARE AGENTS AND SIMULANTS

H Matar[1,2], S C Price[2] and R P Chilcott[1,3]

[1]Health Protection Agency, Porton Down, Salisbury, United Kingdom. [2]Division of Biochemical Sciences, University of Surrey, Guildford, United Kingdom. [3]University of Hertfordshire, Hatfield, United Kingdom.

1 INTRODUCTION

The deliberate release of chemical, biological, radiological and nuclear (CBRN) materials poses a significant threat to civilian populations as exemplified by the 1995 Tokyo sarin incident [1]. A major step in the process of decontamination is the removal of contaminated clothing [2]. Disrobing (the removal of contaminated clothing) is considered to be a central means of limiting the adverse health effects of chemical contamination following exposure to toxic liquids [3]. However, there is likely to be a finite time delay between contamination of clothing and formal instructions to conduct a disrobing procedure during a chemical incident. The Tokyo incident highlighted the importance of removing clothing, as it can prevent continued dermal penetration and potential secondary exposure of emergency medical staff [4].

The purpose of the study was to investigate the effect of common clothing fabrics on the percutaneous absorption of three chemical warfare agents (sulphur mustard (HD), soman (GD) and VX) and a chemical warfare agent simulant (methyl salicylate; MS).

2 MATERIALS AND METHOD

The storage and use of CW agents was in full compliance with the Chemical Weapons Convention (1986). Chemical warfare (CW) agents ('VX'; S-[2-(diisopropylamino)ethyl]-O-ethyl methylphosphonothioate, soman; 'GD'; O-Pinacolyl methylphosphonofluoridate and sulphur mustard; 'HD'; bis(2-chloroethyl)sulphide) and their ([14]C-) radiolabelled analogues were custom synthesised by TNO Defense, Security and Safety (Rijswijk, Netherlands). All were reported to be >97% purity. Methyl salicylate was purchased from the sigma chemical company (Poole, UK) and was reported to be >99% pure. Radiolabelled methyl salicylate (>98% purity) was purchased from ARC (UK) Ltd. (Cardiff, UK). Each radiolabelled CW agent was mixed with 5g of corresponding undiluted agent to provide a stock solution with a nominal activity of ~ 1mCi ml^{-1} and was stored for up to four months at 4°C. Aliquots of each stock solution were diluted with unlabelled CW agent immediately prior to each experiment to provide a working solution with a nominal activity of ~ 0.2 - 0.5 µCi µl^{-1}. Ethanol and isopropanol (both Analytical grade) were purchase from the Sigma Chemical Company (Dorset, UK). Liquid scintillation counting fluid (Ultima Gold), tissue solubliser (Soluene-350™), scintillation counting vials (5ml) and glass vials (20 ml) were purchased from Perkin Elmer LAS (UK) Ltd (Buckinghamshire, UK).

A leading UK clothing retailer provided data on the average type and composition of clothing sold to the UK public (based on 2007/2008 sales figures). Based on this information, samples of relevant materials were purchased according to Table 1.

Table 1 *Summary of clothing materials.*

Garment	Composition	Density (g m^{-2})	Thickness (μm)	Supplier
Overcoat	Nylon (100%)	100	160	Baileyfreerglover, Leics, UK
T-shirt	Cotton (100%)	190	810	Vend Fabrics, Leics, UK
Jumper	Acrylic (100%)	250	1960	Castle knitwear, Leics, UK
Denim Jeans	Cotton (100%)	350	1670	Nova Trimmings, Leics, UK

Full thickness, close-clipped pig skin was obtained post mortem from the dorsal aspect of female animals (*Sus scrofa*, large white strain). The skin from each animal was stored flat between sheets of aluminium foil at -20°C for up to twelve weeks before use. For each experiment, skin from one animal was removed from cold storage and thawed in a refrigerator at 5°C for ~ 24 hours. The skin was then dermatomed (Humeca model D42, Eurosurgical Ltd, Guildford, UK) to a nominal depth of 500μm and cut into squares (3 x 3 cm).

Skin diffusion cells were purchased from PermeGear (Chicago, IL., USA) and comprised an upper (donor) and lower (receptor) chamber with an area available for diffusion of 1.76 cm^2. A section of dermatomed skin was placed between the two chambers (epidermal surface facing the donor chamber) and the ensemble was securely clamped. The receptor chambers were filled with fluid (50% (v/v) aqueous ethanol; 14 ± 0.8 ml) so that the meniscus in the sampling arm was level with the surface of the skin sample. Each diffusion cell was placed in a Perspex™ holder above a magnetic stirrer which constantly mixed the receptor fluid via a (12 x 6 mm) Teflon™-coated iron bar placed within the receptor chamber. The receptor chambers were of the "jacketed" variety through which warm (36°C) water was pumped from a circulating water heater (Model GD120, Grant Instruments, Cambridge, UK) via a manifold to ensure a constant skin surface temperature of ~ 32°C (as confirmed by infrared thermography (FLIR Model P640 camera, Cambridge, UK). Up to 36 diffusion cells were used in each experiment, with six treatment groups (Table 2; each comprising n=6 diffusion cells). Once assembled, the diffusion cells were left *in situ* for an equilibration period of 24 hours.

Table 2 *Summary of treatment groups for both studies. Study one involved prolonged (24 hour) exposure to a liquid droplet of chemical. In study two, the combined clothing layers were removed at 5 – 360 minute intervals post exposure. A total of n=6 replicates were used in each group.*

Study	Treatment Group	Description
1: protective effects of clothing over 24 hours	Control	
	T-shirt	Single layer placed over skin surface.
	Denim Jeans	
	Jumper	
	Overcoat	
2: temporal effects of disrobing	Combination	Overcoat (top layer), jumper and T-shirt (lower layer) placed over skin surface.
	Control	Skin only.
	5	Overcoat (top layer), jumper and T-shirt (lower layer) placed over skin surface at 5, 30, 60, 180 or 360 minutes after exposure.
	30	
	60	
	180	
	360	

Each experiment was started by the addition of 10 µl of ^{14}C-radiolabelled chemical (HD, GD, VX or MS) to the skin (controls) or outermost layer of clothing. Samples of receptor chamber fluid (250 µl) were withdrawn from each diffusion cell at regular intervals up to 24 hours post-exposure and were placed into vials containing 5 ml of liquid scintillation counting (LSC) fluid. Each receptor sample was replaced with an equivalent volume of fresh fluid to maintain a constant volume in the receptor chamber.

The amounts of radioactivity in each sample were quantified using a Perkin Elmer Tri-Carb liquid scintillation counter (Model 2810 TR), using an analysis time of 2 minutes per sample and a preset quench curve specific to the brand of LSC fluid used in this study. The amounts of radioactivity in each sample were converted to amount of ^{14}C-radiolabelled CW agent by comparison to appropriate standards (measured simultaneously).

The relative effectiveness of each treatment was expressed as a protection factor (PF), calculated from equation 1.

$$PF = \frac{QC_{24}}{QT_{24}} \tag{1}$$

Where Q is the total amount of radiolabelled chemical penetrating the skin after 24 hours in treated (T) or control (C) diffusion cells.

3 RESULTS

The protection afforded by single layers of clothing following exposure to GD varied according to fabric (Figure 1). Acrylic and nylon were the least effective whereas a combined layer (Nylon, acylic and cotton t-shirt) was the most effective (13 fold reduction in total amount penetrated). Sulphur mustard was particularly penetrative; no single fabric or combined layers provided any marked level of protection (Figure 1). Clothing provided the most protection against VX (Figure 1), with the combined layers reducing dermal absorption by a factor of 28. However, nylon exhibited the lowest protection.

In general, the protective effect of multiple layers of clothing decreased exponentially with time to disrobing (Figure 2). The actual level of protection afforded by clothing was consistently related to the chemical contaminant, being (in order of highest to lowest protection) VX>GD>MS>HD.

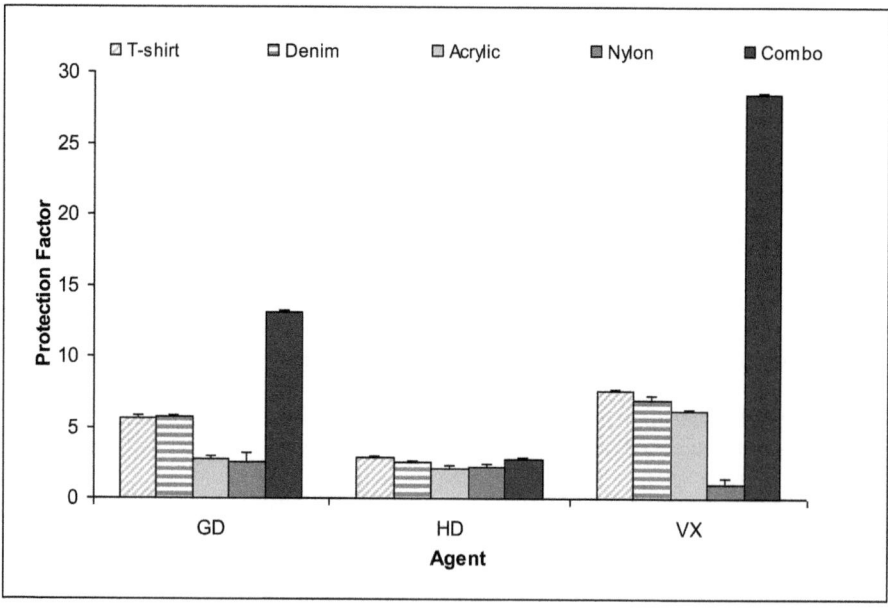

Figure 1: *Protective effect of clothing (expressed as protection factor; see equation 1) afforded by various fabrics over a constant exposure period of 24 hours against the dermal absorption of ^{14}C-radiolabelled soman (GD), sulphur mustard (HD) and VX. All values are average ± standard deviation of n=6 replicates.*

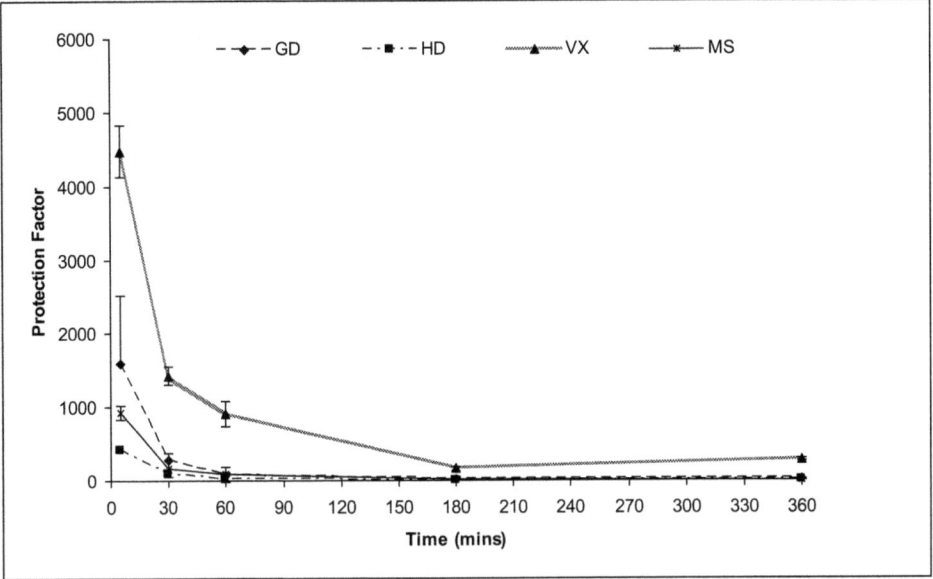

Figure 2 *Protective effect of disrobing (expressed as protection factor; see Equation 1) at 5, 30, 60, 180 and 360 minutes after exposure to ^{14}C-radiolabelled contaminants (GD, HD, VX and MS). All values are average ± standard deviation of n=6 diffusion cells.*

4 DISCUSSION AND CONCLUSIONS

This study has demonstrated that standard civilian clothing may provide protection against the dermal absorption of chemical contaminants under certain circumstances. In particular, these data emphasise the importance of early disrobing. However, the level of protection afforded against sulphur mustard was consistently poor. The relative ease with which mustard penetrates clothing has long been noted [5, 6].

5 ACKNOWLEDGEMENTS

This work was funded by the UK Department of Health as part of the "ORCHIDS 2" project.

6 REFERENCES

1. T. Okumura, N. Takasu, S. Ishimatsu, S. Miyanoki, A. Mitsuhashi, K. Kumada, K. Tanaka and S. Hinohara, Report on 640 victims of the Tokyo subway sarin attack, *Ann Emerg Med,* 1996, **28,** 129–35.

2. S. F. Clarke, R. P. Chilcott, J. C. Wilson, R. Kamanyire, D. J. Baker, and A. Hallett, Decontamination of multiple casualties who are chemically contaminated: a challenge for acute hospitals, *Prehosp Disaster Med,* 2008, **23,** 175–81.

3. R. P. Chilcott, (2010) CBRN Contamination, in *Textbook of Environmental Medicine,* ed. J. Ayres, R. Harrison, G. Nichols and R. L. Maynard, Hodder Arnold, London.

4. T. Okumura, T. Hisaoka, A. Yamada, T. Naito, H. Isonuma, S. Okumura, K. Miura, M. Sakurada, H. Maekawa, S. Ishimatsu, N. Takasu and K. Suzuki, The Tokyo subway sarin attack–lessons learned, *Toxicol Appl Pharmacol,* 2005, **207,** 471–6.

5. Porton, Experiments to ascertain the length of time heavy winter clothing will protect the skin from large drops of liquid mustard gas, *Porton Report 710,* 1929.

6. Porton, Further report on experiments to ascertain the length of time heavy winter clothing will protect the skin from large drops of liquid mustard gas, *Porton Report 727,* 1929.

Subject Index